Building Climate Resilience in Agriculture

Wajid Nasim Jatoi
Muhammad Mubeen • Ashfaq Ahmad
Mumtaz Akhtar Cheema • Zhaohui Lin
Muhammad Zaffar Hashmi
Editors

Building Climate Resilience in Agriculture

Theory, Practice and Future Perspective

Springer

Editors
Wajid Nasim Jatoi
Department of Agronomy
Faculty of Agriculture and Environment
The Islamia University of
Bahawalpur (IUB)
Bahawalpur, Pakistan

Muhammad Mubeen
Department of Environmental Sciences
COMSATS University Islamabad
Vehari Campus, Pakistan

Ashfaq Ahmad
Asian Disaster Preparedness Center
Bangkok, Thailand

Mumtaz Akhtar Cheema
Boreal Ecosystem and Agricultural Sciences
School of Science and the Environment
Memorial University-Grenfell Campus
Corner Brook, NL, Canada

Zhaohui Lin
International Center for Climate
and Environment Sciences
Institute of Atmospheric Physics
Chinese Academy of Sciences
Beijing, China

Muhammad Zaffar Hashmi
Department of Chemistry
COMSATS University Islamabad
Islamabad, Pakistan

ISBN 978-3-030-79407-1 ISBN 978-3-030-79408-8 (eBook)
https://doi.org/10.1007/978-3-030-79408-8

This Springer imprint is published by the registered company Springer Nature Switzerland AG
The registered company address is: Gewerbestrasse 11, 6330 Cham, Switzerland

Preface

Burgeoning world population has put a lot of pressure on food, fuel, fibers, and natural resources. There could be serious challenges of food security in 2050 to feed more than 9 billion people amidst climate change. Agriculture sector has to grow exponentially to produce approximately 70% more food by 2050 to feed the ever-increasing global population. Globally, the agriculture sector contributes 11–14% towards anthropogenic greenhouse gas (GHG) emissions and these emissions are increasing at a rate of approximately 1% per year. Inorganic nitrogen (N) fertilizer and manure application may enhance nitrate (NO_3^-) leaching, which is another significant N loss pathway from the rooting zone, resulting in water pollution and additional indirect nitrous oxide (N_2O) loss through denitrification in water bodies. Emissions of N_2O from the agriculture sector is expected to increase 35–60% by 2030, due to an increased use of N fertilizer and animal manure production. The net emission of GHGs from farming activities can potentially be decreased by changing crop management practices to increase soil organic carbon (SOC) or C sequestration and decrease N_2O emissions. Therefore, it is essential that new agricultural management practices be employed to sequester C, increase soil health, and reduce N and gaseous losses, while maintaining sufficient food and feedstock production. The addition of perennial crops in crop rotation, no tillage, crop residue incorporation, cover cropping, legume-based cropping systems, and biochar amendments have been demonstrated to improve soil health and reduce environmental impacts. Additionally, slow-release fertilizer application, urease inhibitors, and nitrification inhibitors can reduce N losses and improve N use efficiency in different cropping systems.

Any change in climate would eventually disturb agricultural practices. So, it becomes crucial to adopt Climate-Resilient Agriculture (CRA) practices at such a scale so that the disturbing impact of climate change on agriculture sector could be minimized. Adapting and building resilience to climate change means increasing agricultural productivity, farm profitability, and net income and reducing GHGs emissions. A number of strategies can be adopted to achieve sustainable growth in crop production, horticultural crops, livestock, fisheries, and forest under changing climate scenarios.

With increasing concerns over environmental protection, improvement in resource use efficiencies (e.g., efficient use of water and nutrients) has become a prime goal in global agricultural system. Sustainable use of these resources in agricultural production could significantly avoid environmental hazards resulting from their over-utilization. Similarly, salinity of the land has proven to be quite detrimental for agriculture; to study its impacts on the agricultural growth and the ways to deal with this menace has important future implications in the climate change. Modern ways of controlling pests, including insects and weeds, would have to be adopted by curtailing the use of pesticides. The use of crop modeling and remote sensing and adopting innovative crop management practices could make broader future prospects for climate resilience. All these measures would ultimately help for keeping food security challenges to feed ever-increasing global population across the globe.

Bahawalpur, Pakistan Wajid Nasim Jatoi
Vehari, Pakistan Muhammad Mubeen
Bangkok, Thailand Ashfaq Ahmad
Corner Brook, NL, Canada Mumtaz Akhtar Cheema
Beijing, China Zhaohui Lin
Islamabad, Pakistan Muhammad Zaffar Hashmi

Acknowledgments

This book is the outcome of the dedication and efforts of editors Dr. Wajid Nasim Jatoi, Dr Muhammad Mubeen, Prof. Dr. Ashfaq Ahmad Chattha, Prof. Dr. Mumtaz Akhtar Cheema, Prof. Dr. Zhaohui Lin, and Dr. Muhammad Zaffar Hashmi. Their entire support made the completion of this book possible. We would like to sincerely thank our valuable reviewers Prof. Dr. Shakeel Ahmad (Bahauddin Zakariya University Multan, Pakistan), Dr. Shaukat Ali (Global Change Impact Studies Centre (GCISC), Ministry of Climate Change, Pakistan), Dr. Shah Fahad (The University of Haripur, KPK, Pakistan), Dr. Muhammad Adnan Shahid (University of New Hampshire, Durham, USA), Dr. Abdul Ghaffar (The Islamia University Bahawalpur, Pakistan), Dr. Khawar Jabran (Department of Plant Production and Technologies, Faculty of Agricultural Sciences and Technologies, Niğde Ömer Halisdemir University, Niğde 51240, Turkey), Dr. Muhammad Azim Khan (University of Agriculture, Peshawar, Pakistan), Dr. Muhammad Shakeel (South China Agricultural University, Guangzhou, China), Dr. Syed Asif Ali Naqvi (Government College University Faisalabad, Pakistan), Dr. Ishfaq Ahmad (Centre for Climate Research & Development, COMSATS University Islamabad, CUI, Pakistan), Dr. Hafiz Mohkum Hammad (MNS University of Agriculture, Multan, Pakistan), Dr. Shahid Iqbal (MNS University of Agriculture, Multan, Pakistan), Dr. Malik Muhammad Ali Shahid (CUI Vehari-Pakistan), Prof. Dr. Fahim Khokhar (National University of Science and Technology, Islamabad, Pakistan), and Dr. Muhammad Imtiaz Rashid (King Abdul Aziz University, Saudia Arabia). All these reviewers have put enormous effort to improve the quality of the book.

Many eminent researchers and academicians have contributed in the preparation of the proposal of this book. We highly appreciate the kind support and encouragement from Mr. Malik Amin Aslam Khan, Minister for Climate Change/Special Assistant to Prime Minister of Pakistan (on Climate Change), Government of Pakistan, Prof. Dr. Mirza Barjees Baig (Prince Sultan Institute for Environmental, Water & Desert Research, King Saud University, Riyadh, Kingdom of Saudi Arabia), Mr. Afzal Ahmad, Director, Planning & Development, The Islamia University of Bahawalpur, Pakistan., Dr. Abid Shahzad, Director International Linkages, The Islamia University of Bahawalpur, Pakistan., Dr. Muhammad Asif

Naveed Ranjha, Director, Fund Raising and University Advancement, The Islamia University of Bahawalpur, Pakistan., Ms. Farkhanda Tahseen, Principal Officer, Estate Care and Space Management/Additional Director Planning & Development, The Islamia University of Bahawalpur, Pakistan. Dr. Muhammad Ali Raza, Director, National Center for Intercropping, The Islamia University of Bahawalpur, Pakistan. Ayman EL Sabagh, Assistant Professor, Department of Agronomy, Faculty of Agriculture, Kafresheikh University, Egypt. Dr. Syed Tahir Ata-Ul-Karim, Department of Global Agric. Sciences, Graduate School of Agricultural & Life Sciences., The University of Tokyo, Japan. Dr. Khalid Mahmood, IMPACT LAB PROJECT MANAGER, Sustainable Agriculture Sciences, Rothamsted Research, UK. Dr. Michelle Reboita, Natural Resource Institute, Federal University of Itajuba, Av. BPS, 1303 – CEP 37500-903, Itajuba/MG – Brazil. Dr Muhammad Nafees, Chairman, Department of Horticulture, Faculty of Agriculture and Environment, The Islamia University of Bahawalpur, Pakistan. Dr Muhammad Khalil-ur-Rehman, Department of Horticulture, Faculty of Agriculture and Environment, The Islamia University of Bahawalpur, Pakistan. Mr. Nasir Mehmood Elia, CEO, Bristol Pvt Ltd, Pakistan. Dr. Muhammad Nadeem (University of Newfoundland, Corner Brook, Canada), Dr. Hafiz Qaisar Yasin (Punjab Agriculture Department, Lahore, Pakistan), Dr. Syed Aftab Wajid (University of Agriculture, Faisalabad, Pakistan), Dr. Tasneem Khaliq (University of Agriculture, Faisalabad-Pakistan), Dr. Habib ur Rehman (MNS University of Agriculture, Multan-Pakistan), Dr. Ishaq Asif Rehmani (Ghazi University, DG Khan, Pakistan), and Dr. Tauqeer Abbas (North Dakota State University, Fargo, USA). The moral support of Prof. Dr. Nazim Hussain (BZU-Multan-Pakistan), Prof. Dr. Zahid Ata Cheema (UAF-Pakistan), Prof. Dr. Abid Hussain (UAF-Pakistan), Dr. Naeem Khan, (Institute of Food and Agricultural Sciences, University of Florida, Gainesville, USA), and Dr. Muhammad Shahid (CUI, Vehari-Pakistan) may not be ignored.

Furthermore, First Editor (Wajid Nasim Jatoi) is highly thankful to the support and cooperation given by all colleagues Dr. Muhammad Adnan Bukhari, Dr. Muhammad Aurangzaib, Dr. Muhammad Latif, Dr. Muhammad Saqib, Dr. Muhammad Usman Bashir, Dr Rashid Iqbal, Dr Abdul Rehman, Dr Farhan Khalid, Dr Muhammad Asghar Shah, Dr Muhammad Shahzad, Dr Muhammad Usman Aslam and especially Dr. Muhammad Aown Sammar Raza, Chairman, Department of Agronomy, Faculty of Agriculture and Environment, The Islamia University of Bahawalpur, Pakistan, is highly acknowledged and commendable. Dedication and appreciation goes to my Family (Mrs. Muhammad Hanif Khan Jatoi, Mrs. Haji Muhammad Pervaiz Khan and my elder brother Mr. Muhammad Sajid Nasim Khan and their cute families), to whom I am here to be able to do all my professional career. Last but not least, being First Editor (Wajid Nasim Jatoi) my family is my pride, I am very much thankful to my dear wife (Mrs. Neelam Manzoor), my cute kids (Mr. Muhammad Mosa Nasim and Ms. Marium Fatima) without whom, I AM NOTHING. Once again, I thank all my dear and near ones whom I forgot to mention here in my first ever International Book by International Renowned Publisher.

At the end, we are highly thankful to International Publisher (Springer Nature), its administration, and all the respected members who have supported a lot for entire duration from proposal till publication of this book.

Contents

Contributors

Ghulam Abbas Department of Agronomy, Bahauddin Zakariya University, Multan, Pakistan

Mazhar Abbas Department of Management and MIS, College of Business Administration, University of Hail, Hail, Kingdom of Saudi Arabia

Muhammad Abid Department of Soil Science, Bahauddin Zakariya University, Multan, Pakistan

Muhammad Adnan Department of Agriculture, The University of Swabi, Ambar, Pakistan

Department of Soil and Environmental Sciences, the University of Agriculture, Peshawar, Pakistan

Muhammad Afzaal National Agromet Centre, Pakistan Meteorological Department, Islamabad, Pakistan

Obaid Afzal Department of Agronomy, Pir Mehr Ali Shah Arid Agriculture University, Rawalpindi, Pakistan

Ashfaq Ahmad Asian Disaster Preparedness Center, Bangkok, Thailand

Fiaz Ahmad Central Cotton Research Institute, Multan, Pakistan

Hafiz Waqar Ahmad Department of Agronomy, The Islamia University of Bahawalpur, Bahawalpur, Pakistan

Ishfaq Ahmad Climate Resilience Department, Asian Disaster Preparedness Center (ADPC), Islamabad, Pakistan

Mukhtar Ahmed Department of Agronomy, Pir Mehr Ali Shah Arid Agriculture University, Rawalpindi, Pakistan

Department of Agricultural Research for Northern Sweden, Swedish University of Agricultural Sciences, Umeå, Sweden

Naveed Ahmad Department of Zoology, University of Education, Vehari, Punjab, Pakistan

Riaz Ahmad Department of Horticulture, Faculty of Agricultural Sciences and Technology, Bahauddin Zakariya University, Multan, Pakistan

Shakeel Ahmad Department of Agronomy, Bahauddin Zakariya University, Multan, Pakistan

Muhammad Akmal Institute of Soil Science, PMAS-Arid Agriculture University Rawalpindi, Rawalpindi, Pakistan

Rida Akram Department of Environmental Sciences, COMSATS University Islamabad, Vehari, Pakistan

Amjed Ali Department of Agronomy, University College of Agriculture, University of Sorgodha (UoS), Sorgodha, Pakistan

Mazhar Ali Department of Environmental Sciences, COMSATS University Islamabad, Vehari, Pakistan

Muhammad Ali Department of Horticulture, Faculty of Agricultural Sciences and Technology, Bahauddin Zakariya University, Multan, Pakistan

Muhammad Anjum Ali Directorate General Agriculture, Lahore, Punjab, Pakistan

Musaddiq Ali Department of Environmental Sciences, COMSATS University Islamabad, Vehari, Pakistan

Shaukat Ali Global Change Impact Studies Centre (GCISC), Ministry of Climate Change, Islamabad, Pakistan

Khizer Amanet Department of Environmental Sciences, COMSATS University Islamabad, Vehari, Pakistan

Asad Amin Queensland Alliance for Agriculture and Food Innovation (QAAFI), The University of Queensland, Brisbane, QLD, Australia

Muhammad Amjad Department of Environmental Sciences, COMSATS University Islamabad, Vehari, Pakistan

Muhammad Akbar Anjum Department of Horticulture, Faculty of Agricultural Sciences and Technology, Bahauddin Zakariya University, Multan, Pakistan

Muhammad Anwar-ul-Haq Institute of Soil and Environmental Sciences, University of Agriculture, Faisalabad, Pakistan

Muhammad Asif Agricultural Linkages Programme (ALP), Pakistan Agricultural Research Council (PARC), Islamabad, Pakistan

Muhammad Aqeel Aslam Department of Agronomy, Pir Mehr Ali Shah Arid Agriculture University, Rawalpindi, Pakistan

Muhammad Aurangzaib Department of Agronomy, The Islamia University, Bahawalpur, Pakistan

Muhammad Awais Department of Agronomy, University of Agriculture, Faisalabad, Pakistan

Department of Agronomy, The Islamia University, Bahawalpur, Pakistan

Mohamed Ali Awale Department of Agronomy, Pir Mehr Ali Shah Arid Agriculture University, Rawalpindi, Pakistan

Adeem Arshad Awan Department of Agronomy, Pir Mehr Ali Shah Arid Agriculture University, Rawalpindi, Pakistan

Hafiz Usman Ayub Department of Agricultural Engineering, Bahauddin Zakariya University, Multan, Pakistan

Muhammad Farooq Azhar Department of Forestry and Range Management, Bahauddin Zakariya University, Multan, Pakistan

Khuda Bakhsh Department of Management Sciences, COMSATS University Islamabad, Vehari, Pakistan

Rashad Mukhtar Balal Department of Horticulture, College of Agriculture, University of Sargodha, Sargodha, Pakistan

Fatima Bibi Plant Nutrition Section, Mango Research Institute, Multan, Pakistan

Muhammad Bilal Department of Agricultural Engineering, Bahauddin Zakariya University, Multan, Pakistan

Maham Asif Bukari Department of Environmental Sciences, COMSATS University Islamabad, Vehari, Pakistan

Muhammad Adnan Bukhari Department of Agronomy, The Islamia University, Bahawalpur, Pakistan

Hassan Javed Chaudhary Department of Plant Sciences, Quaid-e-Azam University, Islamabad, Pakistan

Muhammad Sami Ul Din Department of Environmental Sciences, COMSATS University Islamabad, Vehari, Pakistan

Mabrouk Elsabagh Department of Animal Production and Technology, Faculty of Agricultural Sciences and Technologies, Niğde Ömer Halisdemir University, Niğde, Turkey

Department of Nutrition and Clinical Nutrition, Faculty of Veterinary Medicine, Kafrelsheikh University, Kafr El-Sheikh, Egypt

Shah Fahad College of Plant Science and Technology, Huazhong Agricultural University, Wuhan, Hubei, China

Department of Agronomy, The University of Haripur, Haripur, Hubei, Pakistan

Hafiz Umar Farid Department of Agricultural Engineering, Bahauddin Zakariya University, Multan, Pakistan

Maham Fatima Department of Environmental Sciences, COMSATS University Islamabad, Vehari, Pakistan

Zartash Fatima Department of Agronomy, Bahauddin Zakariya University, Multan, Pakistan

Abdul Ghaffar Department of Zoology, The Islamia University of Bahawalpur, Bahawalpur, Pakistan

Celina Gomez Environmental Horticulture, Institute of Food and Agriculture, University of Florida, Gainesville, FL, USA

Hafiz Mohkum Hammad Department of Environmental Sciences, COMSATS University Islamabad (CUI), Vehari, Pakistan

Muhammad Zaffar Hashmi Department of Chemistry, COMSATS University Islamabad (CUI), Islamabad, Pakistan

Fayyaz-ul-Hassan Department of Agronomy, Pir Mehr Ali Shah Arid Agriculture University, Rawalpindi, Pakistan

Rifat Hayat Institute of Soil Science, PMAS-Arid Agriculture University Rawalpindi, Rawalpindi, Pakistan

Dilshad Hussain Department of Management Sciences, COMSATS University Islamabad (CUI), Vehari, Pakistan

Khalid Hussain Agro-Climatology Laboratory, Department of Agronomy, University of Agriculture, Faisalabad, Punjab, Pakistan

Nazim Hussain Department of Agronomy, Bahauddin Zakariya University, Multan, Pakistan

Saddam Hussain Department of Agronomy, University of Agriculture, Faisalabad, Pakistan

Sajjad Hussain Department of Environmental Sciences, COMSATS University Islamabad, Vehari, Pakistan

Sajjad Hussain Department of Horticulture, Faculty of Agricultural Sciences and Technology, Bahauddin Zakariya University, Multan, Pakistan

Tajamul Hussain Prince of Songkla University (PSU), Hat Yai, Thailand

Muhammad Zahid Ihsan Cholestan Institute of Desert Studies, The Islamia University, Bahawalpur, Pakistan

Javaid Iqbal Department of Agronomy, Ghazi University, Dera Ghazi Khan, Pakistan

Pakeeza Iqbal Department of Botany, University of Agriculture, Faisalabad, Pakistan

Tasmiya Jabeen Department of Environmental Sciences, COMSATS University Islamabad, Vehari, Pakistan

Khawar Jabran Department of Plant Production and Technologies, Nigde Omer Halisdemir University, Nigde, Turkey

Akash Jamil Department of Forestry and Range Management, Bahauddin Zakariya University, Multan, Pakistan

Hafiz Muhammad Rashad Javeed Department of Environmental Sciences, COMSATS University Islamabad, Vehari, Pakistan

Tahira Kalsoom Department of Horticulture, Pir Mehr Ali Shah Arid Agriculture University, Rawalpindi, Pakistan

Rana Muhammad Asif Kanwar Department of Agricultural Engineering, Bahauddin Zakariya University, Multan, Pakistan

Dildar Hussain Kazmi National Agromet Centre, Pakistan Meteorological Department, Islamabad, Pakistan

Tasneem Khaliq Agro-Climatology Laboratory, Department of Agronomy, University of Agriculture, Faisalabad, Pakistan

Amna Khan Department of Agronomy, University College of Agriculture, University of Sargodha, Sargodha, Pakistan

Amna Khan Department of Agronomy, University College of Agriculture, University of Sargodha, Sargodha, Pakistan

Aziz Khan Key Laboratory of Plant Genetics and Breeding, College of Agriculture, Guangxi University, Nanning, China

Imran Khan Agro-Climatology Laboratory, Department of Agronomy, University of Agriculture, Faisalabad, Punjab, Pakistan

M. Imran Khan Department of Mathematics and Statistics, University of Agriculture, Faisalabad, Pakistan

Naeem Khan Department of Agronomy, Institute of Food and Agricultural Sciences, University of Florida, Gainesville, FL, USA

Zahid Mahmood Khan Department of Agricultural Engineering, Bahauddin Zakariya University, Multan, Pakistan

Saida Komal Department of Agronomy, University of Poonch, Rawalakot, Pakistan

Muhammad Younis Laghari Department of Freshwater Biology and Fisheries, University of Sindh, Jamshoro, Pakistan

Hongyan Liu Hainan Key Laboratory for Sustainable Utilization of Tropical Biosource, College of Tropical Crops, Hainan University Haikou, Hainan, China

Faisal Mahmood Department of Environmental Sciences and Engineering, Government College University, Faisalabad, Pakistan

Mehdi Maqbool Department of Horticulture, University of Poonch, Rawalakot, Pakistan

Nasir Masood Department of Environmental Sciences, COMSATS University Islamabad, Vehari, Pakistan

Neil Mattson Horticulture Section, School of Integrative Plant Science, Cornell University, Ithaca, NY, USA

Muhammad Zeeshan Mehmood Department of Agronomy, Pir Mehr Ali Shah Arid Agriculture University, Rawalpindi, Pakistan

Nosheen Mirza Department of Environmental Sciences, COMSATS University Islamabad (CUI), Abbottabad, Khyber Pakhtunkhwa, Pakistan

Department of Soil and Environmental Sciences, Ghazi University, Dera Ghazi Khan, Punjab, Pakistan

Hussani Mubarak Department of Soil and Environmental Sciences, Ghazi University, Dera Ghazi Khan, Pakistan

Muhammad Mubeen Department of Environmental Sciences, COMSATS University Islamabad, Vehari, Pakistan

M. Akhlaq Muddasir Agro-Climatology Lab, Department of Agronomy, University of Agriculture, Faisalabad, Pakistan

Hassan Munir Agro-climatology Lab, Department of Agronomy, University of Agriculture, Faisalabad, Pakistan

M. Farooq Hussain Munis Department of Plant Sciences, Quaid-e-Azam University, Islamabad, Pakistan

Behzad Murtaza Department of Environmental Sciences, COMSATS University Islamabad, Vehari, Pakistan

Shoaib Nadeem Agro-Climatology Lab, Department of Agronomy, University of Agriculture, Faisalabad, Pakistan

Syed Asif Ali Naqvi Department of Economics, Government College University, Faisalabad, Pakistan

Wajid Nasim Department of Agronomy, Faculty of Agriculture and Environment, The Islamia University of Bahawalpur (IUB), Bahawalpur, Pakistan

Hamid Nawaz Department of Agronomy, The Islamia University of Bahawalpur, Bahawalpur, Pakistan

Sahrish Naz Department of Agronomy, Bahauddin Zakariya University, Multan, Pakistan

Nor Azila Bt Mohd Noor Othman Yeop Abdullah Graduate School of Business, University Utara Malaysia, Bukit Kayu Hitam, Malaysia

Ghulam Qadir Department of Agronomy, Pir Mehr Ali Shah Arid Agriculture University, Rawalpindi, Pakistan

Ihsan Qadir Department of Forestry and Range Management, Bahauddin Zakariya University, Multan, Pakistan

Saeed Ahmad Qaisrani Department of Environmental Sciences, COMSATS University Islamabad, Vehari, Pakistan

Muhammad Rafay Department of Forestry, The Islamia University of Bahawalpur, Bahawalpur, Pakistan

Atta Rasool Department of Environmental Sciences, COMSATS University Islamabad, Vehari, Pakistan

Fahd Rasul Agro-Climatology Lab, Department of Agronomy, University of Agriculture, Faisalabad, Pakistan

Hamid Raza Department of Agronomy, University College of Agriculture and Environmental Sciences, The Islamia University of Bahawalpur (IUB), Bahawalpur, Pakistan

Ramsha Razaq Agro-Climatology Lab, Department of Agronomy, University of Agriculture, Faisalabad, Pakistan

Muhammad Ishaq Asif Rehmani Department of Agronomy, Ghazi University, Dera Ghazi Khan, Pakistan

Ayman El Sabagh Department of Field Crops, Faculty of Agriculture, Siirt University, Siirt, Turkey

Department of Agronomy, Faculty of Agriculture, Kafrelsheikh University, Kafr el-Sheikh, Egypt

Muhammad Sajjad Department of Bio-sciences, COMSATS University Islamabad (CUI), Islamabad, Pakistan

Aashir Sameen Department of Agronomy, Pir Mehr Ali Shah Arid Agriculture University, Rawalpindi, Pakistan

Muhammad Saqib Agricultural and Biological Engineering Department, University of Florida, Gainesville, FL, USA

Department of Horticulture, Faculty of Agricultural Sciences and Technology, Bahauddin Zakariya University, Multan, Pakistan

Ali Sarkhosh Horticultural Sciences Department, Institute of Food and Agriculture, University of Florida, Gainesville, FL, USA

Shah Saud The University of Swabi, Anbar, Khyber Pakhtunkhwa, Pakistan

Muhammad Salman Shabbir Department of Management, College of Commerce and Business Administration, Dhofar University, Salalah, Oman

Ghulam Mustafa Shah Department of Environmental Sciences, COMSATS University Islamabad, Vehari, Pakistan

Adnan Noor Shahi Department of Agricultural Engineering, Khwaja Fareed University of Engineering and Information Technology (KFUIT), Rahim Yar Khan, Punjab, Pakistan

Muhammad Shahid Department of Environmental Sciences, COMSATS University Islamabad, Vehari, Pakistan

Muhammad Sohail Department of Horticulture, Faculty of Agricultural Sciences and Technology, Bahauddin Zakariya University, Multan, Pakistan

Muhammad Adnan Shahid Department of Agriculture, Nutrition, and Food Systems, College of Life Sciences and Agriculture, University of New Hampshire, Durham, NH, USA

Khurram Shahzad Lasbela University of Agriculture, Water and Marine Sciences (LUAWMS), Mango Research Institute, Uthal, Balochistan, Pakistan

Sobia Shahzad Department of Botany, Islamia University of Bahawalpur (IUB), Bahawalnagar Campus, Punjab, Pakistan

Muhammad Shakeel Laboratory of Bio-Pesticide Creation and Application of Guangdong Province, College of Natural Resources and Environment, South China Agricultural University, Guangzhou, China

Aamir Shakoor Department of Agricultural Engineering, Bahauddin Zakariya University, Multan, Pakistan

Muhammad Mudassar Shehzad Department of Forestry and Range Management, Bahauddin Zakariya University, Multan, Pakistan

Muhammad Shehzad Department of Agronomy, University of Poonch, Rawalakot, Pakistan

Ahmad Omid Siddiqui Department of Plant Production and Technologies, Nigde Omer Halisdemir University, Nigde, Turkey

Ajit Singh School of Biosciences, University of Noitingum, Semenyih, Malaysia

Henry Sintim Department of Crop and Soil Sciences, University of Georgia, Tifton, GA, USA

Syeda Refat Sultana Department of Environmental Sciences, COMSATS University Islamabad, Vehari, Pakistan

Syeda Refat Sultana Department of Environmental Sciences, COMSATS University Islamabad, Vehari, Pakistan

Muhammad Tahir Department of Environmental Sciences, COMSATS University Islamabad, Vehari, Pakistan

Muhammad Tariq Agronomy Section, Central Cotton Research Institute (CCRI), Multan, Pakistan

Haseeb Ullah Faculty of Agriculture, University of Agriculture, Faisalabad, Pakistan

Kalim Ullah Department of Meteorology, COMSATS University, Islamabad, Pakistan

Muhammad Habib ur Rahman Institute of Crop Science and Resource Conservation (INRES), Crop Science Group, University Bonn, Bonn, Germany

Department of Agronomy, MNS-University of Agriculture, Multan, Pakistan

Abdul Wahid Department of Environmental Sciences, Bahauddin Zakariya University, Multan, Pakistan

Fazli Wahid The University of Swabi, Anbar, Khyber Pakhtunkhwa, Pakistan

Syed Aftab Wajid Agro-Climatology Lab, Department of Agronomy, University of Agriculture, Faisalabad, Pakistan

Mirza Waleed Department of Environmental Sciences, COMSATS University Islamabad, Vehari, Pakistan

Chao Wu Guangxi Institute of Botany, Chinese Academy of Sciences (CAS), Beijing, China

Wang Xuechun School of Life Science and Technology, South West University of Science and Technology, Mianyang, Sichuan, China

Azra Yasmeen Department of Agronomy, Bahauddin Zakariya University of Bahawalpur, Bahawalpur, Pakistan

Ayşe Yazlik Plant Production Department, Duzce University, Duzce, Turkey

Nosheen Zahid Department of Horticulture, University of Poonch, Rawalakot, Pakistan

Iqra Zakir Department of Agronomy, Faculty of Agricultural Sciences and Technology, Bahauddin Zakariya University, Multan, Pakistan

Muhammad Zamin Department of Agriculture, The University of Swabi, Anbar, Khyber Pakhtunkhwa, Pakistan

Raja Zoq-ul-Arfeen School of Food and Agricultural Sciences, University of Management and Technology, Lahore, Pakistan

Muhammad Zubair Department of Horticulture, College of Agriculture, University of Sargodha, Sargodha, Pakistan

Part I
Basic Concepts of Climate Resilience for Agriculture and Associated Fields

Chapter 1
An Introduction to Climate Change Phenomenon

Sahrish Naz, Zartash Fatima, Pakeeza Iqbal, Amna Khan, Iqra Zakir, Haseeb Ullah, Ghulam Abbas, Mukhtar Ahmed, Muhammad Mubeen, Sajjad Hussain, and Shakeel Ahmad

Abstract Long-term change in the average weather patterns of the earth's local, regional, and global climates is called climate change. These changes have a broad range of observed effects. Climate change is mainly linked to changes, which the scientists observed over the current decade and those that are projected to be happening, mainly as a consequence of human behavior, in all the continents. Other reasons for climate change include greenhouse effect, global warming, urbanization, and deforestation. Due to these impacts of climate change, temperature rise, increase in CO_2 concentration, seawater rise, and ocean acidity also took place in the past. The earth's ecosystem is rapidly changing due to climate change. In this book chapter, we elaborate the reasons for the natural greenhouse gas emissions and the role of various stakeholders in the phenomenon of climate change.

Keywords Greenhouse effect · Global warming · Urbanization · Deforestation · Ozone depletion

S. Naz · Z. Fatima · I. Zakir · G. Abbas · S. Ahmad (✉)
Department of Agronomy, Bahauddin Zakariya University, Multan, Pakistan
e-mail: shakeelahmad@bzu.edu.pk

P. Iqbal
Department of Botany, University of Agriculture, Faisalabad, Pakistan

A. Khan
Department of Agronomy, University College of Agriculture, University of Sargodha, Sargodha, Pakistan

H. Ullah
Faculty of Agriculture, University of Agriculture, Faisalabad, Pakistan

M. Ahmed
Department of Agronomy, Pir Mehr Ali Shah Arid Agriculture University, Rawalpindi, Pakistan

M. Mubeen · S. Hussain
Department of Environmental Sciences, COMSATS University Islamabad, Vehari, Pakistan

W. N. Jatoi et al. (eds.), *Building Climate Resilience in Agriculture*,
https://doi.org/10.1007/978-3-030-79408-8_1

1.1 Introduction

Climate is a long-term fluctuation (statistical expression) of the weather of a particular place. Climate change is known as a change in expected weather. Changes in climate are different at different geographical scales and time period (Ahmad et al. 2019; Abbas et al. 2020). Indicators of climate change are of two types: in which primary indicators are physical which include changes in temperature, rainfall patterns, and weather conditions while the secondary indicators include economic, social, and ecological impacts. Secondary indicators are also known as consequences and vary with the condition (Mendelsohn et al. 2006). Climate change occurs due to the effect of different factors which can be divided into three groups: external, internal, and anthropogenic. External elements include effect of astronomical and orbital activities. Internal elements include earth—geophysical, geological, and geographical (Nikolov and Petrov 2014; Haasnoot et al. 2020).

Global energy balance and total amount of energy stored in the climate system can be used to determine the overall state of global climate; global energy balance can be defined as balance between the amount of solar energy received by the earth and that released back to space (Fig. 1.1) and the regulation of energy balance depends upon the flow of energy in the global climate system (Tariq et al. 2018; Ghahramani et al. 2020). Changes in the energy flow in the climate system and global energy balance are the major causes of changes in climate. Changes in ocean

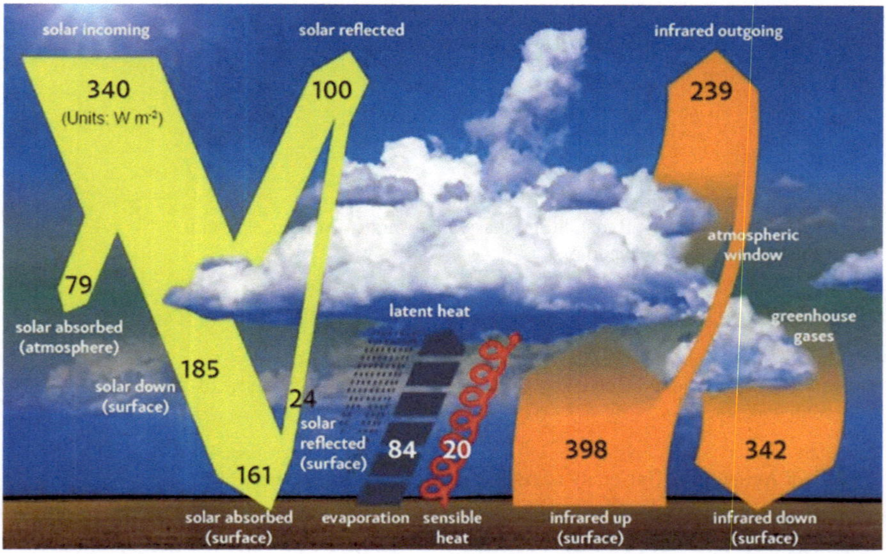

Fig. 1.1 The rates at which energy enters the Earth system from the Sun, and leaves the system. The arrows show global average energy transfer rates in units of Watts per square meter. With more greenhouse gases in the atmosphere, but no other changes, the system must reach a higher temperature to maintain balance

circulation, variations in quantity of energy reaching from the sun, changes in composition of the atmosphere, and changes in earth's orbit are also major causes of climate change (Ahmad et al. 2017a, b; Abbas et al. 2017).

1.2 Greenhouse Effect and Global Warming

Greenhouse gases are causing global warming. Greenhouse gases (GHGs) are the gaseous combinations in the atmosphere which can absorb infrared radiation and trap heat in the atmosphere. By trapping heat in the atmosphere, GHGs cause greenhouse effect that eventually leads to global warming (Ahmad et al. 2017a, b; Awais et al. 2017; Jabran et al. 2017; Rahman et al. 2017*; Nasim et al. 2017; Mahmood et al. 2017; Scoville-Simonds et al. 2020). According to Environmental Protection Agency (EPA) the USA, the most significant GHSs are CO_2, CH_4, N_2O, fluorinated gases, and water vapor. Greenhouse gases enter the atmosphere in two ways: by natural means, i.e., plants and human respiration, and by human activities, i.e., deforestation, use of artificial fertilizers, fossil fuel use, and intensive livestock farming besides industrial processes (Ahmad et al. 2016). The top CO_2 emitter countries are mentioned in Fig. 1.2. Due to climate warming, glaciers are melting and ultimately water level in oceans is rising (Fig. 1.3).

Global warming is the increase of average maximum and minimum air temperature of climate system of earth primarily caused by humans and has been recognized by direct temperature measurements and with the help of measurements of several impacts of climate change (Ahmad et al. 2017a, b; Berrang-Ford et al. 2019; van Valkengoed and Steg 2019). In all the continents, it is a foremost cause of climatic changes which, in addition to enhancing surface temperatures of all regions, also results in changes in rainfall pattern (Ali and Erenstein 2017; Rodriguez et al. 2018). Climate warming happens when greenhouse gases as well as more air contaminants

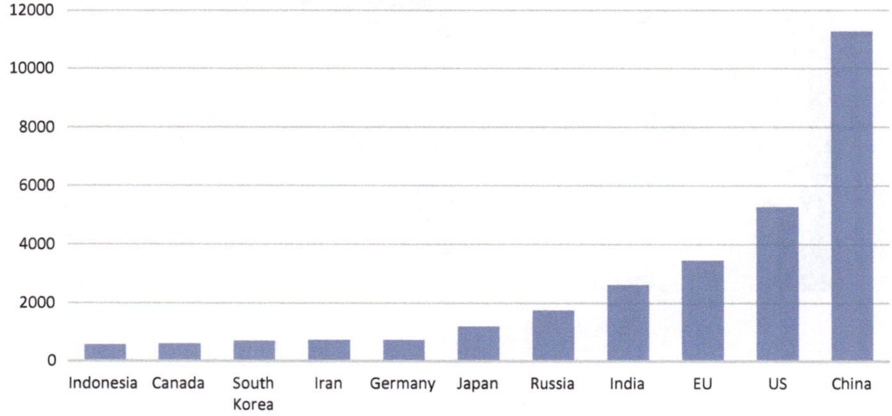

Fig. 1.2 The top CO_2 emitters countries in the world

Fig. 1.3 Effect of climate change on Grinnel Glacier Montana, USA (**Source:** www.nasa.com)

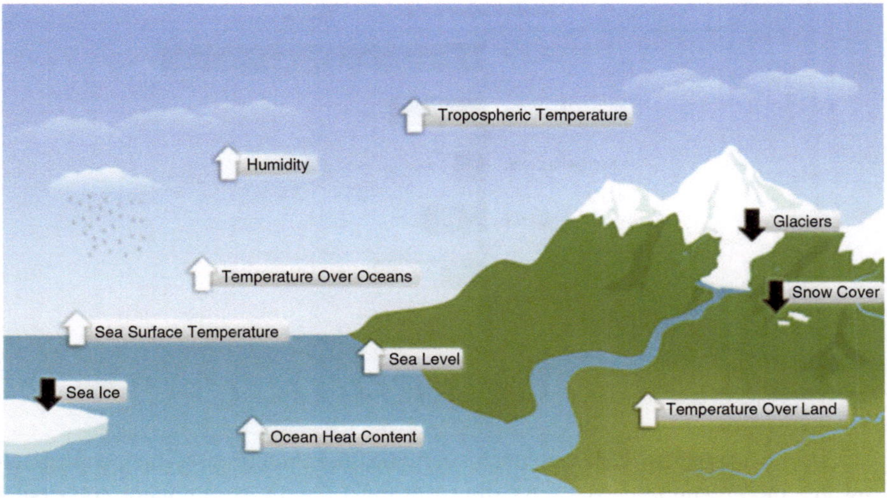

Fig. 1.4 Effect of global warming

accumulate into atmosphere (Adamson et al. 2018). Generally, the radiations can be lost into space, but these pollutants, which can last for years to centuries in the atmosphere, trap the heat and cause the planet to get hotter (Tonmoy et al. 2020). Impacts of global warming comprise rising levels of sea, regional variations in rainfall, more recurrent extreme weather events like heat waves, in addition to deserts expansion (Rippke et al. 2016; Tenzing 2020). Surface temperature rise is highest in Arctic that has contributed to glaciers retreat, permafrost, besides sea ice. Generally, high temperature brings extra precipitation and snowfall; however for certain regions droughts are accompanied by more wildfires. Climate change appears to reduce the productivity of crops, threatening food security; and rising levels of sea may flood coastal infrastructure (Araos et al. 2016). Some of the effects of global warming are shown in Fig. 1.4.

1.3 Ozone Depletion and Climate Change

Ozone is a colorless, irritating, corrosive gas found in the upper atmosphere of Earth. Major causes of Ozone layer depletion include chlorofluorocarbons, global warming, nitrogenous compounds, etc. Ozone present in the atmosphere affects the earth's temperature in two ways. Firstly, the effect of ozone on climate change varies with the ozone concentrations and altitude at which this change occurs. Secondly, it absorbs infrared radiation and traps the heat in troposphere (Lesnikowski et al. 2019). Due to human-produced chlorine and bromine (which produce cooling effect), ozone concentration in lower stratosphere zone has been reduced. Whereas, the ozone amount present in troposphere (which absorbs ultraviolet radiation and

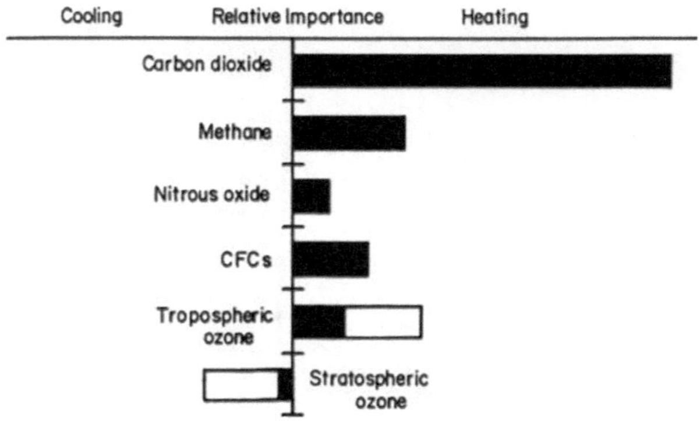

Fig. 1.5 Changes in abundance of various gases in the atmosphere

increases heat) increases due to human activities (i.e., pollution) which contributes towards the greenhouse effect. As shown in Fig.1.5, CO_2 is the major contributor and its concentration increased by burning of oil, natural gas, and coal for transportation and energy. Gases which are causing ozone depletion are also contributing to climate change.

1.4 Effect of Automobiles on Climate Change

Vehicle having its own motor is called automobile. Automobiles were created in the late 1800s and have been causing pollution since their inception (Haida et al. 2019). The effect of automobiles on climate change includes melting of ice on lakes and rivers, shrinking of glaciers, earlier flowering, quick rise in sea level, long heatwaves with more intensity, and smog (Figs. 1.6 and 1.7). Vehicles are one of the major reasons for global warming because of gases emission. Every gallon of vehicular gas contains 24 pounds of carbon dioxide and other global warming gases (Nitrogen oxides, Sulfur dioxide, Hydrocarbons).

1.5 Effects of Deforestation on Climate Change

Forests contribute to the rain pattern and are useful for recycling the rainfall. But, forests are permanently destroyed in order to use the land for other purposes. Causes of deforestation include urbanization, population increase, road construction, over grazing, timber, landslides, fuelwood, forest fires, mining, and forest disease. To quote an example, the annual depletion rate of forest in Pakistan is 1.5% resulting in 39,000 ha of deforestation leaving only 4.8% area with forest. As an international

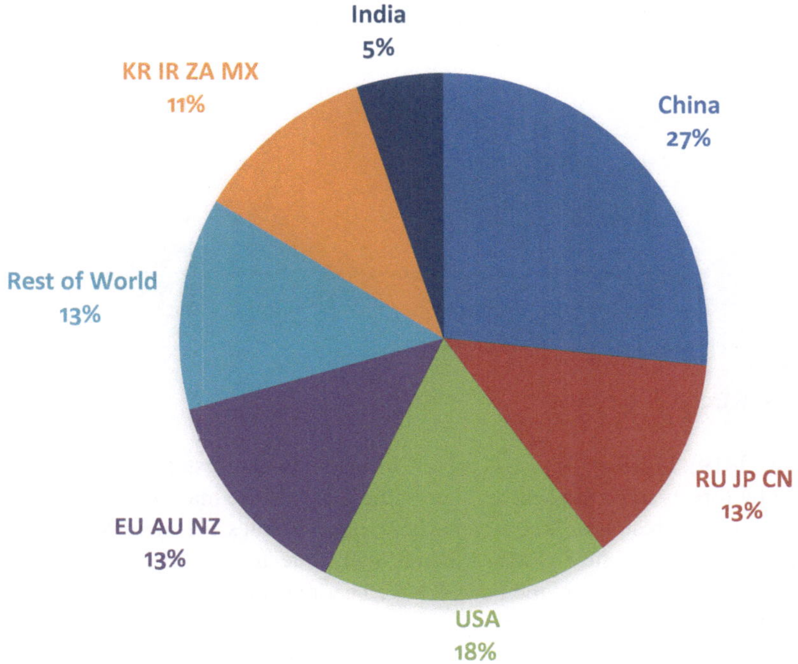

Fig. 1.6 Status of CO_2 emissions due to automobiles

Fig. 1.7 Automobiles source of pollution (**Source:** https://helpsavenature.com/impact-of-car-pollution-on-environment)

standard, 25% area should be under forest cover (Schweikert et al. 2014; Alghabari et al. 2016; Jabran et al. 2016; Fahad et al. 2016c; Nosheen et al. 2016).

Due to deforestation, there is a drastic decrease in the sink of carbon dioxide which results in less CO_2 conversion into oxygen leaving a huge impact on climate

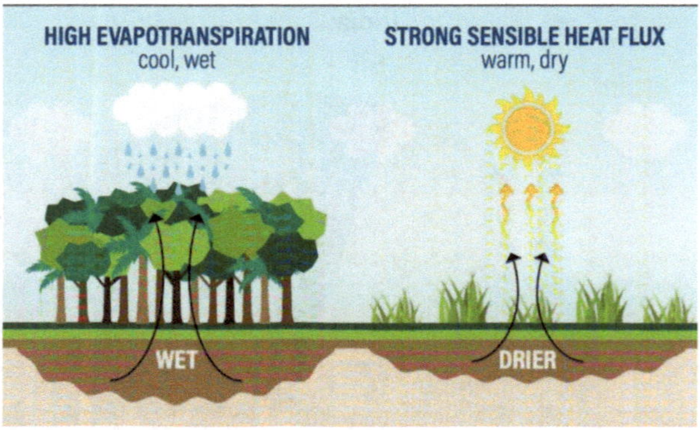

Fig. 1.8 Effect of deforestation

(Juhola 2019). Weather pattern is greatly affected by deforestation water vapor produced by forests traps the heat and keeps the temperature maintained, but due to an increase in deforestation this process is affected besides the increase in earth and atmospheric temperature (Figs. 1.8 and 1.9). Due to the presence of excessive amount of carbon dioxide, the oceans have become acidic leading to change in pH level of the oceans and killing many species of animals and plants. Due to deforestation, the soil (which is sometime contaminated with manmade material such as pesticides and other chemicals) becomes eroded resulting in the contamination of rivers and lakes. So, this water becomes dangerous for animals that drink from these water resources (Nasim et al. 2016a, b, c; Mubeen et al. 2016; Rasool et al. 2016; Fahad et al. 2016a, b; Žurovec and Vedeld 2019).

1.6 Climate Change and Urbanization

Migration of population from rural to urban areas results in the expansion of cities and towns tremendously. Urbanization's effect on climate change include Heat stress, Extreme weather events, Inland flooding, Ocean acidification, Rising temperatures, Sea-level rise, and storm surges. The temperature increase due to urbanization since preindustrial level is from 2.5–4 °C. It also leads to numerous hot days and warm spells. Increase in temperature causes heat-related health issues and pollution.

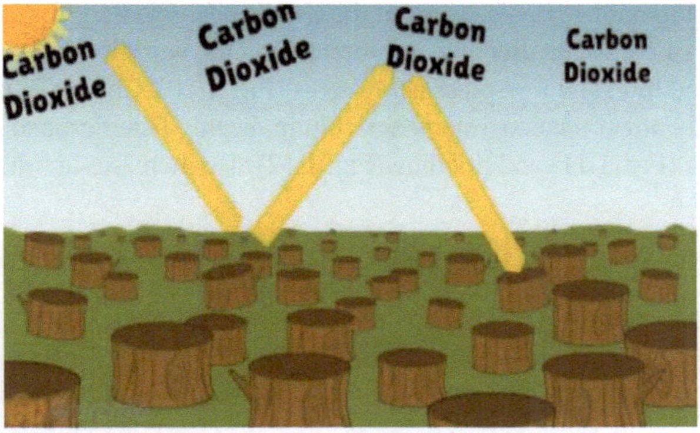

Fig. 1.9 Effect of deforestation (**Source:** https://www.google.com/imgres)

1.7 Human Influence

In earlier human history our effect on the climate was quite simple but with the passage of time it gradually increased. At the end of nineteenth century, the industrial revolution had a great effect on climate. Carbon dioxide amount increased in the atmosphere with burning of fossil fuel and invention of motors which led to the greenhouse effect later (Ahmed et al. 2021; Tariq et al. 2021; Fatima et al. 2020, 2021). Cutting down of trees also contributed towards increased quantity of CO_2 in atmosphere by reducing the absorbance of CO_2 by the trees.

1.8 Role of the Sun

Sun plays a primary role in influencing the earth's climate system. Beer et al. (2000) stated that the sun plays a pivot role in the earth's climate system and is considered as an engine that drives the climate system; however, the variability of results and range of this effect is still not clear. Gray et al. (2010) stated that the sun is the major component among all the components our climate is made of. Whereas, according to IPCC the variation in solar energy (long or short) plays a minute role in climate change as compared to GHGs produced by human.

1.9 El Niño

It is the phenomenon which causes drought and heavy rainfall in different parts of the world by disrupting the weather patterns. It is a complex and natural phenomenon. Break down of trade winds and moving back of water towards the eastern side of Pacific Ocean instead of western side results in rain and storms. Global atmosphere temperature rises because of the excess heat released from the warm ocean water. So, the El Niño can affect the weather around the world by influencing precipitation, winds, and high and low-pressure systems (Fig. 1.10). Change in the weather pattern can greatly affect the ecosystem, agriculture, health, fisheries, and air quality. The risk of wildfire (Fig. 1.11) and flooding (Fig. 1.12) also increases around the globe.

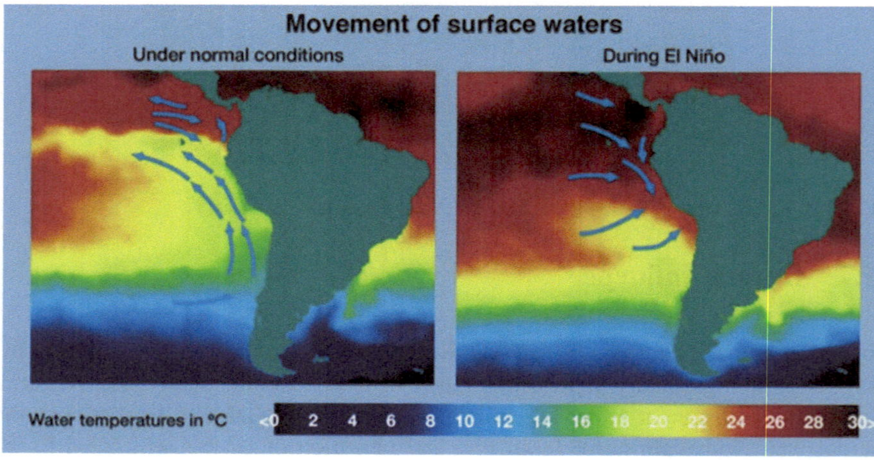

Fig. 1.10 Movement of surface water during El Niño (Source: https://commons.wikimedia.org/wiki/File:Movement_of_surface_waters_during_El_Nino.jpg)

Fig. 1.11 Wildfires in Indonesia

Fig. 1.12 Flooding in Chennai, India

1.10 Conclusion

Climate change is a global phenomenon and is a serious threat to all fields of life. Climate change is increasing rapidly day by day due to the enhancement of concentrations of greenhouse gases at local, regional, and global levels. Main reasons for increasing greenhouse gases are deforestation, use of artificial fertilizers, fossil fuel use, livestock farming, and industrial processes. Ecosystem of the earth is rapidly changing due to climate change. In order to save our future, concentrated efforts are needed at national, regional, and global levels. There are many stakeholders which are playing their respective roles in increasing the chlorofluorocarbons in the environment in the name of development. Development should not be at the cost of our environment. We must save our environment. In this regard, effective measures should be adopted to decrease the release of greenhouse gases in the atmosphere by the developed countries for the benefit of developing and underdeveloped nations.

References

Abbas G, Ahmad S, Ahmad A, Nasim W, Fatima Z, Hussain S, Rehman MH, Khan MA, Hasanuzzaman M, Fahad S, Boote KJ, Hoogenboom G (2017) Quantification the impacts of climate change and crop management on phenology of maize-based cropping system in Punjab, Pakistan. Agr For Meteorol 247:42–55

Abbas G, Fatima Z, Hussain M, Hussain S, Sarwar N, Ahmed M, Ahmad S (2020) Nitrogen rate and hybrid selection matters productivity of maize–maize cropping system under irrigated arid environment of Southern Punjab, Pakistan. Int J Plant Prod. https://doi.org/10.1007/s42106-020-00086-5

Adamson GC, Hannaford MJ, Rohland EJ (2018) Re-thinking the present: The role of a historical focus in climate change adaptation research. Glob Environ Change 48:195-205

Ahmad S, Nadeem M, Abbas G, Fatima Z, Khan RJZ, Ahmed M, Ahmad A, Rasul G, Khan MA (2016) Quantification of the effects of climate warming and crop management on sugarcane phenology. Clim Res 71:47-61

Ahmad S, Abbas G, Fatima Z, Khan RJ, Anjum MA, Ahmed M, Khan MA, Porter CH, Hoogenboom G (2017a) Quantification of the impacts of climate warming and crop management on canola phenology in Punjab, Pakistan. J Agron Crop Sci 203:442-452

Ahmad S, Abbas Q, Abbas G, Fatima Z, Naz S, Younis H, Khan RJ, Nasim W, Rehman MH, Ahmad A, Rasul G (2017b) Quantification of climate warming and crop management impacts on cotton phenology. Plants 6:7

Ahmad S, Abbas G, Ahmed M, Fatima Z, Anjum MA, Rasul G, Khan MA, Hoogenboom G (2019) Climate warming and management impact on the change of phenology of the rice-wheat cropping system in Punjab, Pakistan. Field Crops Res 230:46-61

Ahmed M, Fahad S, Arif Ali M, Hussain S, Tariq M, Ilyas F, Ahmad S, Saud S, Hammad HM, Nasim W, Wu C, Liu H (2021) Hydrogen Sulfide: A Novel Gaseous Molecule for Plant Adaptation to Stress. Journal of Plant Growth Regulation. https://doi.org/10.1007/s00344-020-10284-0

Alghabari, F, et al. (2016) Gibberellin-sensitive Rht alleles confer tolerance to heat and drought stresses in wheat at booting stage. J Cereal Sci 70: 72-78

Ali A, Erenstein O (2017) Assessing farmer use of climate change adaptation practices and impacts on food security and poverty in Pakistan. Clim Risk Manag 16:183-194

Araos M, Berrang-Ford L, Ford JD, Austin SE, Biesbroek R, Lesnikowski A (2016) Climate change adaptation planning in large cities: a systematic global assessment. Environ Sci Policy 66:375-382

Awais, M., et al. (2017) Modeling the water and nitrogen productivity of sunflower using OILCROP-SUN model in Pakistan. Field Crops Res 205: 67-77

Beer J, Mende W, Stellmacher R (2000) Quaternary Science Reviews 19:403–415

Berrang-Ford L, Biesbroek R, Ford JD, Lesnikowski A, Tanabe A, Wang FM, Chen C, Hsu A, Hellmann JJ, Pringle P, Grecequet M (2019) Tracking global climate change adaptation among governments. Nat Clim Change 9:440-449

Fahad, S., et al. (2016a) Exogenously Applied Plant Growth Regulators Affect Heat-stressed Rice Pollens. J Agron Crop Sci 202:139-150

Fahad, S., et al. (2016b) A combined application of biochar and phosphorus alleviates heat-induced adversities on physiological, agronomical and quality attributes of rice. Plant Physiology and Biochemistry, 103:191-198

Fahad S., et al. (2016c) Responses of Rapid Viscoanalyzer Profile and Other Rice Grain Qualities to Exogenously Applied Plant Growth Regulators under High Day and High Night Temperatures. PLoS One: 11(7):1-13

Fatima, Z., Ahmed, M., Hussain, M. Abbas G, Ul-Allah S, Ahmas S, Ahmed N, Ali MA, Sarwar G, ul Haque E, Iqbal P, Hussain S (2020) The fingerprints of climate warming on cereal crops phenology and adaptation options. Sci Rep 10:18013

Fatima Z, Atique-ur-Rehman, Abbas G, Iqbal P, Zakir I, Khan MA, Kamal GM, Ahmed M, Ahmad S (2021) Quantification of climate warming and crop management impacts on phenology of pulses-based cropping systems. Int J Plant Prod 15:107-123

Ghahramani A, Kingwell RS, Maraseni TN (2020) Land use change in Australian mixed crop-livestock systems as a transformative climate change adaptation. Agric Syst 180:102791

Gray LJ, Beer J, Geller M, Haigh JD, Lockwood M, Matthes K (2010) Solar influences on climate. Rev Geophys 48:1–53

Haasnoot M, Biesbroek R, Lawrence J, Muccione V, Lempert R, Glavovic B (2020) Defining the solution space to accelerate climate change adaptation. Reg Environ Change 20(2):1-5

Haida C, Chapagain AK, Rauch W, Riede M, Schneider K (2019) From water footprint to climate change adaptation: Capacity development with teenagers to save water. Land Use Policy 80:456-463

Jabran K., et al. (2016) Economic assessment of different mulches in conventional and water-saving rice production systems. Environmental Science and Pollution Research, 23 (9): 9156-9163

Jabran, K. et al. (2017) Growth and physiology of basmati rice under conventional and water-saving production systems. Archives of Agronomy and Soil Science, 63(10), 1465-1476

Juhola SK (2019) Responsibility for climate change adaptation. Wiley Interdisciplinary Reviews: Climate Change 10:e608

Lesnikowski A, Ford JD, Biesbroek R, Berrang-Ford L (2019) A policy mixes approach to conceptualizing and measuring climate change adaptation policy. Climatic Change 156:447-469

Mahmood F., et al. (2017) Economic and environmental impacts of introducing grain legumes in farming systems of Midi-Pyrénées region (France): A simulation approach. International Journal of Plant Production 1 (1):65-87

Mendelsohn R, Dinar A, Williams L (2006) The distributional impact of climate change on rich and poor countries. Environ Dev Econ 11:159–178

Mubeen M., et al. (2016) Application of CSM-CERES-Maize model in optimizing irrigated conditions. Outlook on Agriculture, 45 (3): 1-12

Nasim, W., et al. (2016a) Modelling climate change impacts and adaptation strategies for sunflower in Punjab-Pakistan. Outlook on Agriculture, 45 (1): 39-45.

Nasim, W., et al. (2016b) Correlation studies on nitrogen for sunflower crop across the agroclimatic variability. Environmental Science and Pollution Research, 23 (4): 3658-3670

Nasim, W., et al. (2016c) Evaluation of the OILCROP-SUN model for sunflower hybrids under different agro-meteorological conditions of Punjab-Pakistan. Field Crops Research, 188: 17-30

Nasim, W., et al. (2017) Response of sunflower hybrids to nitrogen application grown under different Agro-environments. Journal of Plant Nutrition, 40 (1): 82-92

Nikolov T, Petrov N (2014) Main factors influencing climate change: a review Comptesrendusdel 'Academiebulgare des Sciences 67:1455–1475

Nosheen M., et al. (2016) Biochemical and metabolic changes in arsenic contaminated Boehmeria nivea L. BioMed Research International, 2016: 1-8

Rasool A., et al. (2016) Arsenic and heavy metal contaminations in the tube well water of Punjab, Pakistan and risk assessment: A case study. Ecological Engineering, 95: 90-100

Rippke U, Ramirez-Villegas J, Jarvis A, Vermeulen SJ, Parker L, Mer F, Diekkrüger B, Challinor AJ, Howden M (2016) Timescales of transformational climate change adaptation in sub-Saharan African agriculture. Nat Clim Change 6:605-609

Rodriguez RS, Ürge-Vorsatz D, Barau AS (2018) Sustainable development goals and climate change adaptation in cities. Nat Clim Change 8:181-183

Schweikert A, Chinowsky P, Espinet X, Tarbert M (2014) Climate change and infrastructure impacts: comparing the impact on roads in ten countries through 2100. Procedia Eng 78:306–316

Scoville-Simonds M, Jamali H, Hufty M (2020) The Hazards of Mainstreaming: Climate change adaptation politics in three dimensions. World Dev 125:104683

Tariq M, Fatima Z, Iqbal P, Nahar K, Ahmad S, Hasanuzzaman M (2021) Sowing dates and cultivars mediated changes in phenology and yield traits of cotton-sunflower cropping system in arid environment. Int J Plant Prod 15:291-302

Tariq M, Ahmad S, Fahad S, Abbas G, Hussain S, Fatima Z, Nasim W, Mubeen M, Rehman MH, Khan MA, Adnan M (2018) The impact of climate warming and crop management on phenology of sunflower-based cropping systems in Punjab, Pakistan. Agr For Meteorol 256:270-282

Tenzing JD (2020) Integrating social protection and climate change adaptation: A review. Wiley Interdisciplinary Reviews: Climate Change 11:e626

Tonmoy FN, Cooke SM, Armstrong F, Rissik D (2020) From science to policy: Development of a climate change adaptation plan for the health and wellbeing sector in Queensland, Australia. Environ Sci Policy 108:1-13

Van Valkengoed AM, Steg L (2019) Meta-analyses of factors motivating climate change adaptation behaviour. Nat Clim Change 9:158-163

Žurovec O, Vedeld PO (2019) Rural livelihoods and climate change adaptation in laggard transitional economies: A case from Bosnia and Herzegovina. Sustainability 11:6079

Chapter 2
Agro-meteorological Aspect of Climate Change

Dildar Hussain Kazmi, Muhammad Afzaal, Shaukat Ali, and Wajid Nasim

Abstract Climatic conditions have always been a matter of concern for the farming community. Climate change is one of the great challenges faced by today's agriculture system. The industrial revolution has poked the global temperatures through the increment of greenhouse gases. The rainfalls are becoming uncertain and going beyond the normal pattern. Whereas, the global population is on increase in the particular regions and is jointly threatening the underdeveloped states in terms of food security, consequently adding the hardships of the third world countries in providing a fair diet to their population. To address these issues, Food and Agriculture Organization along with other partners have been working on a reform program, "Climate Smart Agriculture." More investments would have to be made for agricultural reforms to meet the new demands and to address the emerging issues like limited water and less available area for production. Mega collaboration by the major stakeholders could enable the world to address the climate-based challenges for agricultural production.

Keywords Climate change · Greenhouse gases · Crop water requirement · Crop production · Reference crop evapotranspiration

D. H. Kazmi (✉) · M. Afzaal
National Agromet Centre, Pakistan Meteorological Department, Islamabad, Pakistan

S. Ali
Global Change Impact Studies Centre (GCISC), Ministry of Climate Change, Islamabad, Pakistan

W. Nasim
Department of Agronomy, Faculty of Agriculture and Environment (FA&E), Islamia University of Bahawalpur (IUB), Bahawalpur, Pakistan

W. N. Jatoi et al. (eds.), *Building Climate Resilience in Agriculture*,
https://doi.org/10.1007/978-3-030-79408-8_2

2.1 Introduction

Climate change has been a great challenge for the research community as well as the policymakers over the globe. Likewise, agricultural production which is the most important segment of life, had largely been affected by climate change (Fig. 2.1). The farmers are growing their crops in the vast uncovered agricultural lands directly under the influence of weather and climate of the specified region. The plant scientists/engineers guide the farmers in accordance with the prevailing conditions of climate and soil features. Most of the farmers practice their activities in the light of those instructions but at times they have to face great losses due to unexpected weather conditions besides other factors. Occasionally, heavy rains along with persistent cloudy conditions trigger some viral or pest attack on standing crops or result in the rapid growth of weeds in the fields. That would prolong the crop season as well and causes delay in the sowing of next season's crops. In the same way, rainfalls right after sowing cause minimized germination that produces a major reduction in the final yield. Rashid and Rasul (2012) stated that fluctuation of rainfall badly affects the agriculture production in rain-fed areas (Amin et al. 2018a, b; Rahman et al. 2018). Also, it becomes problematic for the soil with lower water holding capacity or in the hot and drier seasons with inadequate soil moisture (Pratley 2003).

Abnormal rise in the daytime temperature raises crop water requirement (CWR) at important phenological stages. It causes early completion of significant phases and early maturity of grains as well due to which shriveled grain is obtained. It has been revealed by Elshamy et al. (2006) that the variation in air temperatures triggers the uncertainty in precipitation, rate of evapotranspiration, stream flow, etc., which

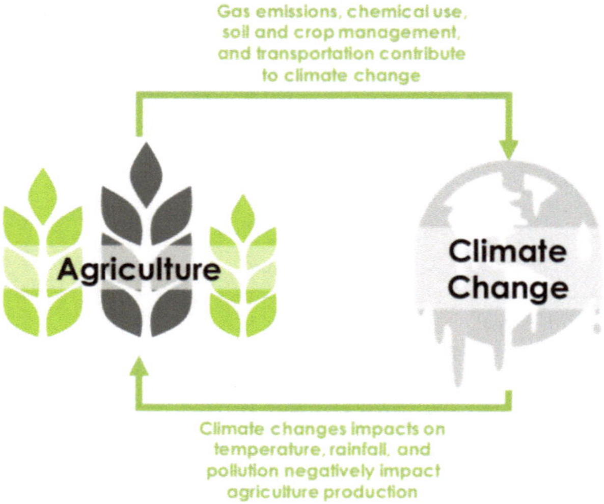

Fig. 2.1 Agriculture and Climate Change Negative Cycle. (Available at https://digital.hbs.edu/platform-rctom/submission/can-vertical-farming-reduce-agricultures-impact-on-climate-change/)

are the most significant factors for crops. Also, the occurrence of extreme events is on increase due to enhanced greenhouse gases (GHG) emissions (Hennessy et al. 1997). Moreover, there are a number of climate-based phenomena including floods, consistent droughts, heat waves, cold waves, etc., which occasionally demolish the crops/orchids over an extended area.

On the other hand, the population of the whole world especially the underdeveloped region is on increase and so is the demand for food and other resources. There are some countries which can manage sufficient food on their own or have the capacity to import it from elsewhere. But generally, the third world states are struggling to provide a fair diet to their people.

2.2 Scope of Agrometeorology

Meteorology is a vast and dynamic scientific field, because the interaction between the atmosphere, hydrosphere, and cryosphere (along with local topographic and orographic features) play an important role in deriving the weather and climate of this planet. In the context of climate impact on agriculture, Agrometeorology is an emerging branch of climate sciences that has got great attention in the recent past. The research and development divisions (R&Ds) serve as a think tank for the advancements in all the realms; therefore, the developed countries have established R&Ds in various scientific, social, and economic arenas including Agrometeorology. The detection of some global phenomena like El-Nino/Southern Oscillation (ENSO) and La-Nina phenomena, global warming, and assessment of their consequent impacts on natural ecosystem, water resources, agriculture, deserts, vegetarian and human health, etc., are the result of dedicated and well-coordinated research activities. These achievements were made possible only with the induction of newly emerged technologies and through employing dedicated and competent professionals. It is revealed by Rasul and Kazmi (2013) that CRW is mainly dependent on air temperature, radiation intensity, cloud cover, air humidity and wind speed, with temperature as the leading actor. It is observed that all of the recent three decades remained warmer based on historical data since 1850. Besides, the period from 1983 to 2012 has been the warmest 30-year period of the last 1400 years in the Northern Hemisphere (IPCC, AR5). In the context of global warming, thermal regime over the globe boosted significantly in the recent era. Consequently, the frequency of extreme events like heat waves, flooding, drought, etc., have been increased, causing threats to crop production.

To address the challenge of climate change-based issues to crop production, Food and Agriculture Organization (FAO) along with other stakeholders have been working on a reform program, "Climate Smart Agriculture" (FAO 2019). A conference was held at Hague in 2010 "The Hague Conference on Agriculture, Food Security and Climate Change" where CSA has been discussed thoroughly (FAO 2010). To achieve the sustainability of agricultural productivity and incomes, CSA scheme focuses on the management of farms, livestock, landscape, etc., and effective services for the farmers.

2.3 Evidences of Climate Change and Consequent Impacts on Crop Production

Just like animals, plants need favorable climate conditions for healthier growth. Soil, water as well as air or atmospheric conditions contribute majorly to the production of biomass. Generally, weather parameters like rainfall, air temperatures, relative humidity, wind velocity, etc., affect the standing crops. Also, their long-term averages are very important for any crop in addition to the geographical features of a particular area. Besides, agrometeorological features like degree days and Reference Crop Evapotranspiration (ETo) have their own significance. Moreover, hydrological factors like CWR are directly linked with climate elements. Breeders introduce a variety after extensive research but they are compelled to modify it shortly due to variation in the atmospheric pattern.

Although, the evidences of climate change and/or global warming are not very new ones but more pronounced in the recent era (Fig. 2.2).

Rasul and Chaudhry (2006) stated that global temperatures are at increase largely due to anthropogenic activities. Frequency of ENSO episodes has been increased so as its impact on the occurrence of major climate features like precipitation in different parts of the world (Nasim et al. 2018; Ali et al. 2019; Hussain et al. 2019). Consequently, the stress represented by agrometeorological elements including ETo, CRW, etc., has been amplified. In a recent study (Naheed and Kazmi 2016) of Pakistan, it has been revealed that air temperatures and corresponding agromet features like Reference Crop Evapotranspiration (ETo) have also been amplified, particularly for summers (Fig. 2.3). It has also been observed that the CRW is enhanced and crop's growth is being affected in many parts of South Asia. On the other hand,

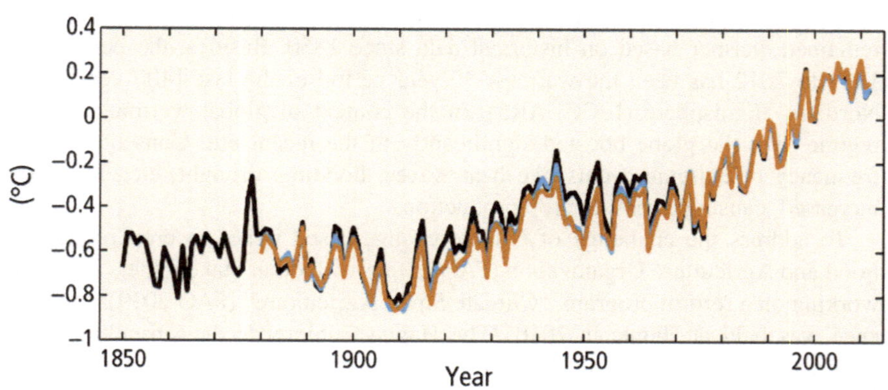

Fig. 2.2 Globally averaged combined land and ocean surface temperature anomaly (IPCC, AR5)

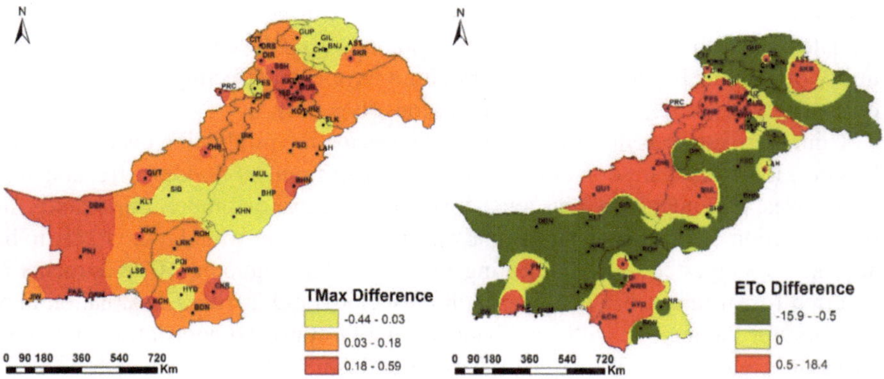

Fig. 2.3 Difference of Max temperature (left) and ET0 (right) during Kharif season (May to September) for the period 1971–2000 and 1981–2010 (Naheed and Kazmi 2016)

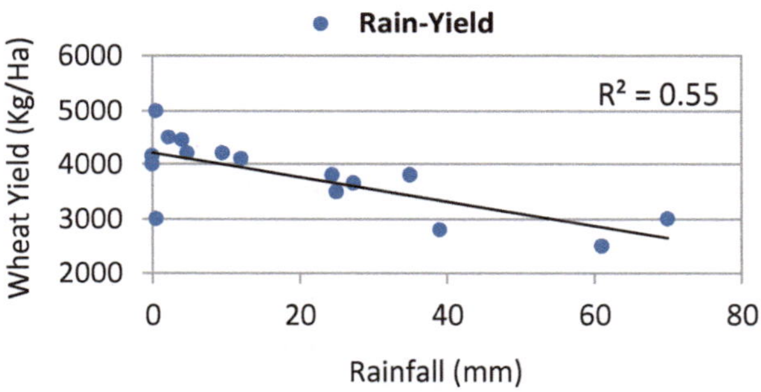

Fig. 2.4 Rainfall (mm) during second decade of January verses Wheat yield (kg/ha) obtained for the period 1994–2011 (Kazmi and Rasul 2012)

in the context of global warming, increasing temperatures may be favorable for the crops in the highlands but the agriculture in the low lands would be affected largely (Hussain and Mudasser 2004). Also, temperature and relative humidity are the major drivers behind CRW and higher rate of relative humidity would project lower efficiency of added nutrients (Rasul 2000; Amin et al. 2018b, c).

In a world where a considerable portion of agriculture production is being met through rain-fed farming, rainfalls are the major source of water for agriculture. Apart from long-term climatic constraints, it has been investigated by Kazmi and Rasul (2012), that more cloudiness or rains during a particular decade (10 days) of January may adversely affect the winter wheat (Fig. 2.4). A possible reason may be the significance of photosynthesis at vegetative stages of the crop, as it needs more sunny days for acquiring higher rate of photosynthesis at that particular time.

There is no doubt that today's agriculture has been equipped with better tools and management systems right from land preparation, fertilizer application, sowing, irrigation scheduling, etc., up to harvesting and post-harvesting facilities but still climate conditions matter. On the other hand, industrial development and enlarged utility of chemicals particularly in agriculture have been affecting crop production over the globe through boosting gas emissions (Fig. 2.5).

Climate change has imposed great challenges to crop production over the world. Besides, long time variation devastating events like frequent flooding, freezing and heat waves, prolonged droughts, etc., are occurring more and more. Importantly, as per projection of climate scenarios, the rain-fed agriculture would more likely to be at risk. As suggested in FAO meeting in 2010, a huge amount of investments is required for sustainability in the agriculture sector (FAO 2010). The situation may be worst for the regions with small landholding farming and poor infrastructure in terms of irrigation, etc (IPCC, 2014; Rashid & Rasul, 2011).

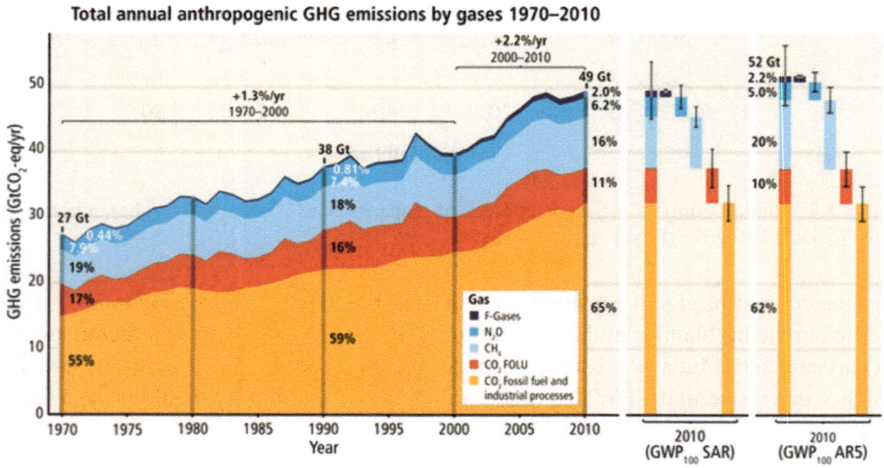

Fig. 2.5 Total annual anthropogenic greenhouse gas (GHG) emissions (gigatonne of CO_2-equivalent per year, $GtCO_2$-eq/year) for the period 1970–2010 by gases: CO_2 from fossil fuel combustion and industrial processes; CO_2 from Forestry and Other Land Use (FOLU); methane (CH4); nitrous oxide (N_2O); fluorinated gases covered under the Kyoto Protocol (F-gases). Right-hand side shows 2010 emissions, using alternatively CO_2-equivalent emission weightings based on IPCC Second Assessment Report (SAR) and AR5 values. Unless otherwise stated, CO_2-equivalent emissions in this report include the basket of Kyoto gases (CO_2, CH_4, N_2O as well as F-gases) calculated based on 100-year Global Warming Potential (GWP100) values from the SAR. Using the most recent GWP100 values from the AR5 (right-hand bars) would result in higher total annual GHG emissions (52 $GtCO_2$-eq/year) from an increased contribution of methane, but does not change the long-term trend significantly (IPCC, ARS)

2.4 Conclusion

The varying climate has been projecting big challenges for crop cultivation from the local to the global level. Huge capital has been invested by the farming community directly and from the governmental side for facilitation purpose which remains at risk due to weather uncertainties. Several industries are mainly dependent on agriculture products, so it is not rather only a matter of food security. Therefore, more detailed and state of the art research for every important region of the world may enable the stakeholders in identifying the root causes of agrometeorological related issues and their optimum solutions. However, it is a fact that crop production is not merely dependent on the climate variables like rainfall or temperatures, there are several other important factors as well including seed quality, fertilizer application, water availability, etc.

Most of the regional-based farming issues may be fixed through local resources with better planning strategies and effective supervision. Besides, keeping in view the importance of this vital sector, more investments would have to be spared for agricultural reforms. Crop production needs to be modernized by addressing the issues like water deficiency and reduced area for cultivation. In this context, the underdeveloped countries would have to be assisted to modify their cultivation system following the climate smart agriculture programs proposed by FAO. As a whole, a joint approach would be worthwhile to address the climate-based challenges over the world.

References

Ali, S., et al. (2019) Assessment of climate extremes in future projections downscaled by multiple statistical downscaling methods over Pakistan, Atmospheric Research, 222: 114–133

Amin, A., et al. (2018a) Evaluation and analysis of temperature for historical (1996–2015) and projected (2030–2060) climates in Pakistan using SimCLIM climate model: Ensemble application. Atmospheric Research, 213: 422–436.

Amin, A., et al. (2018b) Regional climate assessment of precipitation and temperature in Southern Punjab (Pakistan) using SimCLIM climate model for different temporal scales. Theoretical Applied Climatology, 131:121–131.

Amin, A., et al. (2018c) (Simulated CSM-CROPGRO-cotton yield under projected future climate by SimCLIM for southern Punjab, Pakistan. Agricultural Systems, 167: 213–222.

Elshamy, M.E., Wheater, H.S., Gedney, N. and Hunting-ford, C. (2006) Evaluation of the rainfall component of a weather generator for climate impact studies. Journal of Hydrology, 326, 1–24. doi:https://doi.org/10.1016/j.jhydrol.2005.09.017

FAO (2010) The Hague Conference on Agriculture, Food Security and Climate Change. "Climate-Smart" Agriculture Policies, Practices and Financing for Food Security, Adaptation and Mitigation.

FAO (2019) Agriculture and climate change—Challenges and opportunities at the global and local Level—Collaboration on Climate-Smart Agriculture. Rome. 52 pp. Licence: CC BY-NC-SA 3.0 IGO.

Hennessy, K.J., Gregory, J.M. and Mitchell, J.F.B. (1997) Changes in daily precipitation under enhanced green-house conditions. Climate Dynamics, 13, 667–680. doi:https://doi.org/10.1007/s003820050189

Hussain, S. S., Mudasser, M. (2004) Prospects for wheat production under changing climate in mountain areas of Pakistan—An econometric analysis.

Hussain, S., Mubeen, M., et al. (2019) Using GIS tools to detect the land use/land cover changes during forty years in Lodhran District of Pakistan. Environmental Science and Pollution Research, Accepted.

IPCC (2014) Climate Change 2014: Fourth Assessment Report of the Intergovernmental Panel on Climate Change.

Kazmi, D. H., Rasul, G. (2012) Agrometeorological wheat yield prediction in rain-fed Potohar region of Pakistan, Agriculture Sciences, 3, 2, p. 170–177.

Naheed, G., Kazmi, D.H. (2016) ETo Variation in Pakistan in Changing Climate. Pakistan Journal of Meteorology, 13(25).

Nasim, W., Amin, A., et al. (2018) Future risk assessment by estimating historical heat wave trends with projected heat accumulation using SimCLIM climate model in Pakistan. Atmospheric Research, 205:118-133.

Pratley, J. (2003) Principles of Field Crop Production, Oxford University Press, Australia.

Rahman, M. H. R., et al. (2018) Multi-model projections of future climate and climate change impacts uncertainty assessment for cotton production in Pakistan. Agricultural and Forest Meteorology, 253–254: 94–113.

Rashid, K., Rasul, G. (2011) Rainfall Variability and Maize Production over the Potohar Plateau of Pakistan, Pakistan Journal of Meteorology, 8, Issue 15, p 63–74.

Rasul, G. (2000) water requirement of spring sugar cane in Pakistan. Science Vision. 5(4).

Rasul, G., Chaudhry, Q. Z. (2006) Global Warming and Expected Snowline Shift along Northern Mountains of Pakistan. Proc. of 1st Asiaclic Sympos. Yokohama, Japan.

Rasul, G., Kazmi, D. H. (2013) A book titled, "Wheat Yield & Climate Change in Potohar Region of Pakistan", LAP LAMBERT Academic Publishing, ISBN-13: 978-3659-34560-9.

Chapter 3
Impact of Temperature Fluctuations on Plant Morphological and Physiological Traits

Muhammad Aqeel Aslam, Mukhtar Ahmed, Fayyaz-Ul Hassan, Obaid Afzal, Muhammad Zeeshan Mehmood, Ghulam Qadir, Muhammad Asif, Saida Komal, and Tajamul Hussain

Abstract Impact of climate change on plant morphology, physiology, survival, and adaptation is becoming a global concern nowadays. Temperature fluctuations have detrimental effects on many plant morphological and physiological traits. Plant response to warming or chilling temperatures is considered as an important understanding for the agricultural ecosystems. Continuous increase in the global average temperature has significant impacts on the productivity of agricultural crops. Plant growth and development could be affected due to high and low temperature events under changing climate, topography, and land-sea thermal differences. Temperature fluctuations increase the forward and reverse biochemical reactions exponentially resulting in the denaturation of enzymes. Beyond optimal limit, depending on the duration and intensity of temperature, reversible or irreversible changes may lead to plant death. Resilience of plants can be predicted through investigating

M. A. Aslam · F.-U. Hassan · O. Afzal · M. Z. Mehmood · G. Qadir
Department of Agronomy, Pir Mehr Ali Shah Arid Agriculture University, Rawalpindi, Pakistan

M. Ahmed (✉)
Department of Agronomy, Pir Mehr Ali Shah Arid Agriculture University, Rawalpindi, Pakistan

Department of Agricultural Research for Northern Sweden, Swedish University of Agricultural Sciences, Umeå, Sweden
e-mail: mukhtar.ahmed@slu.se

M. Asif
Agricultural Linkages Programme (ALP), Pakistan Agricultural Research Council (PARC), Islamabad, Pakistan

S. Komal
Department of Agronomy, University of Poonch, Rawalakot, Pakistan

T. Hussain
Prince of Songkla University (PSU), Hat Yai, Thailand

W. N. Jatoi et al. (eds.), *Building Climate Resilience in Agriculture*,
https://doi.org/10.1007/978-3-030-79408-8_3

morphological, biochemical, and physiological analysis under different conditions. Plants evolve resilient mechanisms to survive under uncertain temperature stress by limiting the adverse impacts on the metabolic and physiological processes. Knowledge enhancement towards better understanding of various plant processes is needed to cope with detrimental impacts of temperature. This chapter is aimed to illustrate and understand the key plant morphological and physiological traits in response to temperature stress for better management of plants under changing climate.

Keywords Climate change · Temperature · Plant morphology · Physiological responses · Biochemical acclimations

3.1 Introduction

Climate change is adversely impacting the plant physiological mechanisms (Fahad et al. 2017). Climate anomalies are imposing serious threats for the plant's adaptation and survival. Understanding the role of environmental drivers on the plant physiological processes is a key to predict the plant responses under the changing climate. Different studies have been carried out in the past to determine the impacts of climate warming on plants (Gaupp et al. 2017; Nguyen-Huy et al. 2018; Pirttioja et al. 2015). Temperature of the globe might surpass the average of 1.5 °C if current activities of fossil fuel burning is continued (IPCC 2018). According to the Intergovernmental Panel on Climate Change (IPCC), temperature rise of 1.4–5.8 °C till the end of twenty first century has been predicted by the global circulation models while the rise of 1 °C in optimum temperature can cause 10% reduction in rice yield (Fahad et al. 2019a, b). Similarly, both intensity and frequency of high temperature may result up to 40% reduction in rice yield by the end of this century (Fahad et al. 2018). Plants could have positive or negative impacts of climate change depending on their type and geographical position (Liu et al. 2019; Chen et al. 2011; Porter et al. 2017). Over the last couple of decades, climate change research got increased attention due to more noticeable and prominent impacts (Ahmed 2020; Herring et al. 2016; Stott et al. 2013; Ahmed et al. 2019; Ahmad et al. 2017). The negative impacts of elevated temperature on the crop phenotypic, biochemical, shoot, and root features of legume plant is shown in Fig. 3.1.

The best indicator to see the impact of climate change is crop phenology; thus, it is essential to determine the effect of temperature rise on crop phenology by considering seasonal temperature difference. However, temperature change from seasonal shifts, if not considered might lead to biased estimation of warming trends and their corresponding impact on phenology. This concept has been elaborated by Ye et al. (2019) (Fig. 3.2).

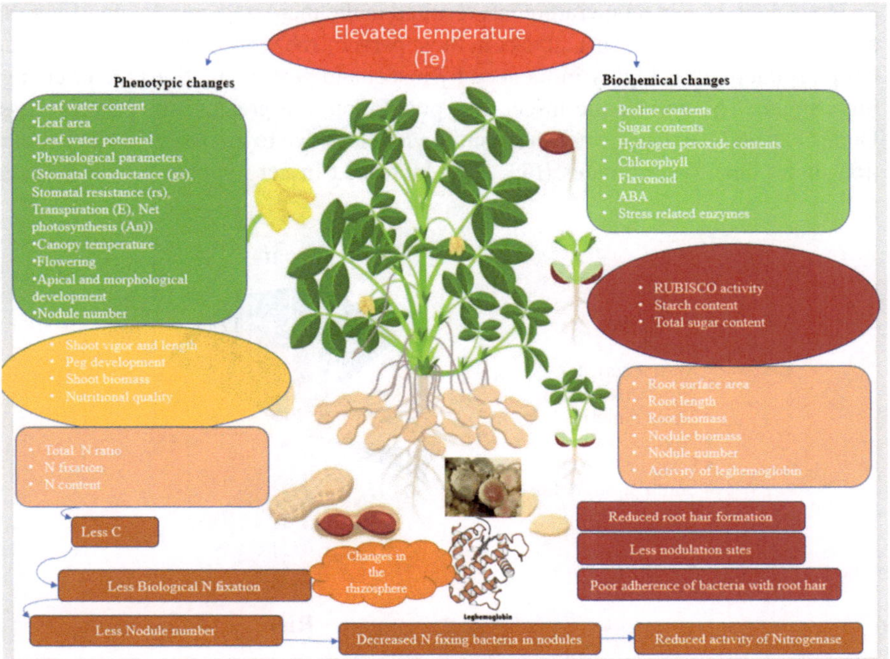

Fig. 3.1 Impact of elevated temperature on phenotypic, biochemical, shoot, and root features of a legume plant

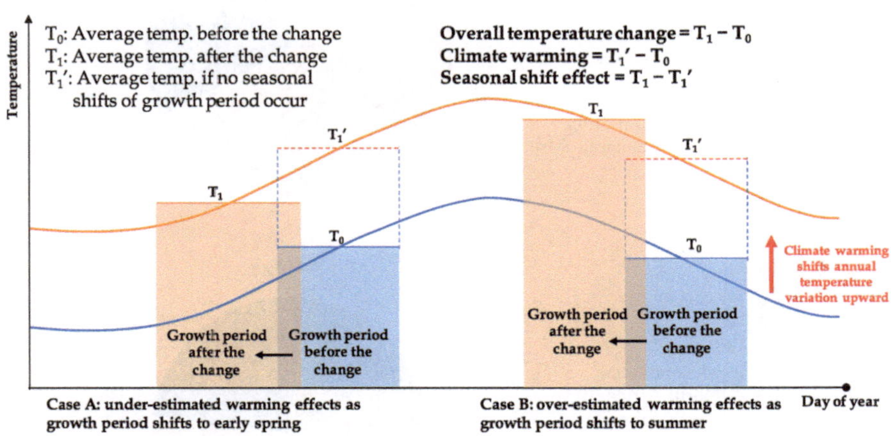

Fig. 3.2 Impact of elevated temperature on cultivar shifts and phonological dates on rice growth period in China

Nowadays, rise in temperature has resulted in the increased population of insects as seen in the form of locust attacks in most parts of the world, particularly in Asia. Crop productivity is highly influenced by insect and pest populations under changing climate. The greater the insect pest population, the greater will be yield loss. Deutsch et al. (2018) projected substantial yield loss due to increased insect population under changing climates (Fig. 3.3). Increased insect population will damage

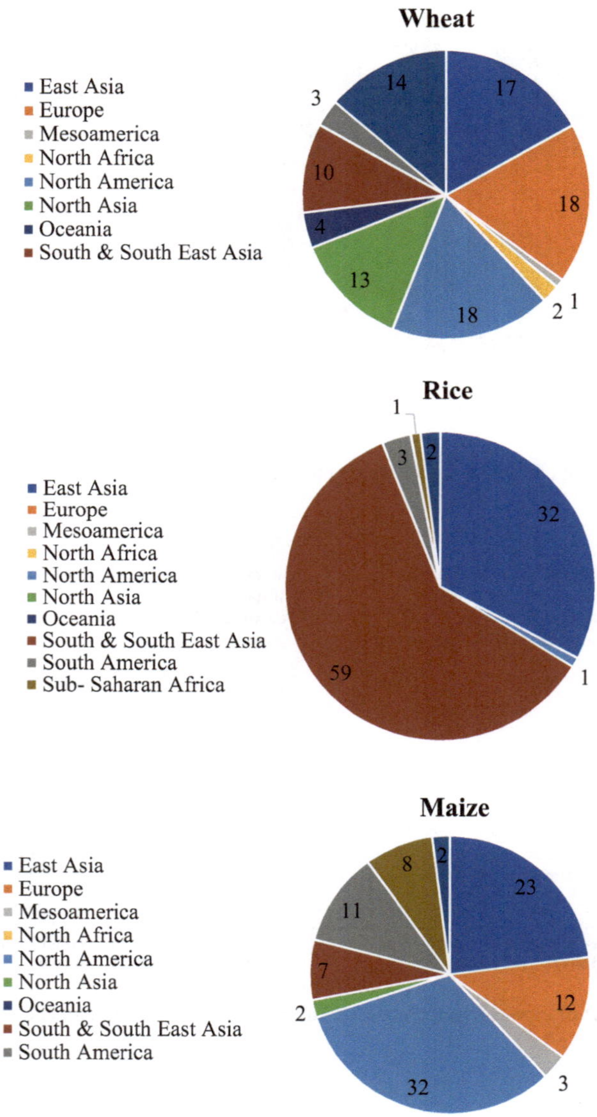

Fig. 3.3 Wheat, maize, and rice substantial yield loss projections due to increased insect population under 2 °C elevated temperature climate (Deutsch et al. 2018)

plant leaves which would ultimately result in the reduced photosynthesis. Most of the high-yielding cereal regions like China and the USA are projected to face serious yield loss by enhanced insect population if the temperature of the globe increases by 2 °C (Fig. 3.3).

Temperature is one of the key environmental factors that affects the physiological and metabolic processes of plants. It has direct impacts on the structures of DNA, proteins, and membranes (Ruelland and Zachowski 2010). Plants can survive under a wide range of temperatures (−10 to 60 °C) depending on the intercellular water freezing, protein and enzyme denaturation (Takai et al. 2008; Fitter and Hay 2012; Taiz et al. 2015). Similarly, plants can quickly detect and respond to temperature stress by triggering specific metabolic and biochemical pathways (Ruelland and Zachowski 2010).

Plants can be classified on the basis of their thermal niches. Classification based on thermal range comprises (1) Psychrophiles, (2) Mesophiles, and (3) Thermophiles. Psychrophiles plants have an optimal temperature range for growth between 0 and −15 °C (e.g., polar region plants), mesophiles plants 10–30 °C (e.g., Higher plants) and the thermopiles plants (30–65 °C) (e.g., Desert plants) (Źróbek-Sokolnik 2012). However, these physiological changes may be reversible or irreversible depending on the intensity and length of prevailing stress. Unlike other organisms, plants do not escape from stress by moving or migrating to favorable environments. However, there are a variety of mechanisms to survive under the prevailing stresses. Such mechanisms may involve the activation and triggering of physiological responses modifying functioning to resist and repair stress-induced damages. Moreover, a plant can also activate such responses by gene expression enabling them to identify the stress conditions and then signaling these responses to the cells (Niu and Xiang 2018). Ability of plants to combat temperature stress mainly depends upon ideal habitats and adaptation capacity (Źróbek-Sokolnik 2012). When plants get exposed to temperature stress, they try to adjust their cell state physiology (homeostasis) known as acclimation (Niu and Xiang 2018). Heat and drought stresses have a significant effect on the plant morphological, physiological, and biochemical responses (Fahad et al. 2017). Similarly, the acclimation capacity of every plant is dependent on species type, habitat, and cardinal temperatures. Plants are becoming more prone to environmental stresses due to the increased intensity of climate change (Shaw and Etterson 2012; Hammad et al. 2020; Hussain et al. 2020). There are suggestions to develop plant evolutionary mechanisms to cope with the unprecedented rate of future climate variability (Anderson et al. 2012). On the other hand, some genotypes may show physiological plasticity to the changing climate but there is a possibility that such kind of species may be pushed beyond their tolerance if changes becomes too extreme in the future.

To predict the plant response to changing climate, enhanced understanding of plant evolutionary morphological and physiological responses are necessary (Medeiros and Ward 2013; Aslam et al. 2017a; Ali et al. 2020; Qasim et al. 2020). Modeling tools are more beneficial to study the impact of climate change under different sets of management (Ahmed 2020; Ahmed et al. 2011, 2013, 2014, 2020; Asseng et al. 2019; Aslam et al. 2017b, c, d; Ijaz et al. 2017; Jabeen et al. 2017;

Ahmed and Hassan 2015; Ahmed and Stockle 2016; Liu et al. 2019; Khan et al. 2020; Wallach et al. 2018). Hence, the evolution of resilience mechanisms in response to climate change is a key factor in determining plant fitness to future environments (Kimball et al. 2012). Research has been carried out in the past to understand the plant morphological and physiological response in the context of thermal changes (Kumudini et al. 2014; Fahad et al. 2016a). In this chapter, we aimed to present the plant biochemical, metabolic, and physiological aspects in response to the elevated temperature stress and some of the adaptive mechanisms to cope with the enhanced temperature.

3.2 Plant Physiological Responses to Temperature Fluctuations

Physiological, molecular, and cellular fluctuations in plants arise when the optimal temperature range is surpassed (Buchanan et al. 2015; Hasanuzzaman et al. 2013b; Sugiyama and Kameshita 2017). Impacts of temperature on the plants are derived by several factors such as temperature level, exposure duration, plant tolerance, and acclimation capacity (Fitter and Hay 2012; Taiz et al. 2015; Nievola et al. 2017; Hasanuzzaman et al 2013b; Hayat and Ahmad 2014). Additionally, these factors also influence the metabolic processes of stressed plants. Severe heat stress results in immediate cell injuries due to protein denaturation within few minutes of exposure, while moderate stress cause injuries to cells during prolonged heat stress (Źróbek-Sokolnik 2012). Metabolites amalgamation in pollens severely declined the pollen fertility and retention, germination, and metabolites synthesis due to an increase in temperature (Fahad et al. 2016b). Climatic predictions revealed that short duration intense cold events will persist in the following decades ranging from few hours to few days (Petoukhov and Semenov 2010; Kodra et al. 2011). Plant metabolic processes may be affected by cold injury depending on the intensity of chilling or freezing (Janská et al. 2010).

3.2.1 *Photosynthesis*

Thermal instability affects photosynthesis; a rise in temperature beyond the optimal range shows declined trend of photosynthesis in plants. Declined photosynthetic activity is attributed to the decreased activity of Rubisco (Ribulose Bisphosphate Carboxylase Oxygenase) enzyme that is mainly responsible for carbon dioxide fixation and carbohydrates synthesis. This condition may even get worsen if the carbon dioxide is prevented to enter due to the decreased stomatal functioning (Hasanuzzaman et al. 2013a; Mathur et al. 2014). Photosystem II is mainly responsible for decline due to its inability to perform efficiently under elevated temperatures (Mathur et al. 2014). More importantly, it is the key determinant of electron transport during photochemical reaction of photosynthesis (O'Sullivan et al. 2017). Moreover, thylakoid membrane is responsible for photosystem II inhibition since it is highly responsive to temperature elevations (Mathur et al. 2014). This membrane may be prone to damage due to reduction in chlorophyll content under heat stress. However, plants can activate resistance mechanisms under mild temperature stresses to recover damage to photosystem (Mathur and Jajoo 2014).

Cold stress reduces plant photosynthetic rates associated with adverse impacts of cold on carbon fixation and thylakoid electron transport. Moreover, in the chloroplast, free phosphate availability for photosynthesis is declined in cold stress due to reduced utilization in chloroplast and accumulation of triose phosphate (Taiz et al. 2015). Ability of plant to maintain levels of photosynthesis is mainly determined by tolerance of thylakoid to cold stress. Depending on the thylakoid tolerance, plants can maintain photosynthesis up to a point where freezing of cellular fluids occur.

3.2.2 Cell Membranes Abnormalities

Heat stress can damage the cell membranes of plants by lipid movements and protein denaturation (Mittler et al. 2012). This results in inhibition of physiological and cellular processes due to increase in membrane fluidity, ion leakage, and cell rupturing (Mittler et al. 2012). Prolonged heat stresses have more prominent impacts on membrane lipids reduction and degradation (Tang et al. 2016; Narayanan et al. 2016). There are significant evidences that extreme temperature may result in thylakoid membrane damage, lipid reduction, greater malondialdehyde, and ox-lipid accumulation by endoplasmic reticulum (Narayanan et al. 2016).

Cold stress significantly affects the fluidity of cell membranes and disrupts the optimal functioning of plants at cellular levels. Membrane lipids are free to move in cellular layers depending on the temperature (Ruelland and Zachowski 2010). Membranes become rigid on exposure to cold temperatures, due to which their fluidity is reduced owing to transition of lipids from crystalline fluid to solid gel state. However, structural differences may influence the cold tolerance of membrane lipids (Ahmad and Prasad 2011). Membrane rigidity result in dysfunctions such as cell solute losses and blocking of transport channels due to increased permeability (Buchanan et al. 2015). Unsaturated fatty acids and malondialdehyde are the good representative of membrane integrity in cold conditions (Ahmad and Prasad 2011). Higher proportion of unsaturated fatty acids and lower malondialdehyde values can serve as a good indicator of cold tolerance to protect the membrane integrity (Guo et al. 2016). Moreover, reduced fluidity of cell membranes is considered as a key signal for enhanced reactive oxygen species ROS and reactive nitrogen species RNS synthesis leading to the injuries like lipid peroxidase and disturbances in antioxidant defense system (Repetto et al. 2012).

3.2.3 Reactive Nitrogen Species (RNS)

Exposure to increased temperature may result in the production of RNS having nitric oxide radical produced from reaction with peroxynitrate, nitrogen dioxide, and peroxynitrous. RNS may lead to toxic impacts on cells by reacting with cellular components such as proteins, lipids, thiols, and nucleotides (Baudouin 2011; Corpas et al. 2011). RNS have same Lipid per oxidation (LPO) effects like ROS specifically for lipids and resulting in denaturation of membranes by nitration reactions (Rubbo and Trostchansky 2014). Despite the deleterious impacts of ROS and RNS on the physiological responses and cellular structures, accumulation of these molecules at initial stages may trigger signaling of plants to activate acclimation responses (Fancy et al. 2017).

RNS induce toxic effects under cold temperatures on cellular proteins and lipids by reacting with cellular components. RNS also react with several nitrogen derivatives and produce nitric oxide radicals (Corpas et al. 2011; Baudouin 2011; Rasool et al. 2019; Zamin et al. 2019). These species also react with the lipids and denature cell membranes during a nitration reaction (Rubbo and Trostchansky 2014). Moreover, RNS can enhance or inhibit antioxidant enzyme depending on nitric oxide-induced translational modifications. Hence, RNS may have dual role either as antioxidant or pro-oxidant. However, this role is highly dose-dependent, as under low concentrations RNS may act as antioxidant and helps in cell survival while under high concentration leads to severe injuries (Begara-Morales et al. 2013, 2015). Proteomic analyses under cold stress suggest that 30% of translational modifications occur in cold responsive signaling proteins (Sehrawat et al. 2013) and nitric oxides mainly target the superoxide dismutase, ascorbate peroxidase, glutathione reductase, and catalase antioxidant enzymes (Begara-Morales et al. 2015; Lin et al. 2012).

3.2.4 Reactive Oxygen Species (ROS)

Heat stresses are associated with disturbing the reactive oxygen species (ROS) balance and increasing ROS specifically in peroxisomes, mitochondria, and chloroplasts (Suzuki et al. 2013) during the photosynthesis and respiration processes. These ROS affect the physiological and metabolic processes by damaging the lipids, proteins, and nucleic acids (Gill and Tuteja 2010; Suzuki et al. 2012). Lipid per oxidation (LPO) is a process of oxidation of PUFAs (Polyunsaturated Fatty Acids) in plants resulting in membrane lipids destruction, bond rearrangement, and radical formation (Repetto et al. 2012). Depending on the intensity of heat stress ROS production may be stimulated and unavailability of plant antioxidants can lead to irreversible oxidative damages affecting plant physiology and survival (Gill and Tuteja 2010). Hence, ability to regulate the ROS is directly interlinked with heat stress resistance (De Pinto et al. 2015). To counter ROS production plants have different antioxidants such as non-enzymatic (like glutathione, carotenoids, tocopherols, etc.) and enzymatic (Like glutathione reductase, catalase, superoxide dismutase, etc.) (Lu et al. 2008; Gill and Tuteja 2010; Hajiboland 2014; Hu et al. 2020).

Synthesis of ROS in cold stress is triggered by decreased fluidity of cell membranes. The consequences of increased synthesis of these ROS are directly linked with injuries to the antioxidant defense system of plants (Repetto et al. 2012). ROS can cause irreversible oxidative damage in the absence of antioxidants and compromise plant survival in stressed environments (Gill and Tuteja 2010). Major increase in ROS may be observed in chloroplast and mitochondria of plants (Suzuki et al. 2013). ROS damage lipids, proteins, and nucleic acids leading to adverse impacts on plant physiological and biochemical processes. To regulate the synthesis of ROS, plants use the enzymatic and non-enzymatic antioxidants while such enzymatic activity may be related to gene expression demonstrating rapid antioxidant responses (Zhang et al. 2010).

3.2.5 Chlorophyll a *Fluorescence*

The process during irradiation when chlorophyll absorbs light to utilize in photochemical reactions of photosynthesis and dissipating excess light is known as chlorophyll *a* fluorescence. Hence, chlorophyll *a* fluorescence is used as an indicator of photosynthetic process efficiency (Li et al. 2014). Any potential changes in photosynthesis process of plants lead to lower values of florescence. Therefore, chlorophyll *a* fluorescence analysis can be used as a tool to assess the performance of plants in a diverse range of cold stresses.

3.2.6 Sugars as Antioxidants

Several kinds of sugars can also act as antioxidants under cold stress (Demidchik 2015). Glucose may hinder ROS synthesis by stimulating ROS scavenging enzymes (Figs. 3.4 and 3.5) as a result of increased NADPH (cofactor for scavenging enzymes) production due to its functioning in pentose phosphate pathway. Fructose can play an important role in plant defense mechanisms under cold stress as fructose is twice more effective in ROS scavenging than glucose (Bogdanović et al. 2008). Similarly, raffinose may have a role in photosystem II stabilization in low temperatures (Knaupp et al. 2011). Furthermore, carbohydrates have a critical role in membrane stabilization to induce cold tolerance (Tarkowski and Van den Ende 2015). Membranes damage and increased permeability by cold stress can be fixed by introducing the different kinds of sugars in lipids polar group in cells (Hincha et al. 2003)

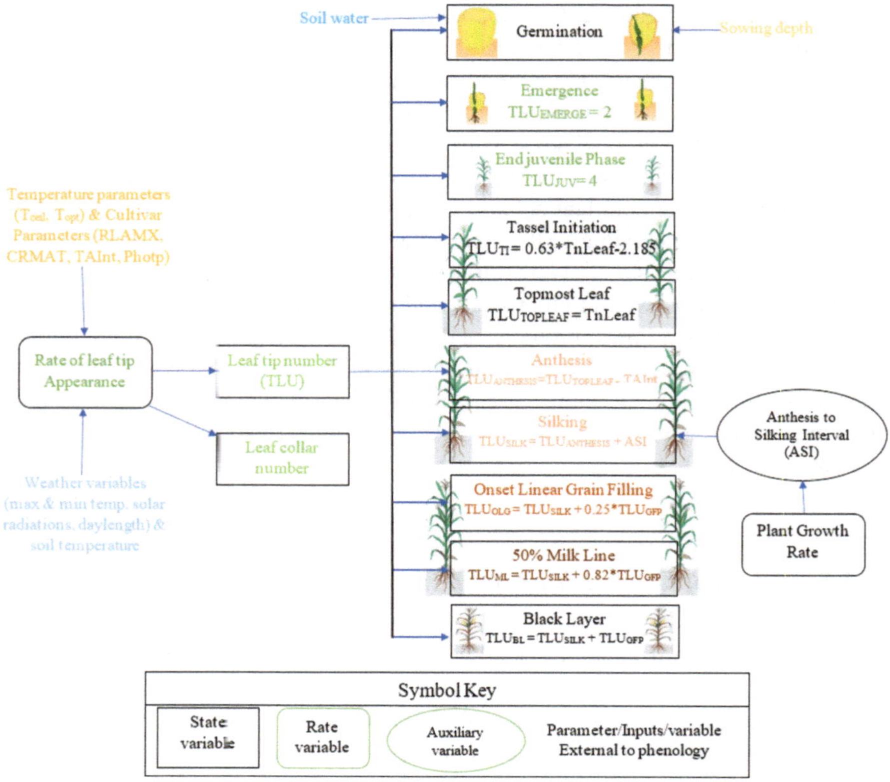

Fig. 3.4 Phenological stages overview in AgMaize and relationship with temperature (Tollenaar et al. 2017). Here TLU; Leaf stage, TLU_{JUV} = End of juvenile phase, TLU_{TI} = Tassel initiation, $TLU_{ANTHESIS}$ = Anthesis stage, T_{nLeaf} = Appearance of topmost leaf, $T_{nLeaf} + T_{AInt}$ = Anthesis, ASI = Anthesis to silking interval, $T_{nLeaf} + T_{AInt}$ + ASI = Silking (SILK), GFP = Duration of grain filling or post flowering period, SILK + 0.25 × GFP = Onset of linear grain filling, SILK + 0.82 × GFP = 50% milk line, SILK + GFP = Black layer

3.2.7 *Temperature and Maize Growth*

Temperature impact on the rate of maize development was elaborated by Tollenaar et al. (2017) using a new process-based maize model (AgMaize) being implemented in the Decision Support System for Agro technology Transfer (DSSAT) (Fig. 3.4 and Table 3.1). Temperature generated rate of leaf appearance (RLA) for maize crop is almost 0.5 leaves per day at 31 °C and expressed in thermal leaf units (TLU), i.e., 1TLU = 1 leaf (°C) which means for 1 leaf appearance plant requires thermal exposure for 2 days at 31 °C temperature. TLU has many advantages to account for heat unit accumulation in maize phenology like maize phenological stage could be determined by using $TLU_n = n$ where nth leaf tip stage is defined as the stage of development when the tip of the nth leaf is first visible from a horizontal plane at the level of whorl. As sowing to flowering duration is determined by number of leaves, time

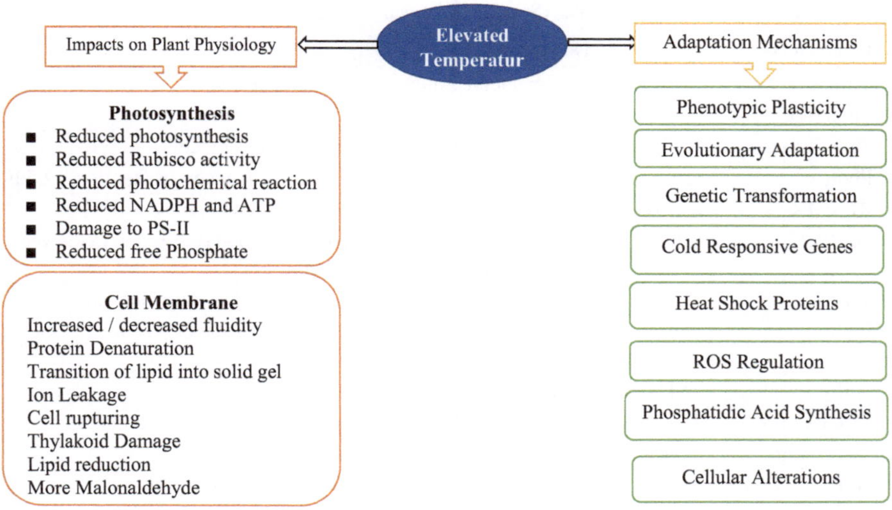

Fig. 3.5 Impact of elevated temperature on plant physiology and Plant adaptation mechanisms to elevated temperature

duration of sowing to silking can also be expressed in terms of TLU. On the other hand, during post flowering or grain filling maize shows different response to temperature than that of pre flowering periods, i.e., morphological development is less responsive to temperature during grain filling period and follows general thermal index (GTI) (Fig. 3.4) (Tollenaar et al. 2017).

3.3 Plant Adaptive Mechanisms and Physiological Responses to Temperature Changes

Altered thermal scenarios significantly affect plant physiological processes (Anderson 2016). However, plants have ability to evolve certain evolutionary mechanisms to survive under stress conditions by limiting the adverse impacts on the metabolic and physiological processes (Franks et al. 2007; Merilä 2012; Mubeen et al. 2019; Wang et al. 2019). Physiological and phenotypic plasticity help the plants to temporarily alleviate the expected adverse effects of thermal changes (Nicotra et al. 2010). There are several acclimation mechanisms in plants to offset the elevated temperature stresses. Several metabolomic, proteomic, and transcriptomic adjustments are responsible for the activation of these mechanisms (Mittler et al. 2012). Plant responses according to their regions to heat and cold stress are elaborated in Table 3.1. Meanwhile, plant adaptation mechanisms are also given in Table 3.2.

Plants undergo a variety of complex signaling processes to tolerate and resist the varying temperatures by adapting different acclimation mechanisms. However,

Table 3.1 CERES-Maize variables (Source: Tollenaar et al. 2020)

Parameters	Unit	Definition
Tmin (0)	°C	Minimal temperature for leaf tip appearance
Tceil (43.7)	°C	Ceiling temperature for leaf tip appearance
Topt (32.1)	°C	Optimum temperature for leaf tip appearance
Photp	leaves h^{-1}	Photoperiod sensitivity for day length values greater than 12.5 h
RLAMX	leaves d^{-1}	Rate of leaf appearance at optimum temperature
GFPYR	GTI	Change in duration of grain filling period (GFP) due to the year of commercial release of the hybrid
TAInt	TLU	Interval between emergence of topmost leaf tip and Anthesis
PGR	g $plant^{-1}$ d^{-1}	Average plant growth rate from growth stages 5 to 7
ASI	TLU	Anthesis to silking interval
TLUBL	TLU	Accumulated TLU from planting to black layer formation
DayLength	Hour	Day length duration (civil twilight)
GFP	TLU	Duration of grain filling or post flowering period
GTI	°cd	General thermal index
GTIGFP	°cd	Duration of GFP in GTIs (°CD)
MeanT	°C	Average air temperature between the end of juvenile phase and tassel initiation
Mrad	MJ m^{-2}	Mean solar radiation during the previous week
RLA	Leaves d^{-1}	Rate of leaf appearance (measured)
RLA (°C)	Leaves (°C) d^{-1}	Computed rate of leaf appearance based on RLA-temperature relationship
RLAFrad	–	Effect of solar radiation on the rate of leaf appearance
TLU	Leaf (°C)	Thermal leaf unit
TLUAnthRM	TLU	Thermal time from planting to Anthesis of hybrid with a RM relative maturity grown under reference conditions
TLUAnthesis	TLU	Thermal time from planting to Anthesis
TLUSilk	TLU	Thermal time from planting to silking
TLUTI	TLU	Thermal time from planting to tassel initiation
TnLeaf	TLU	Computed total number of initiated leaves
TnLeafRM	TLU	TnLeaf of hybrid of RM relative maturity grown under reference conditionsduring the photoperiod/temperature-sensitive phase
ΔTnLeafPhot	TLU	Change in initiated leaf number due to deviation of photoperiod fromreference conditions during the photoperiod/temperature-sensitive phase
ΔTnLeafTemp	TLU	Change in initiated leaf number due to deviation of temperature fromreference conditions during the photoperiod/temperature-sensitive phase

these acclimations to cold stress are attributed to physiological, metabolic, and biochemical changes in the plants. Plants over the globe react differently under varying temperature regimes. Plants have adaptation mechanisms to respond to both temperature extremes (Cold and Heat). Some of the important adaptation mechanisms

Table 3.2 Plant stress (heat and cold) mechanisms over the globe (Source: Nievola et al. 2017)

Native location	Crop Spp	Stress	Temperature treatments	Tolerance mechanisms	References
Temperate region of Tibet, China, Nepal and West Himalaya	Arabis paniculataFranch	Heat	Growth chamber: 35 °C. Duration: 0–22 days	Increased lipid saturation, HSP101 and HSP70 expression, and soluble sugar content	Tang et al. (2016)
Temperate/ Subtropical/Tropical—North, South & Central America	Tomato (Lycopersicon esculentum Mill. cv. Moneymaker)	Cold	Phytotron: 4 °C. Duration: 24 h	Increased NR activity, NR relative expression, NO, ABA, and polyamines content	Diao et al. (2017)
Temperate—France, Great Britain, Spain	Cabbage (Brassica oleracea L. var. capitata L.)	Heat	Growth chamber: exposure to 40, 50, or 60 °C during 20 min every 24 h. Control—constant exposure to 25 °C. Duration: 5 days	Increased isothiocyanate and glucosinolates, ROS scavenging capacity, ascorbic acid, total phenolics content, and antioxidant enzymes activities	Yang et al. (2016)
Arid—South and Southwestern Asia	Rhazya strictaDecne	Heat	Diurnal analysis at the native habitat: max. leaf temperature of 43 °C at 2 PM	Maintenance of net photosynthesis	Lawson et al. (2014)
Tropical—Southeast and South Brazil	Vriesea inflata (Wawra) Wawra	Cold	Growth chamber: 15 and 28 °C (control). Duration: 24 min	Increased cell number of aquiferous parenchyma and maintenance of chlorophyll contents	Pedroso et al. (2010)
Arid and semi-arid—East Africa	Cordeauxia edulis Hemsl.	Heat	Native habitat: leaf temperature of 40–42.4 °C between 1–2:30:00 PM	Increased gene expression of HSPs, chaperones, and aquaporins	Obaid et al. (2016)
			Growth chamber: 32/23 °C (day/night), 37/27 °C, 42/31 °C, or 27/19 °C (control). Duration: 7, 14 or 15 days	Maintenance of net photosynthesis. Increased emission rate of isoprenoids and total phenolics content in leaves	Egigu et al. (2014)

(continued)

Table 3.2 (continued)

Native location	Crop Spp	Stress	Temperature treatments	Tolerance mechanisms	References
Tropical—Southeast Brazil	Nidularium minutum Mez	Cold	Growth chamber: 10, 15, and 25 °C (control). Duration: 3 or 6 months	Increased thickness of aquiferous parenchyma, reducing sugars, and pectin	Carvalho et al. (2013)
Tropical—New Guinea	Sugarcane (Saccharumofficinarum L.)	Heat	Growth chamber: 40/35 °C (day/night) and 28/23 °C (control). Duration: 72 h	Gradual accumulation of free proline, glycinebetaine, and soluble sugars	Wahid and Close (2007)

and physiological responses that may impede or enhance plant metabolic biochemical processes under temperature fluctuations are as follows:

3.3.1 Phenotypic Plasticity of Physiological Processes

It is fundamental mechanism in plants to respond to changing environment. Plants respond by altering and showing the plasticity in their physiological mechanisms (Gunderson et al. 2010; Liancourt et al. 2015). However, plasticity of these physiological processes may show higher fitness level in a certain environment. Moreover, far-reaching plastic responses can be experienced in offspring by evaluating their fitness to the fluctuating climatic conditions (Galloway and Etterson 2007). Such plasticity can regulate immediate imposed effects of stress and can facilitate process of evolutionary mechanisms by providing significantly increased time durations (Chevin et al. 2010).

3.3.2 Evolutionary Adaptation

It is the phenomena in which plant develops varied metabolic and biochemical responses to resist stressed conditions. Genetic variations in physiological traits of any plant population are necessary for the occurrence of significant stress adaptive mechanisms (Johnson et al. 2009). Rapid plant physiological evolution capacity must be improved to explore these mechanisms effectively.

3.3.3 Genetic Transformation

Plant population can promote or restrict evolutionary process under changing climate by gene flow to the successive generations (Kremer et al. 2014). Gene flow may benefit the following generation by introducing alleles that are preadapted to rising temperature (Aitken and Whitlock 2013) or may also lag the adaptive mechanisms by maladapted alleles. However, further adequate examination in context of appropriate ecological and environmental conditions is required to assess the connected gene flows in different generations (Fig. 3.5).

3.3.4 Cellular Alterations at Membranes Level

Plants show the compositional changes and alterations in membrane integrity to cope heat stress (Ruelland and Zachowski 2010; Hasanuzzaman et al. 2013a). Such compositional changes increase thermal melting ranges and avoid fluidization of membranes in response to thermal heating by enhancing the saturated fatty acid contents and changing lipid compositions of membranes (Źróbek-Sokolnik 2012). These changes immediately within a short time period of plant exposure to heat stress. In chloroplast membranes polyunsaturated fatty acids (PUFAs) content is considerably reduced and fluidization may occur under thermal stress and this reduction is overcome by elevating the saturate lipids proportion to prevent membrane fluidization (Tang et al. 2016).

3.3.5 Synthesis of Heat Shock Proteins (HSPs)

The most consolidated and rapid mechanism of plants in response to heat stress is synthesis of HSPs (Vierling 1991). These proteins are highly preserved in plants and during rapid heat stress regular protein synthesis of plants is coupled and dominated with HSPs production to enhance the thermotolerance (Shinozaki et al. 2015). HSPs are categorized based on molecular masses (Ohama et al. 2017). Expression of these HSPs may be induced in few seconds following rise in temperature but maximum expression of their molecular chaperone ability is observed in a couple of hours (Larkindale et al. 2005). However, complete understanding of each HSP functioning is limited but their role is clearly observed in acquisition and maintenance of heat tolerance (Shinozaki et al. 2015; Sedaghatmehr et al. 2016).

3.3.6 Synthesis of Secondary Metabolites

Mainly accumulated plant secondary metabolites in tolerant plants under heat stress environments involve terpenoid and phenolic compounds (Bartwal et al. 2013; Rivero et al. 2001; Zhou et al. 2015). Heat tolerant plants and extreme heat conditions showed large amount of phenolic and carotenoid compounds (Zhou et al. 2015). Recently, steroid's ability to enhance heat stress tolerance has been identified (Wahid et al. 2012). Brassinosteroid has been defined as a phytohormone to suppress the heating adverse impacts on plants (Hayat and Ahmad 2010). Importance of these secondary metabolites under intense heat stress has been significantly exploited in several studies (Singsaas and Sharkey 1998; Yang et al. 2017). Increased emission of isoprenes and improved phenolic content considerably enhanced temperature stress tolerance of plants due to their ability to protect the photosynthesis apparatus and to act as antioxidants (Smirnoff 2005; Mazorra 2011; Ahammed and Yu 2016).

3.3.7 Osmotic Adjustments

Plant tissues experience limited plant water potential as a result of reduced water levels in elevated temperature environments. Hence, compatible osmolytes synthesis may ameliorate problems related to cell's osmotic potential by enabling them to capture the water and maintain water potential levels (Ruelland and Zachowski 2010). Compatible osmolytes are very beneficial for metabolism being having adverse impacts even in larger concentrations. These osmolytes include proline, glycine betaine, sugars, polyols, etc.). Several studies have showed rapid osmotic adjustments during heat stress associated with osmolytes accumulation and improved heat tolerance (Ramani et al. 2017; Wahid and Close 2007). Compatible osmolytes have been reported to act as redox buffer to prevent oxidative damages (Hossain et al. 2014). However, the exogenous application of these osmolytes prior to heat stress can provide a better understanding of these impacts (Hasanuzzaman et al. 2013a). As previously, exogenous applications have shown significant impact on plants under thermal stress conditions (Rasheed et al. 2011).

Cold stress results in cell dehydration due to the freezing of cellular fluid and reduced absorption. Hence, compatible osmolyte accumulation may help in reduced dehydration and maintain cell water contents. Freezing in tissues can be delayed by decreasing the freezing point by accumulation of specific compatible osmolytes and by the process of supercooling to retain water in liquid state even below the freezing point (Vitasse et al. 2014). For instance, sucrose, fructose, and glucose are the most important osmolytes produced in cold stress. Moreover, plants have the ability to produce binding proteins that can bind the ice crystals to reduce ice formation (Dolev and Braslavsky 2017).

3.3.8 Complex Signaling to Trigger Thermal Responses

Plants have a very complex signaling mechanism to trigger acclimation responses against heat stress. Enhanced fluidity of plasma membrane activates affects the signaling process during elevated thermal conditions (Mittler et al. 2012). It also enables calcium ion entry to trigger several heat-tolerant mechanisms such as production of HSPs (Narayanan et al. 2016). Additionally, cytoplasmic sensitivity determines activation rate of adaptation mechanisms (Saidi et al. 2010). Some studies depicted that ROS and NOS can be combined with calcium to trigger acclimation processes (Khan et al. 2015; Niu and Liao 2016). Apoplastic ROS synthesis under thermal stress results in the induction of two peaks and the traveling of these peaks to other plants at larger distances in the form of a systematic signal (Mittler et al. 2012). Similarly, NO synthesis in elevated temperature environments results in transcription of heat shock proteins (Xuan et al. 2010). Moreover, phospholipids (MA also trigger stress signaling. However, these signal carriers respond very immediately to the rising temperature to stimulate acclimation mechanism. Phytohormones can also take part in signaling of stress resistance responses such as salicylic acid, ethylene, and abscisic acid (Bita and Gerats 2013). However, there is a need to explore hormones mediated signaling and complex interaction between the signaling molecules and hormones as well as signal transduction patterns depending on plant type (Ahammed and Yu 2016).

3.3.9 Phosphatidic Acid Accumulation

Cold-induced modifications in membrane fluidity serve as a biological thermometer for signaling different physiological and metabolic plant processes (Rihan et al. 2017). Enzymatic activity on membranal lipids (diacylglycerol kinase and phospholipase D) is regulated by cold-induced rigidity in plasma membranes and results in increased accumulation of phosphatidic acid (Gupta et al. 2011). Phosphatidic acid formation in plants immediately occurs on exposure to cold stress and phosphatase activation occurs due to increased intracellular levels of phosphatidic acid (Gao et al. 2013).

3.3.10 Reactive Oxygen Species (ROS) Regulation

During the cold stress, the production of ROS is increased due to several changes at cellular level and induce negative impacts on physiological processes (Repetto et al. 2012). Increased production of ROS during cold stress is regulated through nitric oxide due to its key role in antioxidant enzyme activation under cold stress (Begara-Morales et al. 2015; Fancy et al. 2017). Nitric oxide is produced by the enzyme called nitrate reductase (Ziogas et al. 2013), whereas mechanisms modulating the activity of nitrate reductase are not clear (Gao et al. 2013). Moreover, accumulation of nitrate reductase and nitric oxide can trigger the expression of cold responsive genes by activating the cold tolerance responses (Gupta et al. 2011).

3.3.11 Cold Responsive Genes Activation

Cold-induced membranal rigidity may enhance the concentration of intracellular calcium due to increased ion influx through plasma membrane via calcium channel and release of vacuole stored calcium. Increased cellular calcium leads to activation of CBF (C-repeat binding) genes within a short duration of cold stress (Thomashow

2010). After CBF-genes activation transcription factors are activated by these CBF-genes and then CBF-proteins result in activation of cold responsive genes (Knight and Knight 2012). After cold responsive gene activation, plants finally attain cold acclimation. This transcription factor controls about 12% of cold responsive genes (Rihan et al. 2017). Hence, membranes fluidity and cytoplasmic calcium are good indicators of cold responsive gene activation and plant responses to cold stress (Rihan et al. 2017; Ruelland and Zachowski 2010).

3.3.12 Cold Tolerance Acquisition

Several signaling processes occurring for acclimation to low temperature may be a prerequisite of freezing tolerance acquisition in plants. This kind of tolerance occurs in plants that are unable to tolerate severe cold temperature such as below freezing point. In such plants, gradual and systematic cold tolerance acquisition starts at a warmer temperature and continues as the temperature falls until the plants become resistant to freezing temperature below 0 °C. Growth is interrupted during this process of acclimation and a variety of tolerance mechanisms such as accumulation of unsaturated lipids and carbohydrates are activated (Figs. 3.5 and 3.6). However, this

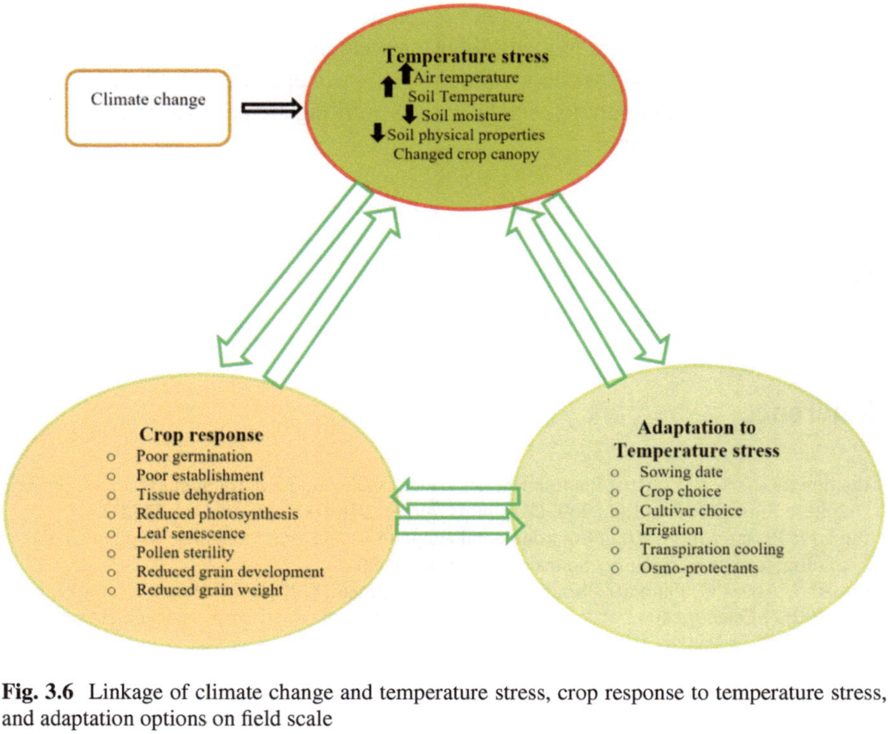

Fig. 3.6 Linkage of climate change and temperature stress, crop response to temperature stress, and adaptation options on field scale

tolerance acclimation process needs gradual thermal changes and sudden or abrupt changes may lead to impairments in physiological functioning and plants may suffer damage (Larcher and Piccioni 1993).

3.4 Conclusion

Temperature extremes or critical thresholds such as heat waves and cold stress alter physiological processes resulting in plant tolerance and susceptibility. Differential physiological response of plants to temperature fluctuations are due to thermosensory network interaction with pedoclimate. Plant response to a range of temperature changes with developmental stage and is critical to evaluate metabolic programming and phytohormones signaling. Current and future abrupt temperature variations pose serious threat to productivity and food security and it can be tackled by different options. Thermo-responsiveness of plants is one of the novel opportunities to assess resilience and promotion of such genotypes to ensure food security globally. Activation of cold responsive genes in response to nitric oxide and nitrate reductase accumulation prevents plants from low-temperature stress. Phenotypic plasticity of plant regulates physiochemical processes and is considered as evolutionary mechanism. Heat shock proteins need to be studied in detail in future to completely understand their function in acquisition of heat tolerance. Modification for emission of phenolic contents and isoprenes in plants can enhance tolerance against thermal stress. Antioxidant defense mechanism of plants has a significant role under abiotic stress. To cope with long-term heat episodes modified agronomic practices, genetic manipulation and hormonal profiling are promising ways forward for holistic understanding.

Acknowledgments Thanks are extended to the Department of Agronomy, Pir Mehr Ali Shah Arid Agriculture University, Rawalpindi, Pakistan and Higher Education Commission, Pakistan.

Conflict of Interest The authors declare no conflict of interest.

References

Ahammed GJ, Yu J-Q (2016) Plant hormones under challenging environmental factors. Springer Nature Singapore Pvt. Ltd., 269. DOI: https://doi.org/10.1007/978-94-017-7758-2

Ahmad P, Prasad MNV (2011) Environmental adaptations and stress tolerance of plants in the era of climate change. Springer Science & Business Media

Ahmad S, Abbas G, Fatima Z, Khan R, Anjum M, Ahmed M, Khan M, Porter C, Hoogenboom G (2017) Quantification of the impacts of climate warming and crop management on canola phenology in Punjab, Pakistan. Journal of Agronomy and Crop Science

Ahmed M (2020) Introduction to Modern Climate Change. Andrew E. Dessler: Cambridge University Press, 2011, 252 pp, ISBN-10: 0521173159. Sci Total Environ 734, 139397. https://doi.org/10.1016/j.scitotenv.2020.139397

Ahmed M, Stockle CO (2016). Quantification of climate variability, adaptation and mitigation for agricultural sustainability. Springer Nature Singapore Pvt. Ltd., pp. 437. doi:https://doi.org/10.1007/978-3-319-32059-5.

Ahmed M, Hassan FU, Aslam MA, Akram MN, Akmal M. (2011) Regression model for the study of sole and cumulative effect of temperature and solar radiation on wheat yield. African Journal of Biotechnology.10(45):9114-21. https://doi.org/10.5897/AJB11.1318

Ahmed M, Asif M, Hirani AH, Akram MN, Goyal A (2013) Modeling for Agricultural Sustainability: A Review. In Gurbir S, Bhullar GS, Bhullar NK (ed) Agricultural Sustainability Progress and Prospects in Crop Research. Elsevier, London

Ahmed, M, M. Asif, M. K. Nawaz Shah, Fayyaz-ul-Hassan and S. K. Basu (2014) Model for integrating traditional agronomic practices and modern biotech applications with omics approach in forage development for uplifting agricultural and rural economy in Pakistan. Omics Technologies and Crop Improvement Published: CRC Press. Editor(s): Noureddine Benkeblia

Ahmed M, Hassan F-U (2015) Response of spring wheat (Triticum aestivum L.) quality traits and yield to sowing date. PloS one 10 (4):e0126097-e0126097. https://doi.org/10.1371/journal.pone.0126097

Ahmed M, Aslam M, Fayyaz ul H, Hayat R, Ahmad S (2019) Biochemical, physiological and agronomic response of wheat to changing climate of rainfed Pakistan. Pakistan Journal of Botany 51. https://doi.org/10.30848/PJB2019-2(10)

Ahmed K, Shabbir G, Ahmed M, Shah KN (2020) Phenotyping for drought resistance in bread wheat using physiological and biochemical traits. Sci Total Environ 729, 139082. https://doi.org/10.1016/j.scitotenv.2020.139082

Aitken SN, Whitlock MC (2013) Assisted Gene Flow to Facilitate Local Adaptation to Climate Change. Annual Review of Ecology, Evolution, and Systematics 44(1): 367–388. https://doi.org/10.1146/annurev-ecolsys-110512-135747

Ali, S. et al., 2020. Assessment of arsenic exposure by drinking well water and associated carcinogenic risk in peri-urban areas of Vehari, Pakistan. Environmental Geochemistry and Health, 42, 121-133

Anderson JT (2016) Plant fitness in a rapidly changing world. New Phytol 210 (1):81-87

Anderson, T J, Inouye, W D, McKinney, M A, Colautti, I R, Mitchell-Olds, Tom (2012) Phenotypic plasticity and adaptive evolution contribute to advancing flowering phenology in response to climate change. Proceedings of the Royal Society B: Biological Sciences 279 (1743):3843-3852

Aslam MA, Ahmed M, Stöckle CO, Higgins SS, Hassan Fu, Hayat R (2017a) Can Growing Degree Days and Photoperiod Predict Spring Wheat Phenology? Frontiers in Environmental Science 5 (57). doi:https://doi.org/10.3389/fenvs.2017.00057

Aslam MA, Ahmed M, Fayyaz-ul-Hassan, Hayat R (2017b) Modeling Nitrogen Use Efficiency Under Changing Climate. In: Ahmed M, Stockle CO (eds) Quantification of Climate Variability, Adaptation and Mitigation for Agricultural Sustainability Springer International Publishing, Cham, 71-90. doi:https://doi.org/10.1007/978-3-319-32059-5_4

Aslam Z, Khattak JZK, Ahmed M, Asif M (2017c) A Role of Bioinformatics in Agriculture. In: Ahmed M, Stockle CO (eds) Quantification of Climate Variability, Adaptation and Mitigation for Agricultural Sustainability, Springer International Publishing, Cham, pp 413-434. doi:https://doi.org/10.1007/978-3-319-32059-5_17

Aslam MU, Shehzad A, Ahmed M, Iqbal M, Asim M, Aslam M (2017d) QTL Modelling: An Adaptation Option in Spring Wheat for Drought Stress. In: Ahmed M, Stockle CO (eds) Quantification of Climate Variability, Adaptation and Mitigation for Agricultural Sustainability, Springer International Publishing, Cham, pp 113-136. doi:https://doi.org/10.1007/978-3-319-32059-5_6

Asseng S, Martre P, Maiorano A, Rötter RP, O'Leary GJ, Fitzgerald GJ, Girousse C, Motzo R, Giunta F, Babar MA, Reynolds MP, Kheir AMS, Thorburn PJ, Waha K, Ruane AC, Aggarwal

PK, Ahmed M, Balkovič J, Basso B, Biernath C, Bindi M, Cammarano D, Challinor AJ, De Sanctis G, Dumont B, Eyshi Rezaei E, Fereres E, Ferrise R, Garcia-Vila M, Gayler S, Gao Y, Horan H, Hoogenboom G, Izaurralde RC, Jabloun M, Jones CD, Kassie BT, Kersebaum K-C, Klein C, Koehler A-K, Liu B, Minoli S, Montesino San Martin M, Müller C, Naresh Kumar S, Nendel C, Olesen JE, Palosuo T, Porter JR, Priesack E, Ripoche D, Semenov MA, Stöckle C, Stratonovitch P, Streck T, Supit I, Tao F, Van der Velde M, Wallach D, Wang E, Webber H, Wolf J, Xiao L, Zhang Z, Zhao Z, Zhu Y, Ewert F (2019) Climate change impact and adaptation for wheat protein. Global Change Biology 25 (1):155-173. doi:https://doi.org/10.1111/gcb.14481

Bartwal A, Mall R, Lohani P, Guru S, Arora S (2013) Role of secondary metabolites and brassinosteroids in plant defense against environmental stresses. J Plant Growth Regul 32 (1):216-232

Baudouin E (2011) The language of nitric oxide signalling. Plant Biol 13 (2):233-242

Begara-Morales JC, Sánchez-Calvo B, Chaki M, Valderrama R, Mata-Pérez C, López-Jaramillo J, Padilla MN, Carreras A, Corpas FJ, Barroso JB (2013) Dual regulation of cytosolic ascorbate peroxidase (APX) by tyrosine nitration and S-nitrosylation. J Exp Bot 65 (2):527-538

Begara-Morales JC, Sánchez-Calvo B, Chaki M, Mata-Pérez C, Valderrama R, Padilla MN, López-Jaramillo J, Luque F, Corpas FJ, Barroso JB (2015) Differential molecular response of monodehydroascorbate reductase and glutathione reductase by nitration and S-nitrosylation. J Exp Bot 66 (19):5983-5996

Bita C, Gerats T (2013) Plant tolerance to high temperature in a changing environment: scientific fundamentals and production of heat stress-tolerant crops. Frontiers in plant science 4:273

Bogdanović J, Mojović M, Milosavić N, Mitrović A, Vučinić Ž, Spasojević I (2008) Role of fructose in the adaptation of plants to cold-induced oxidative stress. Eur Biophys J 37 (7):1241-1246

Buchanan BB, Gruissem W, Jones RL (2015) Biochemistry and molecular biology of plants. John Wiley & Sons,

Carvalho C, Hayashi A, Braga M, Nievola C (2013) Biochemical and anatomical responses related to the in vitro survival of the tropical bromeliad Nidularium minutum to low temperatures. Plant Physiol Biochem 71:144-154. doi:https://doi.org/10.1016/j.plaphy.2013.07.005

Chen C, Lei C, Deng A, Qian C, Hoogmoed W, Zhang W (2011) Will higher minimum temperatures increase corn production in Northeast China? An analysis of historical data over 1965–2008. Agricultural and Forest Meteorology 151 (12):1580-1588

Chevin L-M, Lande R, Mace GM (2010) Adaptation, plasticity, and extinction in a changing environment: towards a predictive theory. PLoS Biol 8 (4):e1000357

Corpas FJ, Leterrier M, Valderrama R, Airaki M, Chaki M, Palma JM, Barroso JB (2011) Nitric oxide imbalance provokes a nitrosative response in plants under abiotic stress. Plant Sci 181 (5):604-611

De Pinto MC, Locato V, Paradiso A, De Gara L (2015) Role of redox homeostasis in thermotolerance under a climate change scenario. Ann Bot 116 (4):487-496

Demidchik V (2015) Mechanisms of oxidative stress in plants: from classical chemistry to cell biology. Environ Exp Bot 109:212-228

Deutsch C, Tewksbury J, Tigchelaar M, Battisti D, Merrill S, Huey R, Naylor R (2018) Increase in crop losses to insect pests in a warming climate. Science 361:916-919. doi:https://doi.org/10.1126/science.aat3466

Diao Q, Song Y, Shi D, Qi H (2017) Interaction of Polyamines, Abscisic Acid, Nitric Oxide, and Hydrogen Peroxide under Chilling Stress in Tomato (Lycopersicon esculentum Mill.) Seedlings. Frontiers in Plant Science 8. https://doi.org/10.3389/fpls.2017.00203

Dolev MB, Braslavsky I (2017) Ice-binding proteins—not only for ice growth control. Temperature: Multidisciplinary Biomedical Journal 4 (2):112

Egigu M, Ibrahim M, Riikonen J, Yahya A, Holopainen T, Julkunen-Tiitto R, Holopainen J (2014) Effects of Rising Temperature on Secondary Compounds of Yeheb (Cordeauxia edulis Hemsley). American Journal of Plant Sciences 05:517-527. doi:https://doi.org/10.4236/ajps.2014.55066

Fahad S, Hussain S, Saud S, Hassan S, Ihsan M, Shah A, Wu C, Yousaf M, Jatoi W, Alharby H, Alghabari F, Huang J (2016a) Exogenously Applied Plant Growth Regulators Enhance the

Morpho-Physiological Growth and Yield of Rice under High Temperature. Frontiers in Plant Science 7. https://doi.org/10.3389/fpls.2016.01250
Fahad S, Hussain S, Saud S, Khan F, Hassan S, Amanullah, Nasim W, Arif M, Wang F, Huang J (2016b) Exogenously Applied Plant Growth Regulators Affect Heat-Stressed Rice Pollens. Journal of Agronomy and Crop Science 202 (2):139-150. doi:https://doi.org/10.1111/jac.12148
Fahad S, Bajwa AA, Nazir U, Anjum SA, Farooq A, Zohaib A, Sadia S, Nasim W, Adkins S, Saud S, Ihsan MZ, Alharby H, Wu C, Wang D, Huang J (2017) Crop Production under Drought and Heat Stress: Plant Responses and Management Options. Frontiers in Plant Science 8 (1147). doi:https://doi.org/10.3389/fpls.2017.01147
Fahad S, Ihsan MZ, Khaliq A, Daur I, Saud S, Alzamanan S, Nasim W, Abdullah M, Khan IA, Wu C, Wang D, Huang J (2018) Consequences of high temperature under changing climate optima for rice pollen characteristics-concepts and perspectives. Archives of Agronomy and Soil Science 64 (11):1473-1488. doi:https://doi.org/10.1080/03650340.2018.1443213
Fahad S, Adnan M, Hassan S, Saud S, Hussain S, Wu C, Wang D, Hakeem KR, Alharby HF, Turan V, Khan MA, Huang J (2019a) Chapter 10—Rice Responses and Tolerance to High Temperature. In: Hasanuzzaman M, Fujita M, Nahar K, Biswas JK (eds) Advances in Rice Research for Abiotic Stress Tolerance. Woodhead Publishing, pp 201-224. doi:https://doi.org/10.1016/B978-0-12-814332-2.00010-1
Fahad S., et al., 2019b. Suppressing photorespiration for the improvement in photosynthesis and crop yields: A review on the role of S-allantoin as a nitrogen source, Journal of Environmental Management, 237: 644-651
Fancy NN, Bahlmann AK, Loake GJ (2017) Nitric oxide function in plant abiotic stress. Plant, Cell Environ 40 (4):462-472
Fitter AH, Hay RK (2012) Environmental physiology of plants. Academic press,
Franks SJ, Sim S, Weis AE (2007) Rapid evolution of flowering time by an annual plant in response to a climate fluctuation. Proceedings of the National Academy of Sciences 104 (4):1278-1282
Galloway LF, Etterson JR (2007) Transgenerational plasticity is adaptive in the wild. Science 318 (5853):1134-1136
Gao H-B, Chu Y-J, Xue H-W (2013) Phosphatidic acid (PA) binds PP2AA1 to regulate PP2A activity and PIN1 polar localization. Molecular plant 6 (5):1692-1702
Gaupp F, Pflug G, Hochrainer-Stigler S, Hall J, Dadson S (2017) Dependency of crop production between global breadbaskets: a copula approach for the assessment of global and regional risk pools. Journal of Risk Analysis 37 (11):2212-2228
Gill SS, Tuteja N (2010) Reactive oxygen species and antioxidant machinery in abiotic stress tolerance in crop plants. Plant Physiol Biochem 48 (12):909-930
Gunderson CA, O'Hara KH, Campion CM, Walker AV, Edwards NT (2010) Thermal plasticity of photosynthesis: the role of acclimation in forest responses to a warming climate. Global Change Biol 16 (8):2272-2286
Guo Y, Liu S, Yang Z, Tian S, Sui N (2016) Responses of unsaturated fatty acid in membrane lipid and antioxidant enzymes to chilling stress in sweet sorghum (Sorghum bicolor (L.) Moench) seedling. J Agric Sci 8 (9):71
Gupta KJ, Hincha DK, Mur LA (2011) NO way to treat a cold. The New phytologist 189 (2):360-363
Hajiboland R (2014) Reactive oxygen species and photosynthesis. In: Oxidative damage to plants. Elsevier, pp 1-63
Hammad, MH., et al. (2020). Comparative effects of organic and inorganic fertilizers on soil organic carbon and wheat productivity under arid region. Communication in Soil Science and Plant Analysis, Accepted
Hasanuzzaman M, Nahar K, Alam M, Roychowdhury R, Fujita M (2013a) Physiological, biochemical, and molecular mechanisms of heat stress tolerance in plants. International journal of molecular sciences 14 (5):9643-9684

Hasanuzzaman M, Nahar K, Fujita M (2013b) Extreme temperature responses, oxidative stress and antioxidant defense in plants. Abiotic stress-Plant responses and applications in agriculture:169-205

Hayat S, Ahmad A (2010) Brassinosteroids: a class of plant hormone. Springer Science & Business Media

Hayat R, Ahmad S (2014) Biochemical, physiological and agronomic response of wheat to changing climate of rainfed Pakistan. Pak J Bot 51:2

Herring SC, Hoell A, Hoerling MP, Kossin JP, Schreck III CJ, Stott PA (2016) Explaining extreme events of 2015 from a climate perspective. Bulletin of the American Meteorological Society 97 (12):S1-S145

Hincha DK, Zuther E, Heyer AG (2003) The preservation of liposomes by raffinose family oligosaccharides during drying is mediated by effects on fusion and lipid phase transitions. Biochimica et Biophysica Acta (BBA)-Biomembranes 1612 (2):172-177

Hossain MA, Hoque MA, Burritt DJ, Fujita M (2014) Proline protects plants against abiotic oxidative stress: biochemical and molecular mechanisms. In: Oxidative damage to plants. Elsevier, pp 477-522

Hu, S., Ding, Y., Zhu, C. (2020) Sensitivity and Responses of Chloroplasts to Heat Stress in Plants. Frontiers in Plant Science, 11.

Hussain, S., et al (2020) Study of land cover/land use changes using RS and GIS: a case study of Multan district, Pakistan, Environmental Monitoring & Assessment, 192:2: 1-15

Ijaz W, Ahmed M, Fayyaz-ul-Hassan, Asim M, Aslam M (2017) Models to Study Phosphorous Dynamics Under Changing Climate. In: Ahmed M, Stockle CO (eds) Quantification of Climate Variability, Adaptation and Mitigation for Agricultural Sustainability, Springer International Publishing, Cham, pp 371-386. doi:https://doi.org/10.1007/978-3-319-32059-5_15

IPCC (2018) An IPCC Special Report on the impacts of global warming of 1.5 C above pre-industrial levels and related global greenhouse gas emission pathways, in the context of strengthening the global response to the threat of climate change, sustainable development, and efforts to eradicate poverty

Jabeen M, Gabriel HF, Ahmed M, Mahboob MA, Iqbal J (2017) Studying Impact of Climate Change on Wheat Yield by Using DSSAT and GIS: A Case Study of Pothwar Region. In: Ahmed M, Stockle CO (eds) Quantification of Climate Variability, Adaptation and Mitigation for Agricultural Sustainability, Springer International Publishing, Cham, pp 387-411. doi:https://doi.org/10.1007/978-3-319-32059-5_16

Janská A, Maršík P, Zelenková S, Ovesná J (2010) Cold stress and acclimation–what is important for metabolic adjustment? Plant Biol 12 (3):395-405

Johnson MT, Agrawal AA, Maron JL, Salminen JP (2009) Heritability, covariation and natural selection on 24 traits of common evening primrose (Oenothera biennis) from a field experiment. J Evol Biol 22 (6):1295-1307

Khan MN, Mobin M, Abbas ZK (2015) Nitric oxide and high temperature stress: a physiological perspective. In: Nitric Oxide Action in Abiotic Stress Responses in Plants. Springer, pp 77-93

Khan MA, Tahir A, Khurshid N, Husnain MIu, Ahmed M, Boughanmi H (2020) Economic Effects of Climate Change-Induced Loss of Agricultural Production by 2050: A Case Study of Pakistan. Sustainability 12 (3):1216. https://doi.org/10.3390/su12031216

Kimball S, Gremer JR, Angert AL, Huxman TE, Venable DL (2012) Fitness and physiology in a variable environment. Oecologia 169 (2):319-329

Knaupp M, Mishra KB, Nedbal L, Heyer AG (2011) Evidence for a role of raffinose in stabilizing photosystem II during freeze–thaw cycles. Planta 234 (3):477-486

Knight MR, Knight H (2012) Low-temperature perception leading to gene expression and cold tolerance in higher plants. New Phytol 195 (4):737-751

Kodra E, Steinhaeuser K, Ganguly AR (2011) Persisting cold extremes under 21st-century warming scenarios. Geophys Res Lett 38 (8)

Kremer A, Potts BM, Delzon S (2014) Genetic divergence in forest trees: understanding the consequences of climate change. Funct Ecol 28 (1):22-36

Kumudini S, Andrade FH, Boote K, Brown G, Dzotsi K, Edmeades G, Gocken T, Goodwin M, Halter A, Hammer G (2014) Predicting maize phenology: intercomparison of functions for developmental response to temperature. Agron J 106 (6):2087-2097

Larcher W, Piccioni M (1993) Ecofisiologia vegetal, vol 350. Edagricole,

Larkindale J, Mishkind M, Vierling E (2005) Plant responses to high temperature. Plant abiotic stress 100:144

Lawson T, Davey PA, Yates SA, Bechtold U, Baeshen M, Baeshen N, Mutwakil MZ, Sabir J, Baker NR, Mullineaux PM (2014) C3 photosynthesis in the desert plant Rhazya stricta is fully functional at high temperatures and light intensities. New Phytol 201 (3):862-873. doi:https://doi.org/10.1111/nph.12559

Li X, Gong B, Xu K (2014) Interaction of nitric oxide and polyamines involves antioxidants and physiological strategies against chilling-induced oxidative damage in Zingiber officinale Roscoe. Scientia Horticulturae 170:237-248

Liancourt P, Boldgiv B, Song DS, Spence LA, Helliker BR, Petraitis PS, Casper BB (2015) Leaf-trait plasticity and species vulnerability to climate change in a Mongolian steppe. Global Change Biol 21 (9):3489-3498

Lin A, Wang Y, Tang J, Xue P, Li C, Liu L, Hu B, Yang F, Loake GJ, Chu C (2012) Nitric oxide and protein S-nitrosylation are integral to hydrogen peroxide-induced leaf cell death in rice. Plant Physiol 158 (1):451-464

Liu B, Martre P, Ewert F, Porter JR, Challinor AJ, Müller C, Ruane AC, Waha K, Thorburn PJ, Aggarwal PK, Ahmed M, Balkovič J, Basso B, Biernath C, Bindi M, Cammarano D, De Sanctis G, Dumont B, Espadafor M, Eyshi Rezaei E, Ferrise R, Garcia-Vila M, Gayler S, Gao Y, Horan H, Hoogenboom G, Izaurralde RC, Jones CD, Kassie BT, Kersebaum KC, Klein C, Koehler A-K, Maiorano A, Minoli S, Montesino San Martin M, Naresh Kumar S, Nendel C, O'Leary GJ, Palosuo T, Priesack E, Ripoche D, Rötter RP, Semenov MA, Stöckle C, Streck T, Supit I, Tao F, Van der Velde M, Wallach D, Wang E, Webber H, Wolf J, Xiao L, Zhang Z, Zhao Z, Zhu Y, Asseng S (2019) Global wheat production with 1.5 and 2.0°C above pre-industrial warming. Global Change Biology 25(4):1428-1444. doi:https://doi.org/10.1111/gcb.14542

Lu P, Sang W-G, Ma K (2008) Differential Responses of the Activities of Antioxidant Enzymes to Thermal Stresses between Two Invasive Eupatorium Species in China. Journal of integrative plant biology 50:393-401. doi:https://doi.org/10.1111/j.1744-7909.2007.00583.x

Mathur S, Jajoo A (2014) Alterations in photochemical efficiency of photosystem II in wheat plant on hot summer day. Physiol Mol Biol Plants 20 (4):527-531

Mathur S, Agrawal D, Jajoo A (2014) Photosynthesis: response to high temperature stress. J Photochem Photobiol B: Biol 137:116-126

Mazorra L (2011) Brassinosteroid action and its relation with heat stress mechanisms in plants. In: Brassinosteroids: A Class of Plant Hormone. Springer, pp 289-307

Medeiros JS, Ward JK (2013) Increasing atmospheric [CO2] from glacial to future concentrations affects drought tolerance via impacts on leaves, xylem and their integrated function. New Phytol 199 (3):738-748

Merilä J (2012) Evolution in response to climate change: in pursuit of the missing evidence. Bioessays 34 (9):811-818

Mittler R, Finka A, Goloubinoff P (2012) How do plants feel the heat? Trends Biochem Sci 37 (3):118-125

Mubeen, M. et al. (2019) Evaluating the climate change impact on crop water requirement of cotton- wheat in semi-arid conditions using DSSAT model. Journal of Water and Climate Change 11 (4): 1661-1675. doi: https://doi.org/10.2166/wcc.2019.179

Narayanan S, Tamura PJ, Roth MR, Prasad PV, Welti R (2016) Wheat leaf lipids during heat stress: I. High day and night temperatures result in major lipid alterations. Plant, Cell Environ 39 (4):787-803

Nguyen-Huy T, Deo RC, Mushtaq S, An-Vo D-A, Khan S (2018) Modeling the joint influence of multiple synoptic-scale, climate mode indices on Australian wheat yield using a vine copula-based approach. European journal of agronomy 98:65-81

Nicotra AB, Atkin OK, Bonser SP, Davidson AM, Finnegan EJ, Mathesius U, Poot P, Purugganan MD, Richards CL, Valladares F (2010) Plant phenotypic plasticity in a changing climate. Trends Plant Sci 15 (12):684-692

Nievola CC, Carvalho CP, Carvalho V, Rodrigues E (2017) Rapid responses of plants to temperature changes. Temperature 4 (4):371-405

Niu L, Liao W (2016) Hydrogen peroxide signaling in plant development and abiotic responses: crosstalk with nitric oxide and calcium. Frontiers in Plant Science 7:230

Niu Y, Xiang Y (2018) An overview of biomembrane functions in plant responses to high-temperature stress. Frontiers in plant science 9:915

O'Sullivan OS, Heskel MA, Reich PB, Tjoelker MG, Weerasinghe LK, Penillard A, Zhu L, Egerton JJ, Bloomfield KJ, Creek D (2017) Thermal limits of leaf metabolism across biomes. Global Change Biol 23 (1):209-223

Obaid AY, Sabir JSM, Atef A, Liu X, Edris S, El-Domyati FM, Mutwakil MZ, Gadalla NO, Hajrah NH, Al-Kordy MA, Hall N, Bahieldin A, Jansen RK (2016) Analysis of transcriptional response to heat stress in Rhazya stricta. BMC Plant Biol 16 (1):252. doi:https://doi.org/10.1186/s12870-016-0938-6

Ohama N, Sato H, Shinozaki K, Yamaguchi-Shinozaki K (2017) Transcriptional regulatory network of plant heat stress response. Trends Plant Sci 22 (1):53-65

Pedroso ANV, Lazarini RAdM, Tamaki V, Nievola CC (2010) In vitro culture at low temperature and ex vitro acclimatization of Vriesea inflata an ornamental bromeliad. Brazilian Journal of Botany 33:407-414

Petoukhov V, Semenov VA (2010) A link between reduced Barents-Kara sea ice and cold winter extremes over northern continents. Journal of Geophysical Research: Atmospheres 115 (D21)

Pirttioja N, Carter TR, Fronzek S, Bindi M, Hoffmann H, Palosuo T, Ruiz-Ramos M, Tao F, Trnka M, Acutis M (2015) Temperature and precipitation effects on wheat yield across a European transect: a crop model ensemble analysis using impact response surfaces. Journal of Climate Research 65:87-105

Porter JR, Xie L, Challinor AJ, Cochrane K, Howden SM, Iqbal MM, Travasso M, Barros V, Field C, Dokken D (2017) Food security and food production systems.

Qasim, M.Z, et al. (2020) The potential applications of picotechnology in biomedical and environmental sciences. Environmental Science and Pollution Research, Accepted

Ramani H, Mandavia M, Dave R, Bambharolia R, Silungwe H, Garaniya N (2017) Biochemical and physiological constituents and their correlation in wheat (Triticum aestivum L.) genotypes under high temperature at different development stages. International Journal of Plant Physiology and Biochemistry 9 (1):1-8

Rasheed R, Wahid A, Farooq M, Hussain I, Basra SM (2011) Role of proline and glycinebetaine pretreatments in improving heat tolerance of sprouting sugarcane (Saccharum sp.) buds. Plant Growth Regulation 65 (1):35-45

Rasool A., et al. (2019) Quantification of Tl (I) and Tl (III) based on microcolumn separation through ICP-MS in river sediment pore water. Environmental Science and Pollution Research, Accepted

Repetto M, Semprine J, Boveris A (2012) Lipid peroxidation: chemical mechanism, biological implications and analytical determination, vol 1. chapter

Rihan HZ, Al-Issawi M, Fuller MP (2017) Advances in physiological and molecular aspects of plant cold tolerance. Journal of Plant Interactions 12 (1):143-157

Rivero RM, Ruiz JM, Garcıa PC, Lopez-Lefebre LR, Sánchez E, Romero L (2001) Resistance to cold and heat stress: accumulation of phenolic compounds in tomato and watermelon plants. Plant Sci 160 (2):315-321

Rubbo H, Trostchansky A (2014) Nitro-fatty acids: synthesis, properties, and role in biological system. In: Nitric Oxide in Plants: Metabolism and Role in Stress Physiology. Springer, pp 153–162

Ruelland E, Zachowski A (2010) How plants sense temperature. Environ Exp Bot 69 (3):225-232

Saidi Y, Peter M, Finka A, Cicekli C, Vigh L, Goloubinoff P (2010) Membrane lipid composition affects plant heat sensing and modulates Ca2+-dependent heat shock response. Plant signaling & behavior 5 (12):1530-1533

Sedaghatmehr M, Mueller-Roeber B, Balazadeh S (2016) The plastid metalloprotease FtsH6 and small heat shock protein HSP21 jointly regulate thermomemory in Arabidopsis. Nature communications 7:12439

Sehrawat A, Gupta R, Deswal R (2013) Nitric oxide-cold stress signalling cross-talk, evolution of a novel regulatory mechanism. Proteomics 13 (12-13):1816-1835

Shaw, RG, Etterson, RJ (2012) Rapid climate change and the rate of adaptation: insight from experimental quantitative genetics. New Phytol 195 (4):752-765

Shinozaki K, Uemura M, Bailey-Serres J, Bray E, Weretilnyk E (2015) Responses to abiotic stress. Biochemistry and molecular biology of plants:1051-1100

Singsaas E, Sharkey T (1998) The regulation of isoprene emission responses to rapid leaf temperature fluctuations. Plant, Cell Environ 21 (11):1181-1188

Smirnoff N (2005) Ascorbate, tocopherol and carotenoids: metabolism, pathway engineering and functions. Antioxidants and reactive oxygen species in plants:53-86

Stott PA, Allen M, Christidis N, Dole RM, Hoerling M, Huntingford C, Pall P, Perlwitz J, Stone D (2013) Attribution of weather and climate-related events. In: Climate science for serving society. Springer, pp 307–337

Sugiyama Y, Kameshita I (2017) Multi-PK antibodies: Powerful analytical tools to explore the protein kinase world. Biochemistry and biophysics reports 11:40-45

Suzuki N, Koussevitzky S, Mittler R, Miller G (2012) ROS and redox signalling in the response of plants to abiotic stress. Plant, Cell Environ 35 (2):259-270

Suzuki N, Miller G, Salazar C, Mondal HA, Shulaev E, Cortes DF, Shuman JL, Luo X, Shah J, Schlauch K (2013) Temporal-spatial interaction between reactive oxygen species and abscisic acid regulates rapid systemic acclimation in plants. The Plant Cell 25 (9):3553-3569

Taiz L, Zeiger E, Møller IM, Murphy A (2015) Plant physiology and development.

Takai K, Nakamura K, Toki T, Tsunogai U, Miyazaki M, Miyazaki J, Hirayama H, Nakagawa S, Nunoura T, Horikoshi K (2008) Cell proliferation at 122 C and isotopically heavy CH4 production by a hyperthermophilic methanogen under high-pressure cultivation. Proceedings of the National Academy of Sciences 105 (31):10949-10954

Tang T, Liu P, Zheng G, Li W (2016) Two phases of response to long-term moderate heat: Variation in thermotolerance between Arabidopsis thaliana and its relative Arabis paniculata. Phytochemistry 122:81-90. doi:https://doi.org/10.1016/j.phytochem.2016.01.003

Tarkowski ŁP, Van den Ende W (2015) Cold tolerance triggered by soluble sugars: a multifaceted countermeasure. Frontiers in plant science 6:203

Thomashow MF (2010) Molecular basis of plant cold acclimation: insights gained from studying the CBF cold response pathway. Plant Physiol 154 (2):571-577

Tollenaar M, Fridgen J, Tyagi P, Stackhouse Jr PW, Kumudini S (2017) The contribution of solar brightening to the US maize yield trend. Nature Climate Change 7(4): 275–278. https://doi.org/10.1038/nclimate3234

Tollenaar M, Dzotsi K, Kumudini S, Boote K, Chen K, Hatfield J, Jones JW, Lizaso JI, Nielsen RL, Thomison P, Timlin DJ, Valentinuz O, Vyn TJ, Yang H (2020) Modeling the Effects of Genotypic and Environmental Variation on Maize Phenology: The Phenology Subroutine of the AgMaize Crop Model. Agroclimatology:173-200. https://doi.org/10.2134/agronmonogr60.2017.0038

Vierling E (1991) The roles of heat shock proteins in plants. Annu Rev Plant Biol 42 (1):579-620

Vitasse Y, Lenz A, Körner C (2014) The interaction between freezing tolerance and phenology in temperate deciduous trees. Frontiers in Plant Science 5:541

Wahid A, Close T (2007) Expression of dehydrins under heat stress and their relationship with water relations of sugarcane leaves. Biol Plant 51 (1):104-109

Wahid A, Farooq M, Hussain I, Rasheed R, Galani S (2012) Responses and management of heat stress in plants. In: Environmental adaptations and stress tolerance of plants in the era of climate change. Springer, pp 135–157

Wallach D, Martre P, Liu B, Asseng S, Ewert F, Thorburn PJ, van Ittersum M, Aggarwal PK, Ahmed M, Basso B, Biernath C, Cammarano D, Challinor AJ, De Sanctis G, Dumont B, Eyshi Rezaei E, Fereres E, Fitzgerald GJ, Gao Y, Garcia-Vila M, Gayler S, Girousse C, Hoogenboom G, Horan H, Izaurralde RC, Jones CD, Kassie BT, Kersebaum KC, Klein C, Koehler A-K, Maiorano A, Minoli S, Müller C, Naresh Kumar S, Nendel C, O'Leary GJ, Palosuo T, Priesack E, Ripoche D, Rötter RP, Semenov MA, Stöckle C, Stratonovitch P, Streck T, Supit I, Tao F, Wolf J, Zhang Z (2018) Multimodel ensembles improve predictions of crop–environment–management interactions. Global Change Biology 24 (11):5072-5083. doi:https://doi.org/10.1111/gcb.14411

Wang, D., et al., 2019. Morphological acclimation to agronomic manipulation in leaf dispersion and orientation to promote "Ideotype" breeding: Evidence from 3D visual modeling of "super" rice (Oryza sativa L.). Plant Physiology and Biochemistry, 135: 499-510

Xuan Y, Zhou S, Wang L, Cheng Y, Zhao L (2010) Nitric oxide functions as a signal and acts upstream of AtCaM3 in thermotolerance in Arabidopsis seedlings. Plant Physiol 153 (4):1895-1906

Yang H, Feng J, Zhai S, Dai Y, Xu M, Wu J, Shen M, Bian X, Koide RT, Liu J (2016) Long-term ditch-buried straw return alters soil water potential, temperature, and microbial communities in a rice-wheat rotation system. Soil and Tillage Research 163:21-31. https://doi.org/10.1016/j.still.2016.05.003

Yang R, Guo L, Wang J, Wang Z, Gu Z (2017) Heat Shock Enhances Isothiocyanate Formation and Antioxidant Capacity of Cabbage Sprouts. J Food Process Preserv 41 (4):e13034. doi:https://doi.org/10.1111/jfpp.13034

Ye T, Zong S, Kleidon A, Yuan W, Wang Y, Shi P (2019) Impacts of climate warming, cultivar shifts, and phenological dates on rice growth period length in China after correction for seasonal shift effects. Climatic Change 155 (1):127-143. https://doi.org/10.1007/s10584-019-02450-5

Zamin M., et al. (2019) Developing the first halophytic turfgrasses for the urban landscape from native Arabian desert grass. Environmental Science and Pollution Research (Accepted)

Zhang S, Jiang H, Peng S, Korpelainen H, Li C (2010) Sex-related differences in morphological, physiological, and ultrastructural responses of Populus cathayana to chilling. J Exp Bot 62 (2):675-686

Zhou R, Yu X, Kjær KH, Rosenqvist E, Ottosen C-O, Wu Z (2015) Screening and validation of tomato genotypes under heat stress using Fv/Fm to reveal the physiological mechanism of heat tolerance. Environ Exp Bot 118:1-11

Ziogas V, Tanou G, Filippou P, Diamantidis G, Vasilakakis M, Fotopoulos V, Molassiotis A (2013) Nitrosative responses in citrus plants exposed to six abiotic stress conditions. Plant Physiol Biochem 68:118-126

Źróbek-Sokolnik A (2012) Temperature stress and responses of plants. In: Environmental Adaptations and Stress Tolerance of Plants in the Era of Climate Change. Springer, pp 113–134

Chapter 4
Infirmity to Climate Change and Regional Impacts

Mazhar Abbas, Muhammad Salman Shabbir, Nor Azila Bt Mohd Noor, Wajid Nasim, and Muhammad Mubeen

Abstract This chapter explains the type of information needed to measure climate change and inconsistency as well as the techniques used for measuring the effects and susceptibility of global climate of various sectors. This chapter is aimed to know the regional and international potential impacts at average elevated temperature up to 5 °C, over 7 indicators that represent temperature, water, agriculture, food security, extreme events, healthcare, coastal zones, and ecosystem. For example, the surface temperature over the last decade had grown about "0.6 °C," which is consistent with increasing global average temperature. Water travels constantly above, on, and below the surface of the planet, interchanging through air, liquid water, and vapor. The growth of agriculture is highly vulnerable to 2 °C (drop) global average temperatures projections with significant impact on both rural and urban agricultural development and food stability. An increase in the severity and the incidents of severe weather is among the most important noticeable effects of global warming; huge flood in different parts of the world affect billions of population and the

M. Abbas (✉)
Department of Management and MIS, College of Business Administration,
University of Hail, Hail, Kingdom of Saudi Arabia
e-mail: m.hussain@uoh.edu.sa

M. S. Shabbir
Department of Management, College of Commerce and Business Administration,
Dhofar University, Salalah, Oman
e-mail: mshabbir@du.edu.om

N. A. B. M. Noor
Othman Yeop Abdullah Graduate School of Business, University Utara Malaysia,
Bukit Kayu Hitam, Malaysia
e-mail: azila@uum.edu.my

W. Nasim
Department of Agronomy, Faculty of Agriculture and Environment,
The Islamia University of Bahawalpur (IUB), Bahawalpur, Pakistan

M. Mubeen
Department of Environmental Sciences, COMSATS University Islamabad,
Vehari Campus, Pakistan

W. N. Jatoi et al. (eds.), *Building Climate Resilience in Agriculture*,
https://doi.org/10.1007/978-3-030-79408-8_4

incidence of Malaria and other infectious disease were enormously reported after the onset of flood. The climatological patterns (rapid change in daytime temperature, mostly in cool climatic season) induce respiration problems characterized as upper and lower respiratory infection. Furthermore, coastal habitats are among the most productive habitats as they are supplemented by land-based and marine minerals but these are much disturbed by climate change. So, an integrated approach is needed to use the elements of nature for getting maximum benefits on a sustained basis.

These seven sectors are being discussed in detail in the following lines.

Keywords Climate change and susceptibility · Temperature · Water · Agriculture · Food security · Extreme events · Ecosystem

4.1 Temperature

4.1.1 Understanding the Role of Temperature in Climate Change

4.1.1.1 Ocean Temperature

Because of the immense ocean temperature, a massive amount of heat energy is needed in order to increase even a small amount of average temperature of the earth's annual global average temperature. The 2 °C significant rise in the worldwide air temperatures has taken place compared with the early period of the modern age (1880–1900). Although it appears tiny, however, this implies that cumulative warm air will rise significantly. This added heat induces high temperatures in the regions and seasons, decreases the amount of ice and snow, increases heavy precipitation, and brings changes in the habitat of animals and plants (Lindsey and Dahlman 2020).

4.1.1.2 Continent Temperature

Asia confirmed its third hottest year, above the "average temperature of 1.68 °C (3.02 °F), in 1910–2000. The five hottest years in Asia have mostly occurred till 2007. The trend in Asia was "+0.16 °C (+0.29 F)" per decade during the period 1910–2019; even so, the pattern in 1981–2019 is multiple times the longer term pattern "(+0.35 °C/+0.63 °F)" (Pidcock 2015).

4.1.1.3 Global Temperature

The Global Change adverse effect studies' center shows that the surface temperature over the last decade had grown to about "0.6 °C", which is consistent with increasing global average temperature. The Prospective global warming forecasts, predicated from all four IPCC-AR5 Scenarios considered, found that the average warming trend at the end of the decade would be around 1 °C above the worldwide average. This rise, particularly in atmospheric pressure, is associated with such an incidence of negative effects, such as the soaring incidence of extreme storms (floods, droughts, heat waves, and cyclonic activity) and the steady decline of most glaciers (Azeem 2019).

4.1.1.4 How Do Scientists Measure Global Temperature?

Researchers should incorporate measurements of terrestrial as well as the surface water acquired from boats, and satellite systems, in order to have a complete picture of the surface climate.

Researchers use four huge databases to explore the atmospheric climate. The United Kingdom Met Office Hadley Center and the University of East Anglia's Climate research group collectively helped in providing Had CRUT4. The GISTEMP range in the US is given by the National Oceanic and Atmospheric Administration (NOAA) via the NASA Goddard Institute for Space Sciences (GISS) and the MLOST database. A fourth dataset is provided by the Japan Meteorological Agency (JMA).

Figure 4.1 shows that how average temperatures over the last 130 years align in the four datasets. You will see that they both display a growing pattern but there are still several variations for year-to-year (Pidcock 2015).

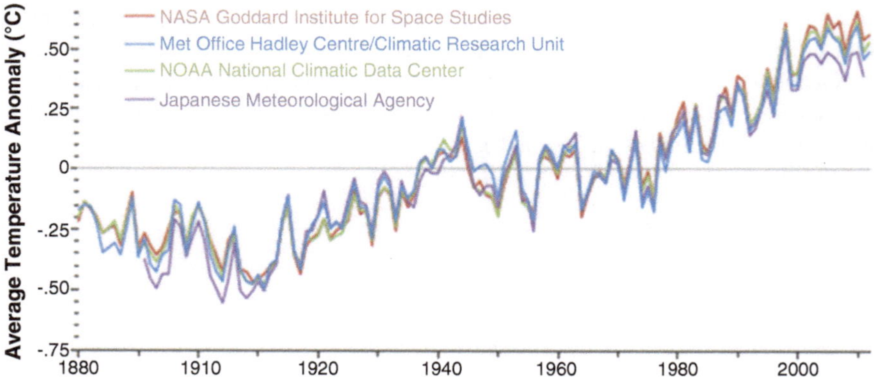

Fig. 4.1 Global average temperature anomaly from 1880 to 2012, compared to the 1951–1980 long-term average. Source: NASA Earth Observatory

4.2 Water

4.2.1 Understanding Role of Water in Climate Change

Water occupies nearly 3/4th of Land. It moves constantly above, on, and below the surface of the planet, alternating through the air, liquid water, and vapor. Water and atmosphere are profoundly intertwined. The higher temperatures cause the water to evaporate from places such as oceans, lakes, streams, and soil. That adds to greater rates of ambient water vapor, which ensures the incidents of precipitation. Similarly, water falling as rain or snow depends on the weather. Thus, while winter conditions in regions such as the North-eastern United States stay below zero, the elevated humidity in the air may indicate heavy snowstorms. (Reports and Multimedia/ Explainer 2010; Vulnerability of Pakistan's 2017).

Increasing water vapor in the air can also increase heating more. Steaming is essentially a greenhouse gas that absorbs moisture in the atmosphere and induces elevated temperatures. Yet unlike many greenhouse emissions that may remain in the environment for years, water vapors normally remains in the air for a matter of days before dropping as precipitation back to Earth (QUEST Staff 2014; Vulnerability of Pakistan's 2017).

4.2.2 Water and Climate Change

Change in climate impacts water supplies and it influences glacier activity, greenhouse gas pollution, and predisposing factors like disasters recurring. The years 1950, 1956, 1957, 1973, 1976, 1978, 1988, 1992, 2010, 2011, and 2012 had extreme floods. The average water supply per unit has now decreased from "5140 m^3 in 1950 to 1000 m^3." It is a fast reaching water shortage. Thus, it is important to develop comprehensive action plans at the state, regional, and local levels addressing scientific, social, economic, institutional, and financial aspects in order to prevent the consequences of climate change on the water issue. Implementation of two parallel mechanisms, i.e., adaptation strategies and mitigation (addressing climate change negativities) is essential (Hussain and Mumtaz 2014; Nasim et al. 2011; Chaudhary et al. 2011; Hammad et al. 2010a, b).

4.2.3 How Do Scientists Measure Water Impact on Climate Change?

4.2.3.1 How Heat Moves

Sunlight is the primary cause of ocean energy. In fact, dust, water vapor, and greenhouse gasses release radiation they have consumed, and some of the thermal pollution spills through the oceans. The water is continually combined by winds, tides, and rivers, shifting heat energy from warmer to colder latitudes and deeper depths (Chaudhary et al. 2011; Chaudhry 2015).

4.2.3.2 Measuring Ocean Heat

Previously, taking temperature from the ocean was possible through ships to drift sensors or sample collections into the sea. But through this technique, we could only know temperatures of a limited portion of the too large oceans spreading on the earth. To gain worldwide visibility, scientists have switched to satellites that determine ocean heights. When the water warms, it spreads, and ocean temperature levels can be measured from the heights of the sea surface. To check that the values they generate make sense, scientists continuously analyze data from rockets, floats, and probes. Every 3 months, the measurement series is processed to estimate the global average ocean heat content. Transforming temperatures into joules (a normal unit of energy) helps us equate heat on the sea as in some areas of the world's climate (Dahlman and Lindsey 2020) (Fig. 4.2).

4.2.4 Understanding Role of Agriculture and Food Security in Climate Change

The main areas for action in the era of climate change are food and agriculture. Growth of agriculture is highly vulnerable to 2 °C (drop) global average temperatures projections for 2100, with a significant impact on both rural and urban agricultural development and food stability.

There are still very uncertain patterns and effects at a number of areas and temporal dimensions; there need to be significant improvements in forecasting how the changes in climate influence food security potential. In light of these uncertainties, adaptation to the extent and rate of expected changes would obviously be required. Adaptation activities in two overlapping areas can be (1) better governance of agricultural threats related to increased climate volatility and events, like strong environment intelligence systems and protection networks; (2) rapid adjustment of agricultural threats to decadal timescales such as advanced infrastructures,

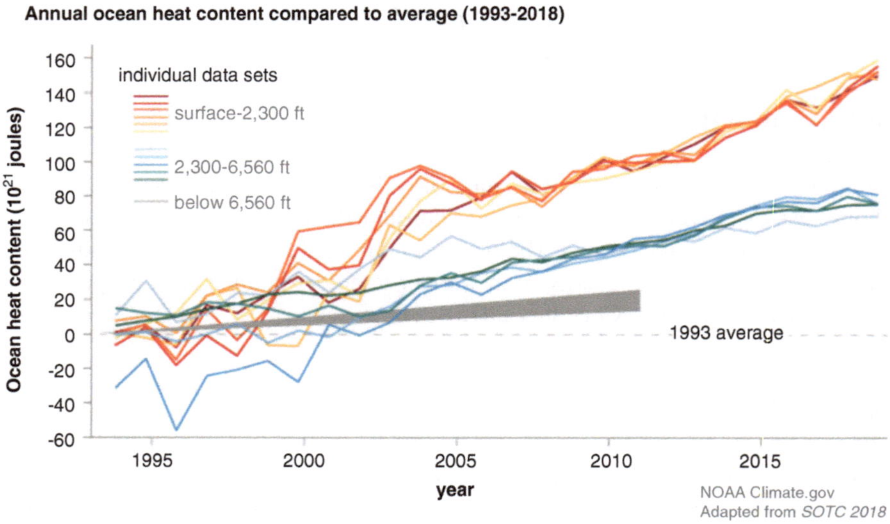

Fig. 4.2 Agriculture Impacts in relation to Food Security

agriculture, and policy options. Investing in technological innovation and agricultural strengthening in connection with increased quality of the inputs will require the production, among other aspects, of incentives and monitoring systems for small-scale farmers (Nasim et al. 2012a, b, c; Munis et al. 2012; Sajjad et al. 2012).

4.2.5 Food Security and Agriculture Role in Climate Change

The growth of domestic and industrial use of chemical products in the form of fertilizers also has a part to be played for a change in the environment, according to new research "Environment Risk and Food Safety Analysis: A Special Report (Islamabad, December 2018)." The high use rate of fertilizers rich in nitrogen affects croplands' heat-storing system (the potential for thermal trapping of nitrogen oxides 300 times higher per unit than carbon dioxide) and the excessive fertilizer run-down in our waters produces "dead zones." High levels of food and nutrition deficiency are compounding the threats it poses. As the atmosphere begins to shift, weather conditions are becoming more common in the face of unpredictable and extreme environment-related events, so food shortage levels will grow much further (Climate Risks and Food Security Analysis 2018).

4.3 Extreme Events

4.3.1 Understanding Role of Extreme Events in Climate Change

In 2019, according to new research "Extreme Weather and Climate Change" an increase in the severity and the incidents of severe weather is among the most important noticeable effects in global warming. The National Climate Evaluation reports that in the United States there have been increasing heat waves, severe droughts, and significant hurricanes and that such occurrences have increased. All types of events, including climate change, have been proven to cause certain environmental events (inundations, tropical storms) and unknown climates (tornados) (Pakistan Droughts 2018).

According to new research 2018 "Global warming is making some extreme weather events worse." Climate change is predicted to increase the length, severity, and effect of certain kinds of severe weather events. For example, the rise in sea-level raises the impacts of coastal storms and warming during droughts may place more stress on water supplies. That's why several cities, states, and industries take action to brace for potentially severe weather conditions.

Researchers classify such severe weather occurrences as dependent on a given region's past background of temperature. Severe weather is known to generate rain or snow, temperature, wind, or some other extremely low or high levels of effect. These incidents are usually called severe because they are unlike 90–95% of comparable weather conditions that happened in the same region previously.

4.3.2 Examples of Some Extreme Events

4.3.2.1 Heat and Drought

The harmful results of heat waves, including death, are induced by temperature and humidity—particularly if these conditions persist for more than 2 days. Due to weather highs being broken month after month, year after year, intense heat conditions are expected to grow more common owing to human-caused global warming. The increased temperatures also enhance evaporation, which in the summer dries the land up—intensifying drought in several regions (Extreme Weather and Climate Change 2019; Extreme Weather 2020).

4.3.2.2 Storms and Floods

When further evaporation from the soil contributes to higher precipitation, rainfall intensifies. We already learn, for instance, the flooding of Hurricane Harvey was 15 years more extreme and 3 times more likely, due to climate change induced by people. We foresee hurricane category 4 and 5 to be more common, as temperatures are also increasing gradually. While scientists do not know if climate change has contributed to more storms, they are sure that sea levels will contribute to higher storm waves and more inundation. Since 1900, about half of the sea-level increase arises from the extension of rising seas, exacerbated by global warming induced by man. (Like all liquids, water normally spreads when it heats up.) The remainder of the growth stems from falling ice sheets and glaciers (Extreme Weather 2020).

4.3.2.3 Snow and Frigid Weather

According to the latest research "Extreme weather gets a boost from climate change," the spike in snowfall during winter storms can be correlated with climate change. Notice the colder weather gets more moisture. Thus, snowfall will smash records when the temperatures are below zero. And scientists in the eastern United States are researching a potential link between the growing Arctic and cold spells. The theory is that the jet stream may be weakened by a rapidly warming Arctic which enables frigid polar air to migrate further south.

4.3.2.4 Extreme Events and Climate Change

According to experts, in the past 20 years, Pakistan (due to climate change) has faced around 150 freak weather incidents resulting in economic losses: winter smog, summer forest fires, glaciers melt, flash floods, freaky heatwaves, landslides, and displaced populations to mention a few. In 2010–2011 nearly 10% of the country's population was affected during massive flooding in two provinces, one north and one south. The year before, the climate-related cost of severe weather was $384 million; over the last 20 years, the plagues of global warming cost nearly 2 billion dollars to the country's economy (Azeem 2019).

4.4 Healthcare

4.4.1 Healthcare and Climate Change

The climatological change pattern induces respiration problems characterized as upper respiratory infection and lower respiratory tract, and it mainly depends upon the rapid change in daytime temperature, mostly in cool climatic season. One of the studies also indicated that the climate variation in season also impact the microflora, and hence increase the risk of a particular type of infection under specific climate condition such as more fungal are present in the environment during summer season (Willarda 2020; Neelam et al. 2017). Figure 4.3 shows some of the impacts of climate change on the health of people.

We would like to quote an example that how healthcare is related to climate change consequences. In Pakistan, the flood of 2010 affected around 14–20 million people with 1700 deaths; nearly 436 health facilities and 1.1 million homes were destroyed. From July 29, 2010 to July 21, 2011, 37,391,802 cases of different illness of flood victims were reported. The predominate illnesses (as shown in Fig. 4.4) are acute diarrhea, respiratory, and skin infection; a small number of malaria was also reported in the flood-affected districts. Other reported data show a few numbers

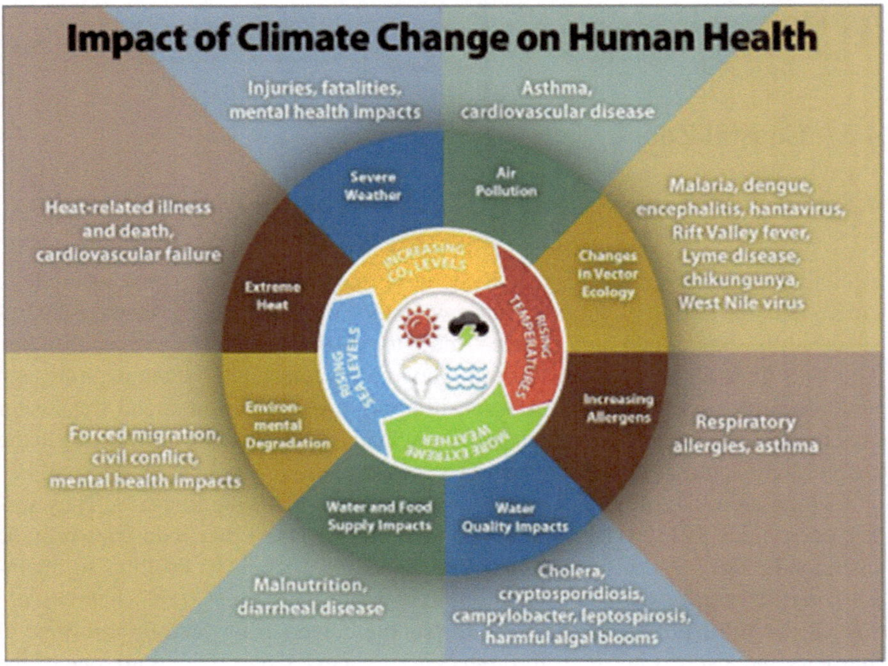

Fig. 4.3 Climate change, health, and health care

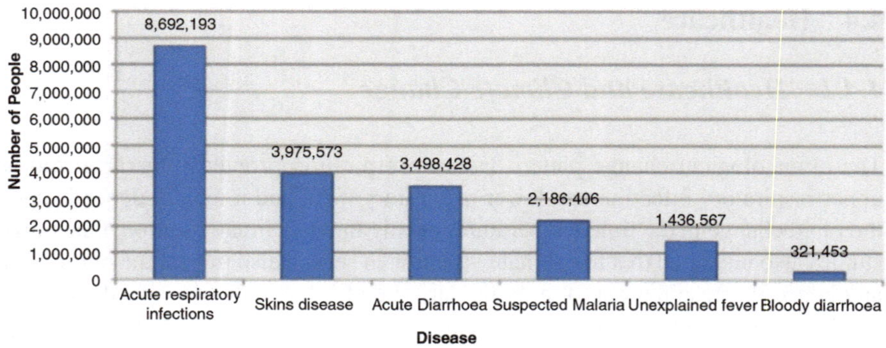

Fig. 4.4 Number of infected cases in flood areas from July 29, 2010 to July 21, 2011

of Meningitis, Viral Hemorrhagic Fever, Diphtheria, and Tetanus (Willarda 2020; WHO 2015).

World is currently implementing climate change adaptation projects in the field of health and is taking action to build institutional and technical capacity to address climate change. In addition, many developing nations are making climate change strategy that considers the health implications of actions to mitigate climate change. This includes the national assessment of climate change effects and vulnerability for health, implementation to increase the climate resilience of health infrastructure, and providing climate information for an integrated disease surveillance and response system.

4.5 Coastal Zones

4.5.1 Understanding Role of Coastal Zones in Climate Change

Among the most vibrant natural ecosystems is really the coastal region where land connects to the ocean. Coast is the place where the three main components—hydrosphere, lithosphere, and atmosphere—meet together and interact with each other. The coastlines are created by shifts in geography, controlled by temperature and geological processes. They form a transition region in which land and freshwater encounter saltwater, and the mutual effects of land and ocean are transformed. In the current fight against climate change, the coastal zones are a crucial battleground.

Coastal areas are of great ecological and socioeconomic significance. They sustain economies and through fisheries, ports, tourism, and other industries provide livelihoods. They also help environmental resources, such as controlling air chemistry, fertilizer and water conservation, and waste disposal. Coastal habitats are among the most productive as they are supplemented from land-driven mineral nutrients and also get nutrients through coastal waters. Coastal areas are one of the densely populated regions as well. Almost half of the world's major cities are located within 50 km of coastline, and the coastal population is 2.6 times greater than the inland population densities (Wajid et al. 2007; Ahmad et al. 2009, Usman et al. 2009, 2010; Pallewatta 2010).

4.5.2 Coastal Zones and Climate Change

It has been observed that people bordering coastal regions face the worst climate impacts. Unconsciousness and over-exploitation of natural capital are some of the factors for worsening the situation; deforestation is one of the key reasons for rising impacts on climate change. For example, deforestation in Pakistan has grown at a rapid rate of 2.1% per annum over the past few years, the fastest in Asia, led by an unprecedented 2.3% annual loss of mangrove forests. Pakistan is endowed with one of the world's largest semi-arid mangroves but has long neglected their ecological importance which has caused great harm. Mangrove cover helps defend coastal areas from extreme climatic weather as they act as a barrier from hurricanes and flooding and often function as a possible refuge for shrimps and sea animals, helping fishing populations economically as well. The severely depleted mangrove forests have made Pakistan's coastal areas vulnerable to harsh climatic conditions, especially cyclones, flooding, sea-level rise, and change impacts (Abu Baker 2014).

4.6 Ecosystem and Climate Change

One of our most precious resources, natural ecosystems, are critical to our survival on the planet. Their benefits are diverse, ranging from marketable goods like pharmaceuticals, leisure activities like camping, to resources such as erosion management and water purification. Terrestrial and coastal ecosystems contain over five times more organic carbon than atmospheric carbon, while net emissions of land cover change and destruction of ecosystem account for about 10% of the overall annual anthropogenic carbon emissions (Epple et al. 2016).

According to new research, "Climate Impacts on Ecosystems," the climatic effect on the ecosystems is significant. Climate change has a variety of effects on ecosystems. For example, warming may cause organisms to migrate to higher latitudes, where temperature enhances their survival. Likewise, intrusions of freshwater into the ecosystem may cause some of the major species to move or die, minimizing vital predators or prey from the food chain (Qureshi and Ali 2011).

Climate change has affected the ecosystem's functioning, which characterizes the flow of chemicals and energy via plants, herbivores, carnivores, and soil organisms (Malcolm and Pitelka 2001; Wajid et al. 2010; Nasim et al. 2010; Hammad et al. 2010).

The management of ecosystems can have a significant role in mitigation and adaptation for climate change; this may be possible by evaluating and improving current practices towards sustainability.

4.7 Conclusion

To get a more consistent analysis of climate change impacts on various sectors like agriculture, food security, and other associated fields, some integrated models should be developed which combine the estimates caused by climate change scenarios, the simulation analysis of agricultural characteristics, and the economic analysis representing the socioeconomic elements of agriculture. Additionally, a more detailed analysis for the economic and policy impacts of each adaptation measures for the agriculture sector. The multiple challenges which the world is facing in terms of climate change, ecosystems degradation, and food insecurity require an integrated approach. This approach should comprise a sustainable, inclusive, and resource-efficient path. The notion that green growth and climate-smart agriculture can help to solve urgent issues could be one of the basis for developing and starting the roadmap for achieving climate resilience.

References

Abu Baker, S. M. Climate Change in coastal areas of Pakistan. Retrieved from https://nation.com.pk/27-Apr-2014/climate-change-in-coastal-areas-of-pakistan. Accessed 2014, April 27

Ahmad, A., et al. (2009) Seasonal growth, radiation interception, its conversion efficiency and aerial dry biomass production of Oryza sativa L. under diverse agro-environments in Pakistan. Pakistan Journal of Botany, 41(3): 1241-1257

Azeem, H. M. Climate change: causes, outcomes in Pakistan and a way forward. https://dailytimes.com.pk/472217/climate-change-causes-outcomes-in-pakistan-and-a-way-forward/. Accessed 2019, September 24

Chaudhary, H.J., et al. (2011) Antimicrobial activities of Sapium sebiferum L. belonging to family Euphorbiaceae. Journal of Medicinal Plant Research, 05(24) 5916-5919.

Chaudhry, Q. Z. (2015). Climate Public Expenditure and Institutional Review Study for Pakistan. Islamabad: United Nations Development Programme.

Climate Risks and Food Security Analysis: A Special Report for Pakistan. https://reliefweb.int/report/pakistan/climate-risks-and-food-security-analysis-special-report-pakistan-islamabad-december. Accessed 2018, December 19

Dahlman, L., Lindsey, R. Climate Change: Ocean Heat Content. https://www.climate.gov/news-features/understanding-climate/climate-change-ocean-heat-content. Accessed 2020, February 13

Epple, C., García Rangel, S., Jenkins, M., Guth, M. (2016) Managing ecosystems in the context of climate change mitigation: A review of current knowledge and recommendations to support ecosystem-based mitigation actions that look beyond terrestrial forests. Technical Series No.86. Secretariat of the Convention on Biological Diversity, Montreal, 55 pages.

Extreme Weather and Climate Change. https://www.c2es.org/category/climate-solutions/. Accessed 2019

Extreme weather gets a boost from climate change. https://www.edf.org/climate/climate-change-and-extreme-weather. Accessed 2020

Global warming is making some extreme weather events worse. https://sites.nationalacademies.org/BasedOnScience/climate-change-global-warming-is-contributing-to-extreme-weather-events/index.htm. Accessed 2018

Hammad, M.H., et al., 2010. Influence of organic manures on weed dynamics and wheat (Triticum aestivum L.) productivity under low rainfed area. Crop & Environ., 1(1): 13-17

Hammad, H. M., et al., 2010a. Optimizing water and nitrogen requirement in maize under semi arid conditions of Pakistan. Pakistan Journal of Botany, 43 (6): 2919-2923

Hammad, H. M., et al., 2010b. Organic farming in wheat crop under arid conditions of Pakistan. Pakistan Journal of Agricultural Sciences, 48 (2): 97-102

Hussain, M., &Mumtaz, S. (2014). Climate change and managing water crisis: Pakistan's perspective. Reviews on environmental health, 29(1-2), 71-77.

Lindsey R., Dahlman, L. Climate Change: Global Temperature. https://www.climate.gov/news-features/understanding-climate/climate-change-global-temperature. Accessed 2020, January 16

Malcolm, J. R., Pitelka, L. F. Ecosystems and Global Climate Change. https://www.c2es.org/document/ecosystems-and-global-climate-change/. Accessed 2001

Munis, M. F. H. et al., 2012. Inheritance pattern of vital post-emergence morphometric and meristic traits of Spring Wheat. Journal of Medicinal Plant Research, 6(16): 3246-3253.

Nasim, W., et al., 2010. Simulation of different wheat cultivars under agro-ecological condition of Faisalabad-Pakistan. Crop & Environ., 1(1): 44-48.

Nasim, W., et al., 2011. Nitrogen effects on growth and development of sunflower hybrids under agro-climatic conditions of Multan. Pakistan Journal of Botany, 43(4): 2083-2092

Nasim, W. et al., 2012a. Impact of Nitrogen and Plant Growth Promoting Rhizobacteria on Yield and Yield Components of Sunflower in Glasshouse Environment. Journal of Crop Science and Biotechnology, 15 (4): 319–324.

Nasim, W., et al., 2012b. Studying the comparative performance of wheat cultivars for growth and grains production. International Journal of Agronomy and Plant Production, 3 (9), 306-312.

Nasim, W., et al. 2012c. Effect of organic and inorganic fertilizer on maize hybrids under agro-ecological conditions of Faisalabad-Pakistan. African Journal of Agricultural Research, 07 (15): 2713-2719.

Neelam, A., et al. (2017). Repercussions of climate change on human health in Pakistan: A Mini Review. Journal of Chemical, Biological and Physical Sciences 7(1)

Pakistan: Drought https://reliefweb.int/disaster/dr-2018-000428-pak. Accessed 2018, July 25

Pallewatta, N. (2010) Impacts of climate change on coastal ecosystems in the Indian Ocean region. Coastal Zones and Climate Change, New York: The Henry L. Stimson Center, 3-16.

Pidcock, R. How do scientists measure global temperature? https://www.carbonbrief.org/explainer-how-do-scientists-measure-global-temperature. Accessed 2015, January 16

QUEST Staff. Water Vapor's Role in Climate Change. https://www.kqed.org/quest/73100/water-vapors-role-in-climate-change. Accessed 2014, December 12

Qureshi, N. A., Ali, Z. (2011) Climate change, biodiversity Pakistan's scenario. J. Anim. Plant Sci, 21(2 Suppl), 358-363.

Reports & Multimedia/Explainer. Water and Climate Change. https://www.ucsusa.org/resources/water-and-climate-change. Accessed 2010, June 24

Sajjad, M., et al., 2012 Association of seed morphology with seedling vigor in wheat (Triticum aestivum L). Research Plant Biology, 2(5): 7-12.

The Vulnerability of Pakistan's Water Sector to the Impacts of Climate Change: Identification of gaps and recommendations for action. https://www.pk.undp.org/content/pakistan/en/home/library/environment_energy/PakistanWaterSectorReport.html. Accessed 2017

Usman, M., et al. 2009. Development and application of crop water stress index for scheduling irrigation in cotton (Gossypium hersutumn L.) under semiarid environment. Journal of Food Agriculture and Environment, 7: (3&4) 386-391.

Usman, M., et al. (2010) Lower and upper baselines for crop water stress index and yield of Gossypium hirsutum L. under variable irrigation regimes in irrigated semiarid environment. Pakistan Journal of Botany, 42(4), 2541-2550.

Wajid, A., et al. (2007) Maize response to yield under semi arid conditions of Pakistan. Pakistan Journal of Agricultural Sciences, 44 (2): 217-220

Wajid, A., et al. (2010) Quantification of growth, yield and radiation use efficiency of promising cotton cultivars at varying nitrogen levels. Pakistan Journal of Botany, 42(3): 1703-1711

WHO. Climate And Health Country Profile Pakistan. https://apps.who.int/iris/bitstream/handle/10665/246150/WHO-FWC-PHE-EPE-15.28eng.pdf;jsessionid=81B3C44CFDA3D5785E462AF48EEFB085?sequence=1. Accessed 2015

Willarda V. E, Climate change, health, and health care. How physicians can help. https://noharm-uscanada.org/sites/default/files/Climate.Physician.Network.pdf. Accessed 2020

Chapter 5
Climate Resilience in Agriculture

Muhammad Shehzad, Noosheen Zahid, Mehdi Maqbool, Ajit Singh, Hongyan Liu, Chao Wu, Aziz Khan, Fazli Wahid, and Shah Saud

Abstract Climate change is a major factor contributing to food insecurity, requiring urgent action and interventions to avoid the risk of hunger and poverty of millions of people. Climate-related threats have a big impact on the life of poor people and subsistent farming communities. Climate changes are forecasted to result in variability in future weather patterns in the form of changes in frequency, quantity, and predictability of precipitation, floods, temperature, drought, and wind storms. In this scenario, promoting climate resilient agriculture (in areas which are highly affected by climate change) is needed for viable food security and valuable income for rural societies. This will enhance the livelihood of rural communities with a focus on people and nature-focused approaches. The best strategies and practices for climate change resilience in agriculture (which address climate risk) may be familiar to some farmers through practices commonly associated with sustainable agriculture. These sustainable agriculture practices include improvement of land preparation such as no-tilling and composting, introduction of disease-resistant and climate hardy seed varieties, crop diversification, integration of crops and livestock,

M. Shehzad (✉)
Department of Agronomy, University of Poonch, Rawalakot, Pakistan

N. Zahid · M. Maqbool
Department of Horticulture, University of Poonch, Rawalakot, Pakistan

A. Singh
School of Biosciences, University of Nottingham, Semenyih, Malaysia

H. Liu
Hainan Key Laboratory for Sustainable Utilization of Tropical Biosource, College of Tropical Crops, Hainan University, Haikou, China

C. Wu
Guangxi Institute of Botany, Chinese Academy of Sciences (CAS), Beijing, China

A. Khan
Key Laboratory of Plant Genetics and Breeding, College of Agriculture, Guangxi University, Nanning, China

F. Wahid · S. Saud
The University of Swabi, Anbar, Khyber Pakhtunkhwa, Pakistan

W. N. Jatoi et al. (eds.), *Building Climate Resilience in Agriculture*,
https://doi.org/10.1007/978-3-030-79408-8_5

soil health improvement through cover crops, management of intensive grazing, reduction in the use of off-farm inputs, development of whole-farm planning, and improvement of agriculture enterprises and marketing. Adaptation of these practices and approaches will ensure agriculture resilience and food security in a changing climate environment.

Keywords Resilient agriculture · Climate change · Smart agriculture · Subsistence agriculture

5.1 Introduction

Cordaid (2016) defined resilience as "systems or communities which bear hazards and are able to anticipate that risk, respond to these disasters, to adapt changing circumstances and risks and have the ability to address the issues and root causes of risks." Resilience literally means "coming back to normal." In the scenario of climate change, the means to reduce the impact of climate change, through adaptations and holistic strategies of mitigation of challenges is considered as climate resilience. This phrasing will be additionally talked about with regard to climate change and the adaptation of sustainable agricultural practices to ensure resilience in crop production and food security in a changing climate (that induced changes in the weather patterns which are unfriendly to agriculture and crop productivity).

Climate resilient agriculture is the key strategy and concept to manage the new risks imposed by unpredictable weather conditions. Christian Aid defined climate resilient agriculture in Time for Climate Justice as "agriculture that reduces poverty and hunger in the face of climate change, improving the resources it depends on for the future generations" (CORDAID 2016). Climate resilient agriculture is the transformation of current agriculture system with the wider perspective of food production at local, regional, and global levels. Along with food production, it takes into account the social, economic, and environmental aspects of agriculture production. In fact, it prepares for the worst to achieve sustainable goals. This approach is a way forward in linking food security with climate change through climate smart agriculture practices. Climate smart agriculture is a new approach that guides the actions needed to reorient the agriculture system in changing climatic scenario and also ensures food security. There are three main objectives of climate smart agriculture: sustainable agriculture with increased productivity and incomes; adaptation of crops with resilience to climate change; and the reduction of greenhouse gas emissions (FAO 2013). The main drawback of climate smart agriculture is that it requires high inputs for food production without proper assessment of social and environmental facts.

5.2 Climate Change and Its Impact on Rural Economy

The major constituent of sustainability of rural communities in the world depends on farming and related activities. It contributes significantly to the overall economy of rural regions in relation to business and employment, environmental quality, and infrastructure. According to FAO (2013), survival of more than 60% of the world's population depends on agriculture. World Bank data shows that 1.0% growth in GDP of the agriculture sector can cover the expenditures of the three poorest families by at least 2.5 times than rest of the economy. This shows the importance of farming and its contribution to rural economic development.

The important factor in agriculture productivity is climate, so changing climate will be the most significant problem affecting food production in the future. Climate change is mainly due to the emission of greenhouse gases such as nitrous oxide, methane, and carbon dioxide. Along with these factors, humans are also responsible for an increase in temperature due to anthropogenic activities and extensive burning of fossil fuels which increases the occurrence of CO_2 in the atmosphere. The concentration of CO_2 has increased from 370 ppm to 412 ppm due to heavy industrialization and massive use of fuels and fossils (Buis 2019). This massive increase in the concentration of gasses will increase global temperature from 1.8 to 4.0 °C with an average of 2.8 °C increase (IPCC 2007). Agricultural sector is the second main contributor of annual increase in anthropogenic greenhouse gas emission (Fig. 5.1). This sector contributes to global warming through carbon dioxide, nitrous oxide, and methane gas emission. These gases allow the transmission of light reaching the earth but block the transmission of heat from escaping into the atmosphere thus trapping the heat as in a "greenhouse" (Smith et al. 2007).

Climate change is responsible for changes in rainfall sequence, drought intensities, and increase in sea levels. Climate change adversely effects agriculture and livestock production, input supplies, hydrologic balances, and other gears of

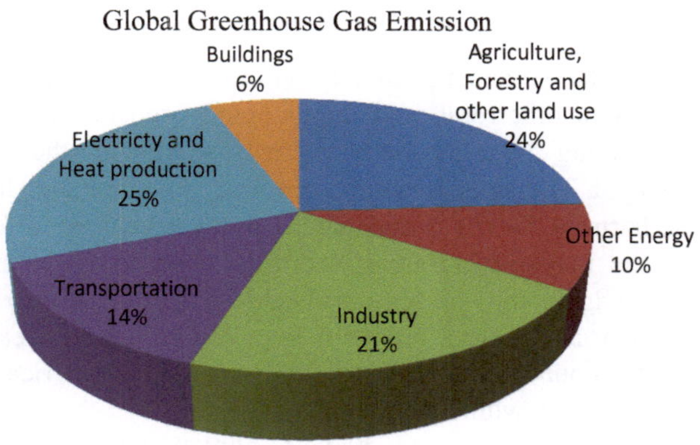

Fig. 5.1 Main contributors in emission of greenhouse gases (Source: IPCC 2014)

agricultural system. This adverse effect of climate change may have negative effects on rural economies.

Economic activities of agriculture are mainly dependent on climatic conditions. Change in climatic conditions mainly threatens the agriculture sector economically because of unpredictability of the changes. Agriculture productivity mainly depends on rainfall, temperature, availability of water, land fertility, date of sowing and harvesting. Effect of change in climatic conditions varies in different regions. Some regions get benefit from these changing conditions while other regions may be adversely affected.

Change in climatic conditions affects agricultural production, while demand is likely to increase due to ever increasing population which causes a change in demand and supply balance of agricultural commodities and also affects the prices of these products which ultimately affect the livelihood of rural society (Masters et al. 2010) (=). Anthropogenic emissions of non-CO_2 greenhouse gases like methane, nitrous oxide, and ozone depleting substances also increased due to climate change and this rise in greenhouse gases affects developing countries more than developed countries (Montzka et al. 2011). Developing countries are more climate sensitive as their economies rely on labor-intensive technologies; whereas, developed countries can cope as adoption of new practices and technologies with payments for mitigation of greenhouse gases (Ogle et al. 2014).

Climate change affects agriculture in different ways across the world. Change in rainfall pattern, change in temperature, and increase in CO_2 production affects global agriculture production. Furthermore, crop productivity patterns change due to these climatic and weather changes. Climatic changes also change the development and distribution pattern of insect pests and diseases. The geographic locations of staple food production may also change with the change in climatic conditions. For the following reasons the modeling of climate change impact on food production is difficult:

1. Uncertainties in the predictions of climatic change.
2. Difficulties of understanding of agricultural responses to increase in CO_2, and the altered pattern of insect pests and disease distribution.
3. Uncertainty in adaptation of certain agricultural practices.

Vulnerable climate change depends mainly on biological, physical, and socioeconomic factors. Low-income rural populations mainly depend on agricultural systems and these changing climatic conditions are responsible for the increase in their hardships and hunger. These people are hardly food sufficient and slight decline in yields could be harmful for the population. The most negative effect of climate change is dominant in low-lying areas where the absence of mountains reduces relief rainfall. Then again arid and semi-arid lands and areas are also highly dependent on rain for their agricultural products. Population of these areas is mostly at risk and most of the countries located in South East Asia and Africa are more vulnerable to these changes (Aydinalp and Cresser 2008).

Most of the Asian regions are vulnerable to climatic changes as an increase in temperature is forecasted in this part of the world. The cooler regions are becoming warmer, while changes in rainfall pattern are less certain making Asia wetter. Heavy rainfalls are expected during the wet seasons increasing the chances of flood, while on other hand the dry seasons are getting drier. These climatic changes have become aggressive on agriculture output resulting in decreasing agricultural productivity and decline in income growth (ADB 2009). Climate change poses severe pressures on farmers of Asia living in marginal areas such as drylands, mountains, and deserts. Impacts of climate change are more devastating in South Asia and may affect wheat productivity up to 50% by 2050 (MoE 2009). Almost 500 million of the rural population in Asia consists of farmers surviving on rain-fed land. Production of maize, wheat, and rice has gone down in the past few decades due to increasing temperatures and water stress (UNFCCC 2007). Considerable damages have also been observed in crop yields which have reduced the earnings of farmers.

5.3 Farmers Perception About Climate Change

Changes in climate system have caused major changes in the growing seasons. A case study is presented in Table 5.1 which shows that the local farming communities of Pakistan are well aware of climate changes (Table 5.1).

There are currently major changes in the planting and harvesting time of different crops in the territory. For example, the planting time of wheat previously began in the first seven-days of December and lasted till the end of December. Currently, wheat is planted in November (beginning to the month's end) showing a 1-month

Table 5.1 Farmers perspective about changing climate

Indicators	Farmers perception
Temperature	Mean temperature of the area has changed. As a result, the summers are warmer and winters are less cold compared to the past 10–30 years
Rainfall variability	Rainfall pattern has changed in the area. The rainfall amount has decreased in both winter and summer seasons and wide uncertainties were found over time
Snowfall	A decline in snowfall has been observed
Soil erosion	Soil erosion has increased due to the increasing flood events
Flooding	Floods have increased in the area compared to past records. The major flood event of 2010 has resulted in major human and economic losses
Landslides	Landslides happen in the study area as it is directly linked with the floods
Summer days	There is an increase in the number of summer days since the past decade. Climate change has affected the length of summer and cold months. More hot days have been observed in the study area compared to the past 10–20 years
Winter days	The number of winter days and intensity has decreased
Extreme weather events	Extreme weather events such as floods, droughts, extreme heat, or heat waves have been increased since the last decade. Now more flash floods and high-intensity cyclones are experienced compared to the past
Early springs	The springs arrive soon now. With change in the average yearly temperatures and extended summers, the springs are experienced in first half of February compared to its arrival in March as observed during the past 10–20 years

Table 5.2 Effect of climate change on livelihood of farmers

Climate indicators	Impact on livelihood sources	Adaptation measures	Potential future risks
Increase in temperature	• Increased irrigation requirements for crops • Negative effects on crops, i.e., loss of seed and crops • Increased mortality rate in poultry and cattle	• Water conservation techniques to reduce water loss from soil • Switching to other jobs • Reducing livestock number	• Livelihood insecurity
Rainfall variability	• Drought conditions, lower agriculture productivity, food scarcity • Problems with natural grazing lands and fodder availability • Drying of springs	• Installation of tube wells or pressure pumps • Delay in sowing different fruits and crops • Purchasing fodder from market, reducing livestock number	• Livelihood insecurity • Increased expenses of livestock rearing
Extreme weather conditions	• Increased flood occurrence • Damage of agricultural lands • Increased soil erosion washing away fertile soils • Cyclones with rise in temperature destroying crops	• Stream/river embankments • Plantation along fields • No coping strategy	• Livelihood insecurity • Decrease in fertile soils/lands • Food security
Warmer winters and less snowfall	• Increased human diseases • Increased livestock diseases	• Hospitalization • Vaccination	• Decreased house-hold welfare • Increased output cost and less income generation

early planting and cultivation period. Likewise, the planting times of rice and maize have also changed. These crops used to be planted in mid-July to the start of August in past but now they are planted in mid-June to early July. The harvesting time for these crops has also changed, for instance, wheat reaping began ahead of schedule in mid-June previously, now reaping is started in mid-May. As a result of these changes, local farming communities have adapted to measures to reduce the impact of climate change on their livelihood (Table 5.2).

5.4 Approaches for Climate Resilient Agriculture

Upgrading strength of agribusiness to climate change is of fundamental significance for ensuring abilities of small landowners to adapt to climate change. With regard to environmental change and fluctuation, small landowners need to adjust rapidly to

Table 5.3 Important physiological traits that confer abiotic stress tolerance in plants

S. No	Abiotic stress	Important physiological traits that confer tolerance
1	Drought	• Improved water use efficiency through plant characters like early closure of stomata (stomatal conductance) • Increased photosynthetic efficiency • Efficient and deeper root system • Waxy leaf surface
2	Heat	• Canopy temperature depression • Membrane thermostability • Chlorophyll fluorescence • Chlorophyll content and stay green • Stem reserve mobilization • Photosynthetic efficiency
3	Salinity	• Osmoregulation (osmolyte production) • K^+/Na^+ selectivity • LEA (late embryogenesis abundant) proteins • Ion homeostasis • Antioxidant enzymes
4	Flooding	Adventitious roots Well-developed parenchyma Plant height and leaf length Quiescence strategy (under flash flooding)

improve their flexibility to adapt to dangers of climatic changeability (Table 5.3), for example, dry seasons, floods, and other extreme weather events.

Major objectives of this initiative are:

1. To enhance the resilience of agricultural crops, livestock, and fisheries to climatic variability and climate change through the development and application of improved production and risk management technologies.
2. To demonstrate site-specific technology packages on farmer's fields for adapting to current climate risks.
3. To enhance the capacity of scientists and other stakeholders in climate resilient agricultural research and its applications.

5.4.1 Physiological Approaches for Climate Resilience Agriculture

Increasing crop production at a time of rapid climate change is going to be challenging for modern agriculture. Thus, there is a need to exploit the genetic potential by diversifying crop cultivation with emphasis on underutilized crops, develop varieties with high yield and tolerance to abiotic stress under climate change scenario. Sharing data among researchers on important physiological traits tolerant to abiotic stress (Table 5.3) and integrating precision agriculture involving modeling and remote sensing could play an important role in achieving global food security. The main drivers of crop production response to changing climate are increase in temperature and carbon dioxide and change in rainfall pattern. Response to these will be positive in some agricultural zones and negative in others (FAO 2013). Some of the physiological responses are highlighted below:

(a) **Effects of temperature** on crop yield will depend on the stage it was exposed to during the growing cycle (CCPSW 2011; Hawkins et al. 2013). For example, flowering and fruiting are sensitive to changes in temperature. Cereal and grain-legume yields are lowered at temperatures above the required optimum (Hatfield et al. 2011). High temperatures reduce the pollen viability, shortens the grain filling period, and reduces the translocation of the photosynthate to the grain (Boote and Sinclair 2006; Hatfield et al. 2011). In addition, rising temperatures will increase the reproductive rate of insects and prolong the breeding season leading to rise in the number of insects and pests. This in turn increases the crop losses.
(b) **Carbon dioxide (CO_2)** gas is a raw material for photosynthetic reaction that eventually results in accumulation of photosynthates, the economic yield of crops. It also increases the water use efficiency of plants. Based on IPCC (2007), in the past 200 years, CO_2 levels in the atmosphere have drastically risen and may reach 450–1000 μmol by 2100. This rise in carbon dioxide level would increase the yield of C3 crops like rice, wheat, and soybean more than the C4 crops like maize (CSSA 2011). However, this increase in yield could be offset by other pressures like insect and pest attacks and the effects of drought. In addition, CO_2 is known to affect crop quality and nutrition as well. For example, increased CO_2 in the atmosphere lowered the protein content in wheat and barley by 4–13% and 11–13%, respectively (Ziska et al. 2004) and raised the carbohydrate level in grain (Erbs et al. 2010).
(c) **Changes in precipitation** affect the water supply of crops. They differ from temperature as they are driven by a wider range of atmospheric processes (Lobell et al. 2008). The existing rainfall pattern and distribution normally required for increased crop production may be disturbed in some regions. This may result in flooding in some parts of the continent and drought in another resulting in negative effects on crop production (CCPSW 2011).

5.4.1.1 Physiological Traits

Increase in temperature and changes in rainfall pattern discussed in the earlier paragraph could lead to drought in some areas and flooding in another. Increased temperature may lead to the development of heat and salinization due to loss of water. Some of the plant traits responsible for the tolerance to these abiotic factors are summarized in Table 5.3. Reynolds et al. (2016) have emphasized the need for identifying important traits and genetic loci that control adaptation to make a greater breeding impact under climate change scenario.

5.4.1.2 Mitigation Strategies

Scientists and farmers across the globe are working to mitigate the climate change effects to increase crop productivity or build resilience through adaptations of production systems. Mitigation strategies work towards reducing the greenhouse gas (GHG) emissions while fixing the carbon in soil in a more stable form (soil organic matter). Increasing the carbon content of soil will improve soil productivity and bring about resilience and adaptability to crop production systems (CCPSW 2011). This is normally achieved by soil conservation using cover crops, crop rotations, and intercropping systems (use of legumes and nitrogen fixation). Supplementing fertilizer with biochar, recycling of crop residues, improved pest control, and integrated soil-crop-water management practices are known to enhance the organic matter content of soil as well as reduce GHG emissions (FAO 2013).

5.4.1.3 Adaptation Strategies

Adaptation strategies work towards adjustment in the production system to withstand climate change for increased performance. Because crop production systems remain the main source of income to the farmers, adaptation to the impact of climate change will be important for the food as well as the social security of farmers (CCPSW 2011; FAO 2012).

Increasing Crop Diversity and Develop New Crops Tolerant to Biotic and Abiotic Stress

An important adaptation to change climate is to diversify crops by developing and promoting crops that are underutilized or underexploited. Currently, we heavily rely on a few crop plants like wheat, rice, maize, soybean, millet, sorghum, and barley, leaving behind a larger number of crops neglected (Massawe et al. 2015). Thus, to address the food security situation in an event of climate change, there is a need to diversify our crop production systems by exploring the underutilized crop species.

Crop diversification can improve resilience by greater ability to tolerate drought and high temperature and to suppress pest and disease outbreaks (Lin 2011). Some of these underutilized legume crops that fix atmospheric nitrogen and are high in protein have been listed by Cheng et al. (2019) which include winged beans (*Psophocarpus tetragonolobus*), lablab beans (*Lablab purpureus*), lima beans (*Phaseolus lunatus*), and Bambara groundnut (*Vigna subterrenea*). Like the underutilized legumes, crops from Chenopodiaceae family such as Quinoa (*Chenopodium quinoa*) and Amaranth spp. could make a significant contribution to the crop production and cropping systems under changing climatic scenarios. Though the impact of underutilized crops will be minimal, regionally, it can play an important role in securing food supply in events of climate change.

To increase crop production under climate change, crops that are new and have not been cultivated before could play an important role. In fact, Challinor et al. (2009) in a metadata analysis suggested that developing new varieties is the most effective solution in climate change adaptation scenarios. Similarly, Noble et al. (2014) also put evolving new crop at the top of the list in different options for adaptation. Scientists have been exploring wild perennial relatives of wheat, maize, rice, sorghum, millet, and sunflower for crossing with their current annual counterparts to evolve new perennial crops that are better adapted to climate change scenarios. However, this will take many years before being tested on the field for farmer's use. In addition, induced mutations could be used to create novel variations using genetic modifications. More and more new cultivars are being developed more efficiently using high-throughput genotyping and phenotyping platforms (FAO 2013).

Development of varieties that could tolerate drought, heat, and waterlogging (abiotic) form an important strategy to address climate stress. Yield of crops normally drops when drought, heat, or waterlogging is experienced at critical growth stages of crops like tillering, flowering, tuber/grain filling during the assimilation of photosynthate (carbohydrate and nutrients) in the sink (grain, fruits, tubers) (CSSA 2011). Heat tolerant varieties of cowpea and corn at pollination and waterlogging at early vegetative growth in soybean and rice are being developed (VanToai et al. 2010; Bailey-Serres et al. 2010). For greater synchronization of pollination and flowering at high temperature and moisture stress, new hybrid maize varieties are being developed (Ribaut et al. 2009). In an event of drought due to climate change, water loving crops like maize (*Zea mays*) could be replaced with drought tolerant or lower moisture needed crops like Bambara groundnut (*V. subterranean*), millet (*Pennisetum glaucum*), lablab beans (*Dolichos lablab*), sorghum (*Sorghum bicolor*), sweet potato (*Ipomoea batatas*), wheat (*Triticum aestivum*), and barley (*Hordeum vulgare*). In a study by Ishimaru et al. (2010), *Oryza sativa* cultivar Koshihikar, was replaced by wild rice (*Oryza officinalis*) to avoid the hottest part of the season that coincided with flowering.

With an increase in the incidence of pest and disease due to rise in temperature, developing cultivars tolerant to insects and pathogens (biotic) will be an important climate resilient adaptation strategy. Significant yield losses have been reported in the US despite the application of chemicals and the use of improved varieties (CSSA 2011). However, interactions between plant pest and disease will be very complex and are presently poorly understood under changing climatic scenarios.

Use of Crop Models in Making Decisions and Predictions

Crop models are dominant tools for simulating the effects of soil, climate, management of pest and disease on production of crops (Zhao et al. 2019). If appropriately used, it could provide a climate resilient agricultural production system. Increasing number of studies have used models to predict yield by recommending climate-change-adaptations like staggering sowing dates, changing sowing densities, varieties and fertilizer regimes, and cropping systems (crop rotations and intercropping) (Waha et al. 2013; Traore et al. 2017). Matthews et al. (2013) also emphasized the use of models to identify suitable management practices for adaptations to climate change. Similarly, the AquaCrop model developed by FAO has become a very popular tool for predicting crop yield under moisture stress conditions (Farahani et al. 2009). This model simulated the growth and yield of sunflower (*Helianthus annuus*) (Todorovic et al. 2009), Bambara groundnut (Karunaratne et al. 2011), and wheat (Farahani et al. 2009) as affected by moisture levels in soil. Decision Support System for Agrotechnology Transfer-Crop Simulation model popularly known as DSSAT-CSM has been widely used for yield prediction of crops under varied climate change scenarios and as well as the adaptation measures to take for increased yield under those conditions (Diasa et al. 2016). Though these models have their own limitation, they present a great potential to develop resilience in our cropping systems and adapt to climate change now and in future.

Adoption of Remote Sensing and Precision Agriculture

Increased crop production requires constant monitoring, soil nutrient, crop growth, crop health, irrigation, and yield. Traditionally, this is achieved by taking direct measurements that may be time-consuming and inaccurate. To meet a 70% increase in food production by 2050 under changing climatic conditions is going to be challenging. To face this challenge, there is a need to make use of technological advances like the use of remote sensing and precision agriculture in the field of agriculture (Radoglou-Grammatikis et al. 2020). Remote sensing is a technique of generating information about the crops/soil on field using drones or popularly known as UAV (Unmanned Aerial Vehicle). It provides a very efficient and effective method of monitoring growth of crop on ground, watering requirements, crop diseases, and crop nutrition (Karthikeyan et al. 2020). It could also be used for foliar fertilizer application as well as spraying pesticides in a mechanized farming (Tellaeche et al. 2008). Precision agriculture refers to the application of production inputs at right time and place and in right quantity (Gebbers and Adamchuk 2010). It aims at optimization of timely use of production inputs like nutrients, pesticides, and water to increase the productivity of crops. This technology has significantly increased the yield of crops. Technological advancement in the field of agriculture has boosted the yield of crops significantly compared with conventional agriculture (Hazel and Wood 2007).

5.4.1.4 Integrated Approach for Climate Resilience in Problem Soils

Problem soils or degraded soils could be defined as soils that are in poor condition (physical, chemical, and biological properties) to support normal plant growth and development leading to low or no crop productivity. Poor conditions could largely be natural or due to human activities like improper farming practices or due to climate change (Yu et al. 2019; FAO 2017). Among the soils classified as problem soils are eroded soils (loss of nutrient and organic matter), saline and sodic soils, acid soils, polluted soils (heavy metal), waterlogged, and compacted soils (FAO 2015). It is estimated that about 33% of the soils on earth are already degraded and this would increase to over 90% by 2050 if reclamation measures are not taken in time (FAO 2019). Therefore, there is an urgent need to improve such soils to provide sustainable supply of food, maintain soil biodiversity, and climate stability (McBratney et al. 2014). It is estimated that restoring the degraded soil can remove up to 51 Gt of CO_2 from the atmosphere and increase food production by 17.6 Mt. per year (FAO 2019).

Literatures addressing the topic directly were very limited but depending on the problem identified in the soil, some remediation measures have been suggested in Table 5.4. Some practices that appeared to be common to all are the development of tolerant varieties, carbon sequestration through biochar application, and addition of organic matter via agroforestry and crop residue recycling practices.

Though each soil will require its own management practice, no one method will achieve desired results. Integrated approach combining the methods suggested by Dagar and Yadav (2017), modeling (Vermue et al. 2013) and remote sensing (Asfaw et al. 2018) would produce desirable results in problem soils in the face of climate change.

5.5 Conclusion

Sustainable agriculture provides a strong foundation upon which to build a resilient agriculture. A rich knowledge base of practical application and innovation informed by a deep understanding of the ecology of agriculture can be used to guide the development of locally adapted, sustainable, and climate resilient systems. Over the last 40 years, farmers, ranchers, and others committed to sustainable food systems have worked together to develop and share information about agricultural sustainability through both formal and informal research, teaching, and learning networks. This accumulated knowledge provides a wealth of potential adaptations, well-tempered by practical, place-based experience, that you can use to enhance the climate resilience of your farm or ranch.

Table 5.4 Problem soils and the approaches taken for remediation

Problem soils	Approaches taken	Reference
Saline soils	• Increase conservation agriculture, judicious water use, improved irrigation system (drip irrigation), develop stress-tolerant crops, and develop database for problem soils	Dagar and Yadav (2017)
Salt-affected soils (saline, sodic)	• Phytoremediation using salt-tolerant crops • Cyanobacteria (Nostoc, Anabaena) for N fixation • Biochar and compost	Li et al. (2019) Jesus et al. (2015) Saifullah et al. (2018)
All problem soils	• Biochar	Haowei et al. (2019)
Heavy metal contamination	• Organic (compost, bio-solids, biochar, etc.) and inorganic soil amendments (lime, gypsum, etc.) • Phytoremediation using hyper-accumulators	Lwin et al. (2018) Ashraf et al. (2019)
All toxic soils (acidic), low fertility	• Development of new cultivars with root adaptations	Rao et al. (2016)
Saline soils (other problem soils)	• SALTMED • APSIM • DSSAT • STICS • SWAP • Modeling for halophytes for optimizing production, understanding of the system • Integration of remote sensing and statistical methods to model and map spatial variations of soil salinity in irrigated areas	Ragab et al. (2005) Keating et al. (2003) Jones et al. (2003) Brisson et al. (2003) Kroes et al. (2009) Vermue et al. (2013) Asfaw et al. (2018)

References

ADB (2009) Building climate resilience in the agriculture sector in Asia and in the Pacific. Asian Development Bank, Annual Development Report, p. 9.

Asfaw E, Suryabhagavan KV, Argaw M (2018) Soil salinity modeling and mapping using remote sensing and GIS: The case of Wonji sugar cane irrigation farm, Ethiopia. Journal of the Saudi Society of Agricultural Sciences, 17(3):250-8.

Ashraf S, Ali Q, Zahir ZA, Ashraf S, Asghar HN (2019) Phytoremediation: Environmentally sustainable way for reclamation of heavy metal polluted soils. Ecotoxicology and Environmental Safety, 174: 714-27.

Aydinalp C, Cresser MS (2008) The effects of global climate change on agriculture. American-Eurasian Journal of Agricultural & Environmental Sciences, 3(5): 672-6.

Bailey-Serres J, Fukao T, Ronald P, Ismail A, Heuer S, Mackill D (2010) Submergence tolerant rice: SUB1's journey from landrace to modern cultivar. Rice, 3(2): 138-47.

Boote KJ, Sinclair TR (2006) Crop physiology: significant discoveries and our changing perspective on research. Crop science, 46(5): 2270-77.

Brisson N, Gary C, Justes E, Roche R, Mary B, Ripoche D, Zimmer D, Sierra J, Bertuzzi P, Burger P, Bussière F (2003) An overview of the crop model STICS. European Journal of agronomy. 18(3-4): 309–32.

Buis, A (2019) The Atmosphere: Getting a Handle on Carbon Dioxide. NASA. Global Climate Change. Retrieved May 1, 2020.
Challinor AJ, Ewert F, Arnold S, Simelton E, Fraser E (2009) Crops and climate change: progress, trends, and challenges in simulating impacts and informing adaptation. Journal of Experimental Botany. 60(10): 2775-89.
Cheng A, Raai MN, Zain NA, Massawe F, Singh A, Wan WA (2019) In search of alternative proteins: unlocking the potential of underutilized tropical legumes. Food Security, 1: 1-1.
Climate Change Position Statement Working Group (2011) Position statement on climate change. Working Group Rep. ASA, CSSA, and SSSA, Madison.
CORDAID (2016) Promoting climate resilient agriculture for sustainable livelihood. Care. Act. Share. Like CORDAID.
CSSA (2011) Position Statement on Crop Adaptation to Climate Change, Crop Science Society of America, Madison.
Dagar JC and Yadav RK (2017) Climate resilient approaches for enhancing productivity of saline agriculture. Journal of Soil Salinity and Water Quality 9(1): 9-29
Diasa MPNM, Navaratnea CM, Weerasinghea KDN, Hettiarachchib RHAN (2016) Application of DSSAT crop simulation model to identify the changes of rice growth and yield in Nilwala river basin for midcenturies under changing climatic conditions. Procedia Food Science, 6: 159-163
Erbs M, Manderscheid R, Jansen G, Seddig S, Pacholski A, and Weigel H (2010) Effects of free-air CO_2 enrichment and nitrogen supply on grain quality parameters and elemental composition of wheat and barley grown in a crop rotation. Agric. Ecosyst. Environ. 136: 59-68
FAO (2012) Economics of plant genetic resource management for adaptation to climate change. A review of selected literature, by S. Asfaw & L. Lipper. (available at http://www.fao.org/docrep/015/an649e/an649e00.pdf)
FAO (2013) Climate Smart Agriculture Sourcebook Food and Agriculture Organization of the United Nations, Rome. http://www.fao.org/3/i3325e/i3325e.pdf
FAO (2015) Status of the World's Soil Resources|Main Report, Rome, Italy
FAO (2017) Handbook of Saline Soil Management, Rome, Italy
FAO (2019) World Soil Day, Rome, Italy
Farahani, H J, Izzi G, Oweis, TY (2009) Parameterization and evaluation of the AquaCrop model for full and deficit irrigated cotton Agronomy Journal, 101(2009): 469-476
Gebbers R, Adamchuk VI (2010) Precision Agriculture and Food Security. 327(5967): 828-831
Haowei Y, Weixin Z, Jianjun C, Hao C, Zebin Y, Jun H, Haoru T, Xiangying W, Bin G. (2019) Biochar amendment improves crop production in problem soils: A review. Journal of Environmental Management, 232: 8–21
Hatfield J, Boote K, Kimball, BA, Izaurralde R, Ort D, Thomson A, and Wolfe D (2011) Climate impacts on agriculture: Implications for crop production. Agronomy Journal, 103: 351–370
Hawkins E, Fricker TE, Challinor AJ, Ferro CA, Ho CK, Osborne TM (2013) Increasing influence of heat stress on French maize yields from the 1960s to the 2030s. Global Change Biology, 19(3): 937-47
Hazel P, and Wood S (2007) Drivers of change in global agriculture. Philos Trans R Soc Lond B Biol Sci. 363(1491): 495–515
IPCC (2007) Climate Change 2007: The physical science basis. Contribution of Work Group I to the Fourth Assessment Report of the Intergovernmental Panel on Climate, Cambridge University Press, United Kingdom.
IPCC (2014) Climate Change 2014: Mitigation of Climate Change. Contribution of Working Group III to the Fifth Assessment Report of the Intergovernmental Panel on Climate Change [Edenhofer, O., R. Pichs-Madruga, Y. Sokona, E. Farahani, S. Kadner, K. Seyboth, A. Adler, I. Baum, S. Brunner, P. Eickemeier, B. Kriemann, J. Savolainen, S. Schlömer, C. von Stechow, T. Zwickel and J.C. Minx (eds.)]. Cambridge University Press, Cambridge, United Kingdom and New York, NY, USA.
Ishimaru T, Hirabayashi H, Ida M, Takai T, San-Oh YA, Yoshinaga S, Ando I, Ogawa T, Kondo M. (2010) A genetic resource for early-morning flowering trait of wild rice Oryza officina-

lis to mitigate high temperature-induced spikelet sterility at anthesis. Annals of Botany, 106(3): 515-20.

Jesus JM, Danko AS, Fiúza A, Borges MT (2015) Phytoremediation of salt-affected soils: a review of processes, applicability, and the impact of climate change. Environmental Science and Pollution Research, 22(9): 6511-6525

Jones JW, Hoogenboom G, Porter CH, Boote KJ, Batchelor WD, Hunt LA, Wilkens PW, Singh U, Gijsman AJ, Ritchie JT (2003) The DSSAT cropping system model. European Journal of Agronomy, 18 (3–4): 235–265

Karthikeyan L, Chawla I, Mishra AK (2020) A review of remote sensing applications in agriculture for food security: Crop growth and yield, irrigation, and crop losses, Journal of Hydrology, 586 124905.

Karunaratne AS, Azam-Ali SN, Izzi G, Steduto P (2011) Calibration and validation of FAO-AquaCrop model for irrigated and water deficient bambara groundnut. Experimental Agriculture, 47: 509-527

Keating BA, Carberry PS, Hammer GL, Probert ME, Robertson MJ, Holzworth D, Huth NI, Hargreaves JNG, Meinke H, Hochman Z, McLean G, Verburg K, Snow V, Dimes JP, Silburn M, Wang E, Brown S, Bristow KL, Asseng S, Chapman S, McCown RL, Freebairn DM, and Smith CJ (2003) An overview of APSIM, a model designed for farming systems simulation. European Journal of Agronomy, 18 (3–4): 267–288

Kroes J, Van Dam JC, Groenendijk P, Hendriks RFA, Jacobs CMJ (2009) SWAP Version 3.2: Theory Description and User Manual. Alterra Report. Alterra, Wageningen.

Li H, Zhao Q, and Huang H (2019) Current states and challenges of salt-affected soil remediation by cyanobacteria. Science of the Total Environment, 669: 258–272

Lin BB (2011) Resilience in agriculture through crop diversification: adaptive management for environmental change. BioScience, 61(3): 183-191

Lobell DB, Burke MB, Tebaldi C, Mastrandrea MD, Falcon WP, and Naylor RL (2008) Prioritizing climate change adaptation needs for food security in 2030. Science, 319: 607–610

Lwin CS, Seo BH, Kim HU, Owens G, Kim KR (2018) Application of soil amendments to contaminated soils for heavy metal immobilization and improved soil quality—a critical review. Soil Science and Plant Nutrition, 64: 156-167

Massawe FJ, Mayes S, Cheng A, Chai HH, Cleasby P, Symonds R, Ho WK, Siise A, Wong QN, Kendabie P, Yanusa Y, Jamalluddinv N, Singh A, Azman R, Azam-Ali SN (2015) The potential for underutilised crops to improve food security in the face of climate change. Procedia Environmental Sciences, 29: 140-141

Masters G, Baker P, and Flood J (2010) Climate change and agricultural commodities. CABI Work Pap, 2, 1-38.

Matthews R, Rivington M, Muhammed S, Newton A, Hallett P (2013) Adapting crops and cropping systems to future climates to ensure food security: the role of crop modelling. Global Food Security, 2(1): 24–28

McBratney AB, Field DJ, Koch, A (2014) The dimensions of soil security. Geoderma, 213: 203-213

MoE (2009) Climate Change Vulnerabilities in Agriculture in Pakistan. Ministry of Environment, Government of Pakistan, Annual Report. pp.1-6.

Montzka SA, Dlugokencky EJ, and Butler JH (2011). Non-CO2 greenhouse gases and climate change. Nature, 476(7358), 43-50.

Noble I, Huq S, Anokhin Y, Carmin J, Goudou D, Lansigan F, Osman-Elasha B, Villamizar A (2014) Adaptation needs and options, in: Field, C., Barros, V., Mach, K., Mastrandrea, M. (Lead authors), Climate Change 2014: Impacts, Adaptation, and Vulnerability. IPCC Working Group II Contribution to AR5. Vol. I: Global and Sectoral Aspects.

Ogle SM, Olander L, Wollenberg L, Rosenstock T, Tubiello F, Paustian K, and Smith P (2014). Reducing greenhouse gas emissions and adapting agricultural management for climate change in developing countries: providing the basis for action. Global Change Biology, 20(1): 1-6.

Radoglou-Grammatikis P, Sarigiannidis P, Lagkas T, Moscholios I (2020) A compilation of UAV applications for precision agriculture, Computer Networks, 172: 107148

Ragab R, Malash N, Gawad GA, Arslan A, Ghaibeh A (2005) A holistic generic integrated approach for irrigation, crop and field management. 2: the SALTMED model validation using field data of five growing seasons from Egypt and Syria. Agricultural Water Management, 78(1–2): 89–107

Rao IM, Miles JW, Beebe SE, Horst WJ (2016) Root adaptations to soils with low fertility and aluminium toxicity. Annals of Botany, 118: 593–605

Reynolds MP, Quilligan E, Aggarwal PK, Bansal KC, Cavalieri AJ, Chapman SC, Chapotin SM, Datta SK, Duveiller E, Gill KS, Jagadish KSV, Joshi AK, Koehler A-K, Kosina P, Krishnan S, Lafitte R, Mahala RS, Muthurajan R, Paterson AH, Prasanna BM, Rakshit S, Rosegrant MW, Sharma I, Singh IR, Sivasankar S, Vadez V, Valluru R, Vara Prasad PV, Yadav OP (2016) An integrated approach to maintaining cereal productivity under climate change. Global Food Security 8: 9–18

Ribaut JM, Betrán, FJ, Monneveux P, and Setter T (2009) Drought tolerance in maize. p. 311–344. In S.C. Hakeand J.L. Bennetzen (ed.) Handbook of maize: its biology. Springer, Netherlands.

Saifullah, DS, Naeem A, Rengel Z, Naidu R (2018) Biochar application for the remediation of salt-affected soils: challenges and opportunities. Science of Total Environment, 625: 320–335

Smith P. Martino D, Cai Z, Gwary D, Janzen H, Kumar P, McCarl B, Ogle S, O'Mara F, Rice C, Scholes B, Sirotenko O (2007) Agriculture. In Climate Change 2007: Mitigation. Contribution of Working Group III to the Fourth Assessment Report of the Intergovernmental Panel on Climate Change [B. Metz, O.R. Davidson, P.R. Bosch, R. Dave, L.A. Meyer (eds)], Cambridge University Press, Cambridge, United Kingdom and New York, NY, USA.

Tellaeche A, BurgosArtizzu XP, Pajares G, Ribeiro A, Fernández-Quintanilla C (2008) A new vision-based approach to differential spraying in precision agriculture. Comput Electronic Agriculture, 60(2): 144–155

Todorovic M, Albrizio R, Zivotic L, Saab MTA, Stöckle C, Steduto P (2009) Assessment of AquaCrop, CropSyst, and WOFOST Models in the Simulation of Sunflower Growth under Different Water Regimes. Agronomy Journal, 101 (3): 509-521

Traore B, Descheemaeker, K, van Wijk, M, Corbeels M, Supit I, Giller K (2017) Modelling cereal crops to assess future climate risk for family food self-sufficiency in southern Mali. Field Crops Research, 201: 133–145

UNFCCC (2007) Climate Change: Impacts, vulnerabilities and adaptation in developing countries. United Nations Framework Convention on Climate Change, UN.

VanToai T, Hoa TTC, Hue NTN, Nguyen HT, Shannon JG, and Rahman MA (2010) Flooding tolerance of soybean [*Glycine max* (L.) Merr.] germplasm from Southeast Asia under field and screen-house environments. Open Agriculture Journal, 4: 38–46

Vermue E, Metselaar K, van der Zee SEATM (2013) Modelling of soil salinity and halophyte crop production. Environmental and Experimental Botany, 92: 186–196

Waha K, Müller C, Rolinski S (2013) Separate and combined effects of temperature and precipitation change on maize yields in sub-Saharan Africa for mid- to late-21st century. Global Planet Change, 106: 1–12

Yu H, Zou W, Chen J, Chen H, Yu Z, Huang J, Tang H, Wei X, Gao B (2019) Biochar amendment improves crop production in problem soils: A Review. Journal of Environmental Management, 232: 8-21.

Zhao C, Liu B, Xiao L, Hoogenboom G, Boote KJ, Kassie BT, Shelia WP, Vakhtang Kim KS, Hernandez-Ochoa IM, Wallach D, Porter CH, Stockle CO, Zhu Y, Asseng S (2019) A SIMPLE crop model. European Journal of Agronomy, 104: 97–106

Ziska LH, Morris CF, and Goins EW (2004) Quantitative and qualitative evaluation of selected wheat varieties released since 1903 to increasing atmospheric carbon dioxide: can yield sensitivity to carbon dioxide be a factor in wheat performance? Global Change Biology, 10: 1810–1819.

Chapter 6
Field Crops and Climate Change

Zartash Fatima, Sahrish Naz, Pakeeza Iqbal, Amna Khan, Haseeb Ullah, Ghulam Abbas, Mukhtar Ahmed, Muhammad Mubeen, and Shakeel Ahmad

Abstract Production of crops and climatic changes are internally linked with each other in several features because changes in climatic conditions are the key reason for abiotic as well as biotic stresses, that have adversative influences on the farming systems at local, regional, and global levels. The yields of major agronomic crops are being negatively impacted by climatic changes in several aspects like disparities in rainfall pattern and intensity, mean temperature, heat waves, changes in weeds infestation, disease causing microorganisms, and pest attack during all growing seasons in major cropping systems. Heat and water shortage stress disturb the crop yield in various ways as response of crop towards these impacts of climatic variables vary. Higher temperature frequently causes a reduction in crop production by reason of the fact that, they generally happen in combination with drought. Crop phenology is negatively affected due to climate change. Yield and yield components are more sensitive under drought condition in comparison to higher temperature in all cropping systems. In this chapter, we summarize the impact of climate change and stresses produced due to climate change on crop production.

Z. Fatima (✉) · S. Naz · G. Abbas · S. Ahmad (✉)
Department of Agronomy, Bahauddin Zakariya University, Multan, Pakistan
e-mail: shakeelahmad@bzu.edu.pk

P. Iqbal
Department of Botany, University of Agriculture, Faisalabad, Pakistan

A. Khan
Department of Agronomy, University College of Agriculture, University of Sargodha, Sargodha, Pakistan

H. Ullah
Faculty of Agriculture, University of Agriculture, Faisalabad, Pakistan

M. Ahmed
Department of Agronomy, Pir Mehr Ali Shah Arid Agriculture University, Rawalpindi, Pakistan

M. Mubeen
Department of Environmental Sciences, COMSATS University Islamabad, Vehari, Pakistan

W. N. Jatoi et al. (eds.), *Building Climate Resilience in Agriculture*,
https://doi.org/10.1007/978-3-030-79408-8_6

Keywords Climate change · Yield · Adaptation · Phenology · Greenhouse gases

6.1 Introduction

Local and regional climatic circumstances are the most important determinants of agronomic crop productivity (Ahmad et al. 2016; De Pinto et al. 2016; Hatfield et al. 2011) because plant catabolism and anabolism physiological processes are controlled by weather variables like maximum and minimum temperature, solar radiation, carbon dioxide concentration as well as availability of water (Fatima et al. 2020, 2021; Ahmed et al. 2021; Tariq et al. 2021). Agronomic cereal crop production can also be influenced due to climatic extreme conditions, like heat waves, storm, drought, salinity, and flooding circumstances (Ahmad et al. 2019; Porter et al. 2019). Local and regional average and extreme climatic situations are influenced by natural impacts which comprise both external to the global climatic systems (including variations in the Sun's intensity and volcanic eruptions), and internal modes of variability (like the multiyear El-Nino Southern Oscillation system). There are also human influences that affect climatic circumstances at the global, regional, and local level, including the release of greenhouse gases into the environment and human influence that have more limitations of geographical impact, for example, changes in land surface characteristics as a result of agriculture activity (Abbas et al. 2020). During the previous century, but predominantly over the most recent decades in this century, the earth has experienced noteworthy climate change, particularly warming trends in most of the regions worldwide (Pironon et al. 2019). The average air temperature has been enhanced by almost 0.95 °C from 1980 to 2018, and it is predicted to increase to almost 3.0–5.0 °C (depending on various regions) by the end of this century (Kaye and Quemada 2017). In the meantime, global population has considerably enhanced and it is expected that the world will need 70% more food by the mid of this century (Tariq et al. 2018).

Due to an abundant increase in the amount of greenhouse gas levels in the atmosphere, a rapid variability in climate trends in different agricultural regions in the world has been recorded. The increase of CO_2 in the future is raising many questions with respect to food security; the worldwide efficiency of agriculture is also included, whether it will be affected or not. Global yield will be affected by the increasing amount of CO_2, as yield per decade will increase 1.8% (Fig. 6.1). Changes in temperature, cost, and availability of mineral fertilizers, levels of funding for research and development (public or private), atmospheric level of O_3 (ozone) and CO_2 (carbon dioxide), and changes in precipitation regimes are included to affect the agricultural productivity.

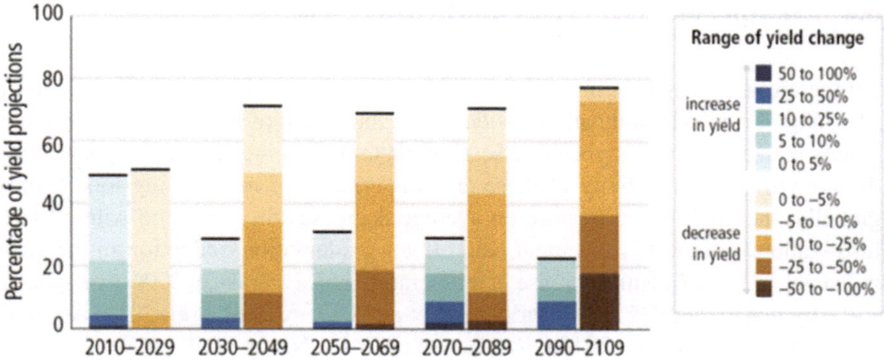

Fig. 6.1 Climate change impact on crop yield

6.2 Crop Response to Global Change

The crop production has been affected and will continue to be effected by the four primary factors:

1. Increase in temperature
2. Increase in CO_2 level
3. Increase in atmospheric O_3
4. Disturbance in hydrological cycle

6.3 Impact on Phenology Change

Phenological stages of the crop plant are set very rigorously to the seasonality of the environment, and therefore are influenced by the changes in the environment. Furthermore, the duration of phenological stages is also associated with CO_2 assimilation, and therefore a shift in the phenology may have effect on crop yields (Ahmad et al. 2017a, b; Abbas et al. 2020). Crop yields are very sensitive to the accumulation of heat units during specific phenological duration, and thus by shifting the phenology, crop yield is also being affected. Several studies reported/projected a reduction in crop yields with climate warming due to shortening of phenological events without considering the management practices (Ahmad et al. 2017a, b; Abbas et al. 2017; Wang et al. 2019; Mousavi-Derazmahalleh et al. 2019).

6.4 Temperature Increase and Crop Yield

Temperature has negative impacts (most likely) on crop yields (Ottman et al. 2012). The major crops which mostly provide 2/3 of the intake of human calories include wheat, maize, rice, and soybean. Assessing the impact of increase in temperature on the growth and productivity of these crops is critical for maintaining food supply globally. Increase in temperature on average decreases the yield and without CO_2 fertilization, genetic improvement, and effective adaptation, reduction in the yield with every degree Celsius increase in temperature is 3.2% by rice, 3.1% by soybean, 7.4% by maize, and 6% by wheat (Zhaoa et al. 2017; Ahmad et al. 2018; Ali et al. 2018a, b; Hammad et al. 2018a; Rahman et al. 2018; Nasim et al. 2018).

To assess the risk of food security and then to develop adaptive strategies to feed the world population, it is necessary to collect the impact of temperature increase on global crop yields; including any spatial variations (Nelson 2010). Wheat production in different countries shows variations towards temperature changes; yield losses for France and the USA were −5.5 to ±4.4% and −6.0 to ±4.2% (per degree Celsius), respectively. With one-degree increase in temperature (global mean temperature) the yield loss would be up to 2.6 ± 3.1% in China (largest wheat producer).

Rice also contributes as a major source of calories in developing countries. Reduction in rice yield with per degree Celsius increase in temperature will be 3.2 to ±3.7 (indicated through analysis of multi method) which is less than wheat and maize. Negative impact of temperature (approximately −6.0% per degree Celsius) was indicated by field warming experiment and grid-point-based simulations but statistical regression suggests no effect. For major rice producing countries, analog differences in estimates between the statistical regressions and other methods are found.

6.5 Climate and Increased CO_2 and O_3 Levels

Rising in the level of CO_2 has been recorded with the start of the industrial era. The concentration of CO_2 was 278 ppm in 1750 and the average increase in its concentration per year is 2 ppm in 2000 (Peters et al. 2011). The increase in CO_2 concentration due to the increasing effect of the industrial revolution is 39% higher than the start with a global average concentration of 390 ppm in 2010. The concentration of O_3 (tropospheric) has also increased due to industrial era from 10–15 ppm to 35 ppm (due to emission of ozone as well). Air pollution has a major contribution and air pollution events can increase concentration to over 100 ppm (Wilkinson et al. 2012). Higher CO_2 concentrations have improved the photosynthesis process in C_3 plants as CO_2 concentration increases fertilization in C_3 crops (wheat, maize, and rice), fruits and vegetable crops (Fig. 6.2). The CO_2 also has a positive effect with respect to water as it plays a vital role in the reduction in stomatal conductance and increased water use efficiency in C_3 and C_4 type crop plants.

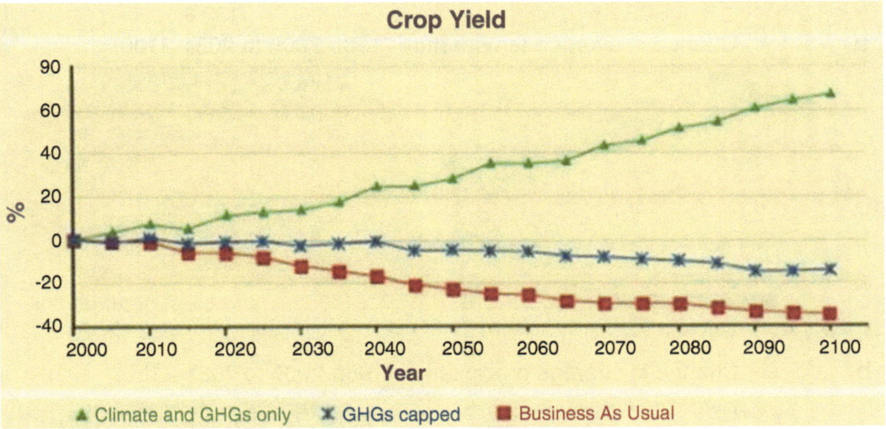

Fig. 6.2 Effect of O_3 and GHGs on crop yield

However, the subject of study also includes the effect of CO_2 concentration which can reduce the nutritional quality of crops (specifically in nutrient poor cropping system) through reduction in protein concentration and nitrate assimilates in harvestable yield (Rahman et al. 2019; Sabagh et al. 2019). Significant fraction of yield growth (loss or gain) of 2–3% is represented per decade (Fig. 6.2).

6.6 Changes in Precipitation Regimes

Precipitation regime has a significant effect on the growth and productivity of crops as 80% of the cropped area is rain-fed where 60% of the world's food is produced. According to the general prediction, the areas with drought will become drier and the areas with high precipitation will receive more precipitation (Liu and Allan 2013). Different soil shows different responses with respect to precipitation regimes. Areas with degraded soil will be greatly affected by seasonal mean precipitation. Water retention will be lower at low moisture potential in the soil with a lower level of organic carbon. Soil with a poor nutrient system recovers slowly from the drought with re-availability of water (Lipiec et al. 2013).

Changes in the precipitation regimes lead to the changes in frequency and length of droughts, changes in the seasonal means, intensity, and timing of individual rainfall events; all these factors are very critical with respect to crop productivity (Fig. 6.3). The effect of rainfall is more vulnerable when it is combined with the temperature changes which ultimately affect the evaporative demand of crops. This problem may lead to the moisture stress of different types with respect to the phonological stage of the crop. It is difficult for the farmers to plan and manage production due to weather patterns and shifting of planting seasons. For example, time

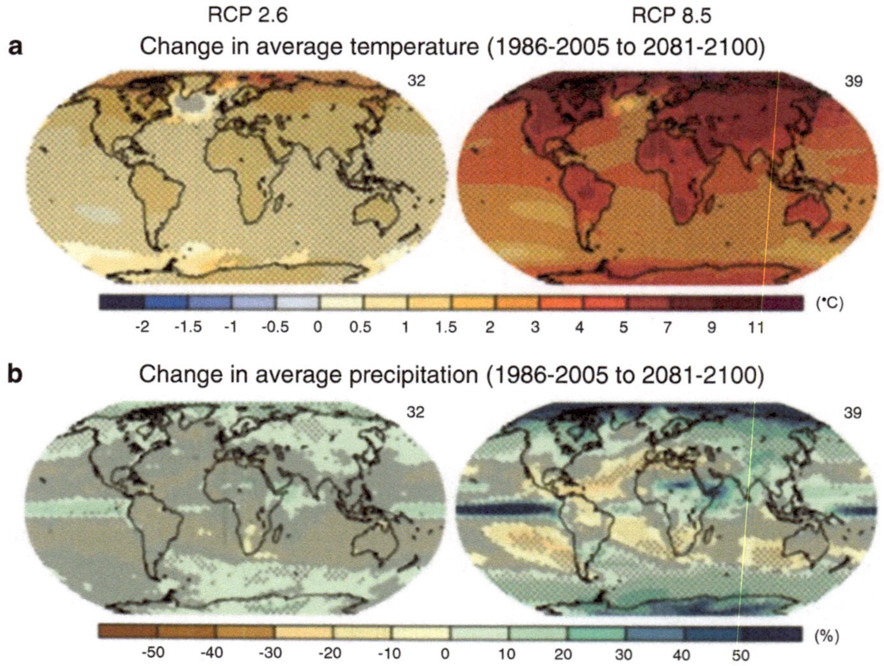

Fig. 6.3 Changes in Temperature and precipitation based on RCP2.6 and RCP8.5 across the world

period of the crop (for completion of growth cycle) is reduced due to late start or earlier end of the rainy season which leads to the reduction in yield.

6.7 Crop Type and Response to Global Change

Various stress factors affect the growth and development of the crop influencing the crop growth linearly or nonlinearly. Crops have different responses towards changes in temperature (Table 6.1) and concentration of CO_2 and O_3 (Fig. 6.4a, b). For example, there is a difference between the crops of different zones as crops of temperate zones include barley and wheat and tropical zones include cassava and sorghum. Average range of temperature has been identified by recent literature as an optimal season average temperature, according to which, for wheat, maize, and soybean, the average temperature is 15 °C, 18 °C, and 22 °C, respectively and for rice and bean (*Phaseolus vulgaris*) suitable temperature is 23 °C and average temperature for cotton and sorghum is 25 °C. Different groups of the crops show different types of responses towards CO_2 sensitivity as C_3 types of plants (grains) are more responsive towards CO_2 as compared to C_4 grains. Root and tuber crops are the most responsive towards it. This can be proved through an example, as with an increase in CO_2 concentration from 385 to 585 ppm, cassava field shows the doubling of dry mass. High

Table 6.1 Climate change impacts on cereals productivity globally and in tropical areas at warming at 1.5 °C and 2 °C beyond the preindustrial levels over the twenty first century

	Increase or decrease in cereal productivity			
	Global		Tropical	
Crops	1.5 °C	2 °C	1.5 °C	2 °C
Wheat (*Triticum aestivum*)	2 (−6 to +17)	0 (−8 to +21)	−9 (−25 to +12)	−16 (−42 to +14)
Maize (*Zea mays*)	−1 (−26 to +8)	−6 (−38 to +2)	−3 (−16 to +2)	−6 (−19 to +2)
Rice (*Oryza sativa*)	7 (−17 to +24)	7 (−14 to +27)	6 (0 to +20)	6 (0 to +24)
Soybean (*Glycine max*)	7 (−3 to +28)	1 (−12 to +34)	6 (−3 to +23)	7 (−5 to +27)

Source: Modified and adapted from Schleussner et al. (2016)

sensitivity can be recorded with high input system (sufficient fertilizer) given that there is no stress of other limiting factors (Ali et al. 2018a, b; Akram et al. 2019; Danish et al. 2019; Iqbal et al. 2019). While, with a high input system, fertilization of CO_2 in C_3 plant will be better with nutritional quality management ().

Increase in CO_2 is helpful for biomass production in areas having drought, but has a drawback of reduction in protein level with high CO_2 level along with no nitrogen inputs into the system (Hammad et al. 2018b; Tariq et al. 2018). Under normal conditions, C_4 plants do not have any benefits from CO_2 level as C_4 plants are capable to increase CO_2 concentration without photosynthesis process. So, C_4 plants will have the lowest reduction in yield under water stress conditions due to less moisture loss (Simpson 2017). The type of damage to crops and seeds will depend upon temperature, and type of plant, developmental stage, and capability to adapt. Highest temperature (above 30 °C) can do permanent damage to plants and seed in storage can also be affected by the temperature above 37 °C (Wahid et al. 2007). The increase in temperature above the threshold level is frequently recorded for maize, wheat, and rice and is predicted to increase worldwide as the climate changes (Gourdji et al. 2013; Amin et al. 2017; Jabran et al. 2017).

6.8 Impact on Pest Infestation

Pests are also influenced by climate change because the body temperature of the pests varies with the temperature of the surrounding and they start moving towards the higher elevations or pole wards (Bebber et al. 2013). Major insect pests of vegetables, fruit crops, pulses, and cereals include pod borers (*Helicoverpa, Spodoptera, and Maruca spp.*), cereal stem borers (*Sesamia, Chilo,* and *Scirpophaga spp.*), whiteflies, and aphids which might travel to temperate areas as changes in cropping patterns are associated with climate change (Sharma 2014). The extent of crop losses will depend on insect biotypes, changes in herbivore-plant interactions, the dynamics of the insect population, species extinctions and the alterations in the

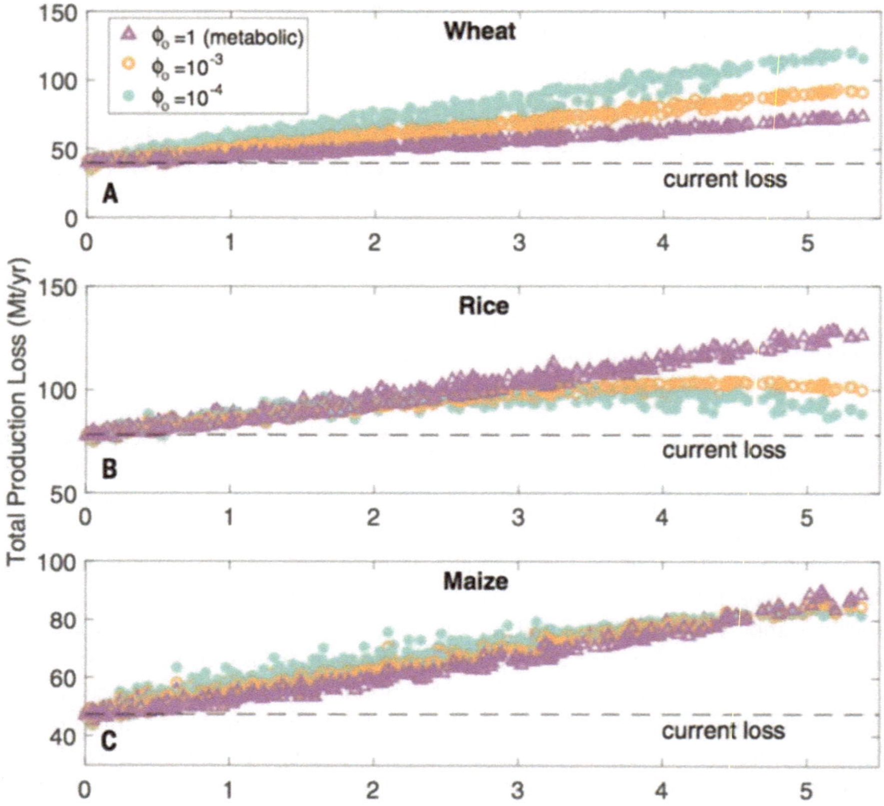

Fig. 6.4 (**a**) Effect of climate change on production of major crops (wheat, rice, and maize). (**b**) Climate change impacts on primary cereals productivity across regions by 2050. *WLD* World, *EAP* East Asia and the Pacific, *EUR* Europe, *FSU* Former Soviet Union, *LAC* Latin America and the Caribbean, *MEN* Middle East and North Africa, *NAM* North America, *SAS* South Asia, *SSA* Sub-Saharan Africa, *NoCC* No climate change, *RCP* Representative Concentration Pathways. **Notes:** Cereals refer to area-weighted average for the following crops: barley, maize, millet, rice, sorghum, wheat, and other cereals

diversity and abundance of arthropods, and the efficacy of crop protection technologies.

6.9 Conclusion

Climatic change is seriously disturbing the farming systems through decreasing production of crops and their products at local, regional, and global levels. Rapid enhancement of concentrations of greenhouse gases are causes of increasing thermal trend, which eventually disrupts the global ecosystem. Overall, in the world,

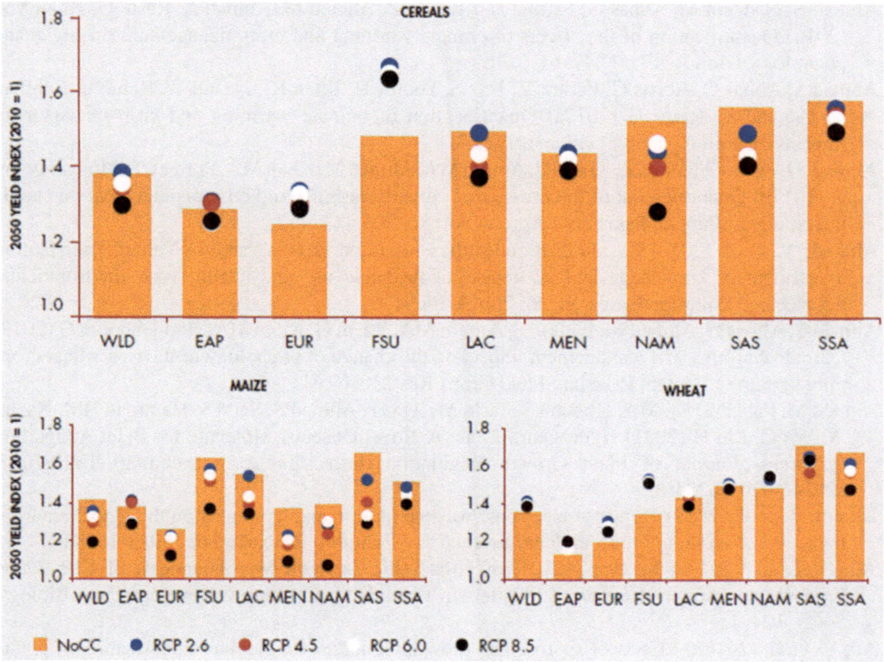

Fig. 6.4 (continued)

climate changes have overwhelming influences on phenological stages and phases, growth, yield attributes, and ultimately crop production in all the farming systems. Climate warming is a major abiotic stress on crops. Abiotic stresses are the foremost category of stresses that crop plants suffer significantly. Major crop yield will be affected more negatively in future scenarios without adaptation strategies. In future, climate change impacts should be studied by using low and high emission scenarios for early, mid, and late centuries. The adaptation strategies should be quantified based on modeling approaches.

References

Abbas G, Ahmad S, Ahmad A, Nasim W, Fatima Z, Hussain S, Rehman MH, Khan MA, Hasanuzzaman M, Fahad S, Boote KJ, Hoogenboom G (2017) Quantification the impacts of climate change and crop management on phenology of maize-based cropping system in Punjab, Pakistan. Agr For Meteorol 247:42–55

Abbas G, Fatima Z, Hussain M, Hussain S, Sarwar N, Ahmed M, Ahmad S (2020) Nitrogen rate and hybrid selection matters productivity of maize–maize cropping system under irrigated arid environment of Southern Punjab, Pakistan, Int J Plant Prod https://doi.org/10.1007/s42106-020-00086-5

Ahmad S, Nadeem M, Abbas G, Fatima Z, Khan RJZ, Ahmed M, Ahmad A, Rasul G, Khan MA (2016) Quantification of the effects of climate warming and crop management on sugarcane phenology. Clim Res 71(1):47–61

Ahmad S, Abbas Q, Abbas G, Fatima Z, Naz S, Younis H, Khan RJ, Nasim W, Habib ur Rehman M, Ahmad A, Rasul G (2017a) Quantification of climate warming and crop management impacts on cotton phenology. Plants 6(1):7

Ahmad S, Abbas G, Fatima Z, Khan RJ, Anjum MA, Ahmed M, Khan MA, Porter CH, Hoogenboom G (2017b) Quantification of the impacts of climate warming and crop management on canola phenology in Punjab, Pakistan. J Agron Crop Sci 203:442–452

Ahmad, A., et al. (2018) Residues of endosulfan in cotton growing area of Vehari, Pakistan: an assessment of knowledge and awareness of pesticide use and health risks. Environmental Science and Pollution Research, 26: 20079-20091

Ahmad S, Abbas G, Ahmed M, Fatima Z, Anjum MA, Rasul G, Khan MA, Hoogenboom G (2019) Climate warming and management impact on the change of phenology of the rice-wheat cropping system in Punjab, Pakistan. Field Crops Res 230:46-61

Ahmed M, Fahad S, Ali MA, Hussain S, Tariq M, Ilyas F, Ahmad S, Saud S, Hammad HM, Nasim W, Wu C, Liu H (2021) Hydrogen Sulfide: A Novel Gaseous Molecule for Plant Adaptation to Stress. Journal of Plant Growth Regulation (https://link.springer.com/article/10.1007/s00344-020-10284-0)

Akram, R. et al. (2019) Trends of electronic waste pollution and its impact on the global environment and ecosystem. Environmental Science and Pollution Research 26: 16923-16938

Ali, S., et al. (2018a) In vitro Effects of Gibberellic Acid on Morphogenesis of CIP Potato Explants and Acclimatization of Plantlets in Field. Vitro Cellular & Developmental Biology, 54(1): 104–111

Ali, S., et al. (2018b) Effects of sucrose and growth regulators on the microtuberization of potato germplasm. Pakistan Journal of Botany, 50(2): 763-768

Amin, A., et al. (2017) Optimizing the Phosphorus Use in Cotton by Using CSM-CROPGRO-Cotton Model for Semi-Arid Climate of Vehari-Punjab, Pakistan. Environmental Science and Pollution Research, 24 (6): 5811-5823

Bebber DP, Ramotowski MAT, Gurr SJ (2013) Crop pests and pathogens move pole wards in a warming world. Nature Climate Change 3:985–988

Danish S., et al. (2019) Alleviation of Cr toxicity in Maize by Fe fortification and Cr tolerant ACC deaminase producing PGPR. Ecotoxicology and Environmental Safety, 185:109706

De Pinto A, Thomas T, Wiebe K (2016) *Synthesis of recent IFPRI research on climate change impacts on agriculture and food security. Background paper prepared for The State of Food and Agriculture 2016.* Washington, DC, IFPRI (International Food Policy Research Institute), (unpublished)

Fatima Z, Ahmed M, Hussain M, Abbas G, Ul-Allah S, Ahmas S, Ahmed N, Ali MA, Sarwar G, ul Haque E, Iqbal P, Hussain S (2020) The fingerprints of climate warming on cereal crops phenology and adaptation options. Sci Rep 10: 18013

Fatima Z, Atique-ur-Rehman, Abbas G, Iqbal P, Zakir I, Khan MA, Kamal GM, Ahmed M, Ahmad S (2021) Quantification of climate warming and crop management impacts on phenology of pulses-based cropping systems. Int J Plant Prod 15: 107–123

Gourdji SM, Sibley AM, Lobell DB (2013) Global crop exposure to critical high temperatures in the reproductive period: historical trends and future projections. Environ Res Lett 8:024041(10pp)

Hammad HM, et al. (2018a) Uptake and toxicological effects of pharmaceutical active compounds on maize. Agriculture, Ecosystem and Environment, 258: 143-148

Hammad, HM, et al. (2018b) Offsetting Land Degradation through Nitrogen and Water Management during Maize Cultivation under Arid Conditions. Land Degradation and Development, 29 (5): 1366-1375

Hatfield JL, Boote KJ, Kimball BA, Ziska LH, Izaurralde RC, Ort D, Thomson AM, Wolfe DW (2011) Climate impacts on agriculture: Implications for crop production. Agron J 103: 351–370

Iqbal, J., et al. (2019) Purple nutsedge (Cyperus rotundus) control through interference by summer crops. International Journal of Agriculture Biology, 21: 1083–1088

Jabran, K., et al. (2017) Water-saving technologies affect the grain characteristics and recovery of fine-grain rice cultivars in semi-arid environment. Environmental Science and Pollution Research, 24: 12971-12981

Kaye JP, Quemada M (2017) Using cover crops to mitigate and adapt to climate change. A review. Agron Sustain Dev 37:4

Lipiec J, Doussan C, Nosalewicz A, Kondracka K (2013) Effect of drought and heat stresses on plant growth and yield: A review. Institute of Agrophysics 2017(27):463–477

Liu C, Allan RP (2013) Observed and simulated precipitation responses in wet and dry region 1850-2100. Environ Res Lett 8:034002(11pp)

Mousavi-Derazmahalleh M, Bayer PE, Hane JK, Valliyodan B, Nguyen HT, Nelson MN, Erskine W, Varshney RK, Papa R, Edwards D (2019) Adapting legume crops to climate change using genomic approaches. Plant Cell Environ 42:6–19

Nasim, W., et al. (2018) Radiation Efficiency and Nitrogen Fertilizer Impacts on Sunflower Crop in Contrasting Environments of Punjab-Pakistan. Environmental Science and Pollution Research, 25 (2): 1822-1836

Nelson GC (2010) Food Security, Farming, and Climate Change to 2050: Scenarios, Results, Policy Options, IFPRI Research Monograph (International Food Policy Research Institute, Washington, DC)

Ottman MJ, Kimball BA, White JW, Wall GW (2012) Wheat growth response to increased temperature from varied planting dates and supplemental infrared heating. Agron J 104:7–16

Peters GP, Marland G, LeQuéré C, Boden T, Canadell JG, Raupach MR (2011) Rapid growth in CO_2 emissions after the 2008-2009 global financial crisis. Nature Climate Change 2:2–4

Pironon S, Etherington TR, Borrell JS, Kühn N, Macias-Fauria M, Ondo I, Tovar C, Wilkin P, Willis KJ. (2019) Potential adaptive strategies for 29 sub-Saharan crops under future climate change. Nat Clim Chang 9:758–63

Porter JR, Challinor AJ, Henriksen CB, Howden SM, Martre P, Smith P (2019) Invited review: Intergovernmental Panel on Climate Change, agriculture, and food—A case of shifting cultivation and history. Glob Change Biol 25: 2518–2529

Rahman, MH, et al. (2018) Multi-model projections of future climate and climate change impacts uncertainty assessment for cotton production in Pakistan. Agricultural and Forest Meteorology, 253-254: 94-113

Rahman HM, et al. (2019) Application of CSM-CROPGRO-Cotton Model for Cultivars and Optimum Planting Dates: Evaluation in Changing Semi-Arid Climate. Field Crops Research, 238: 139-152

Sabagh AEL, et al. (2019) Drought and salinity stresses in barley: Consequences and mitigation strategies. Australian Journal of Crop Science, 13 (6): 810-820 (ISI Indexed).

Schleussner CF, Lissner TK, Fischer EM, Wohland J, Perrette M, Golly A, Rogelj J, Childers K, Schewe J, Frieler K, Mengel M, Hare W, Schaeffer M (2016) Differential climate impacts for policy-relevant limits to global warming: the case of 1.5 °C and 2 °C. Earth System Dynamics 7:327–351

Sharma HC (2014) Climate change effects on insects: Implications for crop protection and food security. Journal of Crop Improvement 28:229–259

Simpson BM (2017) Preparing smallholder farm families to adapt to climate change. Pocket Guide 2: Managing crop resources. Catholic Relief Services: Baltimore, MD

Tariq M, Ahmad S, Fahad S, Abbas G, Hussain S, Fatima Z, Nasim W, Mubeen M, Rehman MH, Khan MA, Adnan M (2018) The impact of climate warming and crop management on phenology of sunflower-based cropping systems in Punjab, Pakistan. Agr For Meteorol 256:270–282

Wahid A, Gelani S, Ashraf M, Foolad M (2007) Heat tolerance in plants: an overview. Environ Exp Bot 61(3):199–223

Wang B, Li Liu D, Evans JP, Ji F, Waters C, Macadam I, Feng P, Beyer K (2019) Modelling and evaluating the impacts of climate change on three major crops in south-eastern Australia using regional climate model simulations. Theor App Clim 138:509–26

Wilkinson S, Mills G, Illidge R, Davies WJ (2012) How is ozone pollution reducing our food supply? J Exp Bot 63:527–536

Zhaoa C, Liub B, Piaoa S, Wanga X, Lobelli DB, Huangj Y, Huanga M, Yaoa Y, Bassuk S, Ciaisl P, Durandm JL, Elliottn J, Ewertp F, Janssensr IA, Lis T, Lint E, Liua Q, Martreu P, Müllerv C, Penga S, Peñuelasw J, Ruaney AC, Wallachz D, Wangg T, Wua D, Liua Z, Zhub Y, Zhua Z, Asseng S (2017) Temperature increase reduces global yields of major crops in four independent estimates. PNAS 114 (35):9326–9331

Tariq M, Fatima Z, Iqbal P, Nahar K, Ahmad S, Hasanuzzaman M (2021) Sowing dates and cultivars mediated changes in phenology and yield traits of cotton-sunflower cropping system in arid environment. Int J Plant Prod 15: 291–302

Chapter 7
Horticultural Crops as Affected by Climate Change

Muhammad Saqib, Muhammad Akbar Anjum, Muhammad Ali, Riaz Ahmad, Muhammad Sohail, Iqra Zakir, Shakeel Ahmad, and Sajjad Hussain

Abstract Climate change is causing substantial effects on horticultural crop productivity as well as quality. The global rise in temperature has forced the farming community for early planting of vegetables and early harvesting of certain fruits. Considering fruit trees, the gradual increase in temperature has disturbed the dormancy breaking process and chilling hours of temperate fruits like apple, peaches, cherry apricots, and many others. This increasing atmospheric temperature has enhanced the evaporative demand, plant transpiration rate, and hence crop water use in certain horticultural crops such as strawberry, blackberry, raspberry, lettuce, spinach, and others. The major elements in climate change such as increasing temperatures, atmospheric CO_2, ozone depletion, UV radiation, heavy metal toxicities, extreme weather events (drought and cold), and changes in the precipitation pattern has markedly affected plant growth and development by modulating various physiological and biochemical mechanisms in horticultural crops. The cultivation of determinate vegetables such as tomato and pepper in the field is becoming difficult and less productive due to early rise in temperature causing flower drop and reduced fruit setting. Vegetable quality may also be affected due to higher O_3 levels, which

M. Saqib
Agricultural and Biological Engineering Department, University of Florida, Gainesville, FL, USA

Department of Horticulture, Faculty of Agricultural Sciences and Technology, Bahauddin Zakariya University, Multan, Pakistan

M. A. Anjum · M. Ali · R. Ahmad · M. Sohail · S. Hussain (✉)
Department of Horticulture, Faculty of Agricultural Sciences and Technology, Bahauddin Zakariya University, Multan, Pakistan
e-mail: sajjad.hussain@bzu.edu.pk

I. Zakir · S. Ahmad
Department of Agronomy, Faculty of Agricultural Sciences and Technology, Bahauddin Zakariya University, Multan, Pakistan

W. N. Jatoi et al. (eds.), *Building Climate Resilience in Agriculture*,
https://doi.org/10.1007/978-3-030-79408-8_7

causes leaf chlorosis and necrosis with premature senescence and subsequent yield reductions. Plantation crops such as tea and coconut are also affected by reduced precipitation and higher temperature. Disease incidence and severity in various horticultural crops such as coffee, cassava, citrus, banana, pineapple, cashew, coconut, and papaya are reported to be affected by climate changing factors. In this chapter, we have discussed different classes of horticultural crops which are affected either positively or negatively by climate change.

Keywords Climate change · Fruits, vegetables · Physiological processes · Pests · Post-harvest quality

7.1 Introduction

Climate change is defined as a change in the statistical parameters of the climate system, when considered over long periods of time, regardless of cause. In contrast, weather is an individual atmospheric condition that occurs at any one time. Both climate and weather affect the cropping area, intensity, and yield in various ways (Iizumi and Ramankutty 2015). Global climate is continuously affected by certain human activities such as burning of fossil fuels, vehicular, and industrial emissions. Since mid-1800s, global average temperature has risen approximately 0.6 °C and there have been severe changes in rainfall pattern, sea levels, and glacier melting. These alterations in climate system are detected to be linked with "greenhouse gases." The major greenhouse gases (GHGs) include methane, ozone, nitrous oxide, and CO_2, the latter being responsible for 70% of the potential of raising temperature on the Earth (Pimentel 2011). The alteration in the concentrations of GHGs disturbs the energy balance of our climate thus causing atmosphere warming. Ozone layer is present in the stratosphere and it is gradually depleting due to the emission of trace gases such as chlorofluorocarbons (CFCs) and nitrous oxides (NOx) thus causing maximum entry of ultraviolet radiations (having wavelength 280–315 nm) in atmosphere. In Europe and North America, different field studies illustrated the detrimental effects of elevated ozone concentration in plants (De Bock et al. 2011). It reduces biodiversity and plays a key role in global warming.

Climate change involves the variation in carbon dioxide (CO_2) concentration, wide and abrupt changes in temperature, and shift in precipitation pattern and intensity. These all factors influence sea levels and salinity, cropped area, soil fertility, and plant diversity (Jackson et al. 2011), resulting in severity in abiotic factors. Agricultural productivity had been threatened by the shifting in rainfall patterns and variation in temperature regimes (Malhotra and Srivastva 2014). The rise in average atmospheric temperature may enhance the risk of drought, thereby limiting various physiological processes including photosynthetic activity and radiation use efficiency in many horticultural crops (including fruits and vegetables).

7.2 Elements of Climate Change Affecting Horticultural Crops

Metabolic mechanisms in horticultural plants are regulated by water availability, temperature, CO_2, and solar radiation, resulting in a decline in crop yield. Different climatic factors induce limiting effect on growth and productivity of horticultural crops (Fig. 7.1). Factors such as tropospheric ozone, drought, excess UV radiation, heat, and soil salinity are causing significant losses of agricultural yield and becoming more prevalent in the coming decades.

7.2.1 *Temperature*

Earth is getting warmer at an unprecedented rate which is clearly evidenced. There are prolonged droughts in arid and semi-arid regions, increased flooding in mid to high latitudes, increase in extreme weather events, etc. Temperature is the most important factor influencing plant growth and development processes. Each plant species has a specific temperature requirement for its optimum growth and development and these temperatures range also depend upon the growth phase of the plant. Beyond these limits, plants are considered under stress. The maximum, minimum, and optimum temperature ranges for agronomic and horticultural crops have already been determined (Hatfield et al. 2011; Hammad et al. 2014; Shahid et al. 2014; Shakeel et al. 2014).

The simulations of rising temperature predicted that the warming trend will be more severe and intense rather than the recent prevailing conditions. High summer temperatures are negatively impacting plant growth and productivity, so there is a dire need for deeper investigations to understand the mechanisms underlying heat tolerance in our commercially important crops (Kumudini et al. 2014).

High temperature and increased intensity of solar radiation is negatively influencing growth, quantity, and quality of fruits and vegetables (Table 7.1). Based on growing season, vegetables are categorized into warm season and cool season categories. Cool season vegetables require 18–25 °C while warm season vegetables need 25–27 °C for optimum growth and yield. Plants are also categorized into C_3 and C_4 plants. The C_4 plants can tolerate more high temperatures as compared to C_3 plants. Most vegetables are C_3 in nature and their optimum range of temperature is 20–32 °C while sweet corn is an example of C_4 plant and it can withstand up to 34 °C (Ruiz-Vera et al. 2015). For tomato, 8–13 days period before flower opening is the most critical developmental phase. In tomato production, higher temperatures about 29 °C for 2 weeks after anthesis has been found to be very critical (Deuter et al. 2012). High temperature at flowering results in flower drop, formation of malformed flowers, and physiological disorder in tomato and pepper (Johkan et al. 2011; Saqib and Anjum 2021).

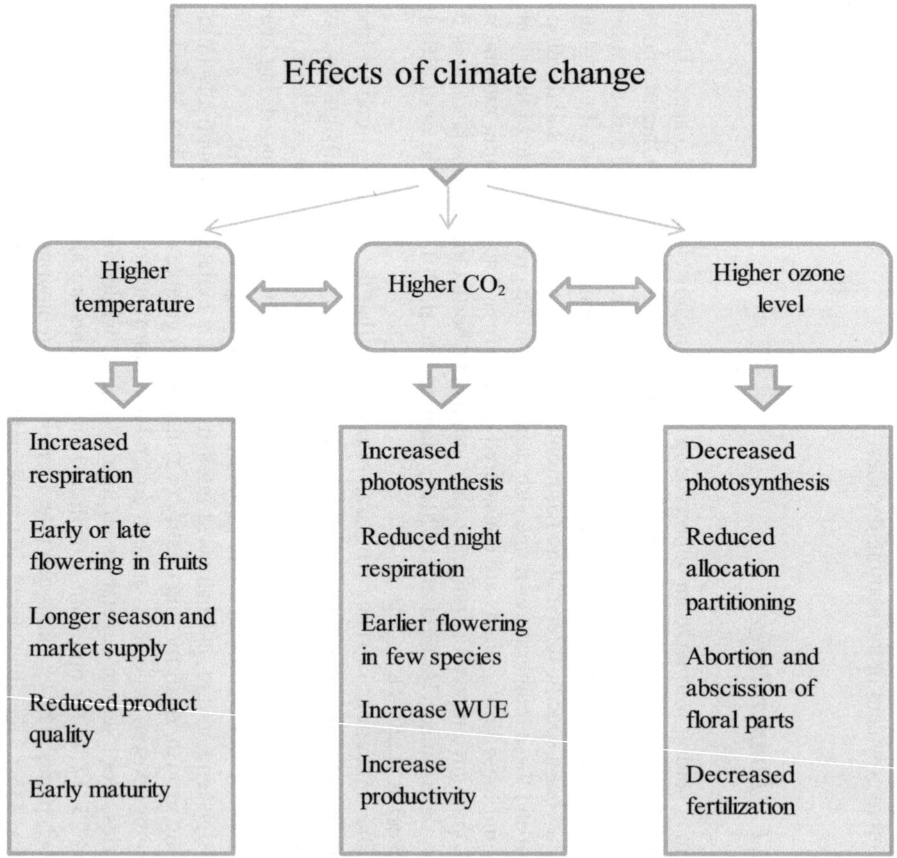

Fig. 7.1 Some positive and negative effects of important climate change elements on horticultural crops

Table 7.1 Negative impact of higher temperature on horticultural crops

Crop	Disorder	References
Lettuce	Tip burn, loose heads	Gruda and Tanny (2014)
Broccoli	Tip burn, bracting, hollow stem, loose heads	Gruda and Tanny (2014)
Pea	Sunburn, reduced fruit and seed size	Lattauschke (2015)
Tomato	Sunburn, reduced fruit and seed size, fruit cracking, blossom-end-rot	Rosales et al. (2011), Gruda and Tanny (2014)
Pepper	Fruit cracking, sunburn, blossom-end-rot	Gruda and Tanny (2014)

In monoecious plants, particularly in cucurbits, male and female flowers are separately produced on a single plant. Temperature has a very important role in regulating flower production in cucurbits because female flowers are favored at low temperature while optimum formation of male flowers is produced at high temperature. Cauliflower performs well in the temperature range of 15–25 °C with highest humidity. Though some varieties have adapted to temperatures over 30 °C, most varieties are sensitive to highest temperatures and delayed curd initiation is observed. In onion, temperature above 40 °C reduced bulb size and yield (Lawande 2010). In potato, heat stress and frost caused 10–20% and 10–50% reduction in marketable yield, respectively.

Fluctuations in temperature also affect flowering in fruit plants such as mango, guava, and citrus. In mango, vegetative and reproductive growth is characterized by flushes and these are principally regulated by temperature. The percentage of hermaphrodite flowers was larger in late emerging panicles, which occurred at higher temperatures (Balogoun et al. 2016). During the peak bloom period, high temperature (35 °C) accompanied by low relative humidity (49%) resulted in highest transpiration and dehydration injury to panicles. Early or late flowering caused fruit set difficulties and further fruit growth and maturity (Rajan et al. 2011; Fahad et al. 2016; Amin et al. 2015).

Likewise, perennial species are also susceptible to high temperature, and their susceptibility is species-dependent. In apple, fruit size and total soluble solids (TSS) increases whereas fruit firmness decreased when the temperature exceeded 22 °C. In cherries, low fruit setting occurs when average optimum temperature rises above 15 °C. In *Citrus sinensis* L. fruit drop is enhanced when temperature increases from 30 °C. High temperature also affects biochemical constituents in citrus fruit.

7.2.2 Carbon Dioxide CO_2

The scientists predicted that CO_2 level will increase up to 700 ppm while temperature will increase by 1.8–4.0 °C by the end of this century. Elevated CO_2 may have a beneficial impact on C_3 plants, but other factors like O_3 concentration and irregular rainfall patterns could have been interfering with it.

Increased CO_2 level is a principal environmental factor which affects the quality of food crops and also stimulates to increase the ozone concentration (DaMatta et al. 2010). The elevated CO_2 effects are more directly linked with the increased temperature. Higher temperature favors more CO_2 enrichment resulting in greater photosynthetic activities. Elevated level of CO_2 has significant impacts on flowering phenology in different plant species. Higher CO_2 concentration promoted earlier flowering in few species; however, in some plant species CO_2 did not influence flowering. Since effects of elevated CO_2 have a strong correlation with climatic temperature, so it is very important to investigate their interactive effect on crop development. The positive responses of elevated CO_2 were noted in sour orange trees when grown under enriched CO_2 concentration than the ambient. Trees exhibited increase in fruit production (upto 70%) and plant biomass. Increment in CO_2 level also improved productivity in peppers, cucumbers, and sugar beet (Kosobryukhov 2009; de Paula et al. 2011).

7.2.3 Ozone and UV Radiation

Among air pollutants, ozone is presumed to be a toxic air pollutant causing severe damage to plants. A gradual increase in ozone concentration has been reported. It is predicted that the concentration of ozone will increase by 20–25% by 2050. These changes are very prevalent in some parts of Asia. This rise in ozone level will reduce the productivity of important crops by 5–20%.

Average monthly ozone concentration of 50 ppb specifically during the growing period of important crops is recorded in different Asian regions. Plants uptake ozone through stomata and it damages cellular structures resulting in leaf senescence, reduced photosynthesis, and disturbances in carbohydrate metabolism. Response of different species to the elevated ozone level varied (Vandermeiren et al. 2012). In tomato, elevated ozone concentration delayed ripening and reduced sugar content in fruits. Ozone caused leaf yellowing, necrosis, and reduction in yield in lettuce, turnip, and spinach depending upon species type and growth stage. C_4 plants observed more ozone tolerant than C_3 ones. Interactive effect of ozone and elevated CO_2 effects foliage quality and starch content in potato tubers.

Sunlight is comprised of wavelengths in the range of 280–1100 nm. Different types of radiations such as UV radiation, photosynthetically active radiation (PAR), far red and near infrared radiations fall in this range. There are research gaps in the effect of different types of irradiance on growth, development, and quality of crops. UV-B radiation is more damaging by altering photosynthetic activity, stomatal functioning, and carbohydrate metabolism. Exposure to UV radiation proved damaging to crops such as tomato, cabbage, potatoes, sugar beets and few others (Ayyogari et al. 2014).

7.3 Climate Change and Physiological Processes

Changing climatic factors influence fruit and vegetable growth and developmental process. Increase in average temperature and onset of early summers may advance the flowering/fruiting in temperate fruits. Physiological processes such as photosynthetic activity, respiration metabolism, membrane stability as well as production of growth hormones and secondary metabolites are responsible for the proper development of fruits and vegetables (Ramírez and Kallarackal 2015). The increased level of CO_2 and O_3 had negative impacts on crop physiological mechanisms in different fruits and vegetables (Tables 7.2 and 7.3). These physiological processes are highly vulnerable to change with changing climate scenarios. Higher temperature reduced or even inhibited the physiological processes involved in seed germination in different species depending upon the stress severity. The optimum temperature range for normal physiological processes is 0–40 °C. However, for different species and ecological zone, this range becomes narrower. It can move towards 0 °C for temperate species such as carrot and lettuce. On the other side, it can be pushed up to 40 °C for tropical species such as cucurbits and cactus species. Suboptimal temperature influences the photosynthetic activity through the alteration in enzymatic activity as well as electron transport chain. Higher temperature also increases the leaf temperature and thereby affect the leaf stomatal conductance.

Produce quality may also be influenced by temperature such as strawberries which had higher antioxidant contents when grown on warmer days (25 °C) and warmer nights (18–22 °C) than berries produced during cooler (12 °C) days. Fruit development, maturity, color development, senescence, and ethylene production in tomato suppresses by exposure to high temperature > 30 °C. Reproductive physiological processes such as number of flowers per plant, pollen tube development, flower fertility, and pollen viability are significantly affected by heat stress (Prasad et al. 2006).

In addition to temperature, CO_2 concentration in the atmosphere also influenced the physiological processes in plant. CO_2 concentration was found to be very influential on physiological mechanism in different vegetables. Ozone concentration had a direct effect on cellular damage, especially in the leaf's cells. This damage may be due to the alterations in membrane permeability and had direct or indirect effects on growth and ultimately leads to reduction in yield. Changes in pigment concentration, leaf chlorosis, and premature senescence are the visible symptoms of low

Table 7.2 Effect of higher O_3 concentration on physiological aspects of fruits and vegetables

Physiological mechanism	Effect	Crop	References
Photosynthesis	↓	Strawberry, french bean	Flowers et al. (2007)
	=	Broccoli	De Bock et al. (2012)
Respiration	↑	Blueberry, broccoli, carrot	Song et al. (2001)
Ripening time	↓	Tomato, cucumber	Calvo et al. (2007)

Adopted from Bisbis et al. (2018)

Table 7.3 Effect of higher CO_2 concentration on physiological aspects of vegetables

Physiological mechanism	Effect	Crop	References
Photosynthesis	↑	Potato; spinach	Katnya et al. (2005), Jain et al. (2007)
Respiration	↓	Asparagus; broccoli; mungbean sprouts; blueberries; tomatoes; pears	Peppelenbos and van't Leven (1996)
Stomatal conductance	↓ ↑	Spinach Chinese cabbage	Jain et al. (2007), Reich et al. (2015)
Ripening	↓	Tomato	Klieber et al. (1996)

ozone concentration in the atmosphere (Hammad et al. 2017; Gillania et al. 2017; Mirza et al. 2017).

Reduced photosynthetic rate, yellowing of leaves, decline in biomass, leaf senescence, and postharvest quality are the main issues in fruits and vegetables associated with higher level of ozone. Ozone accelerates the formation of reactive oxygen species causing lipid peroxidation of membranes which leads to reduced cell membrane permeability.

7.4 Impact of Climate Change on Fruits

Fruit trees are vulnerable to climate change risks in different ways. As fruit trees are categorized into temperate, tropical, and subtropical fruits, they respond to climate change scenario differently.

7.4.1 Temperate Fruits

In temperate fruits, two development phases, viz., dormancy and vegetation are very much climate dependent (especially on temperature). Apple is an important temperate fruit and it ranks third in global fruit production. Flower bud induction and differentiation are highly affected by the climatic condition mainly by temperature. In Germany, the flowering date of apple has advanced up to 2.2 days per decade. Another study indicated that not only temperature but rainfall also influenced full bloom timings of apple (Grab and Craparo 2011). The quality of the apples is also affected by the growing temperature. Fruit softening is dependent upon the number of heat degree days. During the period of 1951–2010, the clear effects of climate change have been observed in the north-western Himalayan region in apple producing areas (Basannagari and Kala 2013). The rise in average and minimum temperature affected the apple productivity through incompletion of chilling period. This was done in Kullu valley (situated at lower altitude) and was famous for apple production and due to this reason farmers started growing vegetables and other fruits requiring low chilling (Rana et al. 2012).

In temperate region, deciduous fruit trees are mostly grown, and they are characterized by the dormant period in autumn. In severe winters trees survive by falling their leaves while buds become dormant. This dormancy is broken by the chilling temperature which is required by the apple and other temperate fruits. It is important for bud break and flowering. If the plant does not receive optimum chilling hours then the result will be partial flowering, poor fruit set, and finally reduced yield. Some species are grown especially in certain locations or regions due to the availability of optimum chilling hours. However, global warming is causing inadequate chilling in that areas, and growing of region-specific fruits becoming difficult.

7.4.2 Tropical Fruits

Diverse fruit trees are found in tropical continent, with about 1200 species in Africa, 500 in Asia (including 300 in the Indian subcontinent), and 1000 fruit species in America. However, only a few species are cultivated and available to mankind as a source of food and nutrition.

Both tropical and subtropical environments are suitable for banana production, and its production is most limited by high temperature and drought. Due to climate change, it is expected that production of banana will be decreased worldwide because of increasing temperature and drought stress during flowering and fruit filling period. However, it may create the opportunities of growing banana in zones characterized by low temperature; though the rainfall pattern is changing and it could limit its production.

Coconut is an important crop in tropical region and it is grown mainly by resource poor farmers in the Philippines, Indonesia, India, Brazil, and Sri Lanka. World's largest coconut producing country is India. Coconut flowering and fruiting are largely affected by temperature and rainfall. For flowering initiation, temperature above 10 °C is required. If temperature falls below 10 °C it causes the nut fall. When temperatures exceed 40 °C in the tropics during April to July it reduces leaf growth as well nut yield (Kumar et al. 2008). Rise in leaf temperature, low water potentials, stomatal and non-stomatal limitations alter the photosynthetic activity resulting in reduced plant biomass and yield. Since temperature and rainfall significantly influence coconut production, it is very important to execute research on coconut in climate change scenario. Papaya is regarded as a valuable fruit crop in the tropical and subtropical region, but water deficit conditions affect its productivity and quality worldwide. Water deficiency is an important issue in papaya cultivation because it reduces stomatal conductance (gs), intervened by hydraulic or non-hydraulic signals resulting in reduced photosynthetic activities.

Technologies such as Open Top Chambers—OTC, temperature gradient tunnels—TGT, Free Air Carbon Dioxide Enrichment—FACE, and Free Atmospheric Temperature Elevation—FATE have been developed to quantify the effect of climate change including elevated temperature and CO_2 level on crop growth, development, and yield. Likewise, simulation models are another approach, which is very useful to improve the productivity of the coconut crop because simulation models analyze changing climate change scenarios and applied under different management practices for assessing the regional impacts.

7.4.3 Subtropical Fruits

Mango is the leading fruit crop and under changing climate scenario its altered flowering phenology affects the productivity and quality of fruit. Early flowering under subtropics may reduce fruit set due to several abnormalities associated with low night temperatures. Delayed flowering also imparts negative impact on fruit set percentage, fruit morphology, and quality parameters. Moreover, high temperature at panicle development stage promotes its growth and reduces the number of days when hermaphrodite flowers are available for effective pollination, which may lead to better crop stand. Severely high temperature results in abnormalities in pollen development and fertilization phenomenon with fruit setting. Though mango is a tree of warmer climates, severe high temperatures markedly influence its morphological and physiological attributes. Leaf scorching and twig dying are commonly observed in young as well as old trees due to extremely high temperatures. Other common symptoms of climate change on mango include abnormal bud differentiation, flowering time variation, and reduced fruit setting and fruit maturity (Rajan et al. 2011).

Citrus is grown worldwide in different tropical and subtropical regions. The major climatic constraints in citrus production are low temperature causing chilling or freezing injury to the plants and may lead to the death of the plant. Low, i.e., below 13 °C, as well as high temperature, i.e., above 37 °C are damaging to the tree and more high temperatures 44–45 °C reduce the growth and may cause fruit abscission.

7.5 Climate Change and Indices of Diseases and Pests in Horticultural Crops

Climate change globally may aggravate the threat to food security posed by crop diseases, which are currently considered to be responsible for 16% of reduction in crop production worldwide. In different crops, such as rice, wheat, barley, maize, potato, soybean, cotton, and coffee, the incidents of crop loss from diseases vary from 9 to 16%. Overall, fungicides worth over US $5 billion are used to control diseases. The food security and crop stability in natural as well as agricultural systems is affected by the climate change either directly affecting the crop yield or through enhancing insect/pests and diseases attack (Beddington 2010). A strong correlation exists between biotic and abiotic factors (Grulke 2011).

The climatic factors responsible for disease severity and spread include changes in precipitation patterns, increased temperature, higher relative humidity, drought, and increased CO_2 and ozone concentrations. Changes in gaseous concentrations impart positive as well as negative effects on disease incidence and severity. On one side, increased CO_2 and O_3 may reduce the resistance expression, increase pathogen attack, leading to ameliorated rates of pathogen evolution whereas on the other hand increased CO_2 was shown to enhance pathogen latent periods (time between infection and sporulation) which ultimately slowed down epidemic rate. However further research is needed to elaborate the effect of elevated CO_2 and O_3 concentration on epidemics of plant diseases (Luck et al. 2011).

In addition, air pollutants such as SO_2, O_3, and acid rain influence plant metabolism and pathogen attack. Various reports have been published which indicate the varying responses of foliar pathogens under polluted air. Higher concentrations, i.e., 200–300 ppb of these pollutants inhibit the life activity of pathogenic fungi whereas lower concentrations (50–100 ppb) of these pollutants may enhance the activity of pathogens and further disease attack.

Potato is a major vegetable grown worldwide. It is vulnerable to different diseases and pathogen attacks during its life span. Studies have been conducted to evaluate the effect of changing climate scenario on disease incidence, severity, and pathogen survival. The attack of potato leaf roll virus was reported to increase under warmer winter climates. Bacterial infections of potato *Pectobacterium carotovorum* and *P. chrysanthemi* spread more rapidly under higher temperatures and humid environments. Scientists considered that if the predictions about gradual increase in average atmospheric temperature actually take place, it will not be feasible to raise the winter season potato crop in lower latitudes due to increased biotic and abiotic diseases. Drought is found to be an important factor affecting the population of insects on plants. In the case of cassava, population of mealybug was found to increase in water stressed plants as compared to well-watered plants (Calatayud et al. 2002; Rasool et al. 2020; Zainab et al. 2020).

Black Sigatoka (*Mycosphaerella fijiensis*) is well-known disease and more damaging for banana crops all over the world. Humidity reduction level is going to increase with passage of time. Therefore, chance of that disease on banana is increased drastically worldwide. Many pathogens, i.e., *Camarotella torrendiella, Camarotella crocomiae*, *Phytomonas* sp., *Phytophthora,* and *Bipolaris incurvata* are severely attacking coconut. The severity of pathogen's attack is increased because of high temperature. Moreover, the pathogen attack was greater because of reduction in precipitation (Ghini et al. 2011). *Hemileia vastatrix* attack increased on coffee plants because of prolonged winter season and increase of CO_2 and temperature. In Mexico, coffee production is economically not viable for producers and going to decrease 34% of current production. Tea cultivation was examined on higher as well as lower elevations in Sri Lanka. Tea cultivation was found to be more vulnerable at lower elevations than higher elevations. The best tea cultivation was recorded at 22 °C. Therefore, viable good actions must be implemented to reduce such adverse effects (Wijeratne et al. 2007).

7.6 Impacts of Climate Change on Postharvest Quality of Fruits and Vegetables

Fruits and vegetables are directly as well as indirectly greatly damaged because of higher solar radiation, CO_2, as well as ozone. The secondary metabolites of fruit and vegetables are greatly influenced because of such constraints. The increase of atmospheric CO_2 level resulted in the reduction of potatoes tuber formation with poor quality. Moreover, chances of common scab and tuber malformation are going to increase (Mattos et al. 2014). Fruits and vegetables have different cultivars which have been developed by breeders. The selection of better cultivars that had potential to stand against different adverse conditions is necessary. The genetic makeup of cultivars varies from one genotype to other. Variability measurements of superior cultivars by using different molecular markers are an essential need of the day. The cultivation of those cultivars having superior traits can be further utilized for higher crop production with the best postharvest quality (Ahmad et al. 2019).

7.7 Conclusion and Future Prospects

Global climate is changing at an unprecedented rate and its constituents such as temperature, precipitation, CO_2, ozone, and UV radiation are imparting their effects on fruits and vegetable in terms of reduced cropped area, decreased region-specific planting, lesser productivity, reduced postharvest quality, and higher incidence of pests and diseases. Higher temperature and increased level of carbon dioxide have shortened the growing season for vegetables such as tomato, peppers, cucumbers, onions, and affected the tuber quality of vegetables. It also affected the tree phenology of fruit trees including apple, cherry, peaches, mango, banana, etc. Rise in temperature and variation in precipitation patterns increased the disease and pest incidence in vegetables and fruits. Vegetable quality may also be affected due to higher O_3 levels, which causes leaf chlorosis, and necrosis with premature senescence and subsequent yield reductions. Further research is needed to elaborate the deep effects of rising temperature and CO_2 on physiological mechanisms of dormancy and vegetative growth in fruit trees. Evaluation of heat tolerant cultivars with a shorter life cycle is needed for attaining higher productivity in vegetables. Planting time should also be adjusted according to the changing local climatic conditions.

References

Ahmad R, Malik W, Anjum MA (2019) Genetic Diversity and Selection of Suitable Molecular Markers for Characterization of Indigenous Zizyphus Germplasm. Erwerbs-Obstbau 4: 1-9

Amin A, Mubeen M, Hammad HM, Nasim W (2015) Climate smart agriculture: an approach for sustainable food security. Agric Res Commun 2(3): 13-21

Ayyogari, K, Sidhya P, Pandit, MK (2014) Impact of climate change on vegetable cultivation-a review. Int J Agri Environ Biotech 7: 145-155

Balogoun I, Ahoton LE, Saidou A, Bello DO, Ezin V, Amadji GL, Ahohuendo BC, Babatounde S, Chougourou DC Ahanchede A (2016) Effect of climatic factors on cashew (*Anacardium occidentale* L.) productivity in Benin (West Africa). J Earth Sci Climatic Change 7:1-10

Basannagari B, Kala CP (2013) Climate change and apple farming in Indian Himalayas: a study of local perceptions and responses. PLoS One, 8:77976

Beddington J (2010) Food security: contributions from science to a new and greener revolution. Philos Trans Roy Soc B: Biol Sci 365:61-71

Bisbis MB, Gruda N, Blanke M (2018) Potential impacts of climate change on vegetable production and product quality–A review. J Clean Prod 170:1602-1620

Calatayud PA, Polania MA, Seligmann CD, Bellotti AC (2002) Influence of water-stressed cassava on Phenacoccus herreni and three associated parasitoids. Entomol Exp Appl 102:163-175

Calvo E, Martin C, Sanz M (2007) Ozone sensitivity differences in five tomato cultivars: visible injury and effects on biomass and fruits. Water Air Soil Pollut 186:167-181

DaMatta FM, Grandis A, Arenque BC, Buckeridge MS (2010) Impacts of climate changes on crop physiology and food quality. Food Res Int 43:1814-1823

De Bock M, de Beeck MO, De Temmerman L, Guisez Y, Ceulemans R, Vandermeiren K (2011) Ozone dose–response relationships for spring oilseed rape and broccoli. Atmos Environ 45:1759-1765

De Bock M, Ceulemans R, Horemans N, Guisez Y, Vandermeiren K (2012) Photosynthesis and crop growth of spring oilseed rape and broccoli under elevated tropospheric ozone. Environ Exp Bot 82: 28-36

de Paula FLM, Frizzone JA, de Paula AL, dos Santos Dias CT, Soares TM (2011) Produção de pimenta tabasco com aplicação de CO2, utilizando-se irrigação por gotejamento. Acta Sci Agron, 33:133-138

Deuter P, White N, Putland D (2012) Critical temperature thresholds case study: Tomato. Agriscience Queensland

Fahad S, Hussain S, Saud S, Hassan S, Ihsan Z, Shah AN, Wu C, Yousaf M, Nasim W, Alharby H, Alghabari F (2016) Exogenously applied plant growth regulators enhance the morpho-physiological growth and yield of rice under high temperature. Front Plant Sci 7: 1250

Flowers MD, Fiscus EL, Burkey KO, Booker FL, Dubois JJB (2007) Photosynthesis, chlorophyll fluorescence, and yield of snap bean (*Phaseolus vulgaris* L.) genotypes differing in sensitivity to ozone. Environ Exp Bot 61: 190-198

Ghini R, Bettiol W, Hamada E (2011) Diseases in tropical and plantation crops as affected by climate changes: current knowledge and perspectives. Plant Pathol 60:122-132

Gillania RA, Shenaza N, Matlooba S, Haqa F, Ngahb WSW, Nasimc W, Munisa MFH, Rehmand A, Chaudharya HJ (2017) Biosorption of Cr (III) and Pb (II) by endophytic *Agrobacterium tumefaciens* 12b3: equilibrium and kinetic studies. Desalin Water Treat 67:206-14

Grab S, Craparo A (2011) Advance of apple and pear tree full bloom dates in response to climate change in the southwestern Cape, South Africa: 1973–2009. Agric For Meteorol 151:406-413

Gruda N, Tanny J (2014) Protected crops. In: Dixon GR, Aldous DE (Eds.), Horticulture: Plants for People and Places, vol. 1. Springer, Netherlands, pp. 327-405.

Grulke N (2011) The nexus of host and pathogen phenology: understanding the disease triangle with climate change. New Phytol 189:8-11

Hammad HM, Saeed S, Ahmad A, Farhad W, Nasim W (2014) Sources of nutrients influence mung bean crop under thal environment. Applied Sci Business Econo 1:44-48

Hammad HM, Farhad W, Abbas F, Fahad S, Saeed S, Nasim W, Bakhat HF (2017) Maize plant nitrogen uptake dynamics at limited irrigation water and nitrogen. Environ Sci Pollution Res 24(3): 2549-2557

Hatfield JL, Boote KJ, Kimball BA, Ziska LH, Izaurralde RC, Ort D, Thomson AM, Wolfe D (2011) Climate impacts on agriculture: implications for crop production. Agron Journal, 103:351-370

Iizumi T, Ramankutty N (2015) How do weather and climate influence cropping area and intensity?. Glob Food Secur 4:46-50

Jackson LE, Wheeler SM, Hollander AD, O'Geen AT, Orlove BS, Six J, Sumner DA, Santos-Martin F, Kramer JB, Horwath WR, Howitt RE (2011) Case study on potential agricultural responses to climate change in a California landscape. Clim Change 109:407-427

Jain V, Pal M, Raj A, Khetarpal S (2007) Photosynthesis and nutrient composition of spinach and fenugreek grown under elevated carbon dioxide concentration. Biologia Plantarum 51:559-562.

Johkan M, Oda M, Maruo T, Shinohara Y (2011) Crop production and global warming. Intech Open Access Publisher.

Katnya MAC, Hoffmann-Thoma G, Schriera AA, Fangmeierd A, Jägerb HJ, van Bel AJE (2005) Increase of photosynthesis and starch in potato under elevated CO_2 is dependent on leaf age. J Plant Physiol 162:429-438

Klieber A, Ratanachinakorn B, Simons DH (1996). Effects of low oxygen and high carbon dioxide on tomato cultivar 'Bermuda' fruit physiology and composition. Scientia Hortic 65: 251-261

Kosobryukhov AA (2009) Activity of the photosynthetic apparatus at periodic elevation of CO_2 concentration. Russ J Plant Physl 56:6-13

Kumar, S.N., Bai, K.K., Rajagopal, V. and Aggarwal, P.K. (2008) Simulating coconut growth, development and yield with the InfoCrop-coconut model. Tree physiol 28(7):1049-1058

Kumudini S, Andrade FH, Boote KJ, Brown GA, Dzotsi KA, Edmeades GO, Gocken T, Goodwin M, Halter AL, Hammer GL, Hatfield JL (2014). Predicting maize phenology: intercomparison of functions for developmental response to temperature. Agron J 106:2087-2097

Lattauschke G (2015) Extreme heat caused massive yield and quality reductions in medium and late peas. In: Versuche im deutschen Gartenbau 2015. S€achsisches Landesamt für Umwelt. Landwirtschaft und Geologie, Dresden (in German)

Lawande KE (2010) Impact of climate change on onion and garlic production. Challenges of climate change in Indian horticulture. Westville Publishing House, New Delhi, 100-103

Luck J, Spackman M, Freeman A, Trebicki P, Griffiths W, Finlay K, Chakraborty S (2011) Climate change and diseases of food crops. Plant Pathol 60:113-121

Malhotra SK, Srivastva AK (2014) Climate smart horticulture for addressing food, nutritional security and climate challenges. Shodh Chintan-Scientific articles, by Srivastava AK et al., ASM Foundation, New Delhi, pp 83–97

Mattos LM, Moretti CL, Jan S, Sargent SA, Lima CEP, Fontenelle MR (2014) Climate changes and potential impacts on quality of fruit and vegetable crops. In Emerging technologies and management of crop stress tolerance (pp. 467-486). Academic Press

Mirza N, Mubarak H, Chai LY, Yang ZH, Mahmood Q, Yong W, Tang CJ, Fahad S, Nasim W (2017) Constitutional tolerance and chlorophyll fluorescence of *Boehmeria nivea* L in response to the antimony (Sb) and arsenic (As) co-contamination. Toxicol Environ Chem 99(2):265-272

Peppelenbos HW, van't Leven J (1996) Evaluation of four types of inhibition for modelling the influence of carbon dioxide on oxygen consumption of fruits and vegetables. Postharv Biol Technol 7:27-40

Pimentel C (2011) Metabolismo de carbono de plantas cultivadas e o aumento de CO2 e de O3 atmosférico: situação e previsões. Bragantia 70

Prasad PV, Boote KJ, Allen Jr LH (2006) Adverse high temperature effects on pollen viability, seed-set, seed yield and harvest index of grain-sorghum [*Sorghum bicolor* (L.) Moench] are more severe at elevated carbon dioxide due to higher tissue temperatures. Agric For Meteorol 139:237-251

Rajan S, Tiwari D, Singh VK, Saxena P, Singh S, Reddy YTN, Upreti KK, Burondkar MM, Bhagwan A, Kennedy R (2011) Application of extended BBCH scale for phenological studies in mango (*Mangifera indica* L.). J Appl Hortic 13:08-114

Ramírez F, Kallarackal J (2015) The effect of increasing temperature on phenology. In Responses of Fruit Trees to Global Climate Change (pp. 11-13). Springer, Cham

Rana RS, Bhagat RM, Singh MM, Kalia V, Singh S, Prasad R (2012) Trends in climate variability over Himachal Pradesh. J Agrometeorol 14:35-40

Rasool A, Nasim W, Xiao T, Ali W, Shafeeque M, Sultana SR, Fahad S, Munis MFH, Chaudhary HJ (2020) Microbial diversity response in thallium polluted riverbank soils of the Lanmuchang. Ecotoxicol Environ Saf 187: 109854

Reich M, Meerakker AN, Parmar S, Hawkesford MJ, De Kok LJ (2015) Temperature determines size and direction of effects of elevated CO_2 and nitrogen form on yield quantity and quality of Chinese cabbage. Plant Biol 18:63-75

Rosales MA, Cervilla LM, Sánchez-Rodríguez E, Rubio-Wilhelmi MDM, Blasco B, Ríos JJ, Ruiz JM (2011) The effect of environmental conditions on nutritional quality of cherry tomato fruits: evaluation of two experimental Mediterranean greenhouses. J Sci Food Agric 91(1): 152-162

Ruiz-Vera UM, Siebers MH, Drag DW, Ort DR, Bernacchi CJ (2015) Canopy warming caused photosynthetic acclimation and reduced seed yield in maize grown at ambient and elevated [CO_2]. Global Change Biol 21:4237-4249

Saqib M, Anjum MA (2021) Mitigation of climate change effect in sweet pepper (Capsicum annuum L.) through adjustment of planting time. Pak J Agri Sci 58(3): 919-927

Shahid M, Austruy A, Echevarria G, Arshad M, Sanaullah M, Aslam M, Nadeem M, Nasim W, Dumat C (2014) EDTA-enhanced phytoremediation of heavy metals: a review. Soil Sediment Contam: An International Journal, 23(4): 389-416.

Shakeel M, Akram W, Ali A, Ali MW, Nasim W (2014) Frequency of aphid (*Aphis gossypii* G.) on brinjal (*Solanum melongena* L.) and farming practices in the agroclimatic conditions of Faisalabad, Pakistan. Int J Agric Innov Res 2: 2319-1473.

Song J, Fan L, Forney CF, Jordan MA, Hildebrand PD, Kalt W (2001) Effect of ozone treatment and controlled atmosphere storage on quality and phytochemicals in high blueberries. Acta Hort 600:417-424

Vandermeiren K, De Bock M, Horemans N, Guisez Y, Ceulemans R, De Temmerman L (2012) Ozone effects on yield quality of spring oilseed rape and broccoli. Atmos Environ 47:76-83

Wijeratne MA, Anandacoomaraswamy A, Amarathunga MKSLD, Ratnasiri J, Basnayake BRSB, Kalra N (2007) Assessment of impact of climate change on productivity of tea (*Camellia sinensis* L.) plantations in Sri Lanka. J Natl Sci Found Sri 35: 12-19

Zainab N, Din BU, Javed MT, Afridi MS, Mukhtar T, Kamran MA, Khan AA, Ali J, Jatoi WN, Munis MFH, Chaudhary HJ (2020) Deciphering metal toxicity responses of flax (*Linum usitatissimum* L.) with exopolysaccharide and ACC-deaminase producing bacteria in industrially contaminated soils. Plant Physiol Biochem 152: 90-99

Chapter 8
Changing Climate Impacts on Forest Resources

Muhammad Farooq Azhar, Ihsan Qadir, Muhammad Mudassar Shehzad, and Akash Jamil

Abstract Forests are the assemblage of natural vegetation dominated by giant plants, i.e., trees having an intrinsic relationship with climate. On one hand, they mitigate climate change effect by capturing and storing carbon dioxide while simultaneously are being affected by climate changes and they are already struggling to survive due to growing human population. Deforestation, cutting, or over-harvesting for burning have become sources of carbon dioxide release which has increased the greenhouse effect. Changing climate has altered seasonal temperature, carbon dioxide concentrations, and rainfall patterns which has brought various changes in all forest types. It has affected snow cover on cliffs which has shifted timberline (Tree line) in subalpine and temperate forests to the northwest. It has resulted in changing forest areas from conifers to scrub, riverine and mangrove forests, so the former has decreased and others have increased. In all forests, vegetation composition is also changing and disturbing ecological balance of indigenous flora and fauna. Low forest out turn, adverse effect on biodiversity, phonological, and biological changes in tree occurred due to change in growing season, are all ultimate results of climate change. Further, it increases insect/pest, disease occurrence in forests which sometimes leads to the extinction of some valuable species and introduction of obnoxious exotic species. Eventually, all these events associated with climate changes had resulted in continuous stress on normal functioning of the forests.

Keywords Climate Mitigation · Forest Resources · Vegetation Composition · Deforestation · Tree Line Shift

M. F. Azhar (✉) · I. Qadir · M. M. Shehzad · A. Jamil
Department of Forestry and Range Management, Bahauddin Zakariya University, Multan, Pakistan

W. N. Jatoi et al. (eds.), *Building Climate Resilience in Agriculture*,
https://doi.org/10.1007/978-3-030-79408-8_8

8.1 Introduction and Background

Forests and climate are interdependent components; if forests are considered to be responsible for balancing and stabilizing the climate then climate is also determinant of forest formation (Rizvi et al. 2015; Inkyin and Woo 2015). A balance of both is necessary for normal functioning of terrestrial life but use of fossil fuel and deforestation has impacted adversely.

The dawn of the industrial revolution triggered large-scale deforestation to cater the ever-increasing needs of fossil fuels. The twentieth century marked intense and systematic deforestation which caught global attention by the start of the twenty first century. Integrated research advancements in the field of ecology, geography, history, and anthropology unfolded the past of forest resources and recast the current forested landscapes. Now, the world is facing problems from global warming to shifting weather patterns with a great threat to food production and public health (Menhas et al. 2016; Newman et al. 2014; Liu et al. 2016).

The interaction between forest ecosystems and climate change is complex and varies in spatial and temporal contexts. Climate change has contributed to obvious impacts on forests (Jandl et al. 2018), some of which are:

- Decline in health and productivity of forests.
- Shifting of forest dynamics and composition.
- Loss of Biodiversity.
- Change in species composition.
- Increased prevalence of pest infections and infestation.
- Increased incidence of biological invasion.
- Higher risks of forest fires.
- Tree phenology and seed disruption due to shift in seasons.

Climate change has not always had a negative impact on forests. Sometimes, changing climate favors forest ecosystems. This extrinsic interaction between climate and forests has much concern for humans. Current swerving of forests into urban and agricultural contexts has further aggravated climatic issues. The capacity of forests to endure and rebound from these threats is variable (Fahey et al. 2016; O'Connor et al. 2017). It becomes complex to discuss forest future in context of climate by neglecting socioeconomic concerns of humans. Many of the problems of climate are directly associated with forests. The potential climate change impact on all forest types must also be assessed.

8.1.1 Climate Change

Climate is the average atmospheric condition that prevails in a specific region for a relatively long period, usually 30 years (Hornsey et al. 2016). It was observed that from 1983 to 2012, the earth had experienced the warmest 30-year period over the last 1400 years. The temperature data both of the earth's surface and ocean surface depicted an upward rise representing an increased temperature of 0.85 °C from the period of 1800 to 2012 (IPCC 2014; Rosenzweig and Hillel 2015a, b; IPCC 2014). Evidence suggests that the foremost driver of global warming is anthropogenic greenhouse gas emission, which largely surged after the onset of the industrial era (Li and Zhang 2011).

Urbanization and industrialization have changed atmospheric and climatic composition in the form of melting glaciers, permafrost, ice sheets, rising sea levels, reducing freshwater availability, shrinking biodiversity, destroying habitats, and disturbing biogeochemical cycles (Amin et al. 2018a, b; Good et al. 2018; Nasim et al. 2018; Pecl et al. 2017). Most disruptive impact of changing climate is the alteration of the growing season length, community composition, and organism's phenology; this, in turn, has triggered increased wildfires, pest outbreak, and extreme events (Richardson et al. 2013; Ali et al. 2019; Hussain et al. 2019). All of these events have directly impacted various ecosystems and inhabiting organisms.

8.1.2 Forestry

Forestry has been the center stage in global climate change debates, owing to the role of forests in sequestering carbon and sustaining biodiversity (Martin et al. 2013). Forests are unarguably the most promising tools to curb carbon emission; there is a strong sociopolitical opinion that forests are pivotal in sustaining and even reversing climate change (Burrascano et al. 2016). Forests are significant in biodiversity, soil and water resource conservation.

Forests are responsible for meeting the world's need for wood and non-wood products (Keča et al. 2013). Forest management and sustainable forest product availability have become the need of the hour. Forests being renewable and having considerable worth on its products have led many countries to manage their forests in such a manner that they could have a stable forest area and at the same time, could increase forest growth and productivity. But many countries due to poor policies and bad management are facing the menace of high deforestation rates.

8.1.3 Forest Biome

During the Silurian period (nearly 420 million years ago) prehistoric arthropods and plants started to inhabit the earth. During the course of the next millions of years, these organisms colonized the earth and laid the foundation for the very first forests that were characterized and dominated by ferns, club mosses, and horse tails that were standing nearly at 40 ft. height.

Earth remained evolved by vegetation with time and dominated with gymnosperms and angiosperms in the Triassic period. Flowering plants appeared in the

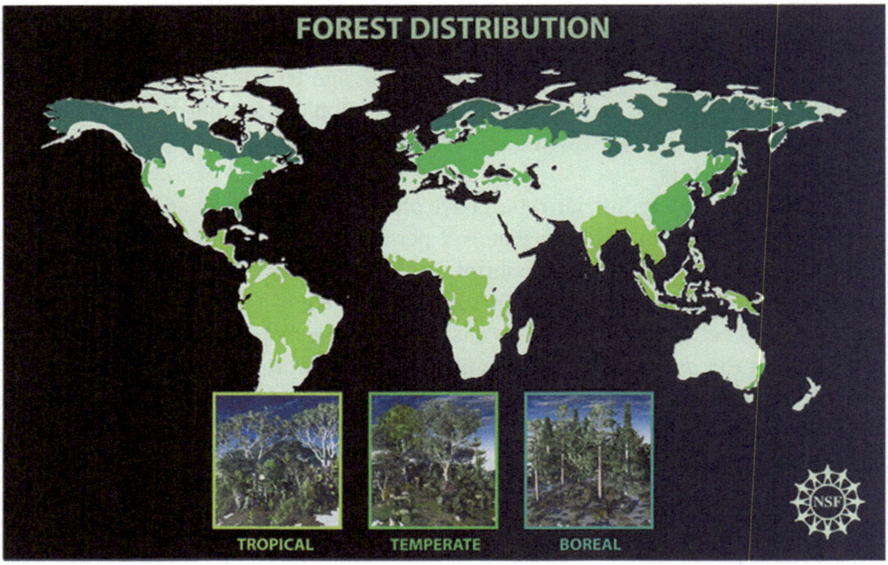

Fig. 8.1 World Forest Distribution

Cretaceous period and started to grow with fauna. These plants started to grow and spread on earth rapidly. The earth went through Pleistocene Ice ages that replaced the earth's long-standing tropical forests with temperate forests that spread all over the Northern Hemisphere. Today, forests spread over one-third of the total earth's terrestrial area.

Forests are classified on the basis of numerous characteristics with various types of forests falling under a broad category. Forest biomes are biological communities of trees (woody vegetation) and are classified on various bases like altitude, latitude, climate, and species composition.

However, on the basis of latitude forests are divided into three main types, i.e., boreal, temperate, and tropical (Fig. 8.1).

8.1.3.1 Tropical Forest

These forests are particularly well known for their rich flora and fauna. They are present in the equatorial belt near the equator along with the latitude by 23.5 N° and 23.5 S°. The unique characteristic of tropical forests is that they do not have a winter season and are characterized by a rainy and dry season. The average temperature during the year is up to 20–25 °C. The precipitation rates are higher (200 cm) and are spread nearly equally all over the year. The vegetation in these layers is capable of retaining much of the nutrients and mineral contents, leaving the undersoil with poor nutrient contents.

Tropical forests are more important because 60–80% of the world's organisms are inhabited in this habitat. The tropical forests are further subdivided according to the distribution of rainfall.

1. Evergreen rainforest
2. Seasonal rainforest
3. Semi-evergreen forest
4. Moist/dry deciduous forest (monsoon)

The biggest danger to these unique and important forests is deforestation. These forests with 7% cover of the land are responsible for supplying a large amount of world's oxygen. If these forests are cut down, not only will the plant and animal biodiversity perished but the indigenous people living here will be displaced.

8.1.3.2 Temperate Forest

These forests occupy the areas of North East Asia, North America, Western Europe, and Central Europe. This biotic community is characterized by a distinct and harsh winter. These forests only experience 140–200 days of growing season in which only 4–6 months are frost-free. The temperature in the temperate forest varies from −30 °C to 30 °C. The soil here is quite fertile due to extensive decaying of the litter. The Temperate forests are further subdivided on the basis of seasonal rainfall distribution.

1. Evergreen broad-leaved and Moist Conifer Forests
2. Mediterranean forests
3. Temperate broad-leaved rainforests
4. Temperate Coniferous forests
5. Dry Conifer forests

8.1.3.3 Boreal Forest (Taiga)

Boreal forests, also known as taiga, are the largest biome of the terrestrial ecosystem. They lie 50° in the North and 60° in the south. These forests occupy the regions of North America and Eurasia. Two third of these forests are present in Siberia while the rest of them are present in Alaska, Canada, and Scandinavia. The seasons here are mainly of two types, i.e., mild warm summers and extensive dry and cold winters having a growing length of only 130 days. The forests are characterizing by low temperature and precipitation is mainly in the form of snow reaching about 40–100 cm/annum.

8.2 Climate Change Effects

Climate change is impinging the world's ecosystems by a notable magnitude. A minor change in different climatic agents like changing temperature and pattern of precipitation alone produces strong multidimensional impacts. Many living species and biotic ecosystems are susceptible in adopting the changing effects of global warming and other associated disturbances. Due to the full effects of climate change, forest ecosystems are also not spared from the effects of rising temperatures, carbon levels, and precipitation fluctuations which have altered phenology, life cycles, geochemical cycles, and pest activity in forest biome (Yang et al. 2017). Another direct effect of changing climate on forest ecosystems is the outset of wildfires. Increasing warm temperatures and prevailing drought conditions have triggered the intensity, frequency, and extent of wildfires all over the world (Primack et al. 2015). In alone 2011, forests of the US experienced wildfires that consumed eight million hectares of forestland (Radeloff et al. 2018). Major projected changes in the forest ecosystems are structural and functional. Furthermore, the ecological relationships of species in different geographical regions with adverse implications for biological diversity are also potential consequences of climate change.

8.2.1 Climate Change Effect on Plant Biodiversity

Biodiversity is known as the number, variability, and variety of living organisms (Meinard et al. 2019). Climate change and global warming are evidently affecting plant biodiversity in world forests and eminent in bringing out considerable influences on the distribution and lifecycles of vegetation types, and plant species from regional to continental scale thus affecting the diversity of species and ecosystem (Mantyka-Pringle et al. 2015). Due to fluctuation and disturbance in climatic factors, the phenotypic plasticity of various life forms has markedly changed making eminent distribution changes of different species (Thom and Seidl 2016). These changes are triggered actually by the disturbances and alteration in the habitat suitability of a specific species thus decreasing the efficacy of the competitiveness and tolerance of that species with other plants in that area (Segan et al. 2016). The frequency and magnitude of changes in rainfall, temperature, occurrence of storm, fire, pest, and disease attacks are other drivers.

Biodiversity disturbance and its ability to adapt to the climate change of the area is largely dependent on the richness and evenness of the species in that particular ecosystem. The evenness and richness are directly dependent upon genetic diversity that plays a major role in contributing to the evolution of an ecosystem. Genetic diversity ensures that the ecosystem is more ready to adapt to Climate-induced disturbance and alterations (Scheffers et al. 2016).

However, biodiversity in forests is not only affected directly by climate change but also has substantial indirect effect via changing forest management practices which are proposed for mitigating adverse effects of changing climate. These practices are associated with species selection in uniform to mixed plantations, rotation age, cultural operations (thinning, pruning), extraction, and felling for bioenergy, fodder, and timber. The existing biodiversity values of world forests are significantly at risk due to such changes; however, new opportunities are being produced for improving biodiversity during changing species composition of forests.

8.2.2 Climate Change Impact on Biomass Production

Global warming due to increased CO_2 levels has made forests to increase their overall production. Coniferous forests due to enhanced growing seasons have increased biomass production (Berndes et al. 2016). The standing forest biomass in the current scenario would likely increase up to 10–30% as compared to the current biomass production (Schadauer et al. 2017). Biomass production of a forest community is largely dependent on the dominant species of the community. Dominant species with an extensive root system and developed foliage are less impacted by climate change and same is the biomass productivity. Both temperature fluctuations and precipitation pattern influence community biomass productivity and stability. The increase in the standing biomass and productivity due to climate change is characteristic of only temperate and boreal ecosystems, but in other tropical and arid ecosystems, there is a negative trend in forest biomass and productivity (Gordon et al. 2018).

8.2.3 Climate Change Impacts the Economy

According to the Fourth National Climate Assessment, if the current rate of carbon in the environment prevails, some serious disruptions in the wood-based world economy could be visible (Broto 2017). Forested areas (mangroves forests) acts as buffer zones to most of the world's coastal cities and much of our society's most important cities and their infrastructure face a risk from flooding. Rapid increasing sea level has the potential to cause damages worth trillions of dollars at the end of this century (Simpson 2017).

Increasing drought conditions and decreasing availability of water have posed a huge threat to manage irrigated plantations that are a valuable source of furniture and fuelwood creating a loss of worth millions of dollars over the globe (Xiao and Xiao 2019). Forests are the hotspots for the bulk of tourism all over the world. Increased warmer temperatures, extreme events, and drought conditions can destroy the scenic attractions of these areas, which in turn is affecting the livelihood of the local communities, as most of these native communities are dependent upon these activities as a source of income (Hoogendoorn and Fitchett 2018).

Table 8.1 Developing countries effected from Climate Change

Serial#	Countries	Deaths per 100,000 inhabitants	Absolute losses (in million US$ PPP)	Losses per unit GDP in %
1	Puerto Rico	90.242	82,315.240	63.2
2	Sri Lanka	1.147	3129.351	1.135
3	Dominica	43.662	1686.894	215.440
4	Nepal	0.559	1909.982	2.412
5	Peru	0.462	6240.625	1.450
6	Vietnam	0.318	4052.312	0.625
7	Madagascar	0.347	693.043	1.739
8	Sierra Leone	6.749	99.102	0.858
9	Bangladesh	0.249	2826.678	0.410
10	Thailand	0.255	4371.160	0.354

Economies of the developing countries are the most vulnerable to the threats of climate change. Various authors have discussed the vulnerability of the developing countries with lower economic, health, and infrastructure facilities to face the wrath of climate change. It has been observed that in the current decade the losses faced by the developing countries in terms of financial and human lives loss is far more than the previous decades (Table 8.1).

8.2.4 Climate Change Impact on Pollination

A synchronization of pollinator activity and time of flowering is important for good pollination. Factors controlling phenology, abundance, and geographic distribution of pollinators and plants are appeared to be greatly impacted by climate change. Climate warming response of plant and pollinator may be different depending on their thermal sensitivity. Climate change effects directly on the structural characters of flowering plants. It has created disruptions of spatial and temporal nature that have the capability of disturbing the interactions of flowers and their essential pollinators. This distraction adds to the mortality of these pollinators and hinders the generation of these specific plants (Haggerty and Galloway 2011). Global warming and fluctuations in precipitation have added frequent dry years, which is putting stress on these organisms as it is hampering pollinator's ability to collect the desired amount of food and limiting the chances of plant species to extend their generations. Studies have shown that the most important genus of honey bees (Apis) is depicted to be the most vulnerable to climate change as a continuous decline in their numbers is being perceived. The causal declining agents are less nectar in the flowers with drought conditions, increased pesticides, and telecommunication signals (Reddy et al. 2013).

8.2.5 Climate Change Impact on Livelihood

There are 240 million inhabitants of forested regions and more than 2 billion are dependent on forest for their livelihood and energy requirements (IPBES 2019). Direct beneficiaries of forests are usually poor. Climate change via floods, hurricanes, sea-level rise, and drought has direct effects on the livelihoods, health, infrastructure, education, and social security of the forest communities like others. These events have resulted in crop failure, water scarcity, and decrease in food, income, and livelihood availability. The decline in food production has effected both income as well as health leaving 600 million people facing malnutrition by 2080 owing to increasing climate change impacts (Clair and Lynch 2010).

Forest-dwelling and depending communities have little adoptive capacity to climate changes due to limited livelihood options. Other risks like activities of climate change mitigation and conservation can also restrict their forest access and forest-related trade. This condition has become much worst due to the eminent impacts of climate change as these people now have to spend part of their income to foster adaptations against climatic changes impacting their existence (Gentle and Maraseni 2012).

8.3 Indicators of Effects of Climate Change

Ecosystems are being affected by Climate change in two ways. One is the direct and short-term effect, which includes the disruption of the life form. The other type of effect depends upon the functioning and behavioral change of the organism in the ecosystem (Walther 2010). These silent impacts on the behavior and phenology of various species, though, are not completely understood, but it reveals specific types of indicators that play a pivotal role in determining these impacts (Maestre et al. 2012). Indicators help in depicting the trends of changing climate and their effects on natural ecosystems and socioeconomic sectors. People involved in forest management and climate mitigation require indicators that help them in monitoring the progress of adoptive approaches and mitigation achievements. These indicators are important in providing early warning signs to the ecosystem managers to develop policy and societal changes to obtain sustainable development. Further, these indicators are applicable in the assessment of possible future changes that could affect global society and the economy. Below are some indicators of climate change effects on the forest ecosystem (Seidl et al. 2017).

8.3.1 Timberline Shift

The boundary of the forests at the altitude from where there is no tree growth is known as the timberline/tree line. In the wake of climate impacts, many studies have predicted and depicted the advancement of the tree line. Timberlines are very sensitive to warm climate and alone temperature is a big timberline shift indicator. The change in the timberline has consequences for the biodiversity of the alpine biome. The immediate effect of an advancing tree line would be on the alpine grasslands as they would start disappearing. The location and cover of various forest types would also come under huge disturbances. Various studies projected that due to global warming, an upward rise of timberline and considerable changes would be observed in forest cover and location of various forest types in a complex way (Abbasi et al. 2015). Temporal and spatial impacts of climate change will be different depending upon the forest types. For example, a study predicted that the highest projected timberline shift is expected in the alpine region, and alpine forests would move 874.29 m north towards alpine pastures (Ahmad et al. 2012; Sheikh et al. 2015). An average of tree line shifting will range from 8.76 to 14.657 m. It is projected that in the next 50 years most of the subtropical forests of the world will be converted into dry subtropical broad leaves forests.

Pakistan, a country in the southeast Asia, though has a little contribution in causing global warming but the effects of climate change in terms of vegetation shift have been eminent here in previous years. These effects could be seen on the timberline of the country constituting alpine forests and pastures above 3350 m. Various studies have shown alterations in the spatio-temporal dynamics of the country's forests from the past decade to have an upward trend. The data assembled from various researches have concluded that various vegetation types of the country have depicted 13–15% of change in the forest cover area for timberline in the previous decade (Table 8.2).

Table 8.2 Status of forest cover and precipitation change along the timberline

Forest types	Elevation change (m)	Forest cover change%	Precipitation change detection (mm)
Subtropical/Moist temperate (Broad leaved)	+285	15	88.5
Moist/Dry Temperate (Conifers)	+233	15	75.5
Alpine/Forests/pastures	+170	13	−3.7

8.3.2 Change in Forest Cover

Global climate change models have predicted an expansion of forest area due to global warming. Forest types are definitely shifting with the rise in temperatures and altering rainfall patterns. It is studied that the sub-alpine forests (as compared to tropical forests) are more vulnerable to climate change. Moisture level accompanied by higher temperatures will bring upon higher productivity in the boreal forests and lead to expansion of the area up to the tundra regions (Ahmad et al. 2012). The present forest types and species composition due to climatic alteration will give way to new forest types and species that have the ability to well adapt to new changing climatic conditions. Studies have projected that in general the total forest area of the south Asian countries will increase to 23.5% up till 2080 (FAO 2010). It was observed that there would be eminent changes in the boundaries of the various forest types.

The riverine forests in the semi-arid region will face heavy receding as about 10.5% of the forest area is projected to be lost during the time period of 2050–2080 (Ahmad et al. 2012). A study projects the loss of mangrove forest in the south, as about 2% of the forest is projected to shrink during 2020–2050. The Scrub forests depicted the potential to experience an expansion; it was projected that these types would be 1.45 times far greater as compared to current area. Furthermore, man-made plantations will enjoy the alteration in the climatic conditions as they will increase by 12 times up till 2080 due to the availability of water and land (Sheikh et al. 2015).

The studies from the past century to the present decade have shown a continuous decreasing trend of forest cover globally (Fig. 8.2). This downward trend in most cases is directly proportional to the rising industrialization accompanied by global warming.

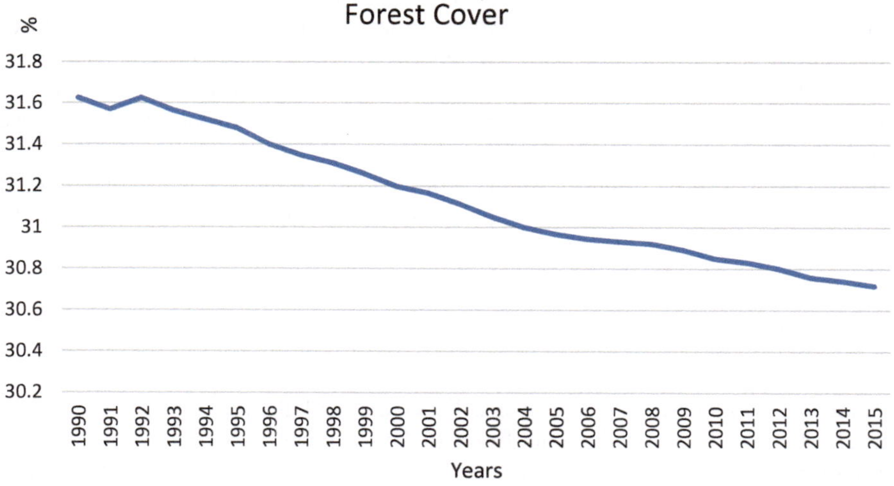

Fig. 8.2 Status of the World's Forest cover (Source: World Bank)

8.3.3 Forest Composition

A Change in temperature as low as 1 °C over a long period of time is capable of altering the species composition and can cause disruption in the distribution of various tree species (Nolan et al. 2018). It is established that different species have different response to climate change due to different climate tolerance (Iverson and McKenzie 2013). So it is projected that the shifting of plant and distribution will be species-specific. It is believed that various plant species will develop competitive abilities and mechanisms which will allow them to avoid same risks of climate change such as dispersal rate and patterns (Benito Garzón et al. 2011). It is studied that as the species reaches its environmental threshold and the synergistic effects of multiple human-induced activities will increase in the future, the responses of the plant species will turn out to be much more divergent as the scale of climate change increases (Iverson and McKenzie 2013).

The projections made in the various studies have shown that the coniferous biomes in the northern region will be shifting to another biome with increased productivity. This change will occur either through stock dieback followed by managed and natural regeneration or in the form of slow conversion of the stock via migration and managed regeneration. The models in the Pakistani scenario have predicted both latitudinal and altitudinal changes in the current forest boundaries. Studies have shown that two types of migrations could occur. Either there would be migration of an individual species or migration of multiple species. The climatic models have depicted that various plant species are directly or indirectly dependent upon certain climatic regimes. This character is important during the redistribution of plant species as they would not be uniform and each species has specific hydrothermal ranges. According to the projected climatic changes in dry temperate scenario, the juniper and Betula species will expand upto the Alpine regions. Further, the stunted evergreen forest species occurring at the timberline will also occupy the alpine grasslands due to reduced frost incidences. The increasing temperatures overall during the various seasons will tend to cause species to move in an upward direction. This will cause the tropical and subtropical forests to move upward into the dry temperate forests resulting in the mortality of some of the species in these regions (Table 8.1). Similarly, the forest species such as *P. wallichiana* and *P. roxburghii* present in the subtropical and moist temperate regions will move upward and dominate higher altitudes in the place of the *A. pindrow* and *P. smithiana.*

8.3.4 Forest Out turn

In the current scenario, it has been established that with the increasing temperatures and carbon levels the production of firewood and timber has overall risen. Further, future projections have shown that the production of the coniferous species has shown a decreasing trend up to 20 and 0% and increase in agricultural lands for food production. The studies also suggested that due to prolonged warm days there is a chance for the provision of longer growing period especially to coniferous forests. Hardwood species are supposed to increase due to the fact that increased atmospheric CO_2 concentration acts like fertilizer and improves water use efficiency of plants. Forest productivity is increased with the increase of CO_2 concentration and rising temperature. Local conditions, like availability of soil moisture and nutrients, strongly temper positive influence of the CO_2 concentration and other climatic factors (Medlyn et al. 2011). With addition of edaphic factors, disproportionate increase in temperature will not necessarily increase the growth of trees rather may reduce it for some species. But high temperatures and longer dry spells definitely would have a negative effect on the riverine and scrub forests and results in less out turn of forests.

8.3.5 *Growing Seasons*

Plants directly depend upon a specific range of seasonal precipitation and temperatures for proper functioning at various life stages. Alterations from climate change have some benefits for some of the species as the lengthening seasons gave them more chance of growing, but for some it brings about greater droughts, pests, and floods. Changing climate will effect precipitation pattern leading to excess water during off-seasons and no or less water during critical growth periods. A prolonged growing season will bear green signal for an increased breeding period for insects meaning multiple generations per year and having greater resistance for pesticides (Linderholm 2006).

8.3.6 *Plant Phenology*

Phenology is defined by the response of organisms against the seasonal appearances and periodicity of life cycle events due to the climatic changes of the ecosystem in which they inhabit (Lieth 2013). Various studies have established the fact of alteration in phenological patterns and disturbance of plant species as a result of global warming accompanied by change in pollinators, predators, and herbivores. Similarly, change in temperature regimes, precipitation pattern, radiation, photoperiod, and droughts will bring about an eminent disturbance in species physical structure. Some scientists have evidenced that a variation in the conventional temperature and precipitation regimes has eminently altered the flowering and fruiting pattern of certain tree species. Climate change has shown to disturb and altered the critical day lengths for various plant species. It has been concluded that these changes in the plant's phenology will bring about markable changes in crop production, species distribution, and subsequently human health. The phenological characteristics of forest trees are effective indicators of climate change (Table 8.2). A few species are highly responsive to even a slight change in any ecosystem.

A study conducted in the scenario of dry scrub forests has depicted that due to seasonal variations there would be definite changes in the flowering seasons thus subsequently effecting the maturity of seeds. It was projected that due to increasing temperatures the coniferous forests will have altered flowering times. It is predicted that there would be considerable variations in the spring and autumn seasons throughout various forest types. This would result in the early flowering of various tree speices such as *A. pindrow, P. wallichiana,* C. *deodara, J. excelsa,* and *P. smithiana*. These changes will bring upon a change in the seasonal rhythms of various members of the food webs thus effecting the functioning of the ecosystem in general (Sheikh et al. 2015).

8.3.7 Adaptivity of Alien Species

Invasive alien/exotic species are regarded as the most important drivers of biodiversity loss worldwide (Masters and Norgrove 2010). The changing climate is providing the most appropriate conditions for invasive species to settle into new conditions. The extreme climatic events inducing disturbances in the ecosystems have provided excellent opportunities in terms of dispersal and growth of alien/exotic species (Early et al. 2016). For example, wildfire incidences increase spaces in the forest canopy, which gives suitable conditions for the invasive species to conduce and spread. Thus, creating implications for the management strategies and biodiversity (Bellard et al. 2018). Invasive pathogens will have a positive effect from climate change as rising temperatures and shrinking winters will provide them conducive conditions for growth, reproduction, and increased transmission (Masters and Norgrove 2010). Replacement of native tree species by alien invasive species is one most marked impact of climate change. With them, these species will bring their unique pathogens new fire regimes and making biodiversity further vulnerable (Early et al. 2016).

A leguminous shrub, *Prosopis juliflora*, (mesquite) was an introduced species in the past due to its high drought tolerance from the southern USA to South Asia and Africa. It has enjoyed the increased temperatures and low rainfall. It is a multipurpose shrub and grows very quickly in arid conditions. However, it has invaded grazing lands and out-competing native grasses and flora. Its eradication is difficult. An increase of temperature as a minimum to 2 °C will favor acacia (*Acacia nilotica*) to invade inland Australia. This expansion is further worsened by farmers shifting livestock husbandry from grazing sheep to browsing cattle as temperature increases. Browsers are better seed dispersers of shrubs and trees (Masters and Norgrove 2010).

8.3.8 *Stunted Growth of Native Species*

Climate change affects biogeochemical/nutrient cycling in forest ecosystems directly through temperature and precipitation impact, or indirectly by altering forest composition, duration of growing season, soil microbial activity, and hydrological cycle. Further, it changes nitrogen and carbon cycles and accelerating land-use changes which create empty niches for invasive new species. Invasive species causing resource loss and devaluation of land is costing a fortune. These changes not only affected the biodiversity but also changes species hierarchies by affecting the growth of native species in forest ecosystems (Table 8.3). As a result of climate perturbation, new dominant tree species may arise through stressing native species. It is established that due to crowding of the invasive species the resources that are to be used by the native plants are decreased; hence, diversity and richness fall considerably (Sathyanarayana and Satyagopal 2013). These plant species when mixing up with native species do not have any problem from the climatic conditions, but after settling, they bring with them the new pathogens, diseases, predators, and above all enhanced competition for the common resource thus leading towards mortality of the native species.

8.3.9 *Insect/Pest Occurrence*

Different components of changing climate affect and interact independently with forest pests. Apart from rising temperatures, increased droughts, and uneven rainfall patterns, another reason for the appearance of new insects and diseases is the introduced exotic tree species (Jactel et al. 2019). The introduced tree species such as Papyrifera, Eucalyptus, Parthenium, Ailanthus, Robinia, and Lantana have brought

Table 8.3 Summary of studies showing impacts of climate change on tree species phenology, distribution, forest composition, and composition

	Land structure		Forest responses to climate change			
Regions	Fragmented	Not reported	Species distribution	Forest composition and structure	Phenology	Total studies
South and Central America	23	3	4	22	–	26
Africa	14	1	3	12	–	15
South Asia	18	6	6	14	4	24
Southeast Asia	7	3	2	7	1	10
Australia	7	3	2	6	2	10
Total	69	16	17	61	7	85

Source: Deb et al. (2018)

with themselves the new stress of insect pests in different biomes. That then becomes the basis for various insect outbreaks that is responsible for weakening and killing native trees. Climatic changes have induced much severity of insect breaks. The rising temperatures will make insects have increased developments with short life cycles (Ramsfield et al. 2016). Warm climate will produce more generations of insects per year, drought spells weekend tree resistance against insects, storms, and elevated CO_2 can substrate for defoliator and bark beetle which damages more trees. Although, these effects are most likely leading to forest damage in particular but can have a possible negative impact on forest insect herbivores. For instance, heat waves may cause more insect mortality and moderate droughts can induce defense.

8.4 Conclusion

Climate changes affect the production and growth of forests directly or indirectly through temperature, rainfall variation, and altering other weather elements. Although, it is more difficult to decide either climate effects are positive for vegetation or may pose abrupt effects which could lead to a diverse change in forests. Occurrence of these changes may be slow and noticed decades after, but are obvious. Various forest ecosystems responded in so many ways. Elevated carbon dioxide levels alone effect plant growth. In some forest types elevated carbon dioxide and temperature increases rainfall which promote vegetation growth. More temperature in boreal forest increases forest productivity and yield, but at the same time forest fire incidents increase. There are also recorded incidents of change in species composition due to climate change. Species composition becomes more diverse but some more valued species can vanish to extinction. Tree's natural ecological regions are likely to proceed north or to higher altitudes to change in timberline. Other indigenous species will no longer be suitable for their natural regions. Some mountain tops tree species may die out with an increase in temperature in subalpine forests.

Climate change in tropical and subtropical areas will likely increase the drought risk. Drought induces risks of dying vegetation and pest infestation.

The solution to all these climate change disasters is simply to increase forest areas. In future, strategies should be focused on afforestation, reforestation, and to better understanding of changing climate effects on trees. Major forest types and grasslands need to be restored according to their natural composition. Similarly, more research is also needed to measure individual species capacity to adopt the climate change or to migrate to new emerging ecosystems in spite of extinction.

References

Abbasi, A. M., Shah, M. H., Khan, M. A. (2015). Pakistan and Pakistani Himalayas. In *Wild Edible Vegetables of Lesser Himalayas* (pp. 1-18). Springer, Cham.

Ahmad, S. S., Abbasi, Q., Jabeen, R., & Shah, M. T. (2012). Decline of conifer forest cover in Pakistan: a GIS approach. *Pakistan Journal of Botany*, *44*(2), 511-514.

Ali, S., H. Eum, et al. (2019). Assessment of climate extremes in future projections downscaled by multiple statistical downscaling methods over Pakistan. *Atmospheric Research*, 222: 114-133

Amin, A., et al. (2018a). Regional climate assessment of precipitation and temperature in Southern Punjab (Pakistan) using SimCLIM climate model for different temporal scales. *Theoretical Applied Climatology*, 131:121–131

Amin, A., et al. (2018b). Evaluation and analysis of temperature for historical (1996–2015) and projected (2030–2060) climates in Pakistan using SimCLIM climate model: Ensemble application. *Atmospheric Research*, 213: 422-436.

Bellard, C., Jeschke, J. M., Leroy, B., Mace G. M. (2018). Insights from modeling studies on how climate change affects invasive alien species geography. *Ecology and evolution, 8(11),* 5688-5700.

Benito Garzón, M., Alía, R., Robson, T. M., & Zavala, M. A. (2011). Intra-specific variability and plasticity influence potential tree species distributions under climate change. *Global Ecology and Biogeography, 20(5),* 766-778.

Berndes G, Abt, B, Asikainen, A, Cowie, A, Dale, V., Egnell, G, Yeh S (2016). Forest biomass, carbon neutrality and climate change mitigation. *From science to policy*, *3*, 3-27.

Broto, V. C., (2017). Urban governance and the politics of climate change. *World development, 93,* 1-15.

Burrascano, S., Chytrý, M., Kuemmerle, T., Giarrizzo, E., Luyssaert, S., Sabatini, F. M., Blasi, C. (2016). Current European policies are unlikely to jointly foster carbon sequestration and protect biodiversity. *Biological Conservation, 201,* 370-376.

Clair, S. B. S., Lynch, J. P. (2010). The opening of Pandora's Box: climate change impacts on soil fertility and crop nutrition in developing countries. *Plant and Soil, 335(1-2),* 101-115.

Deb, J. C., Phinn, S., Butt, N., & McAlpine, C. A. (2018). Climate change impacts on tropical forests: identifying risks for tropical Asia. *Journal of Tropical Forest Science*, 30(2), 182-194.

Early, R., Bradley, B. A., Dukes, J. S., Lawler, J. J., Olden, J. D., Blumenthal, D. M., …, Sorte, C. J. (2016). Global threats from invasive alien species in the twenty-first century and national response capacities. *Nature communications, 7(1),* 1-9.

Fahey RT, Stuart-Haëntjens EJ, Gough CM, De La Cruz A, Stockton E, Vogel CS, Curtis PS. (2016). Evaluating forest subcanopy response to moderate severity disturbance and contribution to ecosystem-level productivity and resilience. *For Ecol Manage*, 376:135–147

FAO (2010). Global Forest Resources Assessment. *Forestry paper no.* 163.

Gentle, P., & Maraseni, T. N. (2012). Climate change, poverty and livelihoods: adaptation practices by rural mountain communities in Nepal. *Environmental science & policy, 21,* 24-34.

Good, P., Bamber, J., Halladay, K., Harper, A. B., Jackson, L. C., Kay, G, Srokosz, M. (2018). Recent progress in understanding climate thresholds: Ice sheets, the Atlantic meridional overturning circulation, tropical forests and responses to ocean acidification. Progress in Physical Geography: *Earth and Environment*, *42(1)*, 24-60.

Gordon, C. E., Bendall, E. R., Stares, M. G., Collins, L., Bradstock, R. A. (2018). Aboveground carbon sequestration in dry temperate forests varies with climate not fire regime. *Global change biology, 24(9),* 4280-4292.

Haggerty, B. P, & Galloway, L. F. (2011). Response of individual components of reproductive phenology to growing season length in a monocarpic herb. *Journal of Ecology, 99(1),* 242-253.

Hoogendoorn, G., Fitchett, J. M. (2018). Tourism and climate change: A review of threats and adaptation strategies for Africa. *Current Issues in Tourism, 21(7),* 742-759.

Hornsey, M. J., Harris, E. A., Bain, P. G. Fielding, K. S. (2016). Meta-analyses of the determinants and outcomes of belief in climate change. *Nature Climate Change, 6(6),* 622.

Hussain, S. et al. (2019). Using GIS tools to detect the land use/land cover changes during forty years in Lodhran District of Pakistan. *Environmental Science and Pollution Research*, Accepted
Inkyin K. & Woo S. Y. (2015) An overview of interrelationship between climate change and forests, Forest Science and Technology, 11:1, 11-18
IPBES (2019). Nature's Dangerous Decline Unprecedented Species Extinction Rates Accelerating. Paris, France.
IPCC (2014). Intergovernmental Panel on Climate Change (IPCC) IPCC Fifth Assessment Report: Climate Change 2014 (AR5) 8.
Iverson L. R. & McKenzie D. (2013). Tree-species range shifts in a changing climate: detecting, modeling, and assisting. Landscape Ecology, 28(5), 879-889.
Jactel, H., Koricheva, J., Castagneyrol, B. (2019). Responses of forest insect pests to climate change: not so simple. *Current opinion in insect science.*
Jandl R, Ledermann T, Kindermann G, Freudenschuss A, Gschwantner T, Weiss P (2018) Strategies for climate-smart Forest Management in Austria. Forests 9(10):592
Keča, L. J., Keča, N., & Rekola, M. (2013). Value chains of Serbian non-wood forest products. *International Forestry Review, 15(3),* 315-335.
Li, J.B., & Zhang, H.P. (2011). Research on spatial-temporal characteristics and affecting factors decomposition of agricultural carbon emission in China. *China Population, Resources and Environment, 21(8),* 80-86.
Lieth, H. (Ed.). (2013). Phenology and seasonality modelling (Vol. 8). *Springer Science & Business Media.*
Linderholm, H. W. (2006). Growing season changes in the last century. *Agricultural and forest meteorology, 137(1-2),* 1-14.
Liu Y, Feng Y, Zhao Z, Zhang Q, Su S. (2016). Socioeconomic drivers of forest loss and fragmentation: a comparison between different land use planning schemes and policy implications. *Land Use Policy*, 54:58–68.
Maestre, F. T., Quero, J. L., Gotelli, N. J., Escudero, A., Ochoa, V., Delgado-Baquerizo, M, García-Palacios, P. (2012). Plant species richness and ecosystem multifunctionality in global drylands. *Science, 335(6065),* 214-218.
Mantyka-Pringle, C. S., Visconti, P., Di Marco, M., Martin, T. G., Rondinini, C., Rhodes, J. R. (2015). Climate change modifies risk of global biodiversity loss due to land-cover change. *Biological Conservation, 187,* 103-111.
Martin, P. A., Newton, A. C., Bullock, J. M. (2013). Carbon pools recover more quickly than plant biodiversity in tropical secondary forests. Proceedings of the Royal Society B: *Biological Sciences, 280(1773)*, 20132236.
Masters, G., & Norgrove, L. (2010). Climate change and invasive alien species. *UK: CABI Working Paper, 1.*
Medlyn, B. E., Duursma, R. A., & Zeppel, M. J. (2011). Forest productivity under climate change: a checklist for evaluating model studies. Wiley Interdisciplinary Reviews: *Climate Change, 2(3)*, 332-355.
Meinard, Y., Coq, S., & Schmid, B. (2019). The Vagueness of "Biodiversity" and Its Implications in Conservation Practice. In *From Assessing to Conserving Biodiversity* (pp. 353-374). Springer, Cham.
Menhas, R., Umer, S., & Shabbir, G. (2016). Climate Change and its Impact on Food and Nutrition Security in Pakistan. Iranian journal of public health, 45(4), 549–550.
Nasim, W., et al. (2018). Future risk assessment by estimating historical heat wave trends with projected heat accumulation using SimCLIM climate model in Pakistan Atmospheric Research, 205-118-133
Newman ME, McLaren KP, Wilson BS. (2014). Assessing deforestation and fragmentation in a tropical moist forest over 68 years; the impact of roads and legal protection in the Cockpit Country, Jamaica, *For Ecol Manage*. 315:138–152.

Nolan, C., Overpeck, J. T, Allen, J R, Anderson, P. M., Betancourt, J. L, Binney, H. A, Djamali, M. (2018). Past and future global transformation of terrestrial ecosystems under climate change. *Science, 361(6405),* 920-923.

O'Connor CD, Falk DA, Lynch AM, Swetnam TW, Wilcox CP (2017) Disturbance and productivity interactions mediate stability of forest composition and structure. *Ecol Appl*. 27:900–915

Pecl, G. T., Araújo, M. B., Bell, J. D., Blanchard, J., Bonebrake, T. C., Chen, I. C., …, Falconi, L. (2017). Biodiversity redistribution under climate change: Impacts on ecosystems and human well-being. *Science, 355(6332),* eaai9214.

Primack, R. B, Laube, J, Gallinat, A. S, Menzel, A. (2015). From observations to experiments in phenology research: investigating climate change impacts on trees and shrubs using dormant twigs. *Annals of botany, 116(6),* 889-897.

Radeloff, V. C., Helmers, D. P., Kramer, H. A., Mockrin, M. H., Alexandre, P. M., Bar-Massada, A., Stewart, S. I. (2018). Rapid growth of the US wildland-urban interface raises wildfire risk. *Proceedings of the National Academy of Sciences, 115(13),* 3314-3319.

Ramsfield, T. D., Bentz, B. J., Faccoli, M., Jactel, H., & Brockerhoff, E. G. (2016). Forest health in a changing world: effects of globalization and climate change on forest insect and pathogen impacts. *Forestry, 89(3),* 245-252.

Reddy, P. R., Verghese, A., Rajan, V. V. (2013). Potential impact of climate change on honeybees (Apis spp.) and their pollination services. *Pest Management in Horticultural Ecosystems, 18(2),* 121-127.

Richardson, A. D., Keenan, T. F., Migliavacca, M., Ryu, Y., Sonnentag, O., Toomey, M. (2013). Climate change, phenology, and phenological control of vegetation feedbacks to the climate system. *Agricultural and Forest Meteorology, 169,* 156-173.

Rizvi, A.R., Baig, S., Barrow, E., Kumar, C. (2015). Synergies between Climate Mitigation and Adaptation in Forest Landscape Restoration. Gland, Switzerland: IUCN.

Rosenzweig, C. and Hillel, D. eds. (2015a). ICP Series on Climate Change Impacts, Adaptation, and Mitigation-Vol. 4. HANDBOOK OF CLIMATE CHANGE AND AGROECOSYSTEMS: The Agricultural Model Intercomparison and Improvement Project (AgMIP) Integrated Crop and Economic Assessment, Part 1. Chapter 8. The AgMIP Coordinated Climate-Crop Modeling Project [C3MP]: Methods and Protocols) Pages 191-221. World Scientific Publishing Co Pt. Ltd.

Rosenzweig, C. and Hillel, D. eds. (2015b). ICP Series on Climate Change Impacts, Adaptation, and Mitigation-Vol. 4. Handbook of climate change and agroecosystems: The Agricultural Model Intercomparison and Improvement Project (AgMIP) Integrated Crop and Economic Assessment, Part 2. Chapter 7. Impact of Climate Change on the Rice–Wheat Cropping System of Pakistan. Pages 191-221. World Scientific Publishing Co Pt. Ltd. isbn:978-1-78326-563-3

Sathyanarayana, N., & Satyagopal, K. (2013). Invasive alien species: Problems and the way forward. Pest Management In *Horticultural Ecosystems, 19(1)*, 85-91.

Schadauer, K., Barreiro, S., Schelhaas, M. J., & McRoberts, R. E. (2017). Future challenges for woody biomass projections. Forest inventory-based projection systems for wood and biomass availability (69-76). Springer, Cham.

Scheffers, B. R., De Meester, L., Bridge, T. C., Hoffmann, A. A., Pandolfi, J. M., Corlett, R. T, and Pacifici, M. (2016). The broad footprint of climate change from genes to biomes to people. *Science, 354(6313).*

Segan, D. B., Murray, K. A., Watson, J. E. (2016). A global assessment of current and future biodiversity vulnerability to habitat loss–climate change interactions. *Global Ecology and Conservation, 5,* 12-21.

Seidl, R., Thom, D., Kautz, M., Martin-Benito, D., Peltoniemi, M., Vacchiano, G, Lexer, M. J. (2017). Forest disturbances under climate change. Nature climate change, 7(6), 395.

Sheikh, M. M., Manzoor, N., Rehman, N., Adnan, M., & Khan, A. M. (2015). Past and Projected Impacts of Climate Change on Forests of Pakistan.

Simpson, M. C. (2017). Quantification and Magnitude of Losses and Damages Resulting from the Impacts of Climate Change: Modelling the Transformational Impacts and Costs of Sea Level Rise in the Caribbean (Key Points and Summary for Policy Makers Document).

Thom, D., & Seidl, R. (2016). Natural disturbance impacts on ecosystem services and biodiversity in temperate and boreal forests. *Biological Reviews, 91(3),* 760-781.

Walther, G. R. (2010). Community and ecosystem responses to recent climate change. Philosophical Transactions of the Royal Society B: *Biological Sciences, 365(1549)*, 2019-2024.

Xiao, Q., & Xiao, Y. (2019). Impact of artificial afforestation on the regional water supply balance in Southwest China. *Journal of Sustainable Forestry, 38(5),* 427-441.

Yang, B., He, M., Shishov, V., Tychkov, I., Vaganov, E., Rossi, S., …, Grießinger, J. (2017). New perspective on spring vegetation phenology and global climate change based on Tibetan Plateau tree-ring data. *Proceedings of the National Academy of Sciences, 114(27),* 6966-6971.

Chapter 9
Climate Change a Great Threat to Fisheries

Muhammad Younis Laghari, Abdul Ghaffar, and Muhammad Mubeen

Abstract For billions of people all around the world, fish is a major source of aquatic animal protein, essential vitamins, fatty acids, and starchy diet. Around 520 million fishers and fish farmers worldwide, mostly in developing countries, depend upon fishing and aquaculture. Climate change influences fish stocks directly and has great impacts on fisheries and aquaculture. There might be unpredictable and surprisingly various impacts of climate change on fishes, which effect in a wide range. Ocean acidification, coral bleaching, and altered river flows are the processes which affect the marine and freshwater ecosystem. Fishes are not only threatened by extreme weather but at the same time these are vulnerable to sea-level rise as well. Meanwhile, the social and economic context of fisheries will be disrupted by impacts on security, migration, transport, and markets. Because of overexploiting natural resources there is a significant decline in fisheries production. In recent years, the fisheries sector has been confronting with numerous challenges (including natural and anthropogenic) such as natural disasters, climate change, industrialization, environmental pollution, and overfishing. Temperature is one of the major factors that cause the physical changes in the aquatic environment such as fluctuation in oxygen in the ecosystem, development of algal blooms, and enhancing the frequency and intensity of disease outbreaks. In Asia and Africa, many regions are dominated by the fishery sector and some regions have greater poverty where fisheries-related livelihood prevails. It is a matter of great concern and needs the utmost attention of policymakers and world aquaculture fisheries authorities to take the matter seriously. Hence, to cope with global climate change and save fisheries resources as well as to increase aquaculture production, a consistent policy is desired in order to find out remedial measures.

M. Y. Laghari (✉)
Department of Freshwater Biology and Fisheries, University of Sindh, Jamshoro, Sindh, Pakistan
e-mail: younis.laghari@usindh.edu.pk

A. Ghaffar
Department of Zoology, The Islamia University of Bahawalpur, Bahawalpur, Pakistan

M. Mubeen
Department of Environmental Sciences, COMSATS University Islamabad, Vehari, Pakistan

W. N. Jatoi et al. (eds.), *Building Climate Resilience in Agriculture*,
https://doi.org/10.1007/978-3-030-79408-8_9

Keywords Climate change · Aquatic protein · Fish threat · Poverty · Socioeconomic condition

9.1 Introduction

Fisheries is one of the vulnerable sectors that is facing widespread and often profound changes for decades. The freshwater, as well as marine water, have been affected by many factors, such as anthropogenic activities and alteration in River flows, coral bleaching, and ocean acidification. Overall, not only fish or aquaculture production is disturbed but fishing communities are vulnerable to such climate change as rise in sea level, and their livelihoods are threatened by extreme weather and storms, etc. On the other hand, due to markets, transport, migration, and security reasons, fisheries will be disturbed in the context of socioeconomic policy. Inland fishery is also at great risk and is a big threat not only to the livelihood of the poorest populations of the world but for food supply too (Smith et al. 2010).

There are millions of people who are dependent on fisheries and aquaculture for their livelihoods throughout the world. At present, it is estimated that about 42 million people work full or part-time as fishers and fish farmers, mainly in Asia. It is expected that by the year 2100, there will be a significant (25%) impact on the inland aquatic ecosystem of Africa, at the same time aquatic culture also will be affected simultaneously.

Through scientific investigation, it has been proved that increases in greenhouse gas emissions have caused global warming. The greenhouse gas emissions have not been recorded as highest at any time since last 650,000 years as it is nowadays (IPCC 2014). The scientific approaches have revealed that globally average increase in air temperature and rise in sea level. While, widespread melting of ice and snow is expected. Sea level is rising due to climate change and causing the salinity increase in rivers and deltaic regions. Hence, this also will have a great impact on fish farming.

9.2 Factors (Related to Climate Change) Affecting Fisheries

Climate change is rapidly proceeding that will have serious impacts on humans, wildlife, aquatic life, and habitat. The present scenario of heat-trapping might lead to a shift in local, regional, and national climate levels. Hence, it will not only disturb the natural processes but will also significantly destroy the ecosystems. That ecosystem degradation certainly will disturb our sustainable natural resources including aquatic and terrestrial. Fishes need a long-term iterative process to adapt to climate change. Habitat loss, water pollution, and diseases are the major threats. The speedy development and expansion of industrial technology is another threat that causes contamination of many freshwater ecosystems (Mashkoor et al. 2013; Ghaffar et al. 2014). The industrial and agricultural processes continuously release wastes into natural water sources and have adverse effects on aquatic biota (Witeska et al. 2014; Shakeel et al. 2017; Amin et al. 2017; Zia et al. 2017; Saud et al. 2017). Management of these wastes is crucial in order to minimize their adverse effects on aquatic ecosystems (Ghaffar et al. 2016). The major challenging factors for their existence are shifts in local climate (temperature and precipitation). The recent years global warming has been accelerated, even it might be rapid in the next century if the current rate of greenhouse gas emissions is not controlled. As a result, most species will adapt to the evolutionary changes (Abbas et al. 2017; Adnan et al. 2017; Awais et al. 2017; Fahad et al. 2017; Rasool et al. 2017).

The major factor of climate change is temperature that is root cause of all other factors affecting the fisheries sector through various aspects.

9.2.1 Temperature

The proper weather record keeping started in the year 1880, since then, the highest earth's surface temperature was recorded in 2019 and it is the largest temperature anomaly of any month. Detailed illustration of temperature can be seen in Fig. 9.1. The global average temperature had increased 0.2 °C per decade up to the year 1970s and average record in 1990 is 1.4 °C (IPCC 2007). It is predicted that the temperature might rise up to 5.8 °C in the year 2100 (WMO 2018). That rise will directly impact the production by affecting all the responsible factors. In these major factors, oxygen, sea level, toxic algal blooms, prevalence of pests, diseases, and predators are included. Based on these factors temperature might influence

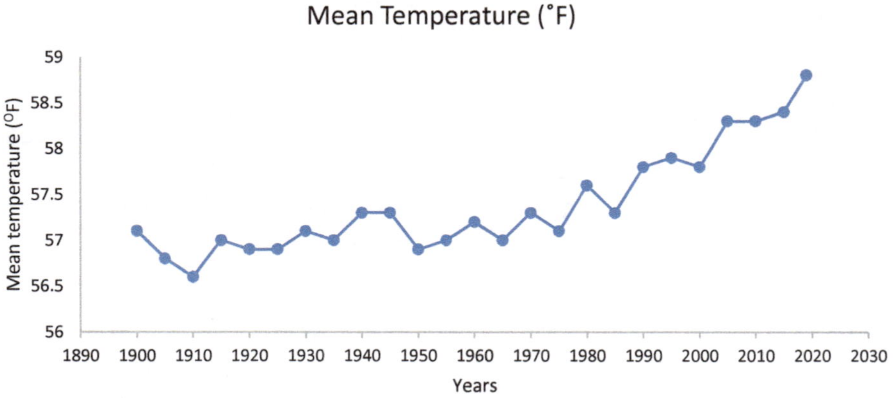

Fig. 9.1 Mean temperature trend from 1900 to 2019

positively or negatively. In general, the increase in temperature boosts up the growth that may increase per unit area production. But, fish stop feeding at above 30 °C temperature and slow down the growth rate (Renee et al. 2019; Eoin et al. 2016).

9.2.2 Oxygen Fluctuation

Oxygen is one of the most important requirements of any living organism. In general, it has been noted that temperature has a direct relationship with the oxygen contents in water. Hence, climate change is the main cause of deoxygenating in the open oceans. The first impact of oxygen fluctuation is on the food chain that plays a primary role in the existence of any living organism. In further, due to stratification on the ocean, the produces are also greatly affected due to the lowest oxygen supply at the bottom level. Hence, such vertical oxygen supply would affect the physical as well as biogeochemical process too. While such changes will also have a great impact on the living style of any aquatic organism. All the physiological processes including respiration, reproduction, digestion, and many others are greatly influenced by the oxygen contents (Tran-Duy et al. 2012; Mohsen et al. 2015). Overall, all these changes ultimately have loss or threat to the fish and cause a decrease in production. A bioindicator fish species, the eelpout (*Zoarces viviparous*), used for environmental monitoring has been decreasing and its growth performance has been affected at North and Baltic Seas. Lakani et al. (2013) showed that mean fish weight was not affected by oxygen levels during the first 2 weeks in large size fish, but after that until the end of the study, fish showed significant differences in mean weight (Fig. 9.2).

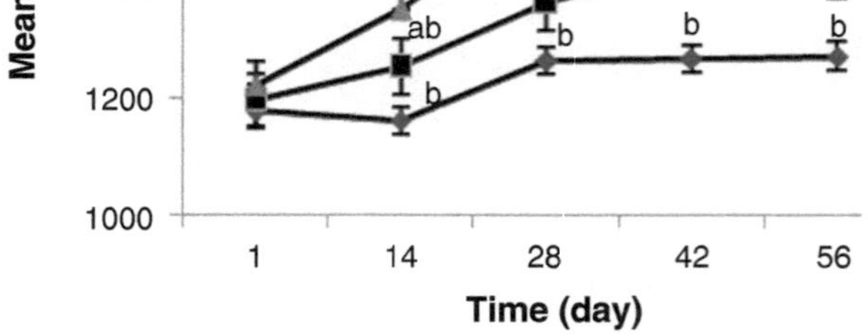

Fig. 9.2 Mean weight of *H. huso* during 8 weeks rearing under different oxygen levels. A significant difference was shown by different letter as determined by ANOVA and Tukey's test at $P \leq 0.05$. [source: Lakani et al. 2013]

9.2.3 Net Primary Productivity (NPP)

The production system of any water body is regulated by the Net Primary Production (NPP). There is a negative as well as positive impact of the climate change in any ecosystem. It will boost up the production of filter feeders by enhancing the NPP and may lower winter mortality. But, there is a huge negative effect of it; as toxins will be released in the water by boosting the algal blooms and dissolved oxygen in the ecosystem will be decreased. That might increase infestation of nuisance species and fouling organisms. Hence, such situation might increase the fish kill. Commonly the water productivity is dependent on planktons, especially the phytoplankton. The phytoplankton population directly depends on the availability of nutrients and light in the water. The atmosphere, ocean mixing, clouds, and the solar cycle are directly effecting both the light and nutrients in the water. It has been recorded that NPP has been altered since the last two decades because of changes in ocean surface temperature, atmospheric iron deposition, and wind (Khan et al. 2018; Hashmi et al. 2018). As we know that temperature plays an important role in thermal stratification, as a result it affects the nutrients in the water and exchange of gases. Finally, it will alter the primary productivity that brings a serious change in the food chain and food web. Hence, that altered food availability might result in alteration of species assemblages because of changes in food availability, species-specific differences in thermal tolerances, disease susceptibility, and shifts in the competitive advantage of species. Due to all these factors, the nutrient supply to upper productivity of the Atlantic and the Pacific oceans is declining.

9.2.4 Ecosystems Alteration

The ecosystem plays a vital role in survival and normal life routine of living organisms of any habitat. Alteration in the ecosystem affects all life activities of an organism from birth to death. The mangroves, seagrass, and coral reefs are considered as basic ecosystems of the marine. These ecosystems are under great threat due to anthropological activities and climate change has added more stress to them. Both factors have brought a complete structural and behavioral change to the ecosystem. The process of the ecosystem is climate-dependent and the influenced process will affect the ecological communities as well as individual species. Recently the concern of many species extinction goes to climate change (Pounds et al. 2006; Amin et al. 2018; Awais et al. 2018; Afzal et al. 2018; Fahad et al. 2018). Climate change might enhance the extreme events and would have rapid changes in any ecosystem in the future. Brander (2007) has briefed the situation as ***"The resilience of species and systems is being compromised by concurrent pressures, including fishing, loss of genetic diversity, habitat destruction, pollution, introduced and invasive species, and pathogens."*** He also added that the pH of ocean water is decreasing due to rise in CO_2 and its consequences are still hidden.

9.2.5 Food Availability

The human population is growing rapidly throughout the globe accompanied by increasing food fish demand. The major source of animal protein is fish for the billions of people all around the world. Fish provides the essential vitamins and fatty acids along with protein complements. About 520 million people, including their dependents, rely on fishing and aquaculture. Majority of them are from developing countries.

There is not only development in modern technology for comfort and luxury life but the food habits also have been changed throughout the world for last few years. Where everyone is demanding food safety and best quality. There are no so far days, when our consumer will demand labeled farmed food, with detailed environmental conditions as greenhouse gas (GHG) emissions per unit of produce, in the market. Hence, based on the consumer consciousness it might be predicted that aquaculture also will meet these aspirations in the future. About 100% mollusks and 70% of finfish are considered minimum emitting GHG. Hence, the aquaculture sector will ensure such demand and increase the fish supply as the most GHG-friendly food source. However, the model change in seafood consumption will be required to achieve it. People get an advantage through fishing during seasonal stock availability when food production and income generation fall down. Fishing communities are dependent on inland fisheries resources that are particularly vulnerable to climate change. Aquaculture is thought to be one of the fast growing food production sectors that have expanded at an average annual rate of 8.9% since 1970. The fishing only could not fulfill the market demand of increasing human population because of overexploiting and climate change. Because, the production of natural sources has been decreasing significantly day by day. Therefore, for the significant nutritional and economic benefits from available land and resources, integrated aquaculture is the best choice. Where as, fish raising in a rice field or using agricultural waste, to fertile the pond, are the best choice of natural resources utilization.

9.2.6 Predation

The temperature and water current leads distribution in fish stock that would cause loss in some area and benefiting other. Only the higher valued fishes are focused in research in this aspect while the potential impact of mass movement of noncommercial or wild species has not been investigated which directly or indirectly has an impact on the fisheries production. We are well aware of the role of such wild species in the food chain. Whenever there is migration of fish due to said factors then the fish predation is effective in two ways. First, when large fishes (mostly carnivores) move to the place of habituating small fishes then those fishes invade them for their food. Secondly, whenever the movement of small fishes to the habitat of large ones then also they become the food of the dominant fishes. Overall the climate change create these circumstances of mass movement where there is a significant impact of predation on fish production. Such type of challenges is significantly observed between climatic regions, especially in the deltaic region, which are mostly impacted by the sea-level rise. Most of the fish habit is to move in schooling; therefore, serious research must be paid to cope with external shock.

9.2.7 *Mass Movement*

There is a strong correlation between temperature on fish migration. The physiology and behavior of fish are also affected by climate change. The fish is a poikilothermic animal and its distribution is controlled by the water temperature because fish cannot maintain its body temperature according to the surrounding degree of hotness/coldness. Hence, for its survival the fish always move to the suitable temperature side from the high temperate water region. It has been reported through the research investigation that fish which normally thrive in the tropics are quickly migrating in an effort to discover cooler seas (Sarah 2019). Such migration of fish reveals that in search of better food and better oxygenated habitat, it moves to the Polar Regions. Hence, due to climate change and marine migration at such speed it might be predicted that a substantial amount of fish species will have evacuated the tropics till the year 2050. Such type of forced migration might have a dangerous effect on the ecosystems, especially the oceans. Such type of migration of temperate to polar region also will have a great impact on the sustainability of that habitat fishes. Rapid dispersal abilities have been found in most of the invasive species because they have a wide range of tolerance and survival in a range of environment. In the present situation of fast climate change, it might be predicted that our oceans will be fully 3 °C warmer globally, in the next 50 years. Such impact of climate change on the marine ecosystem and fish migration would vanish the species from the water once roamed. As there are many records of the history of the planet for such extinction of the species.

Cleaner seas (2019) reported that ***"Once bounding with infinite amounts of underwater life, a small fishing town in the state of Virginia was proudly known as the flounder capital of the world, in recent years however the flounder have struggled with rising sea temperatures and have relocated to cooler waters closer to New York and New Jersey."*** Hence, ocean's ecological stability is affected by the altering producer's chain reaction due to climate change.

9.2.8 *Diseases Distribution*

In aquaculture, where temperature plays a positive role in sense of fish growth there is also an increasing opportunity for bacterial growth, algal blooms, and parasitic development in the ecosystem. Hence, all these might cause various diseases in the fishes. Hence, change in climate that alters the temperature of water could cause slow growth due to above consequences and ultimately causes a decrease in production.

9.2.9 Rise in Ocean Level

Generally, over a long period the mean sea level rises but due to climate change the patterns have been changed and sea level rise earlier. This type of change might cause the fish stock distribution and their migration pattern. While, sea level rise alters the mangroves ecosystem that ultimately affects the fish life. On other hand, the brackish water increase at the coastal side has an impact on the freshwater fishes. Most of the fishes exist at 0–300 m layer all around the world and that layer of the ocean is warming sharply. It is a great concern regarding the prediction of mean sea-level rise may be up to 90 cm in twenty first century. As a result the coastal ecosystem will be destroyed, which is considered most important for the survival of many species. Because the coastal ecosystem is rich in mangroves and marshes and this is not only the shelter for fishes but also provides enough seed supply to aquaculture.

Hence, such significant change in the ocean and climate change may produce various factors including ocean current, temperature, distribution of algae blooms (toxic), predators, and primary productivity will have direct effects on aquaculture and its productivity (Handisyde 2008).

Rise in sea level will increase the salinity in the groundwater that will not only affect the freshwater fishery but agriculture too. At the same time, that saline groundwater will limit domestic as well as industrial consumption. Therefore, it is important to prepare broader policy according to climate change. However, that increasing saline water would lead to develop saline water aquaculture. Where some high-value species including shrimps and crabs culture will be benefited along with the negative consequences of sea rise. Hence, new policies and opportunities will lead the public to get an advantage in such circumstances. Therefore, the fish farmers/fisherman must be trained that how aquaculture can play an important role in diversifying livelihoods in such situations.

9.3 Impacts on Culture

Aquaculture might be affected directly or indirectly through various factors due to climate change. Such factors include fish stock, fish supply, fish consumption, and cost of supply and fish farmers' services.

Rise in sea level will increase saline water intrusion and will reduce the water inflow in the deltaic region will have a significant impact on the tropical aquaculture system. Hence, due to extreme weather conditions inland cage culture, pen culture, and other types of culture could be affected due to increasing upwelling deoxygenated water in the reservoir. As the reproductive cycles of species are dependent of the monsoon season and rain pattern that would affect indirectly on the aquaculture system. Becasue of the seed production and grow out cycle of the species. Especially, the great impact might be observed in the species whose seeds are collected naturally, because of their nature of spawn such as mollusks. While upper tolerance temperature limit of some species, emergent of no pathogen organism and many unknown diseases would make the culture system vulnerable to high temperature. Same time the temperature affects trash fish production, by disturbing the food chain, which in turn has an impact on some culture species, especially the carnivores. While, the basic supply of fishmeal and fish oil, for the preparation of fish feed, also considered to be one of the issues affected by climate change. Hence, all these factors affect the aquaculture practices in the climatic region.

There is a limited research on the climate change impact on aquaculture or vice versa. Overall extreme rains or alteration in monsoon in recent years, more warming post-monsoon, lower number of rainy days, annual mean temperature increase, increase in hot days' frequency, and consequent droughts have a major impact on the aquaculture. Hence, to cope with climate change in such situation, the aquaculture practices must consider the adaptive and mitigate measures. As aquaculture is considered as the backbone in many developing countries, socioeconomic and modern technology must be adopted to address the potential climate change. Hence, small-scale farming, intensive culture, integrated aquaculture system, and recirculatory farming system must be promoted in order to maximize per unit area production at environment-friendly system.

9.4 Impacts on Economics and Community

Climate change varies from region to region. Therefore, the regional productivity and specific species resources may be varied regionally. Ultimately the climate change will impact the effort per unit catch that will lead to go further for harvest. Likewise, it will require much more effort and higher cost. Allison et al. (2009) has reported that out of 132 nations the central, western Africa, and some in Asia have the most vulnerable capture fisheries due to potential climate change impacts. Hence, decrease in harvest capacity in natural water resources has an impact on fisheries production that reduces the access to the market. In such circumstances, the local economies adapt to new conditions in terms of labor and capital mobility that is said to be an indirect economic impact. Aquaculture is growing fast at an annual rate of 8.7% and is thought to be rapidly developing food producing sector of the world. Because of high nutritional value and quality, the aquatic food contributes about 20% per capita animal protein for around 2.8 billion people all over the globe, most of them belong to developing country. Generally, the aquatic environments, freshwater, marine, or brackish water, respond to climate change equally as atmospheric environment. The aquaculture practices represent mostly undefined sources of greenhouse gases (GHGs).

9.5 Conclusion

Fish is a basic source of essential nutrition and income to outpace population growth around the world. It provides a major contribution or might provide livelihood where other food and employment resources are limited. Fisheries and Aquaculture continue to be the fastest growing animal food producing sectors but along with climate change the poor illiterate fisherman/fish farmers are another challenge to adopt changes and have a great impact on the fish resources. In general, the change in temperature, water ecosystem, and precipitation are a major threat to the fisheries. Due to climate change, the storms become frequent and extreme and result in the destruction of infrastructure, affects the livelihood, imperiling habitats and stocks. In such situations, high-level research is required to suggest the best plan and coping strategies to improve the adaptability of the fisher community. Still, the fisheries sector is growing fast in food production sector. Per capita supply from the aquaculture sector increased from 0.7 kg in 1970 to 7.8 kg in 2006. Aquaculture is gradually drawing attention internationally for potential impact of climate change. In the Asian region, there is great poverty and aquaculture-related livelihood prevailing that needs great concern. If we plan properly to challenge the climate change the fishes have property to adapt to climate change as integrating aquaculture and agriculture (Hydroponic, Aquaponic, Recirculatory Aquaculture, Biofloc system, In Raceway Pond System). This type of culture system can help farmers cope with drought while boosting profits and household nutrition. Further, the management system should go forward to seek to maximize yield to increasing adaptive capacity. Therefore, the world aquaculture authorities and policymaker should pay attention to this serious matter. Hence, to cope with climate change the policymakers along with researchers/scientists must prepare policy based on the current scenario and find out the remedial measures.

References

Abbas G et al (2017) Quantification the impacts of climate change and crop management on phenology of maize-based cropping system in Punjab, Pakistan. Agricultural Forest Meteorology, 247: 42-55

Adnan MZ et al (2017) Phosphate-Solubilizing Bacteria Nullify the Antagonistic Effect of Soil Calcification on Bioavailability of Phosphorus in Alkaline Soils. Scientific Reports, 2018: 7-22653

Afzal M et al (2018) Current status and future possibilities of molecular genetics techniques in *Brassica napis*. Biotechnology Letters, 40 (3) : 479-492

Allison EH et al (2009) Vulnerability of national economies to the impacts of climate change on fisheries, Fish and Fisheries. Environmental Information System.

Amin A et al (2017) Comparison of future and base precipitation anomalies by SimCLIM statistical projection through ensemble approach in Pakistan. Atmospheric Research, 194: 214-225

Amin A et al (2018) Simulated CSM-CROPGRO-cotton yield under projected future climate by SimCLIM for southern Punjab, Pakistan. Agricultural Systems, 167: 213-222

Awais M et al (2017) Nitrogen and plant population change radiation capture and utilization capacity of sunflower in semi-arid environment. Environmental Science and Pollution Research, 24 (21): 17511-17525

Awais M et al (2018) Potential impacts of climate change and adaptation strategies for sunflower in Pakistan. Environmental Science and Pollution Research, 25 (14): 13719-13730

Brander KM (2007) Global fish production and climate Change, PNAS.

Cleaner seas (2019) Climate Change: How it is Forcing Mass Fish Migration. 30th May 2019. https://www.cleanerseas.com/climate-change-fish-migration/

Eoin JO et al (2016) Temperature effects on fish production across a natural thermal gradient. Glob Chang Biol. 22(9): 3206-3220.

Fahad S et al (2017) Crop Production under drought and heat stress: Plant responses and management options. Frontier in Plant Sciences. 8: 1147: 1-16

Fahad S et al (2018) Consequences of high temperature under changing climate optima for rice pollen characteristics concept and perspectives. Archives of Agronomy and Soil Science, 64 (11): 1473-1488

Ghaffar A et al (2014) Clinico-hematological disparities induced by triazophos (*organophosphate*) in Japanese quail. Pakistan Veterinary Journal, 34: 257-259.

Ghaffar A et al (2016) Arsenic and Urea in Combination Alters the Hematology, Biochemistry and Protoplasm in Exposed Rahu Fish (*Labeo rohita*) (Hamilton, 1822). Turkish Journal of Fisheries and Aquatic Sciences 16: 289-296.

Handisyde NT (2008) The effects of Climate Change on World Aquaculture: A Global Prospective. 1151.

Hashmi MZ et al (2018) PCB118-Induced Cell Proliferation Mediated by Oxidative Stress and MAPK Signaling Pathway in HELF Cells. Dose-Response January-March : 2018 : 01-08

IPCC (2007) Summary for policymakers. In: M.L. Parry, O.F. Canziani, J.P. Palutikof, P.J. van der Linden and C.E. Hanson (eds.). Climate change 2007: Impacts, adaptation and vulnerability. Contribution of Working Group II to the Fourth Assessment Report of the Intergovernmental Panel on Climate Change. Cambridge, UK: Cambridge University Press. 722.

IPCC (2014) Climate Change 2014: Synthesis Report. Contribution of Working Groups I, II and III to the Fifth Assessment Report of the Intergovernmental Panel on Climate Change [Core Writing Team, R.K. Pachauri and L.A. Meyer (eds.)]. IPCC, Geneva, Switzerland

Khan N et al (2018) Interaction between PGPR and PGR for water conservation and plant growth attributes under drought condition Biologia, 73 (11): 1083-1098

Lakani FB, Sattari M, Falahatkar B (2013) Effect of different oxygen levels on growth performance, stress response and oxygen consumption in two weight groups of great sturgeon Huso huso. Iranian Journal of Fisheries Sciences 12(3) 533-549

Mashkoor J et al (2013) Arsenic induced clinico-hemato-pathological alterations in broilers and its attenuation by vitamin E and selenium. Pakistan Journal of Agricultural Sciences, 50: 131-138.

Mohsen A et al (2015) Effects of dissolved oxygen and fish size on Nile tilapia, *Oreochromisniloticus*(L.): growth performance, whole-body composition, and innate immunity. Aquacult Int. 23:1261-1274

Pounds JA et al (2006) Widespread amphibian extinctions from epidemic disease driven by global warming. Nature 439:161-167.

Rasool, A., et al. (2017) A review of global outlook on fluoride contamination in groundwater with prominence on the Pakistan current situation. Environmental Geochemistry and Health 40 (4) : 1265-1281

Renee M D et al (2019) Warmer and browner waters decrease fish biomass production. Glob Chang Biol. 25(4): 1395-1408.

Sarah Z (2019) Warming pushes lobsters and other species to seek cooler homes. Ecology, Science News for Students. April 11, 2019.

Saud S et al (2017) Effects of Nitrogen Supply on Water Stress and Recovery Mechanisms in Kentucky Bluegrass Plants. Frontier in Plant Sciences. 8:983: 1-18

Shakeel M et al (2017) Environment Polluting Conventional Chemical Control Compared to an Environmentally Friendly IPM approach for control of diamondback moth, *Plutella xylostella* (L), in China: A Review. Environmental Science and Pollution Research, 24: 14537-14550

Smith MD et al (2010) Sustainability and global seafood, *Science* 327 (5967): 784-786.

Tran-Duy A Dam AA and Schrama JW (2012) Feed intake, growth and metabolism of Nile tilapia (*Oreochromis niloticus*) in relation to dissolved oxygen concentration. Aquacul Res 43:730-744

Witeska M et al (2014) The effects of cadmium and copper on embryonic and larval development of ide *Leuciscusidus* L. Fish Physiology and Biochemistry, 40: 151–163.

WMO (2018) World Meteorological Organization, Global temperatures on track for 3-5 degree rise by 2100, News conference at the United Nations in Geneva, Switzerland, November 29, 2018. REUTERS/Denis Balibouse.

Zia Z et al (2017) Effect of water management and silicon on germination, growth, phosphorus and arsenic uptake in rice. Ecotoxicology and Environmental Safety 144: 11–18

Part II
Management of Biotic and Abiotic Stresses in Agriculture Under Changing Climate

Chapter 10
Water Resources in Relation to Climate Change

Hafiz Umar Farid, Zahid Mahmood Khan, Aamir Shakoor, Muhammad Mubeen, Hafiz Usman Ayub, Rana Muhammad Asif Kanwar, and Muhammad Bilal

Abstract In this chapter, the influence of global climate change on precipitation, water availability, drought conditions, runoff, potential evapotranspiration, soil moisture change, and underground water resources (groundwater) is being described for the efficient management of freshwater resources particularly for agriculture purpose (a major freshwater consuming sector). Climate change impacts can be characterized by considering the variability in most important climatic factors such as precipitation and temperature. The timely occurrence of rainfall during the crop growth period helps to increase the crop yield, minimize the cost of production, and maximize the financial benefit. The temperature variation is conducive to increase the frequency of extreme precipitation or drought events. Some studies have reported that precipitation pattern has changed in the past few decades around the globe. This change indicates the uncertainty in precipitation, i.e., occurrence of heavy precipitation or light and moderate precipitation (flood and drought). A study projected that 14% of global land is facing drought events for every 5 years under 3 °C of global warming, as 5–10 times more frequent drought events were observed in most of Africa, Caribbean, central America, central Asia, west Asia, northwestern China, and Oceania. Similarly, strong and frequent drought events were predicted for the southern, large part of the eastern and western Europe, southern and central America. As a result, dryness will increase and put further stress on the agricultural system. On the other hand, the reduction in drought frequency was observed in northern Europe and Russian Federation. These projected scenarios for future climate change showed uncertainties. No doubt, the uncertainties in projected climate change scenarios always need to be taken into consideration. The implementation and regulation of water policies in future water resources projects should consider the projected climate change and uncertainties scenarios. A better understanding of the process of

H. U. Farid (✉) · Z. M. Khan · A. Shakoor · H. U. Ayub · R. M. A. Kanwar · M. Bilal
Department of Agricultural Engineering, Bahauddin Zakariya University, Multan, Pakistan

M. Mubeen
Department of Environmental Sciences, COMSATS University Islamabad, Vehari, Pakistan

W. N. Jatoi et al. (eds.), *Building Climate Resilience in Agriculture*,
https://doi.org/10.1007/978-3-030-79408-8_10

drought and wet spells will lead to better management of agricultural and hydrological activities.

Keywords Climate change · Extreme events projection · Uncertainties · Water resources management

10.1 Introduction

Global freshwater resources like rivers, lakes, and underground aquifers are under stress due to the increasing demand of world populations for drinking, irrigation, and sanitation. Most of the scientific and climate projected studies have proved that climate change is adversely affecting the freshwater resources in different regions of the world (Stagl et al. 2014; Malinowski and Iwona 2018). Change in climate affects the runoff and flood intensity as well as frequency, duration, and intensity of precipitation (Nguyen et al. 2018). These changes have a further impact on the management of water resources and socioeconomic systems. Similarly, global warming and changing climate systems can affect the sea level by melting of snow as well as by thermal expansion of the ocean. Moreover, the northern high latitude glaciers will melt and most of the earth will receive frequent heat waves and heavy precipitation as a result of global warming (Zaffar et al. 2014; Mishra 2019). The glaciers are particularly sensitive to climate change. On the other hand, almost one-sixth population of the world depends on the water flow from mountain glaciers and associated precipitations. The retreating ice and snow will have a remarkable impact on the downstream areas if warming proceeds. Similarly, the drought conditions have increased in some areas due to the more frequent dry spell. Therefore, it is essential to analyze climate change impact on precipitation pattern, water availability, runoff, potential evapotranspiration, soil moisture change, drought condition, and underground water resources for the management of available water resources.

10.2 Temperature Behavior

Climate change can be characterized by the variability of surface air temperature. The earth's temperature has increased by 0.7 °C in the past century alone and it continues to rise. Long-term analysis of historical temperature data indicates that land surface is warming faster than the oceans. Intergovernmental Panel on Climate Change (IPCC) reported that the temperature of the ocean and earth's surface showed a linear trend with warming of 0.85 (0.65–1.06) °C over a period 1980 to 2012 (IPCC 2014). The second half of the twentieth century has turned out to be a particularly warm period and more noticeable changes were observed at high altitude during the spring and winter seasons (Malinowski and Iwona 2018; Sadowski et al. 2013; Majewski and Walczykiewicz 2012). The temperature change trend is conducive to the increase in frequency of extreme precipitation or drought events and occurrence of desertification.

10.3 Global Precipitation Behavior

Precipitation is an important component of weather and climate. It plays an important role in human's everyday life and economy. The timely occurrence of rainfall during the crop growth period helps to increase the crop yield, minimize the cost of production, and maximize the financial benefit. On the other hand, the river flows will increase in response to heavy rainfall. The high river flow contributes to the occurrence of a heavy flood (Almazroui et al. 2017). The flood may become responsible for economic and social losses.

It is particularly important to analyze/understand the current state, change in size, and structure of precipitation at regional, country, and global scales. However, it is difficult to analyze the behavior of precipitation due to unavailability of actual precipitation data for any region with respect to time and space. Some studies have been reported that precipitation pattern has changed in the past few decades around the globe. This change indicates uncertainty in precipitation, i.e., occurrence of heavy precipitation or light and moderate precipitation. Figure 10.1 presents the volumetric change in precipitation over the continents during the last three decades. The analysis indicated that fluctuations in the occurrence of precipitation were

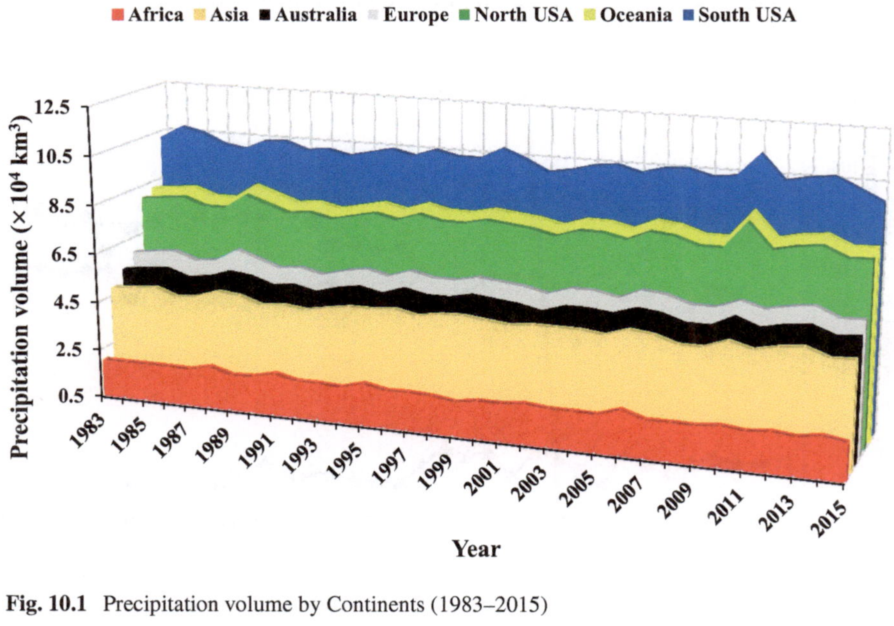

Fig. 10.1 Precipitation volume by Continents (1983–2015)

observed during the whole period but no significant long-term volumetric change in precipitation was observed for either case.

Similarly, Fig. 10.2 showed the comparison of temporal trends of precipitation among the three regions, i.e., north America, south, and east Asia. These regions have an insignificant decreasing trend of precipitation. All the regions with increasing trends cancel out the regions with decreasing trends (northwest to coastal southeast). Figure 10.3 showed the investigation of precipitation trends for management of water resources over the three major basins of the world, i.e., Colorado River Basin, Angola Basin, and Himalayan Basin. The analysis indicated that all three basins showed variability in precipitation over the historical record. The interannual variation in surface temperature also results in variability of global precipitation (Mishra 2019). It was observed that all three basins have been experiencing a decreasing trend for precipitation during the last three decades. Nguyen et al. 2018 analyzed the precipitation behavior over the 237 global basins. The analysis indicated that 20 basins show significant decreasing trends in precipitation whereas 20 basins confirm significant increasing trends. Similarly, statistically insignificant decreasing trends in precipitation were observed over 106 basins and insignificant increasing trends were depicted over 89 basins during the last three decades. This indicated that climate change may occur noticeably especially in terms of precipitation and temperature. If precipitation does not significantly upsurge, there will be more stress on water resources. As a result, dryness will increase and put further stress on the agricultural system.

Tables 10.1 and 10.2 provides the projections of seasonal temperature and precipitation change for the sub-regions of Asia for the three time slices namely 2020s, 2050s, and 2080s using the baseline period of 1961–1990 under the Special Report on Emission Scenarios (SRES) with highest future emission trajectory (A1F1) and

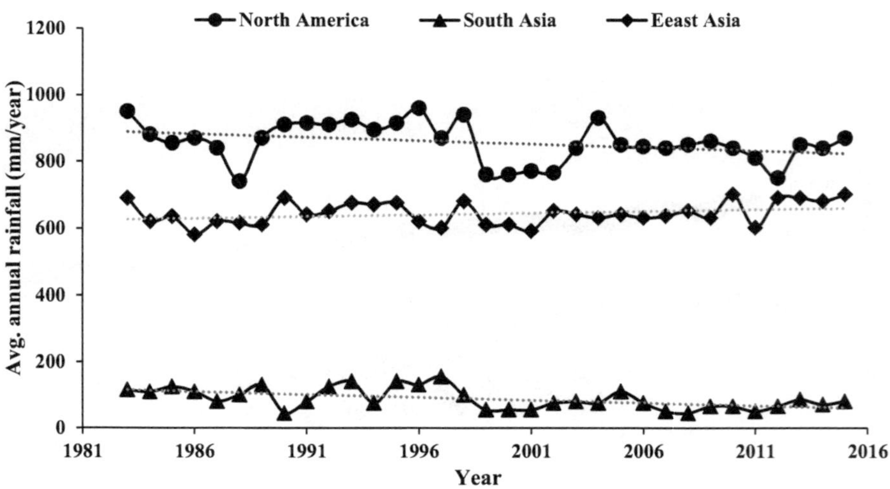

Fig. 10.2 Temporal Trend of rainfall for North America, South, and East Asia

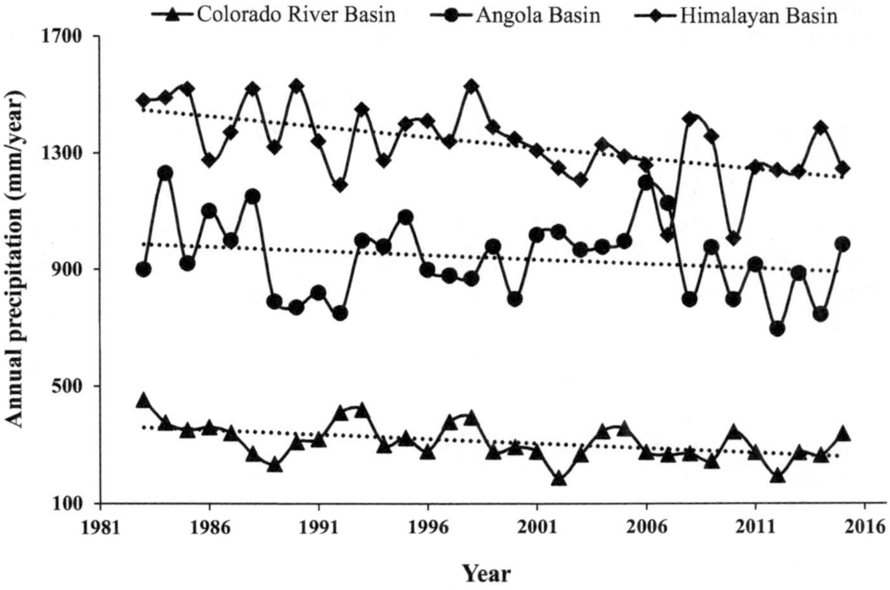

Fig. 10.3 Temporal precipitation trend of three important basins

lowest future emission trajectory (B1) pathways. The annual surface air temperature and precipitation were divided into four seasons based on the 3 months' average such as Summer, Autumn, Winter, and Spring seasons. The higher warming was projected for all the sub-regions of Asia during the northern hemispheric winter than the summer for all the time period. During this century, the maximum noticeable warming was predicted at a high altitude of north Asia and an increasing trend of average annual precipitation was predicted in most of Asia. However, the increasing trend of average annual precipitation was relatively largest and more consistent in the northern and eastern regions of Asia. The projected mean precipitation during the winter will increase in northern Asia and Tibetan Plateau and decrease in central, western, southeastern, western, and eastern regions of Asia. The projected mean precipitation during the summer will increase in northern, southern, southeastern, and eastern regions but it will decrease in western and central regions. It was also observed that due to the very dry summer, the decrease in mean precipitation will be escorted in central Asia. Global warming and precipitation variation may have an impact on crop yield and it will decrease upto 30% by 2050 (Cruz et al. 2007). A recent modeling study confirmed the projection and found a significant warming trend with no change in rainfall in the middle and lower part of the Indus plain (Ahmad et al. 2015). An increasing trend of annual precipitations at four meteorological stations and a decreasing trend at two meteorological stations in the southeastern region of upper Indus plain was observed by Ahmad et al. 2018. Kumar et al. 2016 analyzed the long-term precipitation trends in India and found a significant positive trend during the winter and a significant decreasing trend in monsoon.

Table 10.1 Projected changes in surface air temperature for subregions of Asia under SRES A1F1 (highest future emission trajectory) and B1 (lowest future emission trajectory) pathways for three time slices namely 2020s, 2050s, and 2080s (Cruz et al. 2007)

Subregions	Season	2010–2039		2040–2069		2070–2099	
		A1F1	B1	A1F1	B1	A1F1	B1
North	DJF	2.94	2.69	6.65	4.25	10.45	5.99
Asia	MAM	1.69	2.02	4.96	3.54	8.32	4.69
50N–67.5N	JJA	1.69	1.88	4.20	3.13	6.94	4.00
40E–170W	SON	2.24	2.15	5.30	3.68	8.29	4.98
Central	DJF	1.82	1.52	3.93	2.60	6.22	3.44
Asia	MAM	1.53	1.52	3.71	2.58	6.24	3.42
30N–50N	JJA	1.86	1.89	4.42	3.12	7.50	4.10
40E–75E	SON	1.72	1.54	3.96	2.74	6.44	3.72
West	DJF	1.26	1.06	3.1	2.0	5.1	2.80
Asia	MAM	1.29	1.24	3.2	2.2	5.6	3.0
12N–42N	JJA	1.55	1.53	3.7	2.5	6.3	2.7
27E–63E	SON	1.48	1.35	3.6	2.2	5.7	3.2
Tibetan	DJF	2.05	1.60	4.44	2.97	7.62	4.09
Plateau	MAM	2.00	1.71	4.42	2.92	7.35	3.95
30N–50N	JJA	1.74	1.72	3.74	2.92	7.20	3.94
75E–100E	SON	1.58	1.49	3.93	2.74	6.77	3.73
East	DJF	1.82	1.50	4.18	2.81	6.95	3.88
Asia	MAM	1.61	1.50	3.81	2.67	6.41	3.69
20N–50N	JJA	1.35	1.31	3.18	2.43	5.48	3.00
100E–150E	SON	1.31	1.24	3.16	2.24	5.51	3.04
South	DJF	1.17	1.11	3.16	1.97	5.44	2.93
Asia	MAM	1.18	1.07	2.97	1.81	5.22	2.71
5N–30N	JJA	0.54	0.55	1.71	0.88	3.14	1.56
65E–100E	SON	0.78	0.83	2.41	1.49	4.19	2.17
Southeast	DJF	0.86	0.72	2.25	1.32	3.92	2.02
Asia	MAM	0.92	0.80	2.32	1.34	3.83	2.04
10S-20N	JJA	0.83	0.74	2.13	1.30	3.61	1.87
100E–150E	SON	0.85	0.75	1.32	1.32	3.72	1.90

DJF December-January-February, *MAM* March-April-May, *JJA* June-July-August, *SON* September-October-November

Table 10.2 Projected changes in precipitation for subregions of Asia under SRES A1F1 (highest future emission trajectory) and B1 (lowest future emission trajectory) pathways for three time slices namely 2020s, 2050s, and 2080s (Cruz et al. 2007)

		2010–2039		2040–2069		2070–2099	
Subregions	Season	A1F1	B1	A1F1	B1	A1F1	B1
North	DJF	16	14	35	22	59	29
Asia	MAM	10	10	25	19	43	25
50N–67.5N	JJA	4	6	9	8	15	10
40E–170W	SON	7	7	14	11	25	15
Central	DJF	5	1	8	4	10	6
Asia	MAM	3	−2	0	-2	−11	−10
30N–50N	JJA	1	−5	−7	−4	−13	−7
40E–75E	SON	4	0	3	0	1	0
West	DJF	−3	−4	−3	−5	−11	−4
Asia	MAM	−2	−8	−8	−9	−25	−11
12N–42N	JJA	13	5	13	20	32	13
27E–63E	SON	18	13	27	29	52	25
Tibetan	DJF	14	10	21	14	31	18
Plateau	MAM	7	6	15	10	19	14
30N–50N	JJA	4	4	6	8	9	7
75E–100E	SON	6	6	7	5	12	7
East	DJF	6	5	13	10	21	15
Asia	MAM	2	2	9	7	15	10
20N–50N	JJA	2	3	8	5	14	8
100E–150E	SON	0	1	4	2	11	4
South	DJF	−3	4	0	0	−16	−6
Asia	MAM	7	8	26	24	31	20
5N–30N	JJA	5	7	13	11	26	15
65E−100E	SON	1	3	8	6	26	10
Southeast	DJF	−1	1	2	4	6	4
Asia	MAM	0	0	3	3	12	5
10S–20N	JJA	−1	0	0	1	7	1
100E–150E	SON	-2	0	-1	1	7	2

DJF December-January-February, *MAM* March-April-May, *JJA* June-July-August, *SON* September-October-November

In China, Zhang et al. 2012 determined decreasing trends during the spring and autumn while increasing trends during the winter season.

10.4 Drought Conditions and Wet Spells

The understanding of precipitation behavior is important to analyze the drought conditions and wet spells for planning and efficient management of freshwater resources. The analyses of droughts and wet spells are particularly valuable for arid and semiarid regions due to their high spatial and temporal variability. The drought conditions refer to a period of precipitation deficit resulting in a shortage of water. On the other hand, wet spells refer to a period of intense precipitation resulting in surplus water and occasional floods in semiarid and arid regions (Almazroui et al. 2017). Thus, a better understanding of the process of drought and wet spells will lead to better management of agricultural and hydrological activities as well as raise the quality of life.

The drought can be divided into four categories such as 1. Meteorological 2. Agricultural 3. Hydrological, and 4. Socioeconomic droughts (Fig. 10.4). The direct significant reduction in total precipitation in relation to mean precipitation in each area at the scale of days, weeks, and months is called the meteorological drought. The meteorological drought further leads to other drought types. The deficiency of water from several weeks (6–9 months) for plant growth is called the agricultural drought (Brázdil et al. 2018). Behind the meteorological and agricultural droughts, the hydrological drought can be characterized by a shortage of water in streams, lacks, reservoirs, and aquifers and when negative effects of drought appear in the whole society then it can be characterized into socioeconomic drought.

Different studies in different parts of the world analyzed the occurrence of drought and its impacts on human life. It has been reported that severe drought events have been experienced repeatedly throughout history in the near east region. These drought events have a large number of social and environmental impacts, including mass migration, famines, and human deaths. From 1900 to 2004, drought has been ranked first among all the hazards in terms of number of people killed. During the same time, Africa and Asia among the continents can be ranked first based on the drought-affected people. On the country scale, droughts also affected many countries such as Syria, Jordan, Morocco, Iran, and Iraq. The durations of drought period are increasing in arid regions of the world, i.e., average global drought periods are 2 months/°C below 1.5 °C and 4.2 months/°C when approaching 3 °C. If the current warming rate continues, the water deficits will convert to fivefold size for most parts of Australia, southern America, Africa, southern Europe, Caribbean, and northwest china. A significant increase (5–10 times) in drought frequency was predicted in the Mediterranean basin due to warming of the earth (west and southern Asia, Africa, Oceania, and Central America). Similarly, Naumann et al. 2018 analyzed the increase in drought condition due to warming of the earth

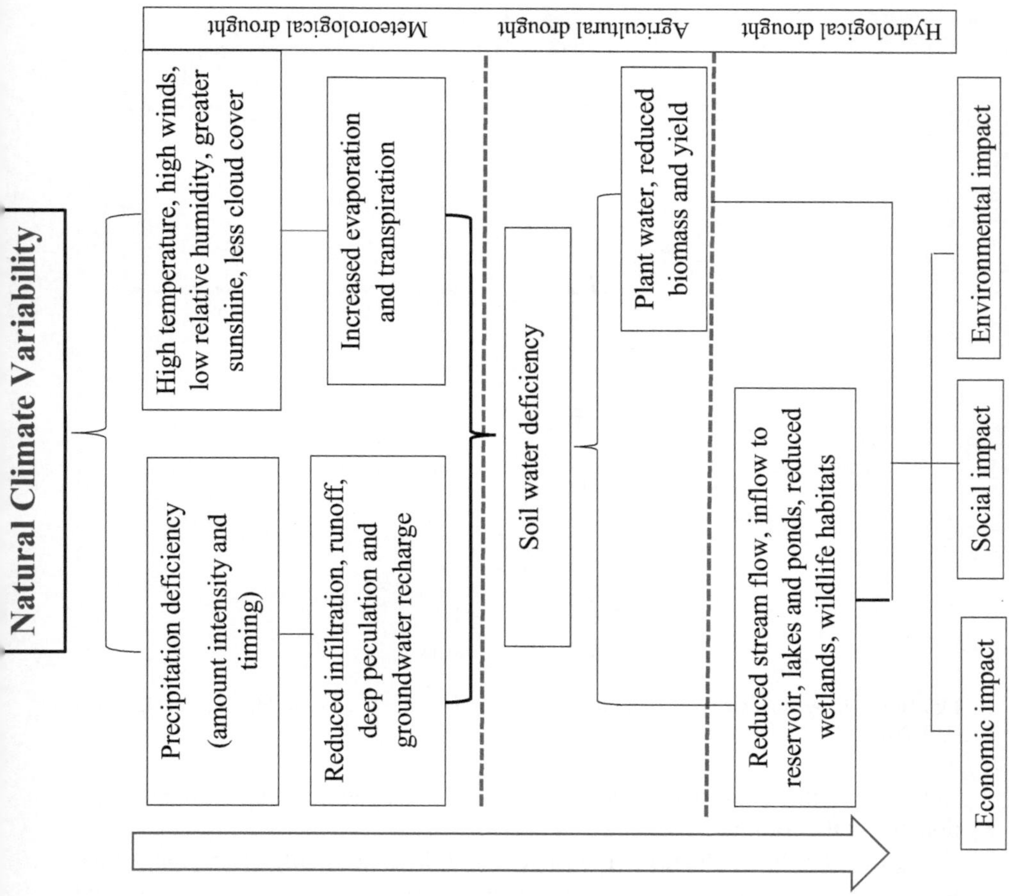

Fig. 10.4 Occurrence and impacts of droughts

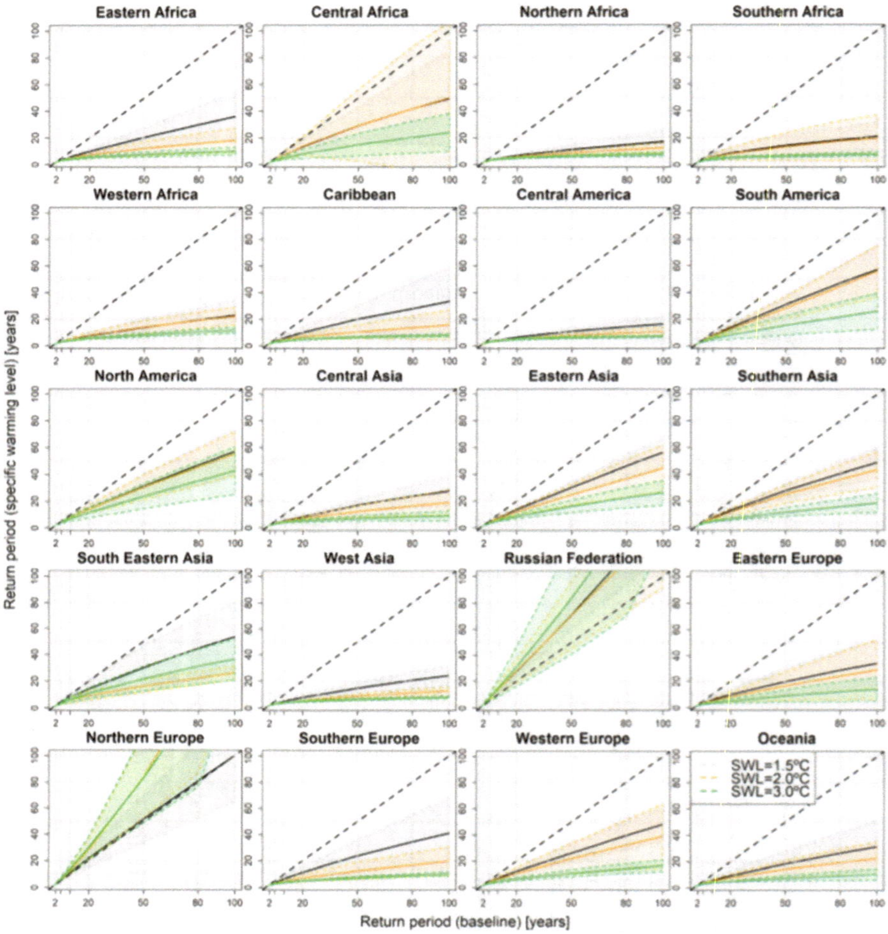

Fig. 10.5 Change in return period of drought in different regions of the world. (Source: Naumann et al. 2018)

and reported that 2/3 of the world's population is being affected by drought condition due to the warming.

Figure 10.5 shows change in return period of drought in different regions of the world. The dashed lines on the graphs (Fig. 10.5) showed no change curve and shaded areas represented the intermodal median absolute deviation for different regions. To a baseline of 100 years, about 15% of the global land may bear droughty events with 5 years interval. As a result, 5–10 times more frequent drought events were projected for most of Africa, Caribbean, central America, central and west Asia, northwest China, and Oceania. Similarly, strong and frequent drought events were predicted for southern, large part of eastern and western Europe, southern and central America. On the other hand, the reduction in drought frequency was observed in northern Europe and Russian Federation (Naumann et al. 2018). These climatic extremes (Drought and flood) are causing considerable economic losses and have a strong social impact on human lives in different parts of the world (Khaliq et al. 2012; Nasim et al. 2012a, b, c).

10.5 Impact of Climate Change on Runoff Change

Climate change and human activities are the most influential factors for global runoff variation including water cycle for any region (Xu et al. 2018). The climate change variation effects on precipitation pattern, which influences urban runoff in such a way that the design of water drainage networks is mainly based on the historical precipitation records and extreme statistics that are assumed to be constant for every year (Zahmatkesh et al. 2018). Moreover, heavy and lower rainfall events increase or decrease the risk of extreme floods and droughts due to climate change in arid or semiarid regions around the globe. Therefore, a better understanding of runoff changes and their potential impacts on water bodies due to climate change is necessary for efficient utilization of freshwater resources. So, Budyko relationship and H08 hydrological model were used across South Asia to simulate the climate elasticity of runoff. The climate elasticity of runoff is a percentage change in a runoff for 1 °C change in temperature or 1% change in potential evaporation or

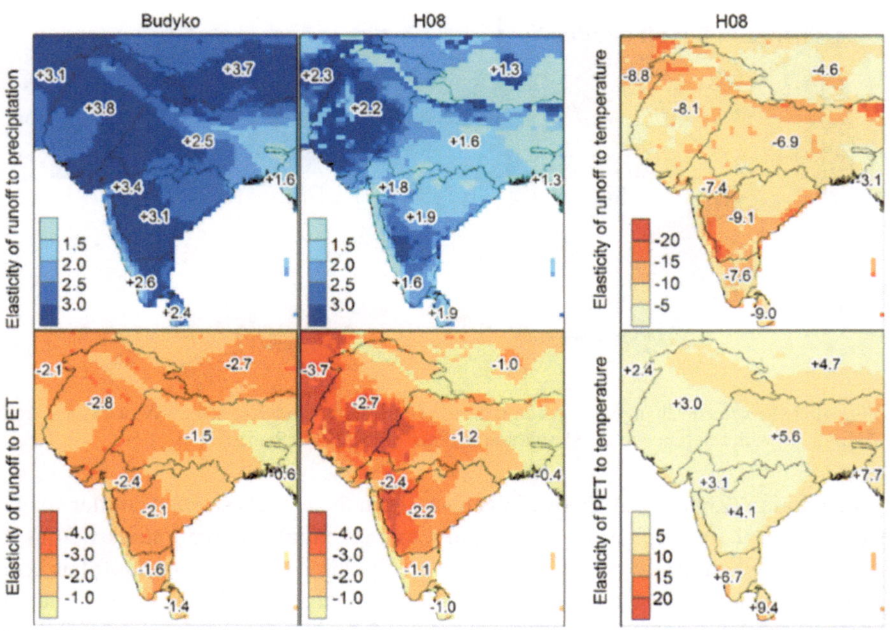

Fig. 10.6 Simulated precipitation, potential evaporation, and temperature elasticities across South Asia (Zheng et al. 2018)

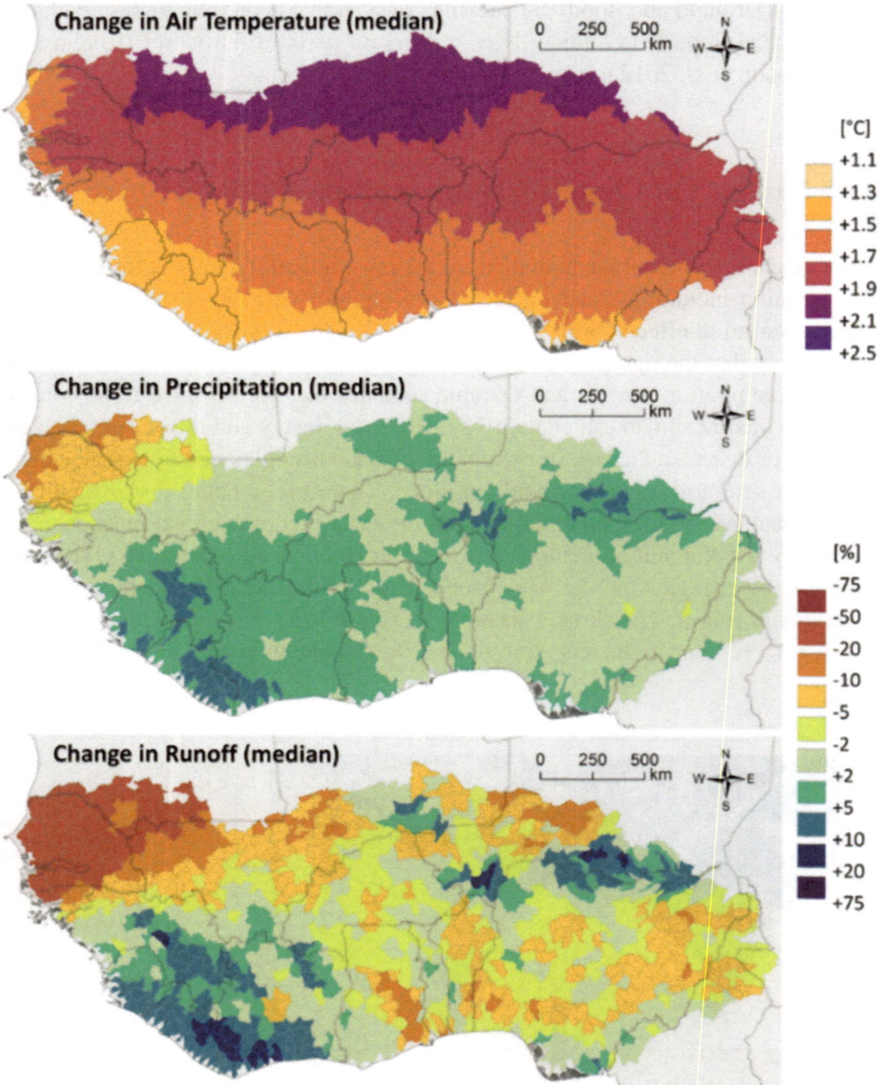

Fig. 10.7 Projected Change for precipitation, temperature, and Runoff in West Africa from 2046–2065 with reference to the simulation data for 1998–2014 (Stanzel et al. 2018)

precipitation (Zheng et al. 2018). In general, both the models showed similar spatial patterns for the precipitation, temperature, and potential evaporation elasticities of runoff. Based on the modeling results, the precipitation elasticity of runoff has been reported in the range of 1.5 or 2.0. This indicated that a 10% change in precipitation can bring a change (15–20%) in the runoff in wetter regions such as northeast, east, and west coast. Similarly, the precipitation elasticity of runoff was found greater than 2 in drier regions such as northwest and inland south. This showed that a 10% change in precipitation can bring 20% change in the runoff in these regions. It was also observed that the spatial pattern for temperature and potential evaporation of elasticities of runoff were found similar. The results showed that a 1 °C rise in temperature can increase the 5% potential evaporation. As a result of an increase in 5% potential evaporation, 5–10% decrease in runoff was observed (Fig. 10.6).

Figure 10.7 shows the projected change for input variables (precipitation and temperature) and for the output variable (runoff) of the water balance model from 2046 to 2065 with reference to the simulation data for 1998 to 2014 in West Africa (Median of 30 model runs). For temperature, the projected median change showed an increase of median change from north to south. The highest warming was projected for arid and semiarid areas in the Sahel region and lowest warming was projected for coastal regions in the middle of this century. For precipitation, most of the area of West Africa showed a median projection in the range of −2 to +2%. The projected median change in precipitation showed a slight increase in precipitation, with values upto 10% in the southwest region of West Africa. The median change in temperature and precipitation exhibits a change in the median runoff. The change in spatial pattern of median runoff is like the precipitation and temperature (Fig. 10.7). The strong warming with lower precipitation showed a strong decrease in the runoff with projected median runoff change −15% in the northwestern region. Similarly, lower warming with substantial precipitation increase showed a considerable increase in runoff in the southwestern coast having median projections >10% (Stanzel et al. 2018). Similarly, several studies have been conducted in other parts of the world to analyze the impact of climate change on runoff variation and reported that water discharge is mainly controlled by the precipitation, temperature, infiltration, and potential evaporation (Naik and Jay 2011; Yao et al. 2015; Xu et al. 2018; Farid et al. 2019).

10.6 Impact of Climate Change on Potential Evapotranspiration (PET)

The role of temperature in the atmosphere is important to quantify the potential change in evapotranspiration (ET) as a result of climate change. The widespread drying in earlier twenty-first century is a clear evidence of increasing PET rate due to climatic variations. The temperature of any region indicates both positive and negative effects on surface and groundwater resources. Global warming trend significantly affects the regional ET pattern, which threatens the utilization of surface water for sustainability (Aladejana et al. 2020). However, very little efforts have been noticed yet to analyze the magnitude and direction of predicted ET from water bodies. It has been studied that the higher temperature in the air increases the ET rate and reduces runoff and soil moisture contents. The minimum temperature in the air decreases the ET rate and increases runoff and soil moisture contents.

Additionally, due to climatic variation, small changes in precipitation and ET may harmfully affect the soil moisture and recharge in arid or semiarid regions around the globe (Ahzegbobor and Kehinde 2017).

10.7 Impact of Climate Change on Soil Moisture (SM) Change

The change of global precipitation and temperature have a significant effect on the SM. Availability of SM and variation in temperature play an important role in terrestrial climate and biochemical reactions in farming lands (Chadha et al. 2019). It has been reported that variability in SM may also affect the groundwater resources and influence atmospheric processes, i.e., cumulus convective rainfall (Thomas and Famiglietti 2019). It has been studied that spatial and temporal patterns in precipitation, ET, and surface water are due to the climatic variables, which affect SM. Other indirect factors that influence the SM contents are land use, land cover, soil texture, slope of the land, and other physiochemical properties of soil (Seneviratne et al. 2010).

10.8 Groundwater

The world's largest fundamental source of freshwater is groundwater (GW), which is a naturally built reservoir under the ground surface and is available for human beings and other sectors (Fig. 10.8). In most of the agrarian countries throughout the world, GW is considered as the foremost source of irrigation (Qazi et al. 2014. Farid et al. 2018). Globally, the 982 km^3/year withdrawal rate of GW was estimated to fulfill drinking and irrigation demands. From the withdrawal rate of GW, about 38% of the land is fortified with GW for irrigation and 60% of worldwide GW withdrawal is used for the agriculture sector. Moreover, from one-third part of all freshwater, GW provides 27% and 36% of water for industrial, and domestic activities all around the globe (Margat and Gun 2013). It is the part of the hydrological cycle and the amount of water which infiltrates into the soil will accumulate on the impermeable strata. The underground layers which both store and transmit groundwater to rivers and sea are named as aquifers. In these GW systems, the water moves slowly in motion from the recharge areas to the discharge areas. Tens or even hundreds of years may pass for this water passage through the hydrological cycle. Water that discharges from the aquifers performs two major roles. First, it benefits the environment by maintaining the river flows. Secondly, it provides water to meet the demands for drinking and industrial use. These aquifers deliver water to rivers during a period of no precipitation. These are natural underground reservoirs and very appropriate resources and can have a massive storage capacity, much greater than even the

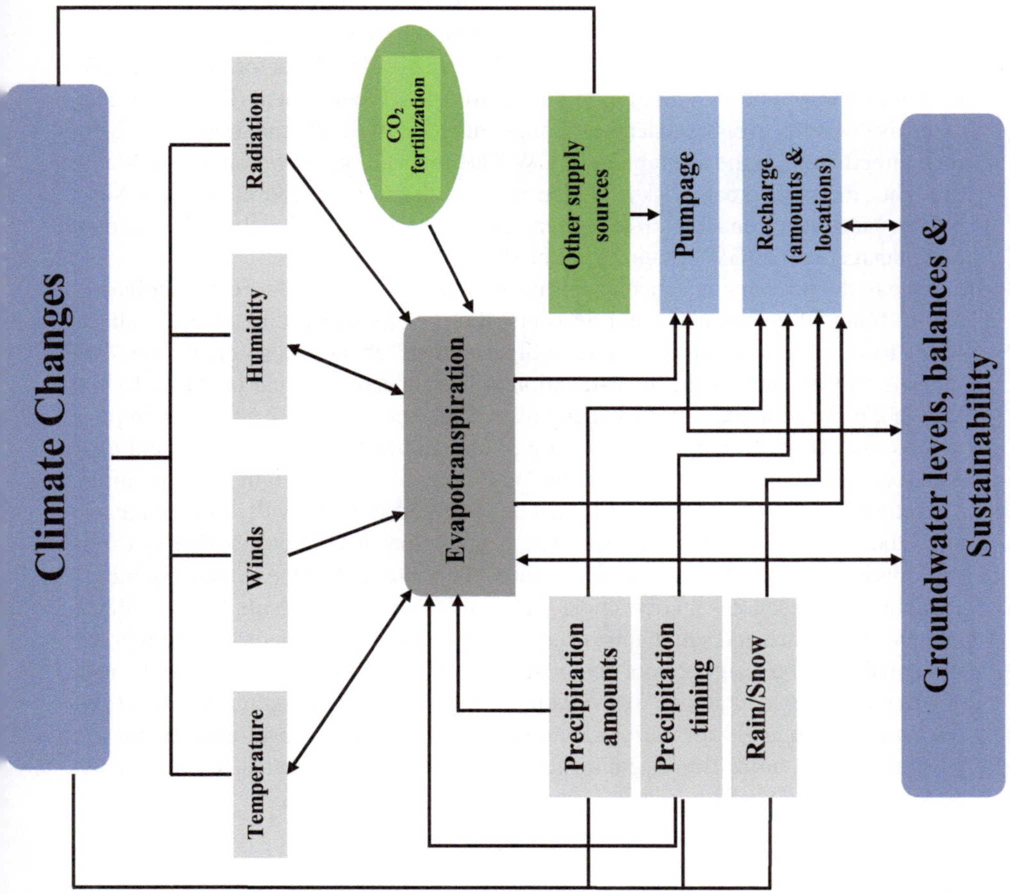

Fig. 10.8 Schematic diagram of groundwater balance and sustainability

largest man-made reservoirs. Due to greenhouse gases, climate change affects the groundwater structures such as it produces stresses to the groundwater recharge and its circulation in time and space. Different researchers have observed that the changing pattern of rainfall and increasing trend in temperature have significant effects on GW recharge and its availability. The increase in temperature means an increase in evaporation and plant transpiration rate. This will give exposure to soil erosion in arid and hot areas. All these pose negative impacts to the groundwater systems. The lowering in GW table due to increasing temperature and changes in rainfall patterns may decrease well efficiency, yield and increase initial pumping cost. The farmer community is extremely affected in those areas, where GW is used as a primary source for irrigation of agricultural lands. Chances of side effects for excessive GW abstraction vary with the situation of the hydrological environment (Fig. 10.8). The extensive weather events can lead to a longer duration of floods and droughts, which then directly affects the availability of GW. Due to the longer duration of floods and droughts, there is a greater risk in depletion of aquifer storage. Therefore, it is essential to quantify the contribution of influencing factors for mastering the GW dynamic and management of GW resources (Li et al. 2020).

Due to the poor maintenance and protection, mostly GW wells are contaminated and in times of emergencies and disasters, these wells will not be able to deliver water to meet freshwater demands. Indirect climate changes also create massive impacts on GW such as the intensification of human activities and land use. Using GW in a more strategic way in a changing climate becomes more and more important. Groundwater recharge is a function of climatic factors, topography, land use, and local geological formation. It has a direct relationship with precipitation. Therefore, it is a vital resource that must be protected so that it will sustain humans and other species for their survival. The degradation in GW occurs due to over-exploitation of GW, which extremely reduces the water table. Due to over-pumping, available GW resources and borehole yields reduce and impart side effects, which include subsidence and saline intrusion. Salinity occurs due to poor irrigation practice and due to the result of poor abstraction. Salinity does not reduce naturally, and it is the main threat to aquifer sustainability. Regardless of its vast importance, GW resources are in crisis for various regions mainly due to the over-exploitation of groundwater to fulfill the irrigation needs and food requirements for increasing population of the world.

10.8.1 Climate Change Impact in Relation to GW Abstraction

The occurrence of GW depletion for a given aquifer can be analyzed when the rate of recharge becomes lower as compared to the discharge rate of GW (Fig. 10.9). Indirectly, changing climatic scenarios in the form of anthropogenic practices of GW pumping, land use pattern, and growing urbanization causes GW depletion. It was analyzed that over-abstraction of GW influences the rise in sea level to some extent. Global warming also causes rise in sea level, if balancing of the system is not

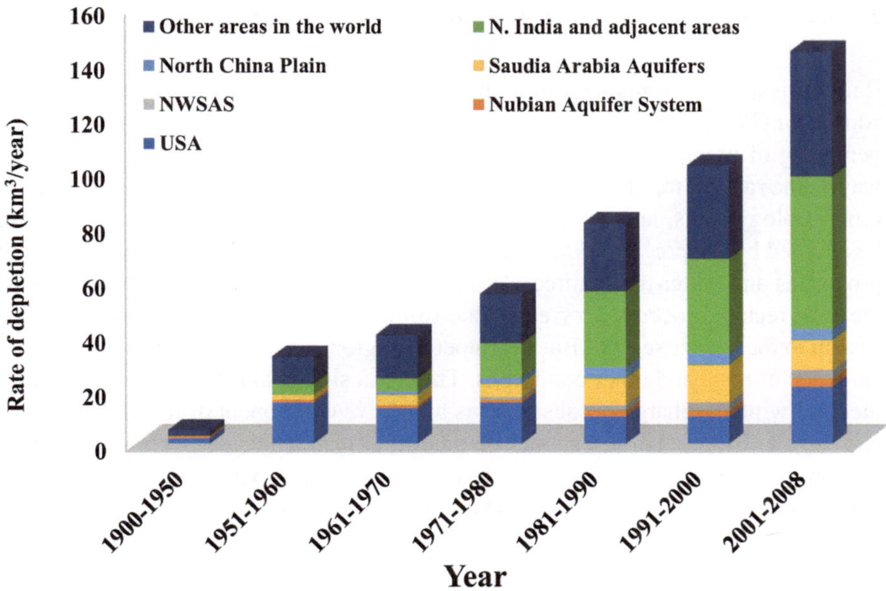

Fig. 10.9 Net groundwater depletion rates around the globe from1900–2008

achieved between the capacity to store water into reservoir and irrigation flow return (Green et al. 2011; Naz et al. 2013; Masood et al. 2013; Mubeen et al. 2013; Saeed et al. 2013; Sultana et al. 2013). The GW resources are being affected by climate change (Aladejana et al. 2020). The occurrence of soil degradation and variability of crop water demand have been influenced by climate change. As a result, the GW discharge/recharge is being affected directly or indirectly (Ho et al. 2016; Foster et al. 2018). A measurable increasing trend in GW depletion has been noticed with increasing population growth and advancement in technology in many regions, where GW resources tend to utilize at many extents. The technological advancement comprises the high-efficiency well pumps, which are used to fulfill water demand that eventually causes the depletion of GW resources (Shakoor et al. 2017; Beck et al. 2018). Furthermore, the quantitive and qualitative gap was found between the GW data collection and analysis which creates difficulty to decide the system for GW resource management). Climatic variations interrupt a GW discharge usually GW discharge increases in arid and semiarid regions. The runoff is a most sensitive hydrological parameter, when it combines with increasing aquifer discharge and winter precipitation, it results in risk of flooding under the wettest climatic condition (Croley and Luukkonen 2013). Correspondingly, an aquifer recharge, which is also a sensitive parameter for the hydrological cycle, decreases in context to dry climatic conditions, resulting in declining the level of GW (Thomas and Famiglietti 2019).

10.8.2 Intrinsic Effect of Climatic Variation on GW Recharge

Rainfall is a basic component of climatic change and it is considered as a primary source for GW recharge. The phenomenon of GW recharge can be analyzed through occurring of precipitation. The GW recharge will not always change in the same way as the rainfall trends change. The factors that affect the recharge are the aquifer type, geology, soils, land cover, and topographic relief. The recharge rate can also be affected by the precipitation type, intensity, and frequency. Also, variation in soil properties and water usage affects groundwater recharge. In high latitude regions, the GW recharge occurs very early due to the spring melt shifts from the spring season to the winter season. But in temperate regions, the annual recharge variation varies on climate and other conditions. There is a significant difference in the summer and winter recharge because GW is a massive component of the global hydrologic cycle. More attention should be paid to climate change effects on recharge. The necessary tools and data required from most of the environments are currently not enough to predict the recharge responses against diverse climate scenarios.

10.8.3 Climate Change Impact on GW Quality

The quality of GW for a given aquifer should be maintained because it provides fresh water to the living population and meet the demands of farming system for irrigation. The change in characteristics of GW at different depths due to increasing air temperature tends to increase the temperature of GW in a shallow aquifer. Likewise, in arid or semiarid regions, due to an increased rate of evapotranspiration, the GW becomes saline. In a region, where the intensity of rainfall is greater, the available pollutants, i.e., pesticides, heavy metals, and inorganic constituents will be washed efficiently on the surface of the soil. At those places, the quality of GW likely to be compromised via the recharging of the aquifer with these surface water bodies. Ultimately, intrusion of poor quality of water takes place in adjacent aquifer (Ho et al. 2016; Shakoor et al. 2017). It has been noticed that comparatively, very few studies have focused on process disturbing the quality of GW. More precisely, entering of water from other resources instead of significantly changing climatic conditions, will affect the GW quality. Moreover, the spatial and temporal variability of precipitation also affects the GW quality. Meanwhile, the rainfall is chemically diluted, which interacts with the dissolved material of an aquifer, causes to change the interaction time, which degrades the quality of GW resources (Ahzegbobor and Kehinde 2017).

Generally, the degradation of GW quality can be explained in three main groups.

1. Injection of contaminated water into the GW during recharging of an aquifer or movement of surface pollutants in the soil and mixing of these impurities in GW, due to deeper percolation.

2. Induction of changing in GW withdrawal produces the movement of water with varying quality. A very common description in this class is seawater intrusion, which increases the salinity of GW and underlies a layer of saline water on fresh GW layer.
3. The quality of GW changes at the mixing stage, where modification in GW regime and inflexing of different contaminants takes place. An example of this group of GW degradation is irrigation return flow in a saline aquifer, which is due to the effect of over fertilizer and pesticide application in the soil.

There is enormous evidence worldwide that explains the vulnerability of GW resources in relation to climate change. However, the effect on surface water resources with changing climatic scenario is recognized to more extent as compared to these effects on GW. The climate system and GW fluctuation, storage, movement, and recharging are not in equilibrium due to which there is a need to assess variation in GW behavior more precisely. The present GW resources in the world are in the position of vulnerability due to spatial and temporal patterns of rainfall, seawater intrusion, over-abstraction and withdrawal of water from an aquifer, rate of evapotranspiration, and interaction of GW surface during recharging or discharging process. In the assessment process, the information relating to climate changes in hydrological variables and their impacts on GW resources are still inadequate or uncertain for different regions in the world.

10.9 Conclusion

The assessment of climate change impacts on worldwide water resources may help for efficient management of these precious resources. Many research studies in different regions of the world analyzed and quantified the effects of climate change on every variable of the hydrological cycle. A better understanding and timely prediction of floods and drought should help to prepare the emergency plan for efficient management of land and water resources. However, a comprehensive plan should be prepared, and action should be taken according to the plan for optimum conservation of freshwater resources. To save the water bodies, a continuous monitoring system is required, and treatment of the domestic wastewater is also required in urban as well as in rural areas to ensure the safe drinking water quality for the global population as well as to save the environment. The relationship between climate change and the quality of GW in relation to GW storage should be established for such places, where contamination and extraction rate is high. More efficient and cost-effective techniques and tools are needed for accurate prediction and quantification of climate change impacts on water resources at the regional and basin-scale. Furthermore, the future water resources development projects, formation, implementation, and regulation of water policies should consider the projected climate change and uncertainties in projection of climate change scenario. This will help to minimize the adverse impact of climate change on water resources and consequently on agricultural production.

References

Ahmad I, Tang D, Wang TF (2015) Precipitation Trends over Time Using Mann-Kendall and Spearman's rho Tests in Swat River Basin, Pakistan. Advances in Meteorology, ID 431860.

Ahmad I, Zhang F, Tayyab M, Anjum MN, Zaman M, Liu J, Farid HU, Saddique Q (2018) Spatiotemporal analysis of precipitation variability in annual, seasonal and extreme values over upper Indus River basin. Atmospheric Research (2018): 134-146.

Ahzegbobor PA, Kehinde D (2017) Oyeyemi and Adebola Adeniran An Overview of the Potential Impacts of Climate Change on Groundwater Resources Journal of Informatics and Mathematical Sciences 9(2): 437–453.

Aladejana JA, Kalin RM, Sentenac P, Hassan I (2020) Assessing the Impact of Climate Change on Groundwater Quality of the Shallow Coastal Aquifer of Eastern Dahomey Basin, Southwestern Nigeria. Water, 12: 224. doi:https://doi.org/10.3390/w12010224

Almazroui M, Balkhair KS, Islam MN, Fen Z (2017) Climate Change Impact on Monthly Precipitation Wet and Dry Spells in Arid Regions: Case Study over Wadi Al-Lith Basin. Advances in Meteorology, Article ID 5132895. https://doi.org/10.1155/2017/5132895

Beck M, Sperlich A, Blank R, Meyer E, Binz R, Ernst M (2018) Increasing Energy Efficiency in Water Collection Systems by Submersible PMSM Well Pumps. Water, 10: 1310. doi:https://doi.org/10.3390/w10101310

Brázdil R, Kiss A, Luterbacher J, Nash DJ, Reznícková L (2018) Documentary data and the study of past droughts: a global state of the art. Climate of the Past 14: 1915–1960. https://doi.org/10.5194/cp-14-1915-2018

Chadha A, Singarayer KF, Chauhan BS, Long B, Jayasundera M (2019) Influence of soil moisture regimes on growth, photosynthetic capacity, leaf biochemistry and reproductive capabilities of the invasive agronomic weed; Lactuca serriola. Water, 12, 224. doi:https://doi.org/10.3390/w12010224

Croley TE, Luukkonen CI (2013) Potential effects of climate change on groundwater in Lansing, Michigan, Journal of the American Water Resources Association 39 (1): 149 – 163.

Cruz RV, Harasawa H, Lal M, Wu S, Anokhin Y, Punsalmaa B, Honda Y, Jafari M, Li C, Huu N (2007) Asia. Climate Change 2007: Impacts, Adaptation and Vulnerability. Contribution of Working Group II to the Fourth Assessment Report of the Intergovernmental Panel on Climate Change, M.L. Parry, O.F. Canziani, J.P. Palutikof, P.J. van der Linden and C.E. Hanson, Eds., Cambridge University Press, Cambridge, UK, 469-506. https://doi.org/10.1080/02626669109492507

Farid HU, Bakhsh A, Ali MU, Mahmood-Khan Z, Shakoor A, Ali I (2018) Field Investigation of Aquifer Storage Recovery (ASR) Technique to Recharge Groundwater—A case study in Punjab Province of Pakistan, Water Science and Technology: Water Supply, 18(1):71-83.

Farid HU, Khan ZM, Ahmad I, Shakoor A, Anjum MN, Iqbal MM, Mubeen M and Asghar M (2019) Estimation of infiltration models' parameters and their comparison to simulate the onsite soil infiltration characteristics. International Journal of Agricultural and Biological Engineering 12(3): 84-91.

Foster S, Pulido-Bosch A, Vallejos Á, Molina L, Llop A, MacDonald AM (2018) Impact of irrigated agriculture on groundwater-recharge salinity: a major sustainability concern in semi-arid regions. Hydrogeology Journal, 26: 2781-2791.

Green TR, Taniguchi M, Kooi H, Gurdak JJ, Allen DM, Hiscock KM, Treidel H, Aureli A, (2011) Beneath the surface of global change: Impacts of climate change on groundwater, Journal of Hydrology 405: 532–560.

Ho M, Parthasarathy V, Etienne E, Russo TA, Devineni N, Lall U (2016) America's water: Agricultural water demands and the response of groundwater, Geophysical Research Letter, 43: 7546-7555.

IPCC (2014) Climate Change 2014: Synthesis Report. Contribution of Working Groups I, II and III to the Fifth Assessment Report of the Intergovernmental Panel on Climate Change [Core Writing Team, R.K. Pachauri and L.A. Meyer (eds.)]. IPCC, Geneva, Switzerland, 151 pp.

Khaliq T, et al., (2012) Effect of Diverse Irrigation Regimes on Growth Parameters and Yield of Cotton under Faisalabad Conditions. Int. Poster J. Sci. Tech. 02 (3): 81-85.

Kumar M, Denis DM, Suryavanshi S, (2016) Long-term Climatic Trend Analysis of Giridih District, Jharkhand (India) using statistical approach, Model. Earth Syst. Environ. 2:116.

Li H, Lu Y, Zheng C, Zhang X, Zhou B, Wu J (2020) Seasonal and Inter-Annual Variability of Groundwater and Their Responses to Climate Change and Human Activities in Arid and Desert Areas: A Case Study in Yaoba Oasis, Northwest China. Water 12: 303; doi:https://doi.org/10.3390/w12010303.

Majewski W, Walczykiewicz T (2012) Sustainable management of water resources and hydrotechnical infrastructure in the light of forecasted climate changes (in Polish). Institute of Meteorology and Water Management – National Research Institute. Warsaw.

Malinowski Ł, Iwona S (2018) Impacts of Climate Change on Hydrological Regime and Water Resources Management of the Narew River in Poland. Journal of Ecological Engineering 19(4): 67–175.

Margat J, Gun J (2013) Groundwater around the World, CRC Press Balkema

Masood N, et al., (2013) Whitefly (Aleurolobus barodensis Mask) Spatio-Temporal Trends In Semi-Arid Agro-Ecological Zones of Pakistan. Pakistan Journal of Agricultural Sciences, 50(1), 69-74

Mishra AK (2019) Quantifying the impact of global warming on precipitation patterns in India. Meteorol Appl. 26:153–160. https://doi.org/10.1002/met.1749

Mubeen, M., et al., 2013. Effect of Growth Stage-Based Irrigation Schedules on Biomass Accumulation and Resource Use Efficiency of Wheat Cultivars. American Journal of Plant Sciences, 04: 1435-1442.

Naik PK, Jay DA (2011) Distinguishing human and climate influences on the Columbia River: changes in mean flow and sediment transport. Journal of Hydrology 404(3-4): 259–277.

Nasim, W., et al. (2012a) Effect of nitrogen on growth and yield of sunflower under semiarid conditions of Pakistan. Pakistan Journal of Botany, 44(2): 639-648

Nasim, W., et al. (2012b) Wheat productivity in arid and semi arid environment of Pakistan using crop simulation model. Int. Poster J. Sci. Tech. 02 (1): 28-35.

Nasim W, et al. (2012c) Effect of nitrogen on yield and oil quality of sunflower (Helianthus annuus L.) hybrids under sub humid conditions of Pakistan. American Journal of Plant Sciences, 3: 243-251.

Naumann G, Alfieri L, Wyser K, Mentaschi L, Betts RA, Carrao H, Spinoni J, Vogt J, Feyen L (2018) Global Changes in Drought Conditions Under Different Levels of Warming. Geophysical Research Letter https://doi.org/10.1002/2017GL076521.

Naz I, et al. (2013) Effect of different Fungicides on the incidence of Maize pathogen Helminthosporium maydis. Jokull Journal, 63. 196-207

Nguyen P, Thorstensen A, Sorooshian S, Hsu K, Aghakouchak A, Ashouri H, Tran H, Braithwaite D (2018) Global Precipitation Trends across Spatial Scales Using Satellite Observations. American Meteorological Society BAMS 689-697.

Qazi MA, Khattak MA, Khan MSA, Chaudhry MN, Mahmood K, Akhter B, Iqbal N, Ilyas S, Ali UA (2014) Spatial distribution of heavy metals in ground water of Sheikhupura District Punjab, Pakistan. Journal of Agricultural Research 52(1): 99–110.

Sadowski M, Romańczak A, Dynakowska M, Kalinowska A, Siwiec E (2013) Sixth Government Report and the first biennial report for the Conference of the Parties to the United Nations Framework Convention on Climate Change. Ministry of the Environment (in Polish). Warsaw.

Saeed HS, et al. (2013) Allelopathic Potential Assessment of Jaman (Syzygium cumini L.) on Wheat. Int. Poster J. Sci. Tech. 3 (1): 09-14.

Seneviratne SI, Corti T, Davin EL, Hirschi M, Jaeger EB, Lehner I, Orlowsky B, Teuling AJ (2010) Investigating soil moisture-climate interactions in a changing climate – a review, Earth-Science Review 99 (3-4):125 – 161

Shakoor A, Khan ZM, Arshad M, Farid HU, Sultan M, Azmat M, Shahid MA, Hussain Z (2017) Regional Groundwater Quality Management through Hydrogeological Modeling in LCC, West Faisalabad, Pakistan. Journal of Chemistry, Article ID 2041648. https://doi.org/10.1155/2017/2041648

Stagl J, Mayr E, Koch H, Hattermann FF, Huang S (2014) Effects of Climate Change on the Hydrological Cycle in Central and Eastern Europe. In: Rannow S., Neubert M. (eds) Managing Protected Areas in Central and Eastern Europe Under Climate Change. Advances in Global Change Research, 58.

Stanzel P, Kling H, Bauer H (2018) Climate change impact on West African rivers under an ensemble of CORDEX climate projections. Climate Services, 11: 36-48.

Sultana SR, et al. (2013) Productivity of some Maize Based Intercropping Systems under Different Planting Geometries. Thai Journal of Agricultural Sciences, 46(2): 65-70.

Thomas F, Famiglietti JS (2019) Identifying Climate-Induced Groundwater Depletion in GRACE Observations Brian. Scientific Reports 9:4124. https://doi.org/10.1038/s41598-019-40155-y

Xu Y, Wang S, Bai X, Shu D, Tian Y (2018) Runoff response to climate change and human activities in a typical karst watershed, SW China. PLOS ONE 13(3): 0193073. https://doi.org/10.1371/journal.pone.0193073.

Yao H, Shi C, Shao W, Bai J, and Yang H (2015) Impacts of Climate Change and Human Activities on Runoff and Sediment Load of the Xiliugou Basin in the Upper Yellow River. Advances in Meteorology. Article ID 481713. https://doi.org/10.1155/2015/481713

Zaffar, M., et al., 2014. Biological assays of plant extract from Araucaria columnaris and Cycas revoluta. Journal of Food Agriculture and Environment, 12 (1): 128-131.

Zahmatkesh Z, Karamouz M, Goharian E, Masce S, Steven J, Burian SJ (2018) Analysis of the Effects of Climate Change on Urban Storm Water Runoff Using Statistically Downscaled Precipitation Data and a Change Factor Approach. Journal of Hydrologic Engineering 20(7): 1-18. https://doi.org/10.1061/(ASCE)HE.1943-5584.0001064

Zhang Q, Singh VP, Peng J, Chen YD, Li J, (2012). Spatial-temporal changes of precipitation structure across the Pearl River basin, China, Journal of Water Resources Research, 440(11):113–122.

Zheng H, Chiew HSF, Charles S, Podgera G (2018) Future climate and runoff projections across South Asia from CMIP5 global climate models and hydrological modelling. Journal of Hydrology: Regional Studies 18: 92-109.

Chapter 11
Water Management in Era of Climate Change

Hamid Nawaz, Nazim Hussain, Muhammad Adnan Shahid, Naeem Khan, Azra Yasmeen, Hafiz Waqar Ahmad, Shah Fahad, Muhammad Rafay, and Wajid Nasim

Abstract Water is a vital and foremost resource that sustains the life at maximum survival level on the earth. Unfortunately, the recent circumstances and sudden temperature fluctuation drastically reduce the availability of usable water especially in the urban cities of many countries. In this way, higher risk of livelihood and negative economic growth under shortage of water become a severe threat to food security. This chapter suggests and recommends the potential water management issues and technologies for mitigating its related problems. The chapter includes the integrated approaches to managing water and climate change, transboundary water management, nexus consideration, water, sanitation, and hygiene, water and health, water and agriculture, water and energy, water and ecosystem, saving water technologies approach.

H. Nawaz (✉) · H. W. Ahmad · W. Nasim
Department of Agronomy, Faculty of Agriculture and Environment, The Islamia University of Bahawalpur, Bahawalpur, Pakistan
e-mail: wajid.nasim@iub.edu.pk

N. Hussain · A. Yasmeen
Department of Agronomy, Bahauddin Zakariya University of Bahawalpur, Bahawalpur, Pakistan

M. A. Shahid
Department of Agriculture, Nutrition and Food Systems, College of Life Sciences and Agriculture, University of New Hampshire, Durham, NH, USA

N. Khan
Department of Agronomy, Institute of Food and Agricultural Sciences, University of Florida, Gainesville, FL, USA

S. Fahad
Department of Agronomy, The University of Haripur, Haripur, Pakistan

M. Rafay
Department of Forestry, Faculty of Agriculture and Environment, The Islamia University of Bahawalpur, Bahawalpur, Pakistan

W. N. Jatoi et al. (eds.), *Building Climate Resilience in Agriculture*,
https://doi.org/10.1007/978-3-030-79408-8_11

Keywords Water management · Water supply system · Water saving technologies

11.1 Introduction

Based on this, since water is of universal importance and significant use, saving water is a major challenge (Shah et al. 2012). Most of the freshwater are found in lakes, rivers, and groundwater aquifers. The burgeoning population and irregular usage of water resources over the last decades have resulted in the imbalance of availability of water. In this condition, water crisis developed in the form of water stress (drought + flood), water quality deterioration, and wastage of freshwater resources (Shah et al. 2020).

The world's climate change crises are increasingly altering the water cycle. During the past decade, heat waves, water scarcity, waterlogging, floods, and other water-related events have led to >90% of natural disasters. The negative impact of climate change has influenced the freshwater system by modifying the streamflow and water quality. Various factors are involved in increasing this risk of climate change including temperature fluctuation, malnutrition, heavy metal pollution through rainfall, drought, flood, etc. (Qureshi et al. 2010). The regions having snowfall have been significantly affected by climate change and streamflow during each season has been continuously altered.

Figure 11.1 shows economic water security index for different countries which are very much prone to water stress. Thus, awareness of climate change in the case of insecurity for water-related problems in the ecosystem becomes the essential

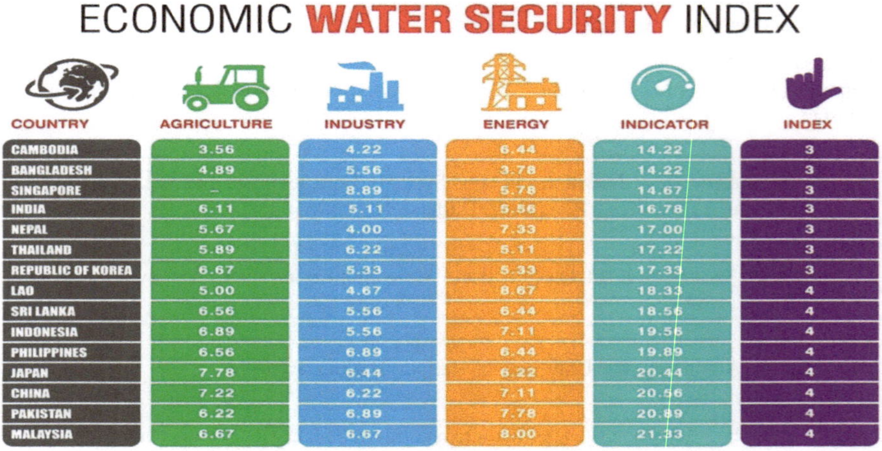

ECONOMIC WATER SECURITY INDEX

COUNTRY	AGRICULTURE	INDUSTRY	ENERGY	INDICATOR	INDEX
CAMBODIA	3.56	4.22	6.44	14.22	3
BANGLADESH	4.89	5.56	3.78	14.22	3
SINGAPORE	–	8.89	5.78	14.67	3
INDIA	6.11	5.11	5.56	16.78	3
NEPAL	5.67	4.00	7.33	17.00	3
THAILAND	5.89	6.22	5.11	17.22	3
REPUBLIC OF KOREA	6.67	5.33	5.33	17.33	3
LAO	5.00	4.67	8.67	18.33	4
SRI LANKA	6.56	5.56	6.44	18.56	4
INDONESIA	6.89	5.56	7.11	19.56	4
PHILIPPINES	6.56	6.89	6.44	19.89	4
JAPAN	7.78	6.44	6.22	20.44	4
CHINA	7.22	6.22	7.11	20.56	4
PAKISTAN	6.22	6.89	7.78	20.89	4
MALAYSIA	6.67	6.67	8.00	21.33	4

Fig. 11.1 Economic water security index (Awais et al. 2017); (Thanks to main author of this image as it was taken from Internet/Google)

target of nations all over the world for nourishing the society's health, balance of nutrients, energy stability, ecosystem, and biodiversity conservation (Qureshi et al. 2010; Khan et al. 2012). In this case, climate change represents a profound threat and an unprecedented opportunity to invest in transforming water governance and management system, so humanity can thrive in an increasingly uncertain and variable future. With regard to intergenerational justice, the global climate crises raise particularly pressing issues, and young people around the world are using their knowledge for innovations to offer solutions, raise awareness, protest for their rights, to address and combat the climate crises (Khan et al. 2012).

This chapter provides an overview of water application approaches and management techniques for sustaining the water resources.

11.2 Integrated Perspectives for Water Managements Under Climatic Change

Integrated management of water resources should be synchronized across crosscutting, sectoral, political, and spatial boundaries (Hamilton et al. 2015; Hammad et al. 2018). There are some modified aspects and outlooks for integrated water managements as given below;

11.2.1 Transboundary Water Management

Globally, various countries dividend water borders across, ponds, waterways and aquifers, river basins, and water-stream cross sovereign borders. It has been estimated that cross-border basins contain about 60% of global freshwater flows and 40% of domestic use in the world's population (Gleeson et al. 2012). The advanced level of transboundary cooperation for alleviation and transformation become a critical importance to avoid the negative impacts of one-sided actions. On the other hand, it also provided reliefs in minimizing the conflicts for the knowledge gaps and enhances regional peace and cohesion along with well-economic progress (Wada et al. 2012).

Maximum transboundary water-sharing contracts are comparatively rigid and motivate in the climate variability. In this climate change condition, the impressive and wider cooperation for climatic adapting approaches for managing water in rivers and aquifers are strongly required among the neighboring countries and states (Shah 2009).

Emphasis should be placed on appropriate cross-border cooperation at all stages of climate change adaptation including information gathering and sharing (Strong Decision Support Systems), joint development of vulnerability assessments, flexible institutional water management and adoption, and development of strategic levels (Jabran et al. 2017a). Different organizations may be able to raise money for adaptation actions in each country. Arrangements for data and information sharing, as well as joint monitoring, are prerequisites for successful cross-border cooperation in the era of climate change (Pittock 2011).

11.2.2 Nexus Attentions for Water Managements

The complicated relationship between the water utilization sectors and economically related lines as energy, agricultural food items, landscaping, environmental ecosystem, and urban structures. It is depicted that rapid increase in urbanization may cause the intensive consumption of land, food, and water application in the world. The things are also debatable which lead to encounter problems related to housing and agricultural productions. So, the dire need to improve the overall water bounciness under this severe threat for living organisms, social attributes, economically, and naturally occurring systems.

This form of sudden climate change addresses the Governance to change the opportunities related to water management systems, approaches, infrastructures, and mechanisms to finance the inherently cross-sectoral nature of water. The portion belongs to transformation added to the top-to-down Governance strategies with comprehensive bottom-to-up community-based welfare making decisions which become the drastic factor for climate (Brunner and Simmons 2012). In these circumstances, these risk-based resilient water management practices for newly developing global community should be practiced.

11.3 Water and Ecosystems

Various studies have been reported by scientists to diminish the disaster of climate changes and also described the different sustainable and ecosystem strategies. In this regard, carbon sequestration, peatlands ecosystem, nutrition managements, vertical (genetic modifications) and horizontal (drought, floods, temperature, air pollution, and sea purification) approaches as per the chronically natural occurring systems (Triana et al. 2010).

Similarly, freshwater consumption always pretended to be a serious problem around the world under facing the reasons belongs to rapidly increasing in the urbanization, decreasing the agricultural cropping intensification and soil degradation, and mining in the groundwater. The interactive impact of these aspects changes carbon sink into carbon sources under water deficit conditions which leads to the unbalancing of the entire ecosystem process (Zhang et al. 2013).

For this, there is a need to increase spending on managing the community-based natural occurring resources, providing green jobs, and Governance opportunities to avoid the negative circumstances for freshwater ecosystem. Governance should make such kind of ecosystem protection measures which are able to entertain all levels of programs, policies, and workshops. One of the best example for maintaining the ecosystem is transboundary basins system worldwide.

11.4 Water and Energy

All the energy production units highly depended directly or indirectly on the water resources, and even water treatment also required energy during the abstraction, transportation vice versa (Hammad et al. 2017). Under these circumstances of rapid population and economic growth, energy and water demand are simultaneously increased (Brodaric and Booth 2010). It is estimated that global energy petition is expected to rise by ≥27% during 2017–2040 as per water demand is ≥55% (Gillani et al. 2017). Moreover, changes in the weather and increasing trend of hydrological variability, such as desalination of long-distance water transport, will likely lead to greater resilience to energy-intensive water supply options (Jabran et al. 2017a, b).

Water and energy sources especially renewable energy represent a significant role in becoming the greater part of supply energy through lower water footmark than their carbon-based substitutions (Hogan et al. 2012). So, to meet these water and energy demands, healthy investments is required to compensate its loads by adopting such strategies as solar photovoltaic, small scale hydropower and wind protection measures, sustainable plannings, suitable structure for applying the rules, and regulations for managing the water and energy sectors both at the national and international level to ensure the safe future perspectives (Dahlhaus et al. 2012).

11.5 Water, Sanitation, and Hygiene

The fluctuation in climate change makes a significantly negative impact on the accessibility and excellence of drinking water. It may lead to the practical performance of water availability, sanitization, and hygiene-related activities. The severe effect of drought and flood create the water flow in the populated areas as well as suburban agricultural lands. In that case, services related to safe and clean water convenience should be adapted in the form of infrastructure and proper policies. It is also necessary to ensure the water recycle processes for public usage in terms of robust WASH systems in the areas where climate change is highly impacted by routine life.

The findings of various research scientists are resulted that this kind of climate change may highly be communicated to the local level. To compensate for these water, sanitation, and hygiene problems, different projects, workshops, training seminars, and international conferences must be conducted in this regard (Schoups and Vrugt 2010).

11.6 Water and Agriculture

Temperature fluctuation in the climate changes in the pattern, intensity, and frequency of rainfall water which may negatively affect the agricultural crop production and ultimately in the food consumption at the end. While food shocks and stress affect everyone, women, indigenous peoples, subsistence farmers, pastoralists, and fishermen are disproportionately transformed.

Climate change management strategies for water resources is significantly at the top priority to avoid the risk of food security issues. For this matter, various mitigation approaches including vertical and horizontal must be increased by adapting the usage of solar pumps for agricultural water supply for crop production, soil conservation activities, waste food recycling, post-harvesting techniques, implementing biofuels, organic agriculture, and biogas. In all the conditions, utilization of food system may improve in the since of nutritional point of view by using the effective usage of land, water, soil and energy sectors (Al Khamisia et al. 2013).

11.7 Saving Water Technologies Approach

The available options and conservation technologies, hereafter we propose the following technologies to be used in order to improve the water management and solve the water issues worldwide. The available options are seepage from canal, skimmed groundwater applications, losses at watercourses level, and increasing the production of water. The conservation technologies for irrigated areas are improvement of watercourses, improvement in farm layout, laser land leveling, improved methods of irrigation, improved cropping patterns, skimming wells, and reuse of wastewater. The saving water options are discussed below.

11.7.1 Irrigation Water as Safe Sources of Drinking Water

Every developing country, in an effort to supply drinking water to their communities, have focused on digging deep tube wells and installing hand pumps to operating bacteriological safe groundwater (Mubeen et al. 2019; Rasool et al. 2020; Ali et al. 2020). But in many of these countries, underground water is not an option because of high AS, F, Fe, or salt levels. Irrigation water is usually the only water in the vast areas as groundwater is too salty for human use, villagers convert canal irrigation water into small mud reservoirs called diggis to meet their domestic needs. In order to improve both quantity and quality, it is suggested that making it possible for people to pump irrigation water into a large storage tank in their houses thereby ensuring a continuous supply of water for drinking, cleaning, and hygiene would greatly reduce the incidence of many diseases, especially when combined with a complain to promote better hygiene. In addition, many pilot projects have been started to develop and check possible interventions including chlorination of irrigation water, low cast water storage containers, and a sewerage scheme (Shah et al. 2012).

11.7.2 Improvement in the Water Management System Efficiency

The efficiency of water management in irrigated and non-irrigated region is not appreciated in the world. Water conversion is therefore critical to achieve the needs (quantity and quality) of all water subsectors (Rasool et al. 2017; Saud et al. 2017; Zia et al. 2017). A lot of exercises have been practiced for water conversion in irrigated land for agriculture purpose which make the best usage of water storage sector. On the other hand, various modern approaches for irrigation effectiveness have resulted in a clear difference in the form of reduced water practices, and utilizations for both the urban and rural domestic supplies (Werner et al. 2013; Awais et al. 2017; Nasim et al. 2016; Rasool et al. 2016; Jabran et al. 2016).

11.7.3 Groundwater Management

Usage of groundwater has reached the maximum limit in most parts of the country. The groundwater tables in most of the freshwater areas are falling and the potential of further groundwater exploitation is there for very limited. Most of the farmers do not have enough awareness of crop water requirements, and crops are irrigated on the basis of physical appearance or as per water availability. For this, awareness seminars and conferences with farmers at their doorsteps are encouraged to make the water conservation measures. Improved and advance cultural practices such as accurate land leveling, tillage, bed, and furrow planting can help a good deal for on-farm water saving (Atta et al. 2017; Khan et al. 2018). Farmers should also be encouraged to use water-efficient irrigation technologies such as drip irrigation, sprinkler irrigation, and other modern irrigation systems. These techniques have been very successful in saving considerable amounts of irrigation water. Surface irrigation systems have been modified to reach high application efficiencies in the United States. Modern agricultural irrigation can reduce water use by between 20 and 30%. The most reliable option includes:

1. Agricultural water should be properly charged for making its importance in utilization.
2. Develop computerize system that can manage schedules of irrigation water
3. Using modern irrigation system and techniques

Fig. 11.2 Various groundwater management in different crops (Al Khamisia et al. 2013); Thanks to Dr. Bashir Ahmad, Climate Change & Geo-Informatics Program Leader, Climate Change, Alternate Energy and Water Resources Institute (CAEWRI), National Agricultural Research Center (NARC), NIH, Shehzad Town, Park Road, Islamabad, Pakistan, for support and help

4. All the agronomic cultural practices especially land preparation should be improved (Fig. 11.2).

11.7.4 Water Harvesting Technique

Farmers in the rain-fed areas should be trained to used water saving and watershed management, including more water storage structures, both small and large purposes. Connecting stormwater drain lines to tanks and rivers can greatly improve the water position of the city with little effort and maintenance (Fig. 11.3).

Fig. 11.3 Various irrigation techniques in various crops (Wada et al. 2012)

11.7.5 Managing Water Wastage by Using Pipelines

Various practices have been worked to minimize the water wastage by using advanced urban pipeline system through replacing and maintenance of pipelines under subjected frequent explosions and leaks. The municipalities in developing countries have succeeded in adopting the improved watering methods for public lawns and gardens, including the practice of nighttime sprinkling (Bhatti and Khan 2012).

11.7.6 Creating Natural Pressure in Open Channels of Irrigation

The regions that belong to paddy surfaces faced various problems related to farm-land consolidation, water leakage, and degraded leveling under the continuous use of superannuated irrigation canals. This canal's water impaired the permeability and disturbed the soil field capacity for crop rotations. In this case, preventive water leakage measures can be achieved by using the following aspects:

1. A natural pressure open channel can be constructed without any pressurization provided that the elevation difference between the water resource basin area and the paddy surface is 20 cm or more.
2. The natural pressure open channels can be erected at the level of the farming system as designed by engineers
3. These methods can be implemented with the agricultural machinery owned by large-scaled agricultural farming system. In addition, it contributes to the efficient utilization of areas with less drainage and the cultivation of abandoned farmlands
4. This method can improve the irrigation system

11.8 Conclusion

The world is blessed with countless water resources, but unfortunately lacks the planning to utilize efficiently. There still exists a scarcity of potable water in most remote areas and desert lands throughout the year. The study found wide gaps between the local people's needs, desires, and expectations and the government policies and possible solutions; between people's practices and historical and proposed institutions; and between local people's and policymakers' understanding of the issues. Water supply projects need to result in improved water services, but with solutions tailor-made to the local culture, together with local actors, rather than being imposed on them.

Acknowledgment The corresponding author (Wajid Nasim Jatoi) is highly thankful to Dr Dr. Bashir Ahmad, Director CEWRI, NARC, Pakistan Agricultural Research Council, Islamabad, Pakistan. Additionally, the support and cooperation given by all colleagues especially Dr. Muhammad Aown Sammar Raza (Chairman), Dr. Muhammad Adnan Bukhari, Dr. Muhammad Aurangzaib, Dr. Muhammad Latif, Dr. Muhammad Saqib, Dr. Muhammad Usman Bashir, Dr Rashid Iqbal, Dr Abdul Rehman, Dr Farhan Khalid, Dr Muhammad Asghar Shah, Dr Muhammad Shahzad, Dr Muhammad Usman Aslam and especially Dr. Muhammad Aown Sammar Raza, Chairman, Department of Agronomy, Faculty of Agriculture and Environment, The Islamia University of Bahawalpur, Pakistan, is highly acknowledged and commendable.

References

Al Khamisia SA, Prathapar SA, Ahmed M. 2013. Conjunctive use of reclaimed water and groundwater in crop rotations. Agric Water Manage 116:228–234

Ali S, Muhammad S, Khalid S, Natasha, SZ, Bakhat HF, Murtaza B, Farooq A, Akram M, Shah GM, Nasim W, Niazi NK. 2020. Assessment of arsenic exposure by drinking well water and associated carcinogenic risk in peri-urban areas of Vehari, Pakistan. Environmental Geochemistry and Health 42:121–133

Atta R, Abida F, Tangfu X,Waqar A, Sifat N, Oyebamiji A, Salar A, Wajid N. 2017. A review of global outlook on fluoride contamination in groundwater with prominence on the Pakistan current situation. Environmental Geochemistry and Health 40: 1265–1281

Awais, M, Aftab W, Wajid N, Ashfaq A, Muhammad FS, Muhammad ASR, Muhammad UB, Muhammad HR, Umer S, Jamshad H, Naveed A, Gerrit H. 2017. Modeling the water and nitrogen productivity of sunflower using OILCROP-SUN model in Pakistan. Field Crops Research 205:67–77

Bhatti AU, Khan MM. 2012. Soil management in mitigating the adverse effects of climate change. Soil and Environment 31:1–10.

Brodaric B, Booth N. 2010. OGC groundwater interoperability experiment final report. Open Geospatial Consortium, Groundwater Data Management, 44 pp. http://portal.opengeospatial.org/files/?artifact_id=43545&version=1

Brunner P, Simmons CT. 2012. HydroGeoSphere: a fully integrated, physically based hydrological model. Ground Water 50:170–176

Dahlhaus PG, MacLeod AD, Thompson HC. 2012. Federating hydrogeological data to visualise Victoria's groundwater. In: Lambert I, and Gordon AC (eds) 34th international geological congress: proceedings, 5–10 Aug 2012, Brisbane, Australian Geoscience Council, 592

Gillani, R.A, Nargis S, Sidra M, Fazal H, Wan SWN, Wajid N, Muhammad FHM, Abdul R, Hassan JC. 2017. Biosorption analysis of Agrobacterium tumefacians 12b3 against Cr (1 III) and Pb(II) by 2 kinetic modeling and equilibrium studies. Desalination and Water Treatment, 67: 206–214

Gleeson T, Wada Y, Bierkens MFP, Van Beek LPH. 2012. Water balance of global aquifers revealed by groundwater footprint. Nature 488:197–200

Hamilton S, El Sawah S, Guillaume JHA, Jakeman AJ. 2015. Integrated assessment and modelling: a review and synthesis of salient dimensions. Environmental Modelling & Software 64:215–229

Hammad, H.M., Wajid F, Farhat A, Shah F, Shafqat S, Wajid N, Hafiz FB. 2017. Environmental Science and Pollution Research 24:2549–2557

Hammad, HM. Farhat A, Shafqat S, Shah F, Artemi C, Wajid F, Chaves CB, Wajid N, Muhammad M, Hafiz FB. 2018. Offsetting Land Degradation through Nitrogen and Water Management during Maize Cultivation under Arid Conditions. Land Degradation and Development 29:1366–1375

Hogan A, Umbrich J, Harth A, Cyganiak R, Polleres A, Decker S. 2012. An empirical survey of linked data conformance. Web Semant Sci Serv Agents World Wide Web 14:14–44

Jabran K, Mubshar H, Shah F, Muhammad F, Ali AB, Hesham A, Nasim W. 2016. Economic assessment of different mulches in conventional and water-saving rice production systems. Environmental Science and Pollution Research, 23: 9156–9163

Jabran, K, Muhammad R, Mubshar H, Nasim W, Umar Z, Shah F, Bhagirath SC. 2017a. Water-saving technologies affect the grain characteristics and recovery of fine-grain rice cultivars in semi-arid environment. Environmental Science and Pollution Research, 24: 12971–12981

Jabran, K. Ehsan U, Nadeem A, Muhammad Y, Umar Z, Nasim W, Muhammad R, Tuba A, Muhammad FA, Mubshar H. 2017b. Growth and physiology of basmati rice under conventional and water-saving production systems. Archives of Agronomy and Soil Science, 63:1465–1476

Khan MA, Ahmed M, Hashmi HS. 2012. Review of available knowledge on land degradation in Pakistan. OASIS Country Report 3, ICARDA. Available at: https://appsicardaorg/asmx/Download File To Local?

Khan N, Asghari B, Muhammad AS, Wajid N, Ali B. 2018. Interaction between PGPR and PGR for water conservation and plant growth attributes under drought condition Biologia, 73:1083–1098

Mubeen M, Ashfaq A, Hafiz MH, Muhammad A, Hafiz UF, Mazhar S, Muhammad SD, Asad A, Amjed A, Shah F, Wajid N. 2019. Evaluating the climate change impacts on water requirements of cotton-wheat in semi-arid conditions using DSSAT model. Journal of Water and Climate Change 179

Nasim W, Hatem B, Ashfaq A, Habib-ur-Rahman M, Khawar J, Kalim U, Shah F, Muhammad S, Gerrit H. 2016. Modelling climate change impacts and adaptation strategies for sunflower in Punjab-Pakistan. Outlook on Agriculture 45:39–45.

Pittock J. 2011. National climate change policies and sustainable water management: conflicts and synergies. Ecology and Society 16:25

Qureshi AS, McCornick PG, Sarwar PG, Sharma BR. 2010. Challenges and prospects for sustainable management in the Indus Basin, Pakistan. Water Resources Management 24:1551–1569

Rasool A. Tangfu X, Abida F, Muhammad S, Sajid M, Salar IA, Shah F, Wajid N. 2016. Arsanic and heavy metal contaminations in the tube well water of Punjab, Pakistan and risk assessment: A case study. Ecological Engineering, 95: 90-100

Rasool A, Tangfu X, Waqar A, Sifat N, Oyebamiji A, Salar A, Wajid N. 2017. A review of global outlook on fluoride contamination in groundwater with prominence on the Pakistan current situation. Environmental Geochemistry and Health 40:1265–1281

Rasool A. Salar A, Waqar A, Wajid N. 2020. Quantification of Tl (I) and Tl (III) based on micro-column separation through ICP-MS in river sediment pore water. Environmental Science and Pollution Research 27:9686–9696

Saud S, Shah F, Chen Y, Muhammad ZI, Hafiz MH, Wajid N, Amanullah Jr, Muhammad A, Hesham A. 2017. Effects of Nitrogen Supply on Water Stress and Recovery Mechanisms in Kentucky Bluegrass Plants. Frontier in Plant Sciences. 983:1-18

Schoups G, Vrugt JA. 2010. A formal likelihood function for parameter and predictive inference of hydrologic models with correlated, heteroscedastic, and non-Gaussian errors. Water Resources Research 46:1–17

Shah T. 2009. Climate change and groundwater: India's opportunities for mitigation and adaptation. Environmental Research Letters 4:035005

Shah T, Giordano M, Mukherji A. 2012 Political economy of the energy-groundwater nexus in India: exploring issues and assessing policy options. Hydrogeol Journal 20:995–1006

Shah AH, Muhammad S, Sana K, Natasha, ZS, Hafiz FB, Behzad M, Amjad F, Muhammad A, Ghulam MS, Wajid N, Nabeel KN.2020. Assessment of arsenic exposure by drinking well water and associated carcinogenic risk in peri-urban areas of Vehari, Pakistan, Environmental Geochemistry and Health. 42:121–133

Triana E, Gates T, badie J. 2010. River GeoDSS for agro environmental enhancement of Colorado's Lower Arkansas River Basin. II: evaluation of strategies. Journal of Water Resources Planning and Management 136:190–200.

Wada Y, van Beek LPH, Bierkens MFP. 2012. Nonsustainable groundwater sustaining irrigation: a global assessment. Water Resource Research 48:W00L06.

Werner AD, Zhang Q, Xue L, Smerdon BD, Li X, Zhu X, Yu L, Li L. 2013. An initial inventory and indexation of groundwater mega-depletion cases. Water Resource Management 27:507–533

Zhang SH, Xia ZX, Wang TW. 2013. A real-time interactive simulation framework for watershed decision making using numerical models and virtual environment. J Hydrol 493:95–104

Zia Z, Saqib ZA, Shah GM, Fahad S, Ashraf MR, Hammad HM, Naseem W, Shahid M. 2017. Effect of water management and silicon on germination, growth, phosphorus and arsenic uptake in rice. Ecotoxicology and Environmental Safety 144:11–18

Chapter 12
Climate Change-Induced Irrigation Water Problems and Resolution Strategies: A Case Study

Muhammad Mubeen, Fahd Rasul, Ashfaq Ahmad, Syed Aftab Wajid, Tasneem Khaliq, Hafiz Mohkum Hammad, Asad Amin, Amjed Ali, Syeda Refat Sultana, Shah Fahad, Khizer Amanet, Musaddiq Ali, Muhammad Sami Ul Din, and Wajid Nasim

Abstract Water requirement is rising day by day, however, the possibilities of improvement of water assets or preserving their sustainable use are failing in Pakistan. Climate Change affects water accessibility during critical times. Irregular droughts, as well as floods caused by climate change, pose a serious threat to many

M. Mubeen (✉) · H. M. Hammad · A. Amin · S. R. Sultana · K. Amanet · M. Ali · M. S. U. Din
Department of Environmental Sciences, COMSATS University Islamabad, Vehari, Pakistan
e-mail: muhammadmubeen@cuivehari.edu.pk

F. Rasul
Washington State University, IAREC, Prosser, WA, USA

Agro-Climatology Laboratory, Department of Agronomy, University of Agriculture, Faisalabad, Pakistan

A. Ahmad
US.-Pakistan Centre for Advanced Studies in Agriculture and Food Security, University of Agriculture, Faisalabad, Pakistan

S. A. Wajid · T. Khaliq
Agro-Climatology Laboratory, Department of Agronomy, University of Agriculture, Faisalabad, Pakistan

A. Ali
University College of Agriculture, University of Sargodha, Sargodha, Pakistan

S. Fahad
College of Plant Science and Technology, Huazhong Agricultural University, Wuhan, Hubei, China

Department of Agronomy, The University of Haripur, Haripur, Pakistan

W. Nasim
Department of Agronomy, Faculty of Agriculture and Environment Sciences, The Islamia University of Bahawalpur (IUB), Bahawalpur, Pakistan

W. N. Jatoi et al. (eds.), *Building Climate Resilience in Agriculture*,
https://doi.org/10.1007/978-3-030-79408-8_12

developing countries including Pakistan which are not capable to develop such infrastructure needed to combat the adverse impacts of floods and droughts. The recent availability of water per capita is about 1066 M^3 which puts the country in the category of "high water stress." Most of the impending water crisis must be overcome through serious political decisions and sustainable water usage. Suitable on-farm water management is required to rise water proficiency for crops and to confirm effective and sustainable crop production. The government should promote agricultural water management practices, strengthen the Department of Autonomous Water Management system as well as encourage the use of computer technologies, such as plant growth models, to combat the effects of problems caused by irrigation in the short and long term.

Keywords Water requirement · Floods · Droughts · Computer technologies

Abbreviations

ADB	Asian Development Bank
CEWRE	Centre of Excellence in Water Resources Engineering
DSSAT	Decision Support System for Agro-technology Transfer
ESCAP	United Nations Economic and Social Commission for Asia and the Pacific
FAO	Food and Agriculture Organization of the United Nations
GOP	Government of Pakistan
IPCC	Intergovernmental Panel on Climate Change
IUCN	International Union for Conservation of Nature
IWMI	International Water Management Institute
MAF	Million acre feet
Mha	Million hectare
PSMD	Potential soil moisture deficit
UNDP	United Nations Development Programme

12.1 Introduction

Water and agricultural sectors are likely to be the most sensitive to climate change. Freshwater availability is expected to be highly vulnerable to the anticipated climate change. Climate change affects the rainfall pattern, especially on water distribution (Qureshi 2011). The variation in rainfall patterns, especially the seasonal shifts, would have likely impacted water availability for irrigation and subsequently on crop growth (Bhatti et al. 2019). The gap between supply and demand for water has widened due to increased agricultural activities and shrinking rivers.

In line with the expected growth in world population (9 billion) by 2050, food demand will increase by 65% in 2050 (Hanjra and Qureshi 2010), while the demand for energy from renewable resources and hydropower will increase by 60%. These difficulties are interlinked, for instance growing agricultural productivity would significantly raise water and energy utilization which would ultimately lead to increased struggle among water users. Water availability is probably decreasing in many regions. It is expected that even the future global use of water in agriculture (both rain-fed and irrigated agriculture) will be increased about 19% by 2050, and will be more due to lack of policy measures and technical development. Recent trends show that abstraction in nonindustrialized countries is predicted to rise by 25%. According to the global consumption of water resources, agriculture accounts for 70% of water resource consumption. Remaining 20% accounts for industrial use and 10% for domestic purposes. According to FAO prediction, irrigation water consumption will increase by 6% by 2050 (Bharathkumar and Mohammed-Aslam 2015; Zhang et al. 2011). This will increase water extraction for irrigation by about 6% from the present (2760 km^3) to 2050 (2926 km^3). Although this appears to be a slight increase, already multiple regions are suffering from water shortage (Alexandratos and Bruinsma 2012).

About 95% of water resources are consumed by agriculture in Pakistan. About 80% of agricultural land is irrigated, and 90% of agricultural production is the result of irrigation (Khalid and Begum 2013). Pakistan is one of the arid countries in the world with less than 240 mm of average rainfall per year. The water shortage situation in Pakistan is still overstressed due to heavy irregular precipitation rainfall. After losing three main rivers Beas, Sutlej, and Ravi to India in accordance with the Indus Waters Treaty (1960), the construction of new dams at Kishanganga and Baghlihar in India is upsetting the continuous downstream flow of water to Pakistan (Wilson 2011). In water deficit benchmark indicator (the Falkenmark Indicator), Pakistan is in the category of "Severe Water Stress" because of the projected existing per capita water accessibility of about 1066 M^3 (Fig. 12.1). In addition, water productivity in Pakistan is low; water efficiency in Pakistan is lower than 0.1 kg/m^3 compared to 0.39 kg/m^3 in India, furthermore water availability in Pakistan decreased by 1.1% between 2014 and 2015 (GOP 2015).

Failure to meet water needs and protect people and property from floods and droughts poses a serious threat to all countries, but developing countries are not capable to develop such infrastructure needed to combat the adverse impacts of floods and droughts. Climate change has serious consequences for precipitation and runoff around the world. Relatively in semiarid and arid areas, ambiguous certainty in rainfall will have huge impacts on water resources. In mountain river basins, an increase in temperature accelerates snow melting rate in the spring and limits the snow season.

The aim of this chapter is to describe some fundamental facts about the irrigation issues in Pakistan (as a case study) caused by climate change and then to devise some suitable on-farm water management strategies which may be required to rise irrigation proficiency in the era of climate change.

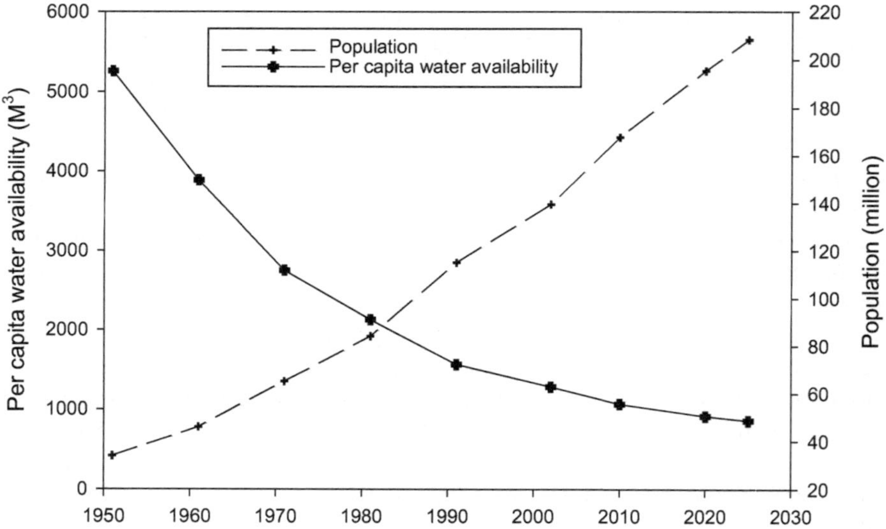

Fig. 12.1 Water availability as against population increase in Pakistan

12.2 Pakistan Location in Asia: Vulnerable to Water Resources Scarcity

Water use efficiency varies with different regions by their management and climatic variations. The annual water use efficiency ratio was predicted to be 41–42% for developed countries and 52–53% for developing countries up to 2050 (Bashir et al. 2016). Out of the total irrigated land of the world (almost 302 million ha), only 22% of the land is occupied by developed countries and 77% of irrigated land is in developing countries (Alexandratos and Bruinsma 2012). Asia-Pacific region has 36% of water resources but 60% population of the world live in this region. Due to the massive population and economic development, the rate of water use is high. On average, it has only 11% of entire renewable water resources (ESCAP 2010). Water availability regarding per capita is very low in this region as compared to other regions of the world (ESCAP and UNDP 2010). A rise of 1 °C in temperature is expected to increase the irrigation demand by at least 10% in arid and semiarid regions of Asia. The Himalayas distribute a massive amount of irrigation water in Asia. Himalayas snow and ice are expected to decrease by 20% by 2030 due to increase in temperature.

The Middle East extends from Kazakhstan to Turkey in the north to Somalia in the south, from the Atlantic Ocean (Mauritania and Morocco) to Pakistan in the west and Kyrgyzstan in the east. The poorest region in the world regarding water resources is the Middle East (due to the prevailing aridity in the region). Pakistan (30.3753° N, 69.3451° E), situated in South Asia, has limited resources of water with the complex terrain. Pakistan is classified as arid and semiarid conditions of climate (Wang et al. 2019) but has near about nine regimes of climate (Ahmed et al. 2019). Pakistan covers 33% irrigated area; this area is nearly about 4% of the region's territory. In Pakistan, 67–75% of the total annual rainfall follows in the summer and 25–33% in the winter. Maximum of the summer heavy rains occur during the monsoon season (from July to September) and it is not accessible for plant growth, as it is wasted due to rapid drainage and infiltration outside the root zone. In other cases, showers might be so mild that water evaporates before it can penetrate to the root zone of vegetation. The annual evaporation rate fluctuates from 150 to 200 cm. The average annual evaporation rates in northern Pakistan ranges from 115 to 195 mm and in southern Pakistan, it varies between 155 and 225 mm (Naheed and Rasul 2010).

Due to limited surface water resources for irrigation, groundwater over-extraction increased to some extent. In 2010, total groundwater extraction worldwide was estimated at about 1000 km^3 per year, of which 67% was used for irrigation purpose (Eurostat 2011; FAO 2011). Groundwater is an important factor for the livelihood of the Pakistan population. All groundwater is not good for irrigation because it carries a high concentration of salts and adds these salts to the soil after irrigation. During the drought, soil becomes salty in nature due to inadequate amount of water for leaching (Qadir et al. 2015).

12.3 Magnitude of the Problem

According to IPCC (2007), global temperature is expected to rise from 1.5 to 5.8 °C at the end of this century and the rainfall is expected to rise from 0 to 14%. Regarding Asia, a forecast of 0.5–2 °C rise in predicted temperature is expected until 2030. Similarly, increased precipitation from the base year is also expected for future forecasts. The average temperature (by Global Circulation Models (GCMs)) for Pakistan will increase 1.3–1.5 °C by 2020s, 2.5–2.8 °C by 2050s, and 3.9–4.4 °C by 2080s (GOP-PC 2010).

The Jhelum, Indus, and Chenab flow from the neighboring region; these are the main rivers of Pakistan and play a significant role to increase irrigation as well as agriculture. To achieve the main goal, it was important to develop and unlock the potential of the Indus River, a leading supplier. River Indus is mainly a snow-fed river; however, the size of the snow cover is rapidly decreasing due to global warming, which can disrupt the flow. The macro-scale hydrological river flow model estimated that Indus flow will decrease by 27% until 2050. This shows that freshwater availability in Pakistan is very vulnerable to changes in climate. A decrease in the mean runoff of snow-feed rivers joined with a rise in peak flows and sedimentation will have a significant impact on hydropower production and agriculture. Water supply for snow-covered rivers may increase in the short term, but has a long-term effect on sustainable water availability. The outflow of rivers that feed on rains may also change in the future.

Changes in hydrology caused by climate affect the magnitude and frequency as well as the cost of severe events that entail the highest economic and social costs for the world population. The floods can become repetitive and more extreme. In 2010, about one-fifth of Pakistan suffered from floods, which affected more than 20 million people in flood-prone areas along the Indus. Flooding also destroyed more than 647,497 hectares of crops (Guha-Sapir et al. 2011). The Intergovernmental Panel on Climate Change reports also indicate that the frequency and extent of drought in some areas may increase due to a decline in annual rainfall and with more repeated periods of drought. Droughts occur frequently in Pakistan and can affect almost one-third of Pakistan (Fig. 12.2). The most drone-prone areas include Cholistan in Punjab, Thar in Sindh, and the Chagai-Kharan region in Balochistan (Khan and Khan 2015). Balochistan is by far the most drought-prone province because of its arid to hyper-arid climate. Balochistan's agricultural sector has experienced major drought losses. It may not be possible to raise high delta crops, such as sugar cane, in the climate change scenario.

12.4 Water Management Strategies to Cope with Changing Climate

Water should be available not only in normal years but also during periods of drought. The overview of risk management indicates that it is recommended to decrease this risk before it can warrant the life of the system. Effective and sustainable water use can significantly maximize the availability of resources. In the longer run, the upcoming water crisis must be overcome through policies and tough political decisions that distribute water to economically and socially beneficial sectors. In Pakistan, almost 50% of water used for irrigation purposes is never accessed by the crop and is lost during runoff and in evaporation. Conveyance losses are also a serious matter of concern because of the huge loss of water from the barrage to the field (Chaudhry 2010). Improvements (by concrete measures) of water channels and watercourses enhance conveyance efficiency and reduce water losses. On average, the economic and environmental costs of developing new reserves will be two to three times higher than present investments (Nasim et al. 2015). The demand for water for irrigation purposes can be reduced if the local government introduces the PSI (pressurized irrigation systems). The basin irrigation efficiency (60–70%) is low as compare to sprinkler and drip irrigation efficiency (70–90%) (Hasanain et al. 2012). After identifying the major reasons behind the water crises, the water losses can be reduced by public awareness about crop selection, modern agriculture techniques, administrative as well as good governance measures which can enable Pakistan to cope with the water crises and irrigation problems (Jamil 2019).

In Pakistan, on-farm productivity is quite low because fields are not laser leveled and farmers use old irrigation methods. About 25% of water is lost through uneven fields. Due to the reasons stated above, the total effectiveness is 35.5%. To ensure

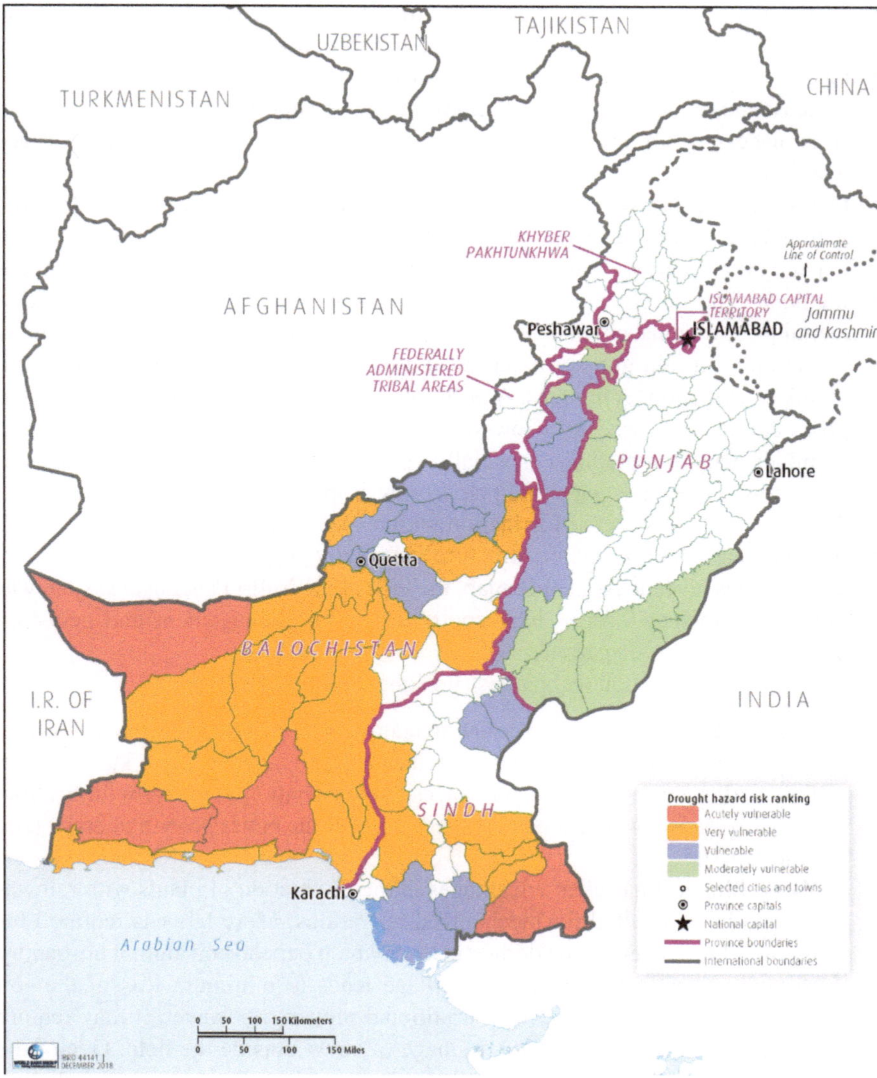

Fig. 12.2 Drought vulnerability of various districts in Pakistan. (*Source: Pakistan Getting More from Water (World Bank Document) December 14, 2018*)

effective and sustainable water consumption for irrigation as well as high productivity, farmers should implement the following measures:

- Choose the highly suitable and commercial crop varieties of the area depending on the amount of water available.
- Choose better quality and drought-tolerant seed varieties.
- Apply green manures and fertilizers efficiently.

- Water frequently and in the precise quantity considering weather conditions and plant growth phase. To save water, avoid over-irrigation.
- Cyclic consumption of ground and surface water is a fascinating option to increase the reliability and flexibility of the water supply.
- It is necessary to install low-cost shallow groundwater and small surface water pumps.
- Overhead irrigation (subsurface irrigation, subsurface exudes, Sprinkle, drip irrigations) are expensive techniques but after they are installed, you do not have to expend every year; we can irrigate small vegetables and fruit orchards with these methods. The developed countries have established high-efficiency irrigation techniques; we must adopt those techniques in our local conditions. For example, irrigation with rain gun sprinklers would be an optimal choice than traditional sprinkler irrigation techniques. A perforated irrigation tube with sufficient sized pores can also be used as a low-stress sprinkler. Also, cyclic use of rain gun sprinkler irrigation methods especially before planting or in the early stages of crop growth will save water and help to sow crops on time.
- Always adopt the system of irrigation with respect to the slope of the soil.
- It is necessary to develop infrastructure, especially large underground water storage facilities in some parts of Punjab, the idea of the Nehri Panchayat system was introduced so that farmers may contribute to the management and decision-making relevant to irrigation.
- Soil water conservation techniques (for long-term sustainability):

  Laser Land Leveling: In surface irrigation performance, laser land leveling plays a crucial role. Land leveling reduces the roughness of the land surface, time and water requirement, enhances the distribution of uniform infiltration, and also leads towards adequate conditions for the better growth of crops (Bai et al. 2010).

  Zero tillage: It has clear advantages in the production of plants with rainwater and is well suited for mechanized operations; More labor is required for the small producers at the domestic level which depends on animal husbandry as well as manual labor. Lowest tillage leads to minimum loss of the soil moisture connected along with traditional plowing; however, it may require weed control (chemicals) and drainage of straw outside the field. One of the best advantages of zero tillage is that it consumes minimum energy including fuel than traditional farming methods. Mulching (with plant debris, dust, and crop remaining, e.g., plastic sheet or straw) periodic tillage to expand the efficient root zone along with raising its porosity strip planting surface formation to improve holding capacity and penetration of runoff and rainfall add soil modifications to increase structure including stability of the root zone in the soil. Transpiration by weeds should be reduced by keeping the lanes between rows dry and, if necessary, taking measures to control weeds.

12.4.1 Overview of On-Farm Water Management Department

A program OFWM (On-Farm Water Management) was introduced in 1976–1977 to enhance the conservation of water resources. The purpose of this project is to minimize water losses during distribution and increase irrigation efficiency. Improving local watercourses is still an important part of all OFWM projects (Government of Punjab 2010). According to GOP (2011), water savings in Pakistan will be provided through a highly efficient irrigation system (drip irrigation system and sprinkler) to improve irrigation to 291,000 hectares by the OFWM department. A brief overview of the various current projects of the department is given in Table 12.1.

12.4.1.1 Functions of the Department

- Board of the water association
- Watercourses reconstruction and restoration (Water Users Association)
- To introduce the multiple resource conservation methods like management of crop yield and zero tillage.

Table 12.1 On-farm water management projects

Project name	Project period	Cost (Million Rs.)	Target
National Program for improvement of Watercourses in Pakistan-the Punjab Component (Supervisory Consultancy)	2004–2005 to 2007–2008 (extended from 2008–2009 to 2009–2010)	28,223.50	Improvement of 30,000 Watercourses/Irrigation Schemes
Water Conservation and Productivity Enhancement through High Efficiency (Pressurized) Irrigation System (Punjab Component)	2008–2009 to 2011–2012	6917.17	• Installation of HEIS – Demonstration Phase-600 acres – Up-scaling Phase-138,788 acres • Training of 14,000 farmers and 155 professionals
Greater Thal Canal (GTC) Command Area Development Project (Phase-I)	2008–2009 to 2012–2013	1994.71	• Earthen construction and lining of 725 Watercourses • Provision of 36 LASER units • Technical assistance for LASER Land Leveling of 36,000 acres

Source: Government of Punjab (2010)

Technology improvements for water conservation, i.e., ground leveling by LASER, effective system of irrigation along with the front bed, sprinkler, and drip irrigation for high-efficiency purpose.

12.4.1.2 Impacts of OFWM Projects

According to the Government of Punjab (2010), improvement of the watercourses according to Punjab Government (2010) can reduce losses during water transportation by up 33%, increase in cropping intensity by 23%, 50% labor, and 28% time saving for the irrigation. Similarly, social effects include:

- Reduce clashes between farmers regarding the distribution of water.
- Saving time filling of the watercourse, which leads to an increase in the duration of field use.
- Less confusion with watercourse exits (Nakkas), particularly in lined sections.
- Smaller passage area for waterways through narrower cross-sections of channels

Similarly, LASER land leveling can:

- Decrease water loss (50%)
- Reduced labor demand (35%)
- Dams and ditches reduction enhance the area (50%)

12.4.2 Adaptation and Mitigation Strategies

Mitigation strategies and adequate adaptations are required to minimize the influence of the worldwide variations in climate and environmental changes on the food system. Here are some of them:

1. Creating plant varieties resistant to heat and drought using conventional breeding and biotechnology approaches.
2. Different plant breeding along with the deep-rooted system.
3. To improve water use efficiency for proper provision, distribution, and use of water for irrigation needs a strict management system along with the administrative measures, and reuse of wastewater for irrigation purposes.
4. Local collection and construction of surface and underground tanks for local needs as well as for agriculture purposes.
5. Farmers should subsidies with services as like free electricity connection and a fixed fee to allow enough groundwater in the weak zone of roots to maintain soil to leave the poor root zone to maintain soil fertility.

12.4.3 Role of PSMD

In the semiarid and arid regions, the dry season cannot be predicted, mostly common in Pakistan, yield depends on the crop variety, because some variety bears the water scarcity (FAO 2008) and soil properties, mainly the water holding capacity of the soil, which will later on provide water to plants. The PSMD (potential soil moisture deficit) model accounts for the unpredictability in the supply of groundwater, water need for the atmosphere through assessing drought in plant growth at any stage as stability between the water supply (rainfall and irrigation) and crop water needs for the PET (potential evapotranspiration) (Mubeen et al. 2013a; Fries et al. 2020). The advantage of PSMD is to summarize the effects of droughts in terms of time and intensity into one index, and it helps to analyze the drought effects on canopy production and ultimately interception of radiation. This concept appears to be more necessary in Pakistan, where farmers do not take into account the requirement for harvesting (due to evaporation) and use water indiscriminately. Department of Agriculture recommended that the water should be supplied on a weekly basis (Government of Punjab 2011). These methods raise production costs and dramatically reduce the efficiency of water consumption.

12.4.4 Role of Irrigation Scheduling

Planning and scheduling irrigation is the decision: when and how much water must be supplied to the field to fully utilize the water. Proper irrigation planning will help to minimize the environmental impacts of irrigation and enhanced sustainability by saving water. When and how much water must be supplied to the field to fully utilize the water. Principally the aim of irrigation scheduling is to achieve a cost-effective water supply, maintaining the proximity of groundwater to field resources (Tariq and Usman 2009).

12.4.5 Role of Crop Growth Modeling

Conventional production techniques can combat climate change using modern tools, such as crop growth models. DSSAT (Decision Support System for Agrotechnology Transfer) model helps researchers to improve the profitability as well as efficiency of the Department of Agriculture system under changing environment through rotational, seasonal, and spatial decision analysis; otherwise, the potential impact of climate change is challenging, not only because of the unpredictability in the magnitude of climate change variables but also because of unpredictability associated with the plant's response to the soil, management and weather, factors. Prof. Dr. Ashfaq and his team have been doing a lot of work in crop growth modeling in

Pakistan. For example, in maize (Mubeen et al. 2013a, b, 2016) and sunflower (Nasim, 2010; Nasim et al. 2011, 2012), irrigation and other management decisions were improved by using the DSSAT model.

12.5 Future Water Needs of Pakistan and High-Level Recommendations

Agriculture takes over nearly about 60% of total withdrawn groundwater for the crops . Out of this water, 80.3% is used for wheat, rice, sugarcane, and cotton. In view of climate change and ever-increasing population, the water demands for urbanization and economic growth drive are also increasing (in addition to agriculture). Industrial and domestic demand will increase by several folds till 2050 due to more industrial activities and household incomes. If we do not consider demand management, faster global warming would cause significant additional increases, and the maximum projected water demand would be 58% higher than now (Amir and Habib 2015). However, agriculture will still continue to lead in water demands (Figs. 12.3 and 12.4). Some recommendations are qualitatively assessed regarding complexity, urgency, and scale of water security impact (bubble size) (Fig. 12.5). In order to manage the water wastages (through seepage) and subsequent waterlogging and salinity, certain concrete measures need to be undertaken, some of which are listed below:

- The first step to reduce seepage is to regularly maintain the canals and other water channels so as to maintain unobstructed water flow.
- Construction of roads and tracks on both sides of the water channels and their permanent use is another way of reducing seepage.

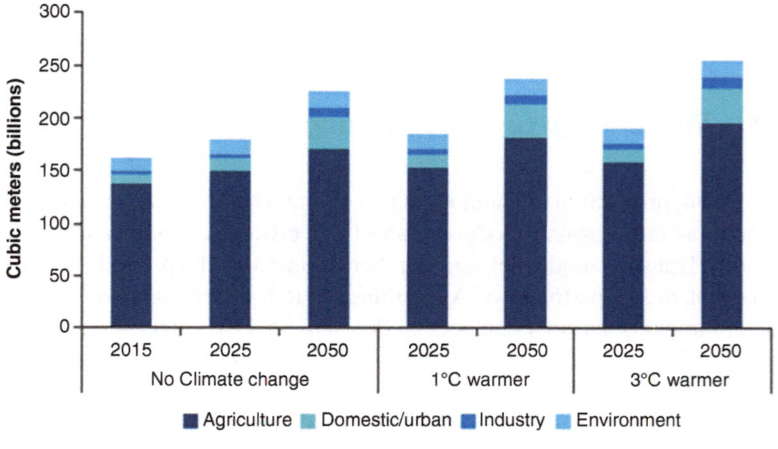

Fig. 12.3 Percentage of District Vulnerable to Drought in Pakistan. (*Source: Amir and Habib* 2015)

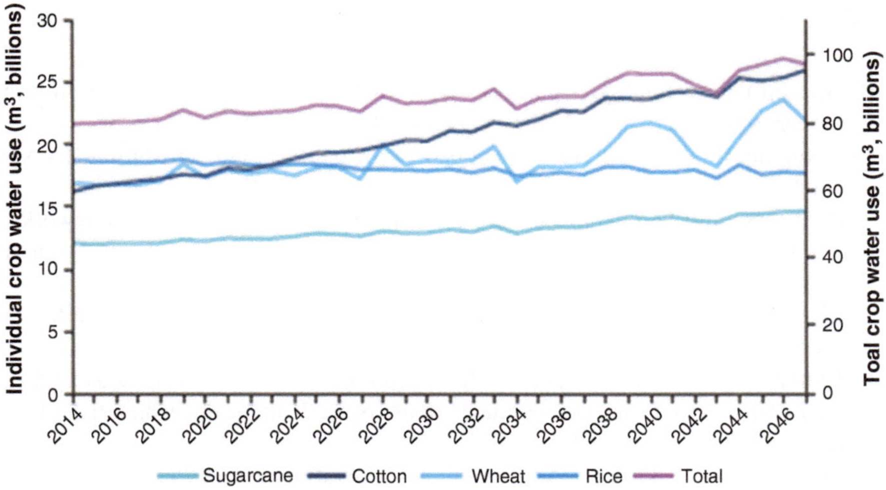

Fig. 12.4 Modeled Annual Crop Water Use in Pakistan, 2014–2047

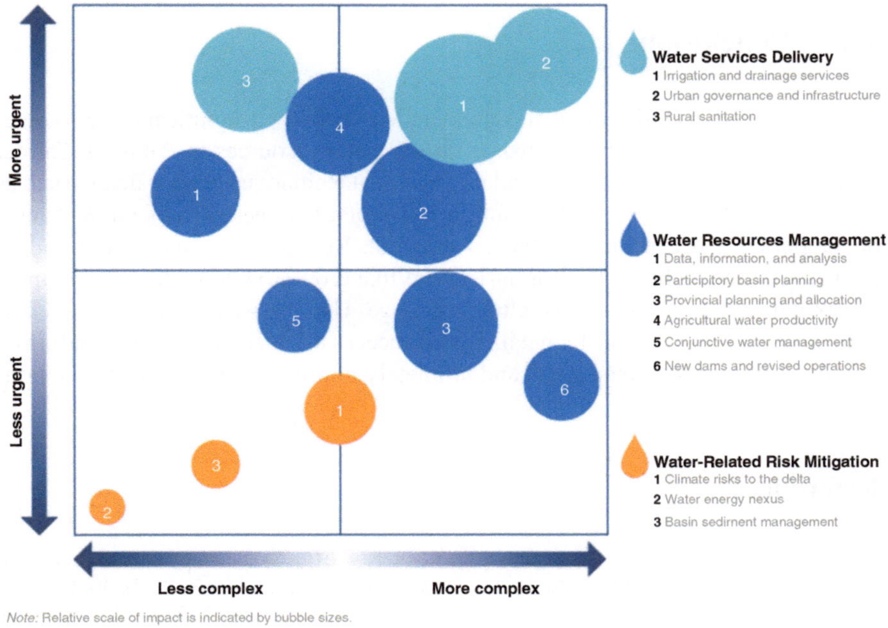

Fig. 12.5 Scale, complexity, urgency of impact, and key recommendations

- Providing an alternate source of water to the grazing animals is another very important step towards protecting the banks of water channels.
- Selective brick linings, particularly in areas which have been affected by rising water levels leading to waterlogging and salinity is a cure to revive the sick land.

12.6 Future Needs

- Appropriate technologies/practices need to be developed/evaluated permitting the use of drainage effluent for crop production.
- Computer models should be developed for irrigation/water scheduling at watercourse command level.
- Studies should be conducted on the mechanism of groundwater framework of regulatory authority; abstraction and development of framework of regulatory authority should be devised.
- More studies on optimizing cropping patterns for different water availability scenarios.
- Bio-saline agriculture to field should be promoted through coordinated efforts by using salt-tolerant varieties, conjunctive use of fresh and saline water, halophytes for fodder production, etc.
- Optimization of skimming well design and operational strategies for the secured extraction of thin freshwater thickness.

12.7 Conclusion

This is the responsibility of all stakeholders to must play a significant role to overcome the influence of climate change on the water resources in Pakistan. Farmer should adopt water conservation and management techniques in their fields (such as high-efficiency irrigation methods and soil conservation techniques). OFWM must be supported and executed by the government. Research institutes should adopt computer-based technologies for agriculture like crop growth modeling to combat short- and long-term impacts of climate change. Use ICTs (Information and communication technologies) which will provide access to timely and accurate information for better water management and ultimately improved agricultural production.

References

Ahmed, K., Shahid, S., Wang, X., Nawaz, N., & Khan, N. (2019). Spatiotemporal changes in aridity of Pakistan during 1901–2016. Hydrology and Earth System Sciences, 23(7), 3081-3096.

Alexandratos, N., & Bruinsma, J. (2012). World agriculture towards 2030/2050: the 2012 revision. *ESA Work. Pap*, *3*.

Amir, P., and Z. Habib (2015). Estimating the Impacts of Climate Change on Sectoral Water Demand in Pakistan. Nottinghamshire, U.K.: ACT. https://cyphynets.lums.edu.pk/images/Readings _concluding.pdf. (accessed June 2018)

Bai, M., Xu, D., Li, Y., & Pereira, L. S. (2010). Stochastic modeling of basins microtopography: analysis of spatial variability and model testing. Irrigation science, 28(2), 157.

Bashir MU, Wajid SA, Ashfaq A, Iqbal M (2016): Potential soil moisture deficit: an alternative approach for irrigation scheduling in wheat. International Journal of Agriculture and Biology 18

Bharathkumar, L., & Mohammed-Aslam, M. A. (2015). Crop pattern mapping of tumkur taluk using NDVI technique: a remote sensing and GIS approach. Aquatic Procedia, 4, 1397-1404.

Bhatti, M. T., Ahmad, W., Shah, M. A., & Khattak, M. S. (2019). Climate change evidence and community level autonomous adaptation measures in a canal irrigated agriculture system of Pakistan. Climate and Development, 11(3), 203-211.

Chaudhry, S.A. (2010). Pakistan: Indus Basin Water Strategy–Past, Present and Future.

ESCAP (United Nations Economic and Social Commission for Asia and the Pacific) (2010). Statistical Yearbook for Asia and the Pacific, 2009. Bangkok.

ESCAP, ADB (Asian Development Bank) and UNDP (United Nations Development Programme) (2010). Achieving the Millennium Development Goals in an Era of Global Uncertainty. Asia-Pacific Regional Report. Bangkok, ESCAP, ADB and UNDP. http://content.undp.org/go/cms-service/stream/asset/?asset_id=2269033.

Eurostat (2011). Online database. Brussels, European Commission (EC). http://epp.eurostat.ec.europa.eu/portal/page/portal/eurostat/home/

FAO (2008). http://faostat.fao.org/site/567/DesktopDefault.aspx?PageID=567.

FAO (2011). AQUASTAT online database. Rome, FAO. http://www.fao.org/nr/water/aquastat/data/query/index.html.

Fries A., et al (2020). Water Balance and Soil Moisture Deficit of Different Vegetation Units under Semiarid Conditions in the Andes of Southern Ecuador. Climate 2020, 8, 30; doi:https://doi.org/10.3390/cli8020030

GOP (2011). Economic Survey. Finance Division, Economic Advisors' Wing, Islamabad.

GOP (2015). Economic Survey. Finance Division, Economic Advisors' Wing, Islamabad. http://www.finance.gov.pk/survey/chapters_15/02_Agricultre.pdf

GOP-PC (2010) Final Report of the Task Force on Climate Change, Planning Commission, Government of Pakistan, Islamabad.

Government of Punjab (2010). http://www.punjabagri.pk (accessed on 24-07-2010)

Government of Punjab (2011). *Production Technology for Maize*. Directorate of Agricultural Information, Punjab, Pakistan.

Guha-Sapir, D. *et al.* 2011. Annual Disaster Statistical Review 2010: The Numbers and Trends. Brussels, CRED. http://www.undp.org.cu/crmi/docs/cred-annualdisstats2010-rt-2011-en.pdf

Hanjra, M. A., & Qureshi, M. E. (2010). Global water crisis and future food security in an era of climate change. *Food Policy*, *35*(5), 365-377.

Hasanain, A., Ahmad, S., Mehmood, M. Z., Majeed, S., & Zinabou, G. (2012). Irrigation and water use efficiency in South Asia. Policy Research Paper: Supporting Policy Research to Inform Agricultural Policy in Sub-Saharan Africa and South Asia. New Delhi.

IPCC (Intergovernmental Panel on Climate Change) (2007). Climate Change 2007: The physical science basis. Contribution of Work Group I to the Fourth Assessment Report of the Intergovernmental Panel on Climate, Cambridge University Press, United Kingdom.

Jamil, M. (2019). Running Dry: Water Scarcity in Pakistan. (Naval Postgraduate School Monterey United States).

Khalid, I., & Begum, I. (2013). Hydro Politics in Pakistan: Perceptions and Misperceptions. *South Asian Studies*, *28*(1), 7-23.

Khan, A. N., and S. N. Khan (2015). Drought Risk and Reduction Approaches in Pakistan. In: Atta-Ur-Rahman, A. Khan, and R. Shaw (eds) Disaster Risk Reduction Approaches in Pakistan. Disaster Risk Reduction (Methods, Approaches and Practices). Springer, Tokyo.

Mubeen, M., A. Ahmad, A. Wajid, T. Khaliq and A. Bakhsh (2013a). Evaluating CSM-CERES-Maize model for irrigation scheduling in semi-arid conditions of Punjab, Pakistan, Int. J. Agric. Biol., 15: 1–10.

Mubeen, M., A. Ahmad, A. Wajid, A. Bakhsh (2013b). Evaluating different irrigation scheduling criteria for autumn-sown maize under semi-arid environment. Pak. J. Bot., 45(4): 1293-1298.

Mubeen, M., A. Ahmad, A. Wajid, T. Khaliq, H. M. Hammad, S. R. Sultana, S. Ahmad, W. Nasim, S. Fahad (2016). Application of CSM-CERES-Maize Model in Optimizing

Irrigated conditions. Outlook on Agriculture. 45(3) 173–184. https://journals.sagepub.com/doi/abs/10.1177/0030727016664464

Naheed G, Rasul G. Projections of crop water requirement in Pakistan under global warming. Pakistan Journal of Meteorology. 2010;7(13):45-51.

Nasim, W. (2010). Modeling the impact of climate change on nitrogen use efficiency in sunflower (*Helianthus Annuus* L.) under different agroclimatic conditions of Punjab-Pakistan. PhD Thesis, Deptt. Agron., Univ. Agric. Faisalabad-Pakistan.

Nasim, W., A. Ahmad, A. Wajid, J. Akhtar and D. Muhammad. 2011. Nitrogen effects on growth and development of sunflower hybrids under agro-climatic conditions of Multan. *Pak. J. Bot.,* Vol 43, No 4, pp 2083-2092.

Nasim, W., A. Ahmad, A. Bano, M. Usman, R. Olatinwo, H.M. Hammad, T. Khaliq, M. Hussain (2012). Effect of nitrogen on yield and oil quality of sunflower (*Helianthus annuus*L.) hybrids under sub humid conditions of Pakistan. Am. J. Plant Sci 3, 243-251.

Nasim W., H. Belhouchette et al. (2015). Correlation studies on nitrogen for sunflower crop across the agroclimatic variability. Environ Sci Pollut Res. https://doi.org/10.1007/s11356-015-5613-1, http://link.springer.com/article/10.1007/s11356-015-5613-1.

Qadir, M., Noble, A. D., Karajeh, F., & George, B (2015). *Potential business opportunities from saline water and salt-affected land resources* (No. H046996). International Water Management Institute.

Qureshi A.S. (2011). Water management in the Indus basin in Pakistan: challenges and opportunities. Mountain Research and Development. (3):252-60.

Tariq J.A., Usman K. (2009). Regulated deficit irrigation scheduling of maize crop. Sarhad Journal of Agriculture, 25(3): 441-450.

Wang, Z., Yang, S., Duan, A., Hua, W., Ullah, K., & Liu, S. (2019). Tibetan Plateau heating as a driver of monsoon rainfall variability in Pakistan. Climate Dynamics, 52(9-10), 6121-6130.

Wilson, S. (2011). Preparation of sub regional plan for Haryana sub-region of NCR-2021. Interim Report–II

Zhang, H., Lan, Y., Lacey, R., Hoffmann, W. C., & Westbrook, J. K. (2011). Spatial analysis of NDVI readings with different sampling densities. Transactions of the ASABE, 54(1), 349-354.

Chapter 13
Morphological, Physiological, and Biochemical Modulations in Crops under Salt Stress

Rashad Mukhtar Balal, Muhammad Adnan Shahid, Naeem Khan, Ali Sarkhosh, Muhammad Zubair, Atta Rasool, Neil Mattson, Celina Gomez, Muhammad Adnan Bukhari, Mirza Waleed, and Wajid Nasim

Abstract Crop plants are affected by biotic and abiotic stresses (including salinity) and such stresses may affect the growth and yield of these crop plants seriously. High temperature (due to climate change) has also changed the pattern of precipitation and caused rise in sea level. These two factors have impacted soil salinization.

R. M. Balal (✉) · M. Zubair
Department of Horticulture, College of Agriculture, University of Sargodha, Sargodha, Pakistan

M. A. Shahid (✉)
Department of Agriculture, Nutrition, and Food Systems, University of New Hampshire, Durham, NH, USA
e-mail: muhammad.shahid@unh.edu

N. Khan
Department of Agronomy, Institute of Food and Agriculture, University of Florida, Gainesville, FL, USA

A. Sarkhosh
Horticultural Sciences Department, Institute of Food and Agriculture, University of Florida, Gainesville, FL, USA

A. Rasool · M. Waleed
Department of Environmental Sciences, COMSATS University Islamabad, Vehari Campus, Pakistan

N. Mattson
Horticulture Section, School of Integrative Plant Science, Cornell University, Ithaca, NY, USA

C. Gomez
Environmental Horticulture, Institute of Food and Agriculture, University of Florida, Gainesville, FL, USA

M. A. Bukhari · W. Nasim
Department of Agronomy, Faculty of Agriculture and Environment, The Islamia University of Bahawalpur (IUB), Bahawalpur, Pakistan

W. N. Jatoi et al. (eds.), *Building Climate Resilience in Agriculture*,
https://doi.org/10.1007/978-3-030-79408-8_13

To address such problems naturally, the crop plants adapt themselves by different mechanisms including changes in morphological, physiological, and biochemical processes. Both ions including sodium and chloride are the main ions, that become the reason for many physio-biochemical modulations inside plant tissues, in a similar way, chloride ion is the most dangerous because NaCl releases around 60% more ions in soil comparatively with Na_2SO_4. An extra amount of such types of salts increases the osmotic potential in soil matrix consequently the water absorbance by plants is reduced that leads towards physiological stresses or drought. This increase of Cl^- relates to salt tolerance that is linked to plant growth, water use efficiency, and transpiration. Increasing salinity in the nutrient solution reduces growth directly and restricts leaf and root mineral fixing. In this chapter, we have discussed insights into various kinds of morphological, physiological, anatomical, and biochemical modulations in plants caused by abiotic stresses especially salinity. In the era of climate change, plant scientists should focus on each shotgun approaches as well as long-term genomic techniques to enhance salt tolerance in commercially important crops to ensure food security and sustainable productivity.

Keywords High temperature · Soil salinization · Morphological modulations · Physico-chemical modulations · Anatomical modulations · Salt tolerant genotypes

13.1 Introduction

In current environmental conditions, saline soils are increasing day by day. Agricultural productivity is greatly degraded by the accumulation of salts in underground water. Quality of water resources and soil fertility are the vital components that fulfill the food, feed, and fiber demand of ever-growing global population. Salinity restricts the production of nearly over 6% of the world's land and 20% of the irrigated land (15% of total cultivated areas) (Munns 2005) and adversely affects agricultural production throughout the world. Many fertile lands are converted into salt affected ones because of underprivileged practices, e.g., irrigation with underground brackish water. Salinization of good arable land results in immense socio-economic losses. The loss of excellent natural resource is another problem because the population depends upon its livelihood on these lands, which are gradually declining through the spread of salinity (Acosta-Motos et al. 2017; Hammad et al. 2018). Salinity is considered as one of the major threats to crop plants which leads to lessen economic productivity by affecting different plant morpho-physiological and biochemical processes (Qazizadah 2016). Salinity tolerance can be elevated as a useful tool to characterize plants in different categories. The plants having the capability to endure high salt concentration and maintain their growth under saline conditions are characterized as halophytes. However, the plant species which face loss in their morphological, physiological, and metabolic processes are identified as glycophytes. The mechanisms adopted by halophyte species to combat salt stress and toxic ions like Na^+ ion toxicity include restriction of Na^+ ions entry into root cells, prohibiting Na^+ ions from leaves surface, movement of Na^+ ions from transpirational xylem stream to roots, compartmentation and sequestering of Na^+ ions in vacuoles and Na^+ exclusion through salt glands (Sun et al. 2009).

Halophytes can uphold their stomatal conductance to a level, that implement minimum effects on photosynthesis; in glycophytes, these effects are more noticeable (López-Climent et al. 2008). Halophytes exhibit high water use efficiency and low internal CO_2 concentration related to glycophytes. Solute accumulation is more prominent in halophytes than glycophytes. Due to various salinity tolerating mechanisms, a lower level of Na^+ and Cl^- ions is observed in halophytic cytoplasm and chloroplast. Halophytes exhibit lower Na^+ ion concentration in roots than upper plant parts (Sun et al. 2009). Sodium partitioning is significantly associated with plant metabolic processes and ion flux at both cellular and whole plant level (Park et al. 2016). Halophytes maintain their survival efficiently in salt shocks. These plants maintain their metabolic activities in a controlled way under saline conditions (Yang and Guo 2018).

The higher salt concentration impose disturbance at the plant cellular level and increase the salt level in soil solution decreases the plant's water uptake ability and results in water scarcity conditions (Munns 2005; Stępień and Kłbus 2006; El Sabagh et al. 2019). Salinity and drought induce similar adverse effects on plant growth; however, salinity have more severe effects on ionic balance in cell through vacuole compartmentation. This high ionic concentration possess restrictions in regular enzymatic activities. As a consequence of this increased ionic concentration in cytoplasm, enzymatic activities become ceased and the plant bends towards decline (Munns 2005). Salt tolerant genotypes regulate toxic ion accumulation (Ryu and Cho 2015) and translocation of these ions in the other plant parts. There are many factors affecting the plant's response to salinity but, irrigation system, edaphic and climatic conditions are the most important ones (Yang and Guo 2018).

13.2 Morphological Responses to Salt Stress

Salinity effects biomass and morphological traits of plants. Roots, shoots, and leaves are the part of plants that are severely affected by salinity. Salt stress in the root zone of plant reduces its root length (Ryu and Cho 2015). Reduced root conductivity is also associated with salt stress (Park et al. 2016). More than 90% of water transport from root to shoot is observed to be declined with increased salt stress. Reduced root hydraulic variables are induced by low soil osmotic potential and high root suberization because of salinity (Rizwan et al. 2015). The morpho-physiological characteristics of roots are directly related to the ability to exclude Cl^- from shoots (Farooq et al. 2017). Size of the root system determines the Cl^- uptake by the plant, which accumulates in the leaves and causes salt stress symptoms (Kaleem et al. 2018). Chloride uptake by roots and accumulation in leaves are linked to transpiration rate and water absorption, respectively. So, root morphological characters, which regulate water use, are of utmost importance in inducing salt tolerance (Farooq et al. 2017). Shoot to root ratio is an important and crucial factor in determining salt tolerance. Salinity also limits the shoot elongation and due to this reason plant canopy volume reduces (Kaleem et al. 2018).

Hence, salt stress adversely affects shoot growth and development. Along with shoot and root, salinity also induced its effects on leaves in terms of reduction in number of leaves, leaf area, and increased necrosis percentage (García-Sánchez et al. 2005). Necrosis/leaf burn symptoms, growth reduction (Rizwan et al. 2015), and leaf drop are identified at medium and high salinity. Sodium accumulation, as a result of salinization induces necrosis in old leaves, which initiate from tips and margins and then extend towards the leaf base. It also decreases the life span of leaves, net productivity, and crop yield (Craine 2005). Biomass reduction and foliar damage become more noticeable with the passage of time, with increasing the salinity levels.

Salinity reduces cell division and thus fresh matter production. As under saline condition, photosynthetic rate imbalances, so production of biomass is greatly decreased (Ryu and Cho 2015). Shoot dry weight decreases significantly with the increase in salinity. A reduction in whole plant dry weight, plant height, root length, root dry weight, and number of nodules was recorded in different crops (Sharifi et al. 2007). Moreover, significant reduction was observed in seed yield per plant, with the increase in salinity (Ashraf et al. 2005). A decreasing trend of plant height and total biomass was also observed in Juncus species (Greenwood and MacFarlane 2009). A salt tolerant plant species show a strong rooting system, which is helpful in osmotic adjustment under stress conditions (Yang and Guo 2018). A stronger rooting system increases the water availability by penetrating deep, below the salinity zone. The direct effect of salinity on roots is its growth retardation by interfering with mineral nutrition of the plant. Under saline growing conditions, decreased concentration of oxygen and internal ethylene accumulation increases, which ultimately leads to decline in root growth and elongation (Ashraf et al. 2005).

13.3 Limitations in Seed Germination, Survival %, and Growth Rate under Salt Stress

Seed germination is significantly affected by increasing salinity in the growing media (Othman et al. 2006). Halophytes and glycophytes vary significantly in their seed germination response to high salinity level. Halophytes follow a characteristic pattern by resisting salinity through normal germination at initial and low salinity levels but afterwards a rapid decreasing trend is observed with rise in salinity level. However, glycophytes show a gradual decrease in germination with rising salinity level. The low solute potential of growing media, because of increased salinity level results in reduced imbibition rate of seeds. This situation negatively affects the metabolic processes within seed like enzymatic imbalances (Ryu and Cho 2015), modifications in Nitrogen metabolism, alterations in plant growth regulator levels, and decline in reserves utilization. The food reserves present in seeds vary significantly in response to salinity. Lipid contents decline while sugar level rises (Yang and Guo 2018). The increment in sugar level is contributed by lipid metabolism and starch and protein breakdown (Acosta-Motos et al. 2017). Salinity cause accumulation of soluble sugars, free protein, and soluble proteins in seeds. These compounds also perform osmotic adjustment and prove to be beneficial for developing embryo.

Plant growth under saline conditions can be determined in terms of relative growth rate (RGR), net assimilation rate on a leaf weight basis (NAR (w)), leaf weight ratio (LWR), and nutrient uptake and utilization. It was found that salt treatment induced negative effects on both RGR and NAR (w), whereas LWR showed no definite trend. Net assimilation rate is correlated with relative growth rate (Nguyen et al. 2015). Salinity induces a negative effect on root mass fraction (RMF) and increase stem mass fraction (SMF) (Acosta-Motos et al. 2017).

13.4 Physiological Responses to Salinity Stress

Salinization disrupts the physiological processes of the plant because of ion toxicity and osmotic effect (Syvertsen and Levy 2005). Salinity proved to be a prominent threat for crop productivity by limiting plant growth processes through increased ion toxicity, inefficiency of photosynthesis, respiration rate, transpiration rate (Yuan 2006), stomatal conductance, membrane instability and permeability, decreased biosynthesis of chlorophyll, nitrite, and nitrate reductase activity

Salinity significantly lowers the stomatal conductance, photosynthesis, transpiration (Liu et al. 2006; Parida and Das 2005), and relative water content (Naumann et al. 2007). Salinity tolerant plant species show higher photosynthetic capacity and more accumulation of organic osmolytes (Hameed and Ashraf 2008). Under salinity conditions, closing of stomata results in decreased photosynthesis as well as transpiration. Chloride accumulation causes the reduction of net assimilation of CO_2 (Nguyen et al. 2015). Sodium is found to have more prominent effect in the reduction of these attributes, which accumulate in excessive amounts under saline conditions. Lower photosynthetic activity is contributed by reduced rubisco activity (Nguyen et al. 2015).

At moderate salinity levels, photosynthesis rate increases, as observed in sugar beet and eggplant. Higher photosynthesis rate is linked to higher stomatal conductance, which results in higher net assimilation of CO_2. This results in higher biomass production and crop yield (Farooq et al. 2017). Gas exchange is inversely related to the concentration of sodium and chloride ions due to salinization. CO_2 assimilation rate is associated with ion toxicity and water relation (Park et al. 2016). Increased sodium concentration leads to lower nitrate reductase activity, photosystem II and chlorophyll degradation (Kaya et al. 2020). It also results in cell membrane leakage due to the replacement of Ca^{2+} with Na^+ (Manchanda and Garg 2008).

Diffusion limitation in respect of mesophyll and stomatal conductance, contribute to photosynthesis inhibition under salt stress. Stomatal limitation with stomatal closure, non-stomatal limitation, or both limitations with stomatal closure at low tissue salt concentration and a disturbance of photosynthetic activity at high tissue salt concentration can be identified for salt-induced low photosynthesis rate. A significant correlation exists between stomatal conductance and photosynthesis rate (Yu et al. 2020). Maintenance of net photosynthetic rate, stomatal conductance, and elevated chlorophyll concentration are considered prerequisite for inducing salinity tolerance.

Elevated CO_2 levels interact with salinity in various ways. In olive (*Olea europea*), elevated CO_2 levels strongly affect the photosystem II and Chlorophyll content, under salinization. Elevated CO_2 does not interfere with Na^+ and Cl^- concentration in leaves and roots of tolerant cultivar whereas decreased these toxic ion accumulations in salt sensitive olive cultivar. Elevated CO_2 increased water use and reduces toxic ion uptake, but not significantly affected plant growth (Yang et al. 2005). Plant growth, shoot/root ratio, net gas exchange, water use, and root Ca^{2+} experience a decreasing trend with the increasing salinity while root N increases.

13.5 Cellular Responses to Salt Stress

Plants tolerate adverse effects of salinity by adjusting their biochemical and molecular processes, accordingly (García-Caparrós and Lao 2018). The plant mechanisms, which contribute to salinity tolerance include ion inclusion or exclusion, controlled uptake and transport of ions into shoots, ion compartmentalization at cellular and whole plant level, compatible solutes synthesis, modifications in photosynthetic mechanism, alterations in membrane structures, and accumulation of antioxidative enzymes (Arbona et al. 2005) and plant hormones (Parida and Das 2005).

Cell division is one of the various metabolic processes which faces serious irregularities. Precisely, it affects the leaf anatomy by inducing epidermal and mesophyll cell thickness, increment in palisade cell length and diameter, spongy cell diameter, reduction in intercellular space, changes in mitochondria and vacuole, decline in plant leaf area, and stomatal density (Yang and Guo 2018).

Ion compartmentalization at cellular level is an essential adaptive mechanism for plant species, to regulate their metabolic activities under salt stress (Parihar et al. 2015). Salt tolerant plant species retain toxic levels of ions in vacuoles and inhibit their interference with cytoplasmic metabolic activities. This adaptation leads to plant survival in adverse conditions. It is known as portioning or compartmentalization of toxic ions. It is independent of the membrane potential (positive or negative) inside the membrane (Farooq et al. 2017). The Na^+ and Cl^- restriction in vacuole induce higher concentrations of K^+ and organic osmolytes in cytoplasm in order to adjust osmotic pressure of the ions in the vacuole (Ashraf et al. 2005). Sodium compartmentation results in vacuolar alkalization, which is found to be partially associated with Na^+/H^+ antiporter activity. The restriction of toxic concentration of sodium ions in vacuole is regulated by salt inducible enzyme Na^+/H^+ antiporter (Padan and Landau 2016). Vacuolar Na^+/H^+ antiporters utilize the proton gradient, produced by H^+-adenosine triphosphatase (H^+-ATPases) and H^+-inorganic pyrophosphates (H^+-PPases). So, salt stress tolerance of plant species is based on the coordinated performance of Na^+/H^+ antiporters, H^+-ATPases, and H^+-PPases (Padan & Landau 2016). The Na^+/H^+ antiporter activity increases with the addition of sodium ions which is very significantly noted in tolerant plant species (Parihar et al. 2015).

13.6 Salinity-Induced Ion Toxicity and Nutrient Imbalance

Ion toxicity leads to reduction in growth due to adverse effects on some essential physiological and biochemical processes (Sun et al. 2009). Chloride, sodium, boron, lithium, etc., are the major ions which interfere with the metabolic processes of the plant when accumulated in excessive amount. As a result of salinization, plants differ in their ion uptake mechanism due to some multiple adaptations to toxic ions operating concurrently within a specific plant. Salt tolerance of a plant species refers to its ability to restrict translocation of toxic ions in shoots (Yu et al. 2020). This ability is regulated by cell specificity of tissues, morphological features, and water use efficiency (Nguyen et al. 2015). These adaptive mechanisms alter the plant response to salinity by inflicting characteristic modifications at both cellular and whole plant level.

Excessive accumulation of sodium and chloride ions results in reduced growth and nutrient imbalance (Liu et al. 2006). High sodium concentration in leaves is found to have a negative effect on net CO_2 assimilation, as indicated in some citrus species (Mishra and Tanna 2017). Chloride ion in high concentration imposes toxic effect on photosynthesis. The chloride uptake is significantly determined by shoot to root ratio through passive transport.

The nutrient imbalance occurs due to high ratios of Na^+/Ca^+, Na^+/K^+, Na^+/Mg^+, Cl^-/NO_3^-, and $Cl^-/H_2PO_4^-$, which leads to reduced yield and growth. The increment in the uptake of Na^+ has an antagonistic effect on the Ca^+ and K^+ uptake. However, Ca^+ and K^+ are of key importance to membrane integrity and proper functioning (Acosta-Motos et al. 2017). Plants adopt a mechanism to maintain adequate Ca^+ and K^+ concentration at cellular level, under saline conditions. Ca^+/Na^+ ratio is maintained in the saline growing media by increasing Ca^+ concentration in order to increase salt tolerance.

13.7 Potassium/Calcium Ions

Under salt stress K^+ concentration decreases (Acosta-Motos et al. 2017). In the presence of excessive Na^+, K^+ leaches out from soil exchange complex. So, a competition develops between Na^+ and K^+ at soil–root interface. Na^+/K^+ ratio can be considered as a source to observe salt tolerance of plant species (Willadino and Câmara 2005). In order to maintain adequate K^+ concentration, plants utilize low affinity and high affinity channels for K^+ uptake. Three low affinity channels, i.e., inward rectifying channels (KIRC), K^+ outward rectifying channels (KORC), and voltage independent cation channels (VIC), and two high affinity transporters are identified for maintaining cellular K^+/Na^+ ratio. The higher K^+ concentration is related to higher biomass production and thus salt tolerant plant species have the ability to retain more concentration of potassium (Ashraf et al. 2005).

Calcium role is significant in new cell wall synthesis, especially the middle lamella, which provides separation in neighboring cells, spindle formation during cell division, and regulating cell membrane integrity (Kaleem et al. 2018). In germinating seeds, calcium concentration with the passage of time decreases with respect to elevated levels of salt stress. Salt tolerant species experience more accumulation of K^+ and Ca^{2+}, which helps in maintaining optimal growth (Ryu and Cho 2015). Salinity tolerant grass species *Cynodon dactylon* (L.) exhibited this attribute through less Na^+ accumulation in roots and more accumulation of K^+ and Ca^{2+} in roots as well as leaves (Hameed and Ashraf 2008).

13.8 Magnesium, Nitrogen, and Nitrate Ions

Salt affected trees have low leaf Mg^{2+} concentrations because of low Mg^{2+} concentration in the exchange complex. Increasing Ca^{2+} (by addition of Ca^{2+} as gypsum $CaSO_4$) implements an antagonistic effect by displacing Mg^{2+} from the soil complex hence reduce Mg^{2+} concentration. Moreover, high sodium accumulation shows an antagonistic relationship with nitrogen (N), potassium (K^+), and manganese (Mn) uptake. Salinization results in high concentration of phosphorous and low Mg content (Min et al. 2014). However, the N and Ca^{2+} percentages in the roots do not vary significantly with the increase in salinity while a significant reduction in leaf N and Ca^{2+} concentrations is observed in salinity affected seedlings. Chloride uptake negatively influences the N uptake, while positively affects the concentration of Ca^{2+} and Na^+. Salinity interferes with the translocation mechanism of these elements. These findings displayed that under saline conditions, leaf mineral contents are least affected by root mineral status.

An inhibitory effect is observed on NO_3 (Sun et al. 2009) and phosphorous uptake, due to accumulation of high concentration of Cl^- ions under saline conditions. High leaf chloride contents, as a result of increased salinity caused reductions in chlorophyll contents, decreased the photosynthesis rate (Sun et al. 2009), and induced bleaching or bronzing of in leaves while high leaf sodium contents lowered the gas exchange rates. Salinity also induced progressive depletions of carbohydrates in leaves and roots thus inhibiting root growth and development (Iqbal et al. 2006).

13.9 Micronutrients

The micronutrient (Cu, Fe, Mn, Mo, and Zn) availability is affected by salinity as evidenced in saline and sodic soils. However, plant type, plant tissue, growing conditions, micronutrient concentration, salinity level, and its composition determine the effect of salt stress on the availability of micronutrients (Acosta-Motos et al. 2017). Manganese deficiency is reported in barley, with increasing salinity level, which can be compensated by manganese addition in soil. In contrast, few reports also suggested no effect and increased manganese concentration in plant shoots. Similarly, contrasting opinions are given by scientists, related to Zn, Fe, Mo, and Cu concentration. At different salinity levels, plants respond by decreasing the concentration of molybdenum, magnesium, iron, and zinc in their leaves. Copper is reported to remain unaffected under salt stress (Yang and Guo 2018).

13.10 Toxic Ions Inclusion and Exclusion Mechanism Under Salinity Stress

Salt exclusion at whole plant level involves ion partitioning, which constrains the salt movement towards shoots (Sun et al. 2009). Cellular level exclusion involves the inhibition of entry of toxic ions into the cell or directs their outward flow if get entered. Exclusion mechanism may also involve toxic ion extrusion through salt glands present on the leaf surface (Ashraf et al. 2005). Toxic ion exclusion is an adaptive strategy by plants experiencing salt stress. Roots maintain salinity levels by extrusion to soil or transport to shoots via xylem transpirational stream (Davenport et al. 2005). Phloem is reported as a chief source of toxic ion transport from shoots to roots. As Na^+ and Cl^- are the major toxic ions salt tolerant species are either Cl^- excluder or Na^+ excluder (Farooq et al. 2017). High concentration of Na^+ and Cl^- ions were recorded in shoots of salt tolerant species. Plant's ability to exclude sodium or chloride ions decreases with the increase in salinity level (Ghosh et al. 2016).

Some halophytes and some salt tolerant plant species like barley are also identified as Na^+ accumulators, which are signified by Na^+ concentration in their roots (Ghosh et al. 2016). The ionic distribution trend and vacuolar compartmentalization determine the extent of these strategies in glycophytes (Yang and Guo 2018). Sodium accumulation under salinity can also be explained through Na^+-ATPases concept. The absence of Na^+-ATPases (García-Sánchez et al. 2005) restrict the efflux of Na^+. The sodium efflux occurs against electrochemical gradient, which requires energy. Consequently, the sodium efflux decreases and sodium accumulates in plant tissues (Ashraf et al. 2005). Exclusion of 97% of Na^+ present in the soil at root surface is considered essential for all plants to maintain safe Na^+ level in shoots (Acosta-Motos et al. 2017).

13.11 Salinity and Biochemical Attributes

13.11.1 Leaf Pigments

Leaf pigments experience noticeable changes in their concentrations under salinity. Carotenoid contents are reported to decline in response to salinity however anthocyanin pigment increased as a result of salt stress. A decreasing trend of total chlorophyll, as well as Chl a and Chl b has been reported as a result of salinity, which can be contributed by the absolute concentration of chloride and/or sodium in the leaves. However, the increase in chlorophyll a, b, and total chlorophyll was reported on weight basis. As salinity causes a reduction in leaf size therefore one gram of salt affected plant showed a greater number of leaves (Weisany et al. 2011).

13.11.2 Sugars, Protein, and Lipid

Under saline conditions, sugar concentration becomes high in mesophyll cells of leaves, which induces feedback inhibition to photosynthesis process. The disturbance in normal sugar utilization process causes the increment in sugar concentrations in growing tissues (Skorupa et al. 2019). The concentration of total soluble carbohydrates increased in the leaves and roots of the seedlings, grown under saline conditions (NaCl or Cl is osmotic condition). The increment in carbohydrate level may be contributed by high chloride concentration in plant tissue or starch degradation as salt tolerant species have less starch accumulation (Poór et al. 2011). The carbohydrate accumulation rate varies among salt tolerant species (Skorupa et al. 2019). A decline in starch content and starch phosphorylase activity and increment in reducing, non-reducing, and sucrose phosphate synthase activity is observed under saline conditions. Soluble protein decreases in response to salinity (Park et al. 2016). Lipids play a key role in the protection of delicate organs. Energy is stored in the form of lipids and is a constituent of cellular membranes. Lipids play a significant role in inducing tolerance against stress conditions. Lipid content declines at higher levels of salinity.

13.11.3 Osmoprotectants

Physiological tolerance involves compartmentation and osmotic adjustment, utilizing inorganic and organic compounds (Acosta-Motos et al. 2017). This osmotic adjustment in plants occurs through the accumulation of compatible solutes which may be inorganic like essential elemental ions such as K^+ and organic compounds including sugars (glucose and fructose, mainly), sugar alcohols (glycerol, methylated inositol), and complex sugars (trehalose, raffinose, and fructans) (Zhu et al. 2016). Other important compounds include quaternary amino acid derivatives (proline, glycinebetaine, β alaninebetaine, and proline betaine), tertiary amines (1,4,5,6-carboxyl pyrimidine), and sulfonium compounds (choline-o-sulfate, dimethyl sulfonium propionate) (Ashraf et al. 2005). The compatible solutes lower the osmotic potential of the cell and maintain water status of plant (Parihar et al. 2015). They are hydrophilic in nature and act to replace water molecules present on the surface of proteins, protein complexes, and membranes and thus serve as an osmoprotectant (Zhu et al. 2016). They do not disturb the enzymatic activity of the cell and pH of the cytosol. These are produced as a result of some specific modifications in biochemical reactions, which occur only under stress conditions (Parida and Das 2005).

Compatible solutes regulate the enzyme activity under salt stress and do not interfere with the metabolic activities of the cell even if accumulated in high concentration. These organic solutes protect proteins and ribosome structures from the adverse effects of a toxic concentration of ions. These organic compounds mainly accumulate in leaves, with maximum concentration in salt tolerant species. These osmoprotectants scavenge the reactive oxygen species (ROS), which poses damage to cell functioning (Ashraf and Foolad 2005).

The osmolyte synthesis is related to basic metabolic activities. The amino acid biosynthesis pathways lead to origin of proline (glutamic acid), aspartate (ecotine), glycinebetaine (choline metabolism), and pinitol (myo-inositol synthesis). Mass action to the same extent can be regarded as a mechanism, adopted by osmolytes for regulating cytoplasmic osmotic potential (Parihar et al. 2015).

13.11.4 Glycinebetaine

The accumulation of nitrogen-containing compounds varies with plant species under salt stress (Grieve et al. 2007). These compounds perform the specific functions of osmotic adjustment, cell macromolecule protection, nitrogen storage, buffering, cell detoxification, and ROS scavenging to induce salt tolerance in plant species (Parida and Das 2005). Glycinebetaine (GB) (a quaternary compound) is identified as one of the prominent compounds, which is utilized for osmotic adjustment (Yang et al. 2005). Glycinebetaine restores the integrity of thylakoid membrane in salt and drought stress (Ashraf and Foolad 2005). It improves salt tolerance by protecting photosynthetic protein complexes (Nguyen et al. 2015) and reducing lipid peroxidation of cellular membranes. Glycine, betaine indirectly enhances the photosynthetic activity of plant cells, experiencing salt stress by regulating photosynthetic machinery by concentrating in chloroplast, regulating chloroplast metabolism, and protecting thylakoid membranes. It stabilizes the extrinsic proteins of PSII and hence increases the efficiency of PSII under salt stress (Yang and Guo 2018).

13.12 Free Amino Acids, Total Soluble Proteins, and Proline

Accumulation of some free amino acids and GB in osmoregulation is found to be a prominent stress tolerant strategy in plants. Amides (glutamine and asparagines) accumulation is also reported in plants experiencing salt stress (Cui et al. 2018). A considerable accumulation of proline, asparagine, and glutamine is also reported in cultivars of strawberry grown under salinity (Keutgen and Pawelzik 2008). The most prominent amino acid which accumulates under salt stress conditions in plants is proline. Proline accumulation because of salinity stress increased its importance as a compatible solute, which aids in osmotic adjustment (Cui et al. 2018). Salt tolerant plant species show maximum amino acid accumulation in leaves, as observed in Sunflower, Safflower, *Eruca sativa*, *Lens culinaris,* and *Phragmites australis* (Parida and Das 2005).

Proline accumulation, under salt stress is more profound in monocots. However, proline accumulation was insignificant in salt stressed barley seedlings (Parihar et al. 2015). Proline is not accumulated specifically for salinity stress, but also occur under drought stress. Proline synthesis is contributed by low water potential of growing media. Proline regulates the membrane stability and alleviates the salinity effects on cell membrane disruption by controlling the osmotically active useable N accumulation. Proline enhances the salt tolerance by protecting the protein turnover machinery against stress damage and up-regulating stress protective proteins in *Pancratium maritimum* L..

Few reports also displayed the opposite picture, i.e., more proline accumulation in salt sensitive varieties of tomato as compared to tolerant species (Yu et al. 2020). An inverse relation is reported between proline accumulation and salt tolerance in *Vigna mungo (*Win and Oo 2017*)*. Soluble proteins also play a key role in osmotic adjustment during salt stress. Salt tolerant species show high content of soluble protein, as evidenced in barley (El-Esawi et al. 2018), Sunflower (Zeng et al. 2016), rice (Ghosh et al. 2016), and finger millet (Acosta-Motos et al. 2017). However, a decreasing trend of soluble protein was also noted with the increase in salinity (Mishra and Tanna 2017). Increased level of protein synthesis in response to salt stress is found in many plant species. For example, osmotins and dehydrins, regulate the protein structure and activity. Late embryogenesis proteins (LEA) are synthesized excessively under salt stress which safeguards the adverse osmotic effects of NaCl (Saha et al. 2016).

Proline accumulation can be considered as a representative of severe saline conditions. Soluble protein accumulation cannot be considered as related to salt tolerance mechanism (Sun et al. 2009). Proline stabilizes subcellular structures, scavenge ROS, and protect cell membranes by stimulating antioxidant activity (Kaleem et al. 2018). It may be an inhibiting agent for plant growth if high concentration of proline is applied. It is of utmost importance to optimize the effective concentration of exogenously applied proline. High proline concentration applied reduced seedling growth and lowered leaf Na^+/K^+ ratio (Saha et al. 2015). The exogenously applied proline concentration varies with the plant species and plant developmental stage (Cui et al. 2018). Proline is found to have a protective role in photosynthesis by provision of regenerated NADP through transcriptional activation of the NADPH-dependent P5C-Synthetase. This NADP provision prevents photo-inhibition.

13.12.1 Polyamines

Accumulation of polyamines like putrescene, spermidine, spermines, etc., is also related with plants growing under saline conditions. The major role of polyamines is in cell elongation, root formation (Saha et al. 2015), cell division, organogenesis, and plant senescence. However, polyamines contribute little to osmotic adjustment. Extent of polyamine accumulation varies in a single species. In *Brassica compestris*, small alterations in polyamine level are noticed upon exposure to long-term salinity stress, whereas plant experienced significant increment in polyamine level and enzymatic activities when subjected to a short duration of salt stress (Ke et al. 2018). More polyamine accumulation was observed in salt sensitive cultivars of rice and tomato. The salt tolerant plant species synthesize excessive concentration of polyamines such as putrescene and spermine (Sequera-Mutiozabal et al. 2017). However, increased concentration of putrescene and tyramine was observed in roots of salt sensitive plant species of rice (Ghosh et al. 2016).

Polyols contribute to osmoregulation in salt stress. These are classified as acyclic (manitol) and cyclic (pinitol) (Parida and Das 2005). The high concentration of ions in vacuole results in osmotic disturbance. Polyols concentrate in cytoplasm to counteract these changes. The salt tolerating ability of tobacco is found to be related to high accumulation of polyols (Saha et al. 2015).

13.13 Summary and Future Research Prospects

Salinity is restraining the crop yield by causing modulations at molecular, cellular, morphological, physiological, and biochemical levels. It is depicted that plants try to cope with salt stress by regulating nutrient uptake, maintaining water status, osmotic adjustment, and through antioxidant defensive system. All these mechanisms are species-dependent. Salinity is limiting the plant growth, production, and quality not only in agronomic crops but in horticultural crops too. So, there is a dire need to improve the salt tolerance potential of crops. Plant scientists should focus on each shotgun approaches as well as long-term genomic techniques to enhance salt tolerance in commercially important crops to ensure food security and sustainable productivity. Identification of tolerant genotypes by using various physiological and biochemical indicators of stress tolerance, exogenous application of stress-inducing compounds like silicon, new generation growth hormones, and growth promoting microbes are the potential strategies to mitigate salinity-induced drastic effects of product quality and quantity within shorter span on time.

References

Acosta-Motos, J. R., Ortuño, M. F., Bernal-Vicente, A., Diaz-Vivancos, P., Sanchez-Blanco, M. J., & Hernandez, J. A. J. A. (2017). Plant responses to salt stress: adaptive mechanisms. *7*(1), 18.

Arbona, V., Marco, A. J., Iglesias, D. J., López-Climent, M. F., Talon, M., & Gómez-Cadenas, A. (2005). Carbohydrate depletion in roots and leaves of salt-stressed potted Citrus clementina L. *Plant Growth Regulation, 46*(2), 153-160.

Ashraf, M., Alvi, A., Sarwar, G., Qureshi, M., Ashraf, M., & Hussain, M. (2005). Effect of ammonium chloride on the growth and nutrient uptake by cotton grown in alkaline soil. *Agrochimica, 49*(3-4), 153-164.

Ashraf, M., & Foolad, M. R. (2005). Pre-sowing seed treatment—A shotgun approach to improve germination, plant growth, and crop yield under saline and non-saline conditions. *Advances in agronomy, 88*, 223-271.

Craine, J. M. (2005). Reconciling plant strategy theories of Grime and Tilman. *Journal of ecology, 93*(6), 1041-1052.

Cui, F., Sui, N., Duan, G., Liu, Y., Han, Y., Liu, S., . . . Li, G. J. F. I. P. S. (2018). Identification of metabolites and transcripts involved in salt stress and recovery in peanut. *9*, 217.

Davenport, R., James, R. A., Zakrisson-Plogander, A., Tester, M., & Munns, R. (2005). Control of sodium transport in durum wheat. *Plant Physiology, 137*(3), 807-818.

El-Esawi, M. A., Alaraidh, I. A., Alsahli, A. A., Ali, H. M., Alayafi, A. A., Witczak, J., & Ahmad, M. J. M. (2018). Genetic variation and alleviation of salinity stress in barley (Hordeum vulgare L.). *23*(10), 2488.

Farooq, M., Gogoi, N., Hussain, M., Barthakur, S., Paul, S., Bharadwaj, N., . . . Biochemistry. (2017). Effects, tolerance mechanisms and management of salt stress in grain legumes. *118*, 199-217.

García-Caparrós, P., & Lao, M. T. J. S. H. (2018). The effects of salt stress on ornamental plants and integrative cultivation practices. *240*, 430-439.

García-Sánchez, F., Botia, P., Fernández-Ballester, G., Cerdá, A., & Lopez, V. M. (2005). Uptake, transport, and concentration of chloride and sodium in three citrus rootstock seedlings. *Journal of Plant Nutrition, 28*(11), 1933-1945.

Ghosh, B., Md, N. A., & Gantait, S. J. R. R. O. A. (2016). Response of rice under salinity stress: a review update. 1-8.

Greenwood, M., & MacFarlane, G. (2009). Effects of salinity on competitive interactions between two Juncus species. *Aquatic Botany, 90*(1), 23-29.

Grieve, A., Prior, L., & Bevington, K. (2007). Long-term effects of saline irrigation water on growth, yield, and fruit quality of 'Valencia'orange trees. *Australian Journal of Agricultural Research, 58*(4), 342-348.

Hameed, M., & Ashraf, M. (2008). Physiological and biochemical adaptations of Cynodon dactylon (L.) Pers. from the Salt Range (Pakistan) to salinity stress. *Flora-Morphology, Distribution, Functional Ecology of Plants, 203*(8), 683-694.

Hammad HM et al., 2018. Offsetting land degradation through nitrogen and water management during maize cultivation under arid conditions. Land Degradation and Development 2018; 1366-1375

Iqbal, N., Ashraf, M., Javed, F., Martinez, V., & Ahmad, K. (2006). Nitrate reduction and nutrient accumulation in wheat grown in soil salinized with four different salts. *Journal of Plant Nutrition, 29*(3), 409-421.

Kaleem, F., Shabir, G., Aslam, K., Rasul, S., Manzoor, H., Shah, S. M., . . . biotechnology. (2018). An overview of the genetics of plant response to salt stress: present status and the way forward. *186*(2), 306-334.

Kaya, C., Ashraf, M., Alyemeni, M. N., Ahmad, P. J. E., & safety, e. (2020). The role of nitrate reductase in brassinosteroid-induced endogenous nitric oxide generation to improve cadmium stress tolerance of pepper plants by upregulating the ascorbate-glutathione cycle. *196*, 110483.

Ke, Q., Ye, J., Wang, B., Ren, J., Yin, L., Deng, X., & Wang, S. J. F. I. P. S. (2018). Melatonin mitigates salt stress in wheat seedlings by modulating polyamine metabolism. *9*, 914.

Keutgen, A. J., & Pawelzik, E. (2008). Quality and nutritional value of strawberry fruit under long term salt stress. *Food chemistry, 107*(4), 1413-1420.

Liu, N.-Y., Ko, S.-S., Yeh, K.-C., & Charng, Y.-Y. (2006). Isolation and characterization of tomato Hsa32 encoding a novel heat-shock protein. *Plant Science, 170*(5), 976-985.

López-Climent, M. F., Arbona, V., Pérez-Clemente, R. M., & Gómez-Cadenas, A. (2008). Relationship between salt tolerance and photosynthetic machinery performance in citrus. *Environmental and Experimental Botany, 62*(2), 176-184.

Manchanda, G., & Garg, N. (2008). Salinity and its effects on the functional biology of legumes. *Acta Physiologiae Plantarum, 30*(5), 595-618.

Min, W., Guo, H., Zhou, G., Zhang, W., Ma, L., Ye, J., & Hou, Z. (2014). Root distribution and growth of cotton as affected by drip irrigation with saline water. *Field Crops Research, 169*, 1-10.

Mishra, A., & Tanna, B. J. F. I. P. S. (2017). Halophytes: potential resources for salt stress tolerance genes and promoters. *8*, 829.

Munns, R. (2005). Genes and salt tolerance: bringing them together. *New phytologist, 167*(3), 645-663.

Naumann, J. C., Young, D. R., & Anderson, J. E. (2007). Linking leaf chlorophyll fluorescence properties to physiological responses for detection of salt and drought stress in coastal plant species. *Physiologia Plantarum, 131*(3), 422-433.

Nguyen, H. T., Stanton, D. E., Schmitz, N., Farquhar, G. D., & Ball, M. C. J. A. O. B. (2015). Growth responses of the mangrove Avicennia marina to salinity: development and function of shoot hydraulic systems require saline conditions. *115*(3), 397-407.

Othman, Y., Al-Karaki, G., Al-Tawaha, A., & Al-Horani, A. (2006). Variation in germination and ion uptake in barley genotypes under salinity conditions. *World Journal of Agricultural Sciences, 2*(1), 11-15.

Padan, E., & Landau, M. J. T. A. M. I. T. R. F. L. (2016). Sodium-proton (Na+/H+) antiporters: properties and roles in health and disease. 391-458.

Parida, A. K., & Das, A. B. (2005). Salt tolerance and salinity effects on plants: a review. *Ecotoxicology and Environmental Safety, 60*(3), 324-349.

Parihar, P., Singh, S., Singh, R., Singh, V. P., Prasad, S. M. J. E. S., & Research, P. (2015). Effect of salinity stress on plants and its tolerance strategies: a review. *22*(6), 4056-4075.

Park, H. J., Kim, W.-Y., Yun, D.-J. J. M., & cells. (2016). A new insight of salt stress signaling in plant. *39*(6), 447.

Poór, P., Gémes, K., Horváth, F., Szepesi, A., Simon, M., & Tari, I. (2011). Salicylic acid treatment via the rooting medium interferes with stomatal response, CO2 fixation rate and carbohydrate metabolism in tomato, and decreases harmful effects of subsequent salt stress. *Plant Biology, 13*(1), 105-114.

Qazizadah, N. A. (2016). *Response of wheat varieties to nitrogen under saline water irrigation.* HARYANA AGRICULTURAL UNIVERSITY HISAR,

Rizwan, M., Ali, S., Ibrahim, M., Farid, M., Adrees, M., Bharwana, S. A., . . . Research, P. (2015). Mechanisms of silicon-mediated alleviation of drought and salt stress in plants: a review. *22*(20), 15416-15431.

Ryu, H., & Cho, Y.-G. (2015). Plant hormones in salt stress tolerance. *Journal of Plant Biology, 58*(3), 147-155. doi:https://doi.org/10.1007/s12374-015-0103-z

EL Sabagh A, et al. (2019) Drought and salinity stresses in barley: Consequences and mitigation strategies. Australian Journal of Crop Science 13(06):810-820

Saha, B., Mishra, S., Awasthi, J. P., Sahoo, L., Panda, S. K. J. E., & Botany, E. (2016). Enhanced drought and salinity tolerance in transgenic mustard [Brassica juncea (L.) Czern & Coss.] overexpressing Arabidopsis group 4 late embryogenesis abundant gene (AtLEA4-1). *128*, 99-111.

Saha, J., Brauer, E. K., Sengupta, A., Popescu, S. C., Gupta, K., & Gupta, B. J. F. I. E. S. (2015). Polyamines as redox homeostasis regulators during salt stress in plants. *3*, 21.

Sequera-Mutiozabal, M., Antoniou, C., Tiburcio, A. F., Alcázar, R., & Fotopoulos, V. J. C. M. B. R. (2017). Polyamines: emerging hubs promoting drought and salt stress tolerance in plants. *3*(1), 28-36.

Sharifi, M., Ghorbanli, M., & Ebrahimzadeh, H. (2007). Improved growth of salinity-stressed soybean after inoculation with salt pre-treated mycorrhizal fungi. *Journal of plant physiology, 164*(9), 1144-1151.

Skorupa, M., Gołębiewski, M., Kurnik, K., Niedojadło, J., Kęsy, J., Klamkowski, K., . . . Tyburski, J. J. B. P. B. (2019). Salt stress vs. salt shock-the case of sugar beet and its halophytic ancestor. *19*(1), 1-18.

Stępień, P., & Kłbus, G. (2006). Water relations and photosynthesis in Cucumis sativus L. leaves under salt stress. *Biologia Plantarum, 50*(4), 610.

Sun, J., Chen, S.-L., Dai, S.-X., Wang, R.-G., Li, N.-Y., Shen, X., . . . behavior. (2009). Ion flux profiles and plant ion homeostasis control under salt stress. *4*(4), 261-264.

Syvertsen, J., & Levy, Y. (2005). Salinity interactions with other abiotic and biotic stresses in citrus. *HortTechnology, 15*(1), 100-103.

Weisany, W., Sohrabi, Y., Heidari, G., Siosemardeh, A., & Ghassemi-Golezani, K. (2011). Physiological responses of soybean ('Glycine max'L.) To zinc application under salinity stress. *Australian Journal of Crop Science, 5*(11), 1441.

Willadino, L., & Câmara, T. (2005). Aspectos fisiológicos do estresse salino em plantas. *R. Custodio, E. Aráujo, L. Gómez, and U. cavalcante (eds.). Estresses ambientais: Danos e benefícios em plantas. MXM. Gráfica e editora. Recife, Pernambuco, Brasil*, 127-137.

Win, K., & Oo, A. J. A. P. A. R. (2017). Salt-stress-induced changes in protein profiles in two blackgram (Vigna Mungo L.) varieties differing salinity tolerance. *7*(1), 00239.

Yang, H., Yuan, X., Zhou, Y., Mao, Y., Zhang, T., & Liu, Y. (2005). Effects of body size and water temperature on food consumption and growth in the sea cucumber Apostichopus japonicus (Selenka) with special reference to aestivation. *Aquaculture Research, 36*(11), 1085-1092.

Yang, Y., & Guo, Y. J. J. O. I. P. B. (2018). Unraveling salt stress signaling in plants. *60*(9), 796-804.

Yu, Z., Duan, X., Luo, L., Dai, S., Ding, Z., & Xia, G. J. T. I. P. S. (2020). How plant hormones mediate salt stress responses.

YUAN, X.-T. (2006). Salinity effect on respiration and excretion of sea cucumber Apostichopus japonicus (Selenka). *Oceanol Limnol Sinica, 37*(4), 354-360.

Zeng, W., Xu, C., Wu, J., & Huang, J. J. F. C. R. (2016). Sunflower seed yield estimation under the interaction of soil salinity and nitrogen application. *198*, 1-15.

Zhu, Y., Guo, J., Feng, R., Jia, J., Han, W., Gong, H. J. P., & Soil. (2016). The regulatory role of silicon on carbohydrate metabolism in Cucumis sativus L. under salt stress. *406*(1), 231-249.

Chapter 14
Weed Management and Climate Change

Ahmad Omid Siddiqui, Ayşe Yazlık, and Khawar Jabran

Abstract Weeds and their management are being affected by warming of the earth and rise in carbon dioxide (CO_2) levels in the atmosphere. Evidence from the recent literature confirms that high carbon dioxide in the atmosphere will increase the growth of weeds particularly the C_3 weeds. Warming on the other hand will support several weeds to expand their range and establish into new areas. Further, high levels of warming and CO_2 may also have a negative effect on the efficacy of several herbicides. Under this situation, use of cultural and mechanical weed control will be important. No doubt, these and other nonchemical weed control methods can be integrated with herbicides to achieve effective control of weeds under changing climate.

Keywords Weeds · Weed control · Climate change · Global warming

14.1 Introduction

Climate change is a state of change in global climatic patterns that persists for long periods, probably decades or even longer. Frequent climate change is attributed to a rapid rise in greenhouse gas (GHG) concentrations, which can be the results of many factors including volcanic explosions, deforestation, and fossil fuel burning (Hussain et al. 2020). The development of industries and human activities has resulted in a tremendous escalation in global warming through the emission of GHG, i.e., carbon dioxide (CO_2), nitrous oxide (N_2O), and methane (CH_4) (Ramesh et al. 2017). Continued change in climate may lead to higher global temperature,

---

A. O. Siddiqui (✉) · K. Jabran
Department of Plant Production and Technologies, Niğde Ömer Halisdemir University, Niğde, Turkey

A. Yazlık
Plant Production Department, Düzce University, Düzce, Turkey

W. N. Jatoi et al. (eds.), *Building Climate Resilience in Agriculture*,
https://doi.org/10.1007/978-3-030-79408-8_14

irregular wind patterns, precipitation, evapotranspiration, unusual weather, i.e., drought and fluctuation in temperature, salinity, floods, and storms (Karkanis et al. 2018). It is predicted until 2100 that CO_2 concentration will reach ~1000μmol mol^{-1} followed by 2–4 °C rise in the earth's annual surface temperature (IPCC 2014).

Climatic changes affect plant life and agricultural practices including weed management (Singer et al. 2013; Shahzad et al. 2018). Potential significant effects of climatic changes on agriculture are now known throughout the world (Kizildeniz et al. 2015, 2018; Ferrero et al. 2018; Hammad et al. 2018). Weeds interfere with the crop growth, agronomic operations, and harvesting of crops. Despite large number of benefits, the role of some weeds as alternate hosts of insect pests and poisoning effects of some weeds to humans and animals are reported (Panter et al. 2017). The yield losses caused by weeds in the crops are generally more than 30% and this high level of yield decrease by weeds necessitated the development of advanced weed control technologies (Clements et al. 2014; Jabran et al. 2017). Weeds are a serious issue in agriculture because they are strong competitors and can tolerate harsh environmental conditions. It is predicted that an increase in CO_2 might support the growth of weeds and pose greater weed control problems in the future (Jabran et al. 2015a; Jabran and Doğan 2018). Climate change affects the weeds, their management, and cost of their control. High levels of carbon dioxide, air temperature, soil temperature, levels of precipitation, and evaporation affect the weed biology, ecology, and their distribution. Climate change also affects weed–crop interactions and weed management. This chapter is aimed at explaining the effects of climatic changes on weeds, the herbicides, and their efficacies against weeds. Further, this chapter also proposes some control methods that are effective to control weeds under changing climate.

14.2 Response of Weeds to Increasing CO_2

Elevated atmospheric CO_2 concentration has been described as a key component of climatic change. Globally, CO_2 continues to increase and will hit 550μmol mol^{-1} by the year 2050. Availability of CO_2 to plants has been significantly increased over the last two decades from 280 to 400 ppm. Such a major increase in the atmospheric CO_2 will have a severe impact on physiology and biochemistry of crops and weeds (Ainsworth and Long 2005).

Bajwa et al. (2019) reported that at a higher CO_2 concentration, parthenium weed (*Parthenium hysterophorus* L.) growth and biomass accumulation were improved significantly. Similarly, Navie et al. (2005) reported an increase in the plant biomass, plant height, and number of seeds due to elevated CO_2 level of 480 ppm in comparison to CO_2 level of 360 ppm. Response of C_3 and C_4 plants to elevated CO_2 concentrations varies for photosynthetic rate, owing to different mechanism of ribulose-1,5-bisphosphate carboxylase (Rubisco) activity; weeds with C_3 photosynthetic pathway benefit from elevated CO_2 and exhibit an increase in water use efficiency and net photosynthetic rates, though in case of C_4 weeds elevated CO_2

concentrations only improve water use efficiency. Elevated CO_2 concentrations enhanced the growth and water use efficiency of C_3 as well as C_4 species due to reduced stomatal conductance (Amin et al. 2017; Jabran and Doğan 2020).

Under elevated CO_2 concentration, a C_3 weed common lambsquarters (*Chenopodium album* L.), showed significant increase in the biomass and reduced the yield of soybean by 39%, while increasing CO_2 concentration exhibited no variation in biomass of a C_4 weed redroot pigweed (*Amaranthus retroflexus* L.).

14.3 Response of Weeds to Increasing Temperature

Under the global warming conditions, some crop plants suffer from high temperatures and fluctuations in temperature that cause stress to plants (Shahid et al. 2014). The response of plants to temperature varies among C_3 and C_4 plants due to physiological variations in photosynthetic biochemistry. Temperature increase in C_3 plants often inhibits growth by suppressing CO_2 assimilation and increase in photorespiration; while in C_4 plants, increase in temperature stimulates meristematic regions, and maximizes canopy growth and root proliferation (Morgan et al. 2001; Khaliq et al. 2012). Higher temperatures as a result of global warming would improve the C_4 weed's growth rates (Ramesh et al. 2017). Because of the wider pool of genes compared to crops, weeds have a higher amplitude of response to elevated temperatures, making them adaptable to diverse environmental conditions. As a result, weeds may migrate to new lands damaging the world's cropping systems.

14.4 Response of Weeds to Increasing Rainfall and Drought

Changes in rainfall patterns will alter weed seed germination, plant size, and seed dispersal. In the future, increase in temperature is anticipated with the rise of CO_2 level (Wajid et al. 2010). Higher temperature results in higher evaporation rate and affects rainfall pattern of the monsoon zones making the crops in the region more prone to drought (Giannini et al. 2008). Weeds response to water stress can differ because of their wider range of physiological and developmental mechanisms (Karkanis et al. 2018). Various weed species have adapted the mechanism of reducing plant leaf area in response to water stress. Velvetleaf (*Abutilon theophrasti* Medik.) adopts senescence and drops its oldest leaves, which allows the younger leaves to retain higher leaf water potential in response to drought stress (Schmidt et al. 2011). Moreover, several weeds decrease their water requirements by shortening their life cycles. Karkanis et al. (2011) observed that *A. theophrasti* plants (compared with well-watered plants) flowered earlier. Similarly, Volis et al. (2009) observed early reproduction for wild barley (*Hordeum spontaneum* K. Koch). Under prolonged drought spells C_4 weeds have the capability to produce more seeds, more biomass, and robust roots than C_3 weeds (Rodenburg et al. 2011). Zand et al. (2006)

reported that drought stress enhanced the growth of *A. retroflexus* (C4 weed) compared to C. album (C_3 weed). In addition, Hyvönen (2011) reported the response of C_4 weed *A. retroflexus* for carbon fixation pathway and concluded that the weed efficiently fixes carbon to cope with drought stress.

14.5 Climate Change and Crop-Weed Competition

Weeds quickly adapt to changes in nature as they have the potent ability for their survival under harsh climatic conditions. Weeds flourish in a better way as compared to the cultivated plants. Climate change may create higher competition between crops and weeds if proper weed management control would not be adopted, and it is predicted to cause great yield losses of crops (Valerio et al., 2013).

Several species of weeds would benefit from higher rates of CO_2 and warmer temperatures, and weeds grow better and quicker than major cultivated plants. Photosynthesis in the plant will be directly affected by increasing levels of CO_2 and may influence the crop's ability to compete with weeds (Chandrasena, 2009; Hatfield et al. 2011). Agriculture depends heavily on specific climate conditions, so the effect of climate change on the production of crops and agricultural output is vital to understand; increased CO_2 levels, changes in earth's mean temperature, extreme weather condition are a major context of climate change that affect crop-weed competition (Ramesh et al. 2017; Amanet et al. 2019).

Effects of climatic changes, for example, increased level of CO_2, temperature fluctuations, and associated moisture stress on crop-weed competition will influence the morphology, photosynthesis, and physiology of both crops and weeds and will vary greatly depending on the nature of weeds and crops (Chongtham et al. 2019). Previous studies indicate that increased CO_2 level will benefit C_3 crops as compared to C_4 plants; although most of the crops fall under C_3 category and the majority of the notorious weeds are using C_4 photosynthesis type, it may not reduce crop-weed competition under higher CO_2 (Ramesh et al., 2017). At increased CO_2 concentrations C_3 weeds like canary grass (*Phalaris minor* Retz.) and wild oats (*Avena ludoviciana* Dur.) in wheat (C_3) would aggravate due to climate change (Naidu and Ojha 2015). Currently, there are over 450 problematic species of weeds (C_3 and C_4) associated with around 50 major crops worldwide. This means that if a C_4 species of weed does not respond to higher CO_2, there is a high chance that C_3 weed species will respond efficiently (Chongtham et al. 2019). C_3 weeds giant ragweed (*Ambrosia trifida* L.), *A. theophrasti,* common ragweed (*Ambrosia artemisiifolia* L.) will be benefitted by elevated CO_2 level due to abridged stomatal aperture, improved water use efficiency, shorter life cycle, and easily disseminated seeds and due to all these characteristics, the weeds would become more aggressive and difficult to control (Miri et al. 2012). Ziska (2000) stated that at higher CO_2 concentrations *C. albuma* (a C_3 weed) had a 65% increase in the biomass which reduced the seed yield of soybean by 39%; on the other hand, loss of soybean yields decreased from 45% to 30% in the presence of *A. retroflexus*, a C_4 weed. Furthermore, there was an increase in biomass and seed yield of wild types of weedy rice (*Oryza sativa* f. *spontanea*) compared to cultivated rice types under higher CO_2 levels, which suggests a greater decrease in the yield of cultivated rice types in the presence of C_3 weeds and the future atmospheric CO_2 concentrations (Ziska et al. 2010). High CO_2 amount can be beneficial for various crops, i.e., wheat (*Triticum aestivum* L.) and soybean (*Glycine max* L.), and it can also increase yield up to 30%, while other plants, such as maize (*Zea mays* L.), may reveal a trivial response (<10%). Nevertheless, if elevated CO_2 levels are associated together with water deficiency and unfavorable temperature, yield and quality of crops may reduce (CCSP 2008). In addition, changing climate can alter weed, pest, and incidence of diseases in different pathways and can be a serious concern for agricultural pests and weed control.

Fluctuation in temperatures as the result of climate change also alters the crop-weed rotation and increases the competition between them. The most important effect of increasing temperature is the extension of growing period in plants; increase in temperature may favor weeds with C_4 photosynthesis pathway as compared to crops with C_3 pathways (Yin and Struik 2009). Lee (2011) reported that under 4 °C higher temperature, the flowering time of white goose-foots (*Chenopodium album* L.) and green foxtail (*Setari aviridis*L.) reduced by 50–31.5 days and increased emergence time by 26–35 days. Studies have shown that C_4 weeds can be more benefited than C_3 crops by an increase in temperature accompanied by elevated CO_2 levels. Alberto et al. (1996) revealed that rice (C_3 crop) under higher CO_2 level alone performed much well as compared to C_4 weed barnyard grass (*Echinochloa glabrescens* Munro ex Hook. f.) but when CO_2 and temperature increased concurrently, C_4 weed emerged as a better competitor. Furthermore, in another study, the growth rate of itchgrass (*Rottboellia exaltata* L. f.), a C_4 weed, was increased by rising 3 °C in temperature. Similarly, a rise in temperature (26/18 °C; daytime maximum/night time minimum) and higher CO_2 (800μmol mol^{-1}) level was found to increase competition for tomato crops against C_4 (*A. retroflexus*) and C_3 (*C. album*) weeds (Valerio et al. 2013).

Climate change will alter the rainfall pattern and will result in occurrences of extreme drought and flood events. This might provide new habitats and a higher competitive advantage for weeds (Rodenburg et al. 2010, 2011). The meteorological data shows that the average precipitation per year in the last 20 years has risen while the rate of annual evaporation has decreased in the Tamil Nadu State, India; the study also reveals that two alien weed species, i.e., Chinese sprangletop (*Leptochloa chinensis* (L.) Nees) and water clover (*Marsilea quadrifolia L.*) have become major weeds in the area as compared to other native weed types, such as *Echinochloa* spp. (*Echinochloa crus-galli* and *Echinochloa colona*) in the rice fields due to adaptation to changing flood and soil moisture levels (Yaduraju and Rao 2013; Kathiresan et al. 2006). Under prolonged drought, the competitive capability of native plants gets reduced and new weeds may get their place (Karkanis et al. 2018). Moreover, different weeds exhibit higher shoot/root ratio to cope with drought conditions (Xu et al. 2006). Heschel et al. (2004) reported that under drought conditions lady's thumb (*Polygonum persicaria* L.) improved their WUE (water use efficiency) and root biomass. Similarly, Travlos (2013) reported that silver leaf nightshade (*Solanum elaeagnifolium* Cav.) weed showed an elevated root/shoot ratio in drought conditions.

14.6 Herbicide–Climate Interactions

Herbicides are important substances for weed control because of their easier application, reliability, and cost-effectiveness than any other methods of weed management. Climate change factors may affect the chemical properties of herbicides in a similar way as they affect crop growth and development. Herbicide efficiency will probably decrease in future climate due to global warming and aggressive growth of weeds and this may demand increased use of herbicides and exploit other potential measures for weed control. Climate factors, i.e., light, CO_2, fluctuation in temperature, wind velocities, changing humidity, rainfall, and soil water level may disturb the physical and chemical characteristics of herbicides and alter their absorption, penetration, and translocation into plants (Keikotlhaile 2011). The impact of climate change on the application patterns of foliar and soil-applied herbicides relies on the basic chemistry of herbicide and specific environmental conditions. Soil moisture along with temperature influences the effect of herbicides that are applied to soil, while several other factors affect the foliar-applied herbicides (Jugulam et al. 2018).

14.7 Effects of Elevated CO_2 and High Temperature on Herbicides

Higher CO_2 concentrations and temperature will have a substantial influence on weed biology and the sustainability of the chemical properties of herbicides (Gutierrez et al. 2008). One important influential effect of high CO_2 concentration is the decrease of stomatal conductivity, which may raise by up to 50% in various plants. Low stomatal conductance may prevent the uptake of soil and foliar-applied herbicides (Jackson et al. 2011). Moreover, high CO_2 concentration causes leaf thickness and a decrease in the number of open stomata. Decreased stomatal activity often results in decreased transpiration, which further decreases the absorption of herbicides used in the soil (Ziska 2008). Glyphosate absorption at a high CO_2 concentration of (720μmol mol^{-1}) compared to an ambient CO_2 concentration of 380μmol mol^{-1} was found to decrease in quackgrass (*Elymus repens* L.) due to a reduction in stomatal and leaf thickness (Ziska and Teasdale 2000). Moreover, elevated CO_2 levels can stimulate root over shoot growth and diluting the efficacy of foliar-applied herbicides absorbed in weeds. Ziska and George (2004) reported that elevated CO_2 concentration stimulated root to shoot growth in Canada thistle [*Cirsium arvense* (L.) Scop.] and minimized the efficiency of glyphosate.

Contrarily the effect of temperature is complex and has direct and indirect influences on herbicide efficiency. High temperature was found to decrease the thickness of cuticular lipids hence elevating penetration and translocation of herbicides via the cuticle. Higher temperature of 35 °C (compared to 15 °C) increased the ^{14}C-glyphosate injury to meristematic tissues of roundup ready soybean (Pline et al. 1999). In another study, extreme temperatures reduced herbicide activity on target plants by inducing rapid metabolism of herbicide and enhanced the action of some antioxidant enzymes that helped to detoxify reactive oxygen molecules (Godar et al. 2015). Furthermore, high temperatures appear to alleviate the absorption and translocation ability of most foliar herbicides; in various cases, higher air temperatures possibly cause rapid metabolism, which in turn reduces herbicide efficacy in the plants. A rise in temperature from 25–40 °C significantly reduced the efficiency of mesotrione on palmer amaranth (*Amaranthus palmeri* S. Watson) demonstrating a probable increase in mesotrione metabolism (Godar et al. 2015). Furthermore, plant germination and seedling development can be affected by the rise in air temperature which depicted plant's sensitivity to herbicide application. Temperature also affects uptake and translocation of herbicides in plants by biomass accumulation, cuticle thickness, and water absorption and movement (Rodenburg et al. 2011). Warmer temperature and a rise in CO_2 concentration would have notable effects on the herbicide due to damage to Photosystem II and pigment inhibition hindering photosynthesis. Linuron potency in wild buckwheat (*Polygonum convolvulus* L.) at high CO_2 concentrations was reduced by 15% (Archambault et al. 2001). However, the efficacy of atrazine was maximized in the control of *A. theophrasti* and *A. artemisiifolia* at high air temperature (Stewart et al. 2009).

14.8 Effects of Solar Radiation on Herbicides

Light is one of the main environmental factors which directly influences anatomy, morphology, and plant growth and development (by inducing photosynthesis), that in turn affects herbicide activity. High light intensity is followed by an increase in photosynthesis and phloem translocation that in turn improves the efficacy of foliar-applied herbicides. High light intensity maximized the efficacy of bentazon, clethodim, and talkoxydim on weeds. Occasionally due to photo-degradation, light may directly affect the efficacy of herbicides, for instance, at the low light intensity the efficacy of cyclohexanedione (CHD) herbicides (ACCase inhibitors) was minimized due to photo-degradation. Light intensity is important for seed, leaf cuticle development, and stomatal conductance. Foliar-applied herbicide efficacy improves at high light intensity due to open stomata; these modifications in plant growth and development may affect the ability of herbicide absorption. For instance, post-emergence herbicides absorption by the plant was increased in higher plant branching while herbicides absorption was reduced in thicker leaves. (Ohadi et al. 2010; Hatterman-Valenti et al. 2011)

14.9 Influence of Precipitation and Relative Humidity on Herbicides Efficacy

Precipitation level and RH (relative humidity) are interdependent climate factors that affect herbicide efficacy by influencing herbicide retention inside and outside of the plant. Rainfall often affects the amount of soil moisture required for plant growth. Irregular rainfall usually followed by warmer temperatures may result in extreme weather, i.e., droughts and floods, affecting the total growth and development of crops (Clements et al. 2014). In these cases, weeds are supposed to show a stronger survival tendency and resistance to herbicides applications. Intense rainfall instantly after herbicide application makes herbicide less effective by washing the herbicide from the plant surfaces and less rainfall also affects herbicide efficacy by decreasing the amount of transpiration and translocation of herbicides on the leaf surface (Keikotlhaile 2011). Under a low rainfall scenario, the efficacy of Acetyl CoA Carboxylase (ACCase) inhibitors was less effective when applied in plantain [*Brachiaria plantaginea* (Link)] (Pereira et al. 2010). Most preemergence herbicides require optimal soil moisture to pass through the soil and be absorbed via plant root. Jursík et al. (2013) explained that under less soil moisture, the efficiency of pethoxamid (applied as preemergence) decreased significantly. These studies suggest that irregular rainfall may decrease herbicide efficacy and persistence. Relative humidity affects the action of foliar-applied herbicides via herbicide spray droplet absorption and translocation inside the plant. Glufosinate ammonium uptake significantly increased on *A. ludoviciana* plants after exposure to high RH (Ramsey et al. 2002). Weeds under low RH are harder to control with postemergence herbicides compared to the normal growing plants. For instance, systemic herbicides require normal plant growth and soil moisture to effectively translocate into the target site in weed (Naidu and Ojha 2015).

14.10 Nonchemical Weed Control Options Under Climate Change

Considering the effects of climate change, the application rate and overall effectiveness of chemicals may be affected by temperature and precipitation. Cultural methods (e.g., narrow row spacing, crop rotation, etc.) can be useful practices to control weeds under climate change. In addition, some physical methods of control (e.g., solarization, mulching, etc.) can be listed among the effective methods that can be proposed considering climate change. Benefits of mulching in maintaining sustainability of agricultural systems and controlling weeds have been explained by Jabran (2019). Cover crops provide several potential benefits for any agro-ecosystem (Jabran and Chauhan 2018). This method is used to protect the surface that may be bare against erosion and avoid nutrient losses by covering the soil. Cover crops are also important for their role in controlling weeds (Tursun et al. 2018; Mennan et al.

2020). The effectiveness of cover crops can be used to control weeds under climate change. Soil solarization can be practiced in areas that have high solar radiation. Effective soil solarization is dependent on proper methodology (use of appropriate soil covering), timing, and temperature. Temperatures above 50 °C will kill the weeds completely (Lee and Thierfelder 2017). Crop rotation exerts positive ecological effects in agricultural systems and can be used to control weeds under changing climate without any extra expenses (Shahzad et al. 2016; Jabran 2017). Using allelopathy for weed control under climate change should also be considered because allelopathic means of controlling weeds have gained great attention by researchers worldwide (Jabran et al. 2015b; Jabran 2017).

14.11 Conclusion

Climate change has a wide range of effects on weeds and their control. Some changes are expected in the efficacy of herbicides against weeds under climate change. The researchers' community is desired to investigate and understand those changes critically for devising satisfactory weed control ways under the changing climate. Nonchemical weed control methods can have a greater scope under the changing climate.

References

Ainsworth, E.A. and Long, S.P., 2005. What have we learned from 15 years of free-air CO2 enrichment (FACE)? A meta-analytic review of the responses of photosynthesis, canopy properties and plant production to rising CO2. New Phytologist, 165(2), 351-372.

Alberto, A.M., Ziska, L.H., Cervancia, C.R. and Manalo, P.A., 1996. The influence of increasing carbon dioxide and temperature on competitive interactions between a C3 crop, rice (Oryza sativa) and a C4 weed (Echinochloa glabrescens). Functional Plant Biology, 23(6), 795-802.

Amanet K., M. Mubeen et al. (2019). Cotton Production in Africa. In: John Wiley &Sonss, Ltd. Pages 359-369. (https://onlinelibrary.wiley.com/doi/pdf/10.1002/9781119385523#page=370)

Amin, A., W. Nasim, M. Mubeen, et al., 2017. Optimizing the Phosphorus Use in Cotton by Using CSM-CROPGRO-Cotton Model for Semi-Arid Climate of Vehari-Punjab, Pakistan. Environmental Science and Pollution Research, 24 (6): 5811-5823

Archambault, D.J., Li, X., Robinson, D., O'Donovan, J.T. and Klein, K.K., 2001. The effects of elevated CO2 and temperature on herbicide efficacy and weed/crop competition. Rept. Prairie Adapt. Res. Collab, 29.

Bajwa, A.A., Wang, H., Chauhan, B.S. and Adkins, S.W., 2019. Effect of elevated carbon dioxide concentration on growth, productivity and glyphosate response of parthenium weed (Parthenium hysterophorus L.). Pest Management Science, 75(11), 2934-2941.

CCSP (Climate Change Science Program). 2008. The effects of climate change on agriculture, land resources, water resources, and biodiversity in the United States. A report by the US Climate Change Science Program and the Subcommittee on Global Change Research, Washington, DC: US Department of Agriculture.

Chandrasena, N., 2009. How will weed management change under climate change? Some perspectives. Journal of Crop and Weed, 5(2), 95-105.
Chongtham, S.K., Devi, E.L., Singh, R.P., Jat, A.L., Lhungdim, J., Bhupenchandra, I. and Rowndel, T.B.S.K., 2019. Crop-weed interactions and their management under climate change: A review. IJCS, 7(3), 4498-4505.
Clements, D.R., DiTommaso, A. and Hyvönen, T., 2014. Ecology and management of weeds in a changing climate. In Recent Advances in Weed Management (13-37). Springer, New York, NY.
Ferrero, A., Milan, M., De Palo, F., Fogliatto, S. and Vidotto, F., 2018. Weed control in rice grown with plastic mulching and drip irrigation system. In18th European Weed Research Society Symposium" New approaches for smarter weed management" (218-218). Kmetijski inštitut Slovenije.
Giannini, A., Biasutti, M., Held, I.M. and Sobel, A.H., 2008. A global perspective on African climate. Climatic Change, 90(4), 359-383.
Godar, A.S., Varanasi, V.K., Nakka, S., Prasad, P.V., Thompson, C.R. and Mithila, J., 2015. Physiological and molecular mechanisms of differential sensitivity of Palmer amaranth (Amaranthus palmeri) to mesotrione at varying growth temperatures. PLoS One, 10(5).
Gutierrez, A.P., Ponti, L., d'Oultremont, T. and Ellis, C.K., 2008. Climate change effects on poikilotherm tritrophic interactions. Climatic Change, 87(1), 167-192.
Hammad HM., et al. 2018. Uptake and toxicological effects of pharmaceutical active compounds on maize. Agriculture, Ecosystem and Environment, 258: 143-148
Hatfield, J.L., Boote, K.J., Kimball, B.A., Ziska, L.H., Izaurralde, R.C., Ort, D., Thomson, A.M. and Wolfe, D., 2011. Climate impacts on agriculture: implications for crop production. Agronomy Journal, 103(2), 351-370.
Hatterman-Valenti, H., Pitty, A. and Owen, M., 2011. Environmental effects on velvetleaf (Abutilon theophrasti) epicuticular wax deposition and herbicide absorption. Weed Science, 59(1), 14-21.
Heschel, M.S., Sultan, S.E., Glover, S. and Sloan, D., 2004. Population differentiation and plastic responses to drought stress in the generalist annual Polygonum persicaria. International Journal of Plant Sciences, 165(5), 817-824.
Hussain, S., et al. 2020. Study of land cover/land use changes using RS and GIS: a case study of Multan district, Pakistan. Environmental Monitoring & Assessment, 192:2: 1-15
Hyvönen, T., 2011. Impact of temperature and germination time on the success of a C4 weed in a C3 crop: Amaranthus retroflexus and spring barley. Agricultural and Food Science, 20(2), 183-189.
IPCC, 2014: Climate Change 2014: Synthesis Report. Contribution of Working Groups I, II and III to the Fifth Assessment Report of the Intergovernmental Panel on Climate Change[Core Writing Team, R.K. Pachauri and L.A. Meyer (eds.)]. IPCC, Geneva, Switzerland, 151.
Jabran, K. and Chauhan, B.S. (eds.) 2018. Non-Chemical Weed Control. Academic Press. USA.
Jabran, K. and Doğan, M.N., 2018. High carbon dioxide concentration and elevated temperature impact the growth of weeds but do not change the efficacy of glyphosate. Pest Management Science, 74(3), 766-771.
Jabran, K. and Doğan, M.N., 2020. Elevated CO2, temperature and nitrogen levels impact growth and development of invasive weeds in the Mediterranean region. Journal of the Science of Food and Agriculture, 100(13), pp.4893-4900.
Jabran, K., 2017. Manipulation of allelopathic crops for weed control. Cham: Springer International Publishing.
Jabran, K., 2019. Role of Mulching in Pest Management and Agricultural Sustainability. Springer International Publishing, Switzerland, AG.
Jabran, K., Dogan, M.N. and Eren, Ö., 2015a. Effect of ambient and simulated CO_2 on the growth invasive weed Potentilla recta L. Agriculture and Forestry, 61: 107-112.
Jabran, K., Mahajan, G., Sardana, V. and Chauhan, B.S., 2015b. Allelopathy for weed control in agricultural systems. Crop Protection, 72, 57-65.
Jabran, K., Mahmood, K., Melander, B., Bajwa, A.A. and Kudsk, P., 2017. Weed dynamics and management in wheat. Advances in Agronomy 145: 97-166.

Jackson, L.E., Wheeler, S.M., Hollander, A.D., O'Geen, A.T., Orlove, B.S., Six, J., Sumner, D.A., Santos-Martin, F., Kramer, J.B., Horwath, W.R. and Howitt, R.E., 2011. Case study on potential agricultural responses to climate change in a California landscape. Climatic Change, 109(1), 407-427.

Jursík, M., Kočárek, M., Hamouzová, K., Soukup, J. and Venclová, V., 2013. Effect of precipitation on the dissipation, efficacy and selectivity of three chloroacetamide herbicides in sunflower. Plant, Soil and Environment, 59(4), 175-182.

Jugulam, M., Varanasi, A.K., Varanasi, V.K. and Prasad, P.V.V., 2018. Climate Change Influence on Herbicide Efficacy and Weed Management. Food Security and Climate Change, 433-448.

Karkanis, A., Bilalis, D. and Efthimiadou, A., 2011. Cultivation of milk thistle (Silybum marianum L. Gaertn.), a medicinal weed. Industrial Crops and Products, 34(1), 825-830.

Karkanis, A., Ntatsi, G., Alemardan, A., Petropoulos, S. and Bilalis, D., 2018. Interference of weeds in vegetable crop cultivation, in the changing climate of Southern Europe with emphasis on drought and elevated temperatures: a review. The Journal of Agricultural Science, 156(10), 1175-1185.

Kathiresan, A., Lafitte, H.R., Chen, J., Mansueto, L., Bruskiewich, R. and Bennett, J., 2006. Gene expression microarrays and their application in drought stress research. Field Crops Research, 97(1), 101-110.

Keikotlhaile, B.M., 2011. Influence of the processing factors on pesticide residues in fruits and vegetables and its application in consumer risk assessment (Doctoral dissertation, Ghent University).

Khaliq, T., M. Mubeen, A. Ali, A. Ahmad, A. Wajid, F. Rasul, W. Nasim. 2012. Effect of Diverse Irrigation Regimes on Growth Parameters and Yield of Cotton under Faisalabad Conditions. Int. Poster J. Sci. Tech. 02 (3): 81-85.

Kizildeniz, T., Irigoyen, J.J., Pascual, I. and Morales, F., 2018. Simulating the impact of climate change (elevated CO2 and temperature, and water deficit) on the growth of red and white Tempranillo grapevine in three consecutive growing seasons (2013–2015). Agricultural Water Management, 202: 220-230.

Kizildeniz, T., Mekni, I., Santesteban, H., Pascual, I., Morales, F. and Irigoyen, J.J., 2015. Effects of climate change including elevated CO2 concentration, temperature and water deficit on growth, water status, and yield quality of grapevine (Vitis vinifera L.) cultivars. Agricultural Water Management, 159: 155-164.

Lee, J.S., 2011. Combined effect of elevated CO2 and temperature on the growth and phenology of two annual C3 and C4 weedy species. Agriculture, Ecosystems and Environment, 140(3-4),484-491.

Lee, N. and Thierfelder, C., 2017. Weed control under conservation agriculture in dryland smallholder farming systems of southern Africa. A review. Agronomy for Sustainable Development, 37(5), p.48.

Mennan, H., Jabran, K., Zandstra, B.H. and Pala, F., 2020. Non-Chemical Weed Management in Vegetables by Using Cover Crops: A Review. Agronomy, 10(2), .257.

Miri, H.R., Rastegar, A. and Bagheri, A.R., 2012. The impact of elevated CO2 on growth and competitiveness of C3 and C4 crops and weeds. European Journal of Experimental Biology, 2(4),1144-1150.

Morgan, J.A., Lecain, D.R., Mosier, A.R. and Milchunas, D.G., 2001. Elevated CO2 enhances water relations and productivity and affects gas exchange in C3 and C4 grasses of the Colorado shortgrass steppe. Global Change Biology, 7(4), 451-466.

Naidu, Y.R. and Ojha, A.K., 2015. A hybrid version of invasive weed optimization with quadratic approximation. Soft Computing, 19(12), 3581-3598.

Navie, S.C., McFadyen, R.E., Panetta, F.D. and Adkins, S.W., 2005. The effect of CO2 enrichment on the growth of a C3 weed (Parthenium hysterophorus L.) and its competitive interaction with a C4 grass (Cenchrus ciliaris L.). Plant Protection Quarterly, 20(2), 61.

Ohadi, S., Rahimian Mashhadi, H., Tavakkol-Afshari, R. and Beheshtian Mesgaran, M., 2010. Modelling the effect of light intensity and duration of exposure on seed germination of Phalaris minor and Poa annua. Weed research, 50(3), 209-217.

Panter, K.E., Colegate, S.M., Davis, T.Z., Stegelmeier, B., Welsh, S.L., Gardner, D., Lee, S.T., Cuneo, P.S. and Stonecipher, C.A., 2017. Fiddleneck (Amsinckia intermedia Lehmann Boraginaceae): Toxicity in Cattle Potentiated by Burrow Weed (Isocoma acradenia). International Journal of Poisonous Plant Research, 4(1), 16-24.

Pereira, M.R.R., Martins, D., Silva, J.I.C.D., Rodrigues-Costa, A.C.P. and Klar, A.E., 2010. Efeito de herbicidas sobre plantas de Brachiaria plantaginea submetidas a estresse hídrico. Planta Daninha, 28(SPE), 1047-1058.

Pline, W.A., Wu, J. and Hatzios, K.K., 1999. Effects of temperature and chemical additives on the response of transgenic herbicide-resistant soybeans to glufosinate and glyphosate applications. Pesticide Biochemistry and Physiology, 65(2), 119-131.

Ramesh, K., Matloob, A., Aslam, F., Florentine, S.K. and Chauhan, B.S., 2017. Weeds in a changing climate: vulnerabilities, consequences, and implications for future weed management. Frontiers in Plant Science, 8, 95.

Ramsey, R.J.L., Stephenson, G.R. and Hall, J.C., 2002. Effect of relative humidity on the uptake, translocation, and efficacy of glufosinate ammonium in wild oat (Avena fatua). Pesticide Biochemistry and Physiology, 73(1), 1-8.

Rodenburg, J., Meinke, H. and Johnson, D.E., 2011. Challenges for weed management in African rice systems in a changing climate. The Journal of Agricultural Science, 149(4), 427-435.

Rodenburg, J., Riches, C.R. and Kayeke, J.M., 2010. Addressing current and future problems of parasitic weeds in rice. Crop Protection, 29(3), 210-221.

Schmidt, J.J., Blankenship, E.E. and Lindquist, J.L., 2011. Corn and velvetleaf (Abutilon theophrasti) transpiration in response to drying soil. Weed science, 59(1), .50-54.

Shahid, M., A. Austruy, G. Echevarria, M. Arshad, M. Sanaullah, M. Aslam, M. Nadeem, W. Nasim, C. Dumat, 2014. EDTA-Enhanced Phytoremediation of Heavy 1 Metals: A Review. Soil and Sediments Contamination, 23:389–416

Shahzad, K., A. Hussain, W. Nasim et al. (2018). Tillage and biochar effects on wheat productivity under arid conditions. Crop Science, 59: 1-9

Shahzad, M., Farooq, M., Jabran, K., Hussain, M., 2016. Impact of different crop rotations and tillage systems on weed infestation and productivity of bread wheat. Crop Protection 89, 161-169.

Singer, A., Travis, J.M. and Johst, K., 2013. Interspecific interactions affect species and community responses to climate shifts. Oikos, 122(3), 358-366.

Stewart, C.L., Nurse, R.E. and Sikkema, P.H., 2009. Time of day impacts postemergence weed control in corn. Weed Technology, 23(3), 346-355.

Travlos, I.S., 2013. Competition between ACCase-inhibitor resistant and susceptible sterile wild oat (Avena sterilis) Biotypes. Weed Science, 61(1),26-31.

Tursun, N., Işık, D., Demir, Z. and Jabran, K., 2018. Use of living, mowed, and soil-incorporated cover crops for weed control in apricot orchards. Agronomy, 8(8), 150.

Valerio, M., Tomecek, M., Lovelli, S. and Ziska, L., 2013. Assessing the impact of increasing carbon dioxide and temperature on crop-weed interactions for tomato and a C3 and C4 weed species. European Journal of Agronomy, 50, 60-65.

Volis, S., Uteulin, K. and Mills, D., 2009. Russian dandelion (Taraxacum kok-saghyz): one more example of overcollecting in the past?. Journal of Applied Botany and Food Quality, 83(1), 60-63.

Wajid, A., A. Ahmad, T. Khaliq, S. Alam, A. Hussain, K. Hussain, W. Nasim, M. Usman and S. Ahmad. 2010. Quantification of growth, yield and radiation use efficiency of promising cotton cultivars at varying nitrogen levels. Pakistan Journal of Botany, 42(3): 1703-1711

Xu, X., Eng, M., Zheng, Y. and Eng, D., 2006. Comparative study of torsional and bending properties for six models of nickel-titanium root canal instruments with different cross-sections. Journal of Endodontics, 32(4), 372-375.

Yaduraju, N.T. and Rao, A.N., 2013. Implications of weeds and weed management on food security and safety in the Asia-Pacific region. Proceedings of 24th Asian-Pacific Weed Science Society Conference October 22-25, 2013, Bandung Indonesia

Yin, X. and Struik, P.C., 2009. Theoretical reconsiderations when estimating the mesophyll conductance to CO2 diffusion in leaves of C3 plants by analysis of combined gas exchange and chlorophyll fluorescence measurements. Plant, Cell & Environment, 32(11), 1513-1524.

Zand, E., Soufizadeh, S. and Eskandari, A., 2006. Water stress and nitrogen limitation effects on corn (Zea mays L.) competition with a C3 and a C4 weed. Communications in Agricultural and Applied Biological Sciences, 71(3 Pt A), 753-760.

Ziska, L.H. and George, K.A.T.E., 2004. Rising carbon dioxide and invasive, noxious plants: potential threats and consequences. World Resource Review, 16(4), pp.427-447.

Ziska, L.H. and Teasdale, J.R., 2000. Sustained growth and increased tolerance to glyphosate observed in a C3 perennial weed, quackgrass (Elytrigia repens), grown at elevated carbon dioxide. Functional Plant Biology, 27(2),159-166.

Ziska, L.H., 2000. The impact of elevated CO2 on yield loss from a C3 and C4 weed in field-grown soybean. Global Change Biology, 6(8),899-905.

Ziska, L.H., 2008. Rising atmospheric carbon dioxide and plant biology: the overlooked paradigm. DNA and Cell Biology, 27(4),165-172.

Ziska, L.H., Tomecek, M.B. and Gealy, D.R., 2010. Competitive interactions between cultivated and red rice as a function of recent and projected increases in atmospheric carbon dioxide. Agronomy Journal, 102(1),118-123.

Chapter 15
Insect Pest Management Under Climate Change

Nasir Masood, Rida Akram, Maham Fatima, Muhammad Mubeen, Sajjad Hussain, Muhammad Shakeel, Naeem Khan, Muhammad Adnan, Abdul Wahid, Adnan Noor Shah, Muhammad Zahid Ihsan, Atta Rasool, Kalim Ullah, Muhammad Awais, Mazhar Abbas, Dilshad Hussain, Khurram Shahzad, Fatima Bibi, Ishfaq Ahmad, Imran Khan, Khalid Hussain, and Wajid Nasim

Abstract Insect responses to climate change are vital for knowing the response of agroecosystems to climate change. Although numerous insect species are pests in crops, yet they also play critical roles as parasitoids and predators for other key pest species. Changes in an insect population's biochemistry, physiology, population dynamics, and biogeography may occur due to alterations in their distribution, among crop types and among the growing seasons . The response of an insect population to a quickly changing climate may also be inconsistent when insects have to interact with diverse competitors, parasitoids, and predators, and impose variable costs at a no. of life stages. The overall influence is on food production systems

N. Masood · R. Akram · M. Fatima · M. Mubeen · S. Hussain · A. Rasool
Department of Environmental Sciences, COMSATS University Islamabad, Vehari, Punjab, Pakistan

M. Shakeel
Laboratory of Bio-Pesticide Creation and Application of Guangdong Province, College of Natural Resources and Environment, South China Agricultural University, Guangzhou, Guangdong Province, China

N. Khan
Department of Agronomy, Institute of Food and Agricultural Sciences, University of Florida, Gainesville, FL, USA

M. Adnan
Department of Agriculture, The University of Swabi, Ambar, Swabi, Khyber Pakhtunkhwa, Pakistan

Department of Soil and Environmental Sciences, The University of Agriculture, Peshawar, Khyber Pakhtunkhwa, Pakistan

A. Wahid
Department of Environmental Sciences, Bahauddin Zakariya University, Multan, Punjab, Pakistan

W. N. Jatoi et al. (eds.), *Building Climate Resilience in Agriculture*,
https://doi.org/10.1007/978-3-030-79408-8_15

which can be already at acute risk from the influences of climate change. A significant limitation in improving crop production is the massive yield loss due to diseases, insect pests, and weeds all around the world. An unwise application of pesticides on crops has produced resistance among the insects and other pests and caused a severe effect on the economy of any country. This condition demands the need to endorse the idea of integrated pest management (IPM) among the farmers. IPM techniques are highly environment-sensitive that depend on the reasonable blend of physical, social, and biochemical control strategies utilized to control the pests, to minimize the economic loss and hazardous impact on the environment.

Keywords Climate change scenario · Pest outbreak · Insect pest management · IPM

A. N. Shah
Department of Agricultural Engineering, Khwaja Fareed University of Engineering and Information Technology (KFUIT), Rahim Yar Khan, Punjab, Pakistan

M. Z. Ihsan
Cholestan Institute of Desert Studies, The Islamia University Bahawalpur, Bahawalpur, Punjab, Pakistan

K. Ullah
Department of Meteorology, COMSATS University Islamabad, Islamabad, Pakistan

M. Awais · W. Nasim (✉)
Department of Agronomy, Faculty of Agriculture and Environment, The Islamia University Bahawalpur, Bahawalpur, Punjab, Pakistan
e-mail: wajid.nasim@iub.edu.pk

M. Abbas
Department of Management and MIS, College of Business Administration, University of Hail, Hail, Kingdom of Saudi Arabia

D. Hussain
Department of Management Sciences, COMSATS University Islamabad (CUI), Vehari, Punjab, Pakistan

K. Shahzad
Lasbela University of Agriculture, Water and Marine Sciences (LUAWMS)), Uthal, Balochistan, Pakistan

F. Bibi
Plant Nutrition Section, Mango Research Institute, Multan, Punjab, Pakistan

I. Ahmad
Climate Resilience Department, Asian Disaster Preparedness Center (ADPC), Islamabad, Pakistan

I. Khan · K. Hussain
Agro-Climatology Laboratory, Department of Agronomy, University of Agriculture, Faisalabad, Punjab, Pakistan

15.1 Introduction

The increasing global warming highlights the need for a comprehensive understanding of its results (Smith et al. 2015). Variations due to anthropogenic activities, in precipitation, increasing rate of severe meteorological events, the melting of ice-caps, and expanding ocean levels are very significant (Nicholls and Cazenave 2010; Maclean and Wilson 2011). Alongside the damaging impacts of climate change, the increasing temperature and carbon dioxide (CO_2) cons. may raise the rate of photosynthesis in medium to higher latitude areas thus improving agricultural production (McMahon et al. 2010; Asseng et al. 2015). While higher CO_2 conc. may bring a more significant increase in photosynthetic rate, it might decrease the quality of foliage as the concentration of defensive compounds in plants is increased. The increase in C-N proportion may influence C3 plants more as compared to C4 plants. Variations in vegetation characteristics influence the insects associated with them as well as affect the competitiveness between plants, the rate of plant infections, and higher order interactions of predation and parasitism. Accordingly, the impacts of climate change will proliferate all through food webs (Hoekman 2010).

Crop domestication started approximately a thousand years ago; usually, farmers have been plagued by huge numbers of pathogens (we shall call these pests later on) causing starvation and social disturbance (Woods 2011). Exemplary models are the 1840s Irish potato starvation brought about by oomycetes (*Phytophthora infestans)* and the 1943 Great Bengal famine because of fungi *(Helminthosporium oryzae)* (Strange and Scott 2005). Somewhere in the range of 10% and 16% of yield is lost due to pests (Chakraborty and Newton 2011). The multifarious crop pests (parasites, microscopic organisms, infections, viroids, oomycetes, bugs, and nematodes) keep on growing through the development and spread of new pathotypes (Fisher et al. 2012). Newly developed strains of the rusts *Puccinia graminis* and *Puccinia striiformis* are among the most harmful and quickly spreading pathogens (Singh et al. 2011; Cooke et al. 2012). The greater part of every single rising infection of plants is spread by introduction. Climate is the second most significant factor (Anderson et al. 2004). For instance, fusarium head blight of wheat has reappeared in the USA, supported by a warm, wet climate at anthesis. Warm conditions stimulate the insect pest's herbivory at higher scopes, principally through expanded winter endurance (Bale et al. 2002), as found in mountain pine creepy bugs (Dendroctonus ponderosae) episodes in the US Pacific Northwest. The impacts of climate are reliant on both host and pest reactions. For instance, dry season pressure can diminish plant obstruction as reported by Ansar et al. (1994), Mauch-Mani and Mauch (2005), Ali et al. (2014a), however disease likelihood is lower in dry conditions.

The impact of climate on crop infection has prompted hypothesis about the impacts of anthropogenic climate change on worldwide food security (Gregory et al. 2009). Projections are complicated by the collaborating impacts of expanding environmental CO_2 conc., changing climatic systems, frequency of severe climate events, and contrasting reactions of the plant and its competitors (Shaw and Osborne 2011). Variation in species abundance and variety because of climate change may bring about a decrease in the viability of IPM approach as described by Dhaliwal et al. (2010), Ewald et al. (2015).

Current sensitivities on environmental contamination, human health risks, and pest resurgence are a result of inappropriate utilization of manufactured pest sprays (Sharma 2016). A few naturally and organically based items are utilized as eco-friendly items. A significant number of these strategies for pest management are profoundly sensitive to nature. Increment in temperatures and UV radiation, and abatement in relative humidity may reduce a large number of these control strategies to be incapable as mentioned by Sharp et al. (1986) and Niziolek et al. (2012). Consequently, there is a need to create proper procedures for pest control that will be successful under circumstances of a worldwide temperature change in the future. Host-plant resistance, biopesticides, regular adversaries, and agronomic practices offer a possibly feasible alternative for IPM. However, the overall viability of a significant number of these control measures is probably going to change because of climate change (Ali et al. 2019).

15.2 Effects of Climate Change on Insect/Pest

In agriculture, pest outbreaks occur by a number of variables as well as insect-pest biology, synchronization of resources in host-plant, and natural competition among species (Letourneau 2012). These components are completely affected by climate. Increments in temperature, GHGs (particularly CO_2), and precipitation because of worldwide climate change will without a doubt keep on worsening many insect-pest issues in cereal and other crops. Shifting, prior spring excursions of aphids in European countries (Hulle et al. 2010), disruption of the occurrence patterns of the multivoltine tea tortrix moth in Japan (Nelson et al. 2013), and prior appearance of the potato leafhopper over the USA particularly in hotter years have all been due to change in climatic conditions (Baker et al. 2015). Eigenbrode et al. (2015) described that there are different manners by which pests may straightforwardly react to climate change (Fig. 15.1). Abiotic environments are influenced by different structures such as rocks, topsoil, landscape, and plant canopy. Biotic environments are influenced by nearby organisms, such as common herbivores insect, which are mostly influenced by change in leaves surface's temperature and humidity by opening of stomata (Akram et al. 2019; Zahoor et al. 2019). Both abiotic and biotic environments can be influenced to some extent by organisms to find their most favorable climate and make it difficult to predict the change in climate (Gia and Andrew 2015; Akram et al. 2018a, b).

Behavioral adaptation to climate change by insects is not fully studied but is a critical aspect of an insect's response to climate change (Andrew et al. 2013). Population abundances of pests, beneficial insects, competitors, and symbionts may go through substantive changes with a changing climate. For example, if a pest species is released from competitive interactions, its abundance may increase with a changing climate and it may become more invasive and impact on a wider number

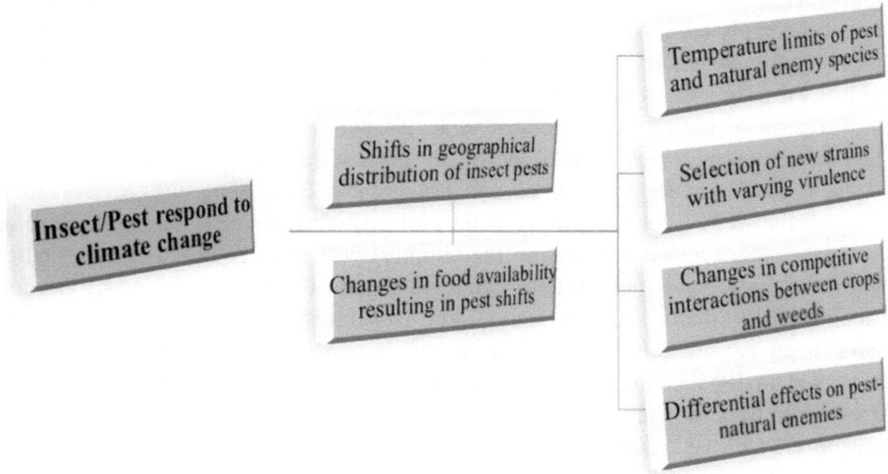

Fig. 15.1 Various ways in which pest directly respond to climate change

of species within its realized niche (Bolnick et al. 2010). If crops are planted in a new environment to keep within their climatic envelope (Nasim et al. 2016, 2018), then natural enemy and pest species may show different responses, as outlined by Gilman et al. (2010) for native species along a climatic gradient (Nasim et al. 2011, 2012).

15.2.1 Climate Change Scenario and Pest Outbreak

All international organizations working on climate change are convinced that different countries would face worst effects of global climate change. Obviously, pathosystem would also be affected by this climate change. Tree decline and dieback are emerging problem in tree plantations in these areas. It started from guava and shisham decline in 90s and now similar phenomenon is observed in trees such as mango, citrus, and loquat. Though, symptoms are more or less similar but reported causal organisms differ tremendously. It includes pathogens such as *Fusarium oxysporum* and *Colletotrichum gloeosporioides*, *Phytophthora cinnamon*, *Ceratocystis fimbriata*, pathological complex include bacteria, fungi, mollicutes (Ali et al. 2014b), and *Lasiodiplodia theobromae* (Naz 2017). This confusion exists because all authors more or less worked on the etiology of disease but there is no concrete effort to corelate the occurrence of tree disease with changing climate. The unwarranted and unprecedented increase in automobile units has increased carbon footprint in the atmosphere. Tree being perennial plants suffered most by this micro-climate change. However, no interdisciplinary effort is made to tackle this menace of plant growth. Bajwa et al. (2015) pointed out that the rise in temperature and precipitation had a profound impact on blue pine (*Pinus wallichiana*) in the forest of the study area. An educated guess could be drawn that climate change made these perennials more vulnerable against pathogens against whom in past these plants were showing resistance.

Invasion of okra crop by means of lepidopterous borers, Earias vittella, and Helicoverpa armigera has been one of the present-day troubles of the farmers in African and Asian areas (Munthali and Tshegofatso 2014). The study was planned to assess an IPM (bio-in-depth) module evolved by means of selecting through in situ assessment and incorporating the simplest pest manage alternatives together with the biological and cultural strategies against okra lepidopterous borers. According to effects, maximum shoot and fruit infestations by using okra borers have been recorded in control (unsprayed) module, while minimal have been observed in IPM module. So, IPM module is suggested to the indigenous okra growers to fight infestations of E. vittella and H. armigera and different lepidopterous borers (Aziz et al. 2012; Nawaz et al. 2019).

If weather modifications result in an intensification in pest outbreaks, farmers can also respond with the aid of making use of extra pesticides to decrease the volume of pest damage (Ziska 2014; Rosenblatt and Schmitz 2014). Determining the likely influences of climate trade on pest invertebrates, and quantifying the effects of those impacts, is needed to offer advice to farmers concerning adaptive responses. Table 15.1 shows various insects and their influences on exclusive crops. Such responses may also consist of adjustments to pesticide use, improved pest tracking technologies, modifications to crop rotation sequences, and modifications to tillage and stubble retention practices. Some of those responses are inexpensive and constitute a shift towards greater sustainable pest control practices. Others are steeply priced to put into effect. For instance, shifting from one crop kind to a distinct crop type that has a decreased susceptibility to pest damage can also involve modifications to seeding and harvesting device, modifications to crop rotation practices and adjustments to buyers, and advertising and marketing of the grain. Conversely, converting to a specific crop range that permits for earlier or later sowing may additionally require only minimal changes (Sutherst et al. 2011).

15.3 Insect Pest Management to Mitigate the Effects of Climate Change

IPM rose after WWII following the acknowledgment that aimless utilization of pesticide spray would be biologically hazardous. From that point forward, it has been expressed that IPM has become the predominant yield insurance worldview (Parsa et al. 2014). Viable IPM focuses on the guideline of conveying numerous corresponding techniques for pest, weed, and infection control (Fig. 15.2). IPM has been characterized as a "choice based procedure for organizing different strategies for control of all classes of pest in a naturally and financially stable manner." This wide scope of choices considers numerous understandings of IPM (Gadanakis et al. 2015).

Table 15.1 Various types of insects and their impacts on differnet crops

Insect name	Scientific name	Crop (s)
American bollworm	*Helicoverpa armigera* (Hubner)	Cotton, chickpea, pigeonpea, sunflower, tomato
Whitefly	*Bemisia tabaci* (Gennadius)	Cotton, tobacco
Brown planthopper	*Nilaparvata lugens* (Stal)	Rice
Green leafhopper.	*Nephotettix* spp	Rice
Serpentine leaf	miner *Liriomyza trifolii* (Burgess)	Cotton, tomato, cucurbits, several other vegetables
Fruit fly	*Bactrocera* spp.	Fruits and vegetables
Wheat aphid	*Macrosiphum miscanthi* (Takahashi)	Wheat, barley, oats
Pink stem borer	*Sesamia inferens* (Walker)	Wheat
Gall midge	*Orseolia oryzae* (Wood-Mason)	Rice
Diamondback moth	*Plutella xylostella* (Linnaeus)	Cabbage
Hoppers Several species	Pyrilla *Pyrilla perpusilla* (Walker)	Sugarcane or rice at times
Tomato leaf Miner	*Tuta absoluta* (Meyrick)	Tomato
Coconut eriopyhid mite	*Aceria guerreronis*	Coconut
Papaya mealybug	*Paracoccus marginatus*	Papaya, cotton and mulberry
Coconut leaf Beetle	*Brontispa longissima* (Gestro)	Coconut
Coffee berry Borer	*Hypothenemus hampei* (Ferrari)	Coffee
Western flower Thrips	*Frankliniella occidentalis*	Fruits and vegetables
Serpentine leaf Miner	*Liriomyza trifolii*	Burgess Cotton, tomato and cucurbits
Mealy bugs	*Paracoccus marginatus* and *Phenacoccus solenopsis*	Field and horticultural crops
Thrips Several species	*Scirtothrips dorsalis Frankliniella, schultzei* Trybom, *Thrips tabaci* L., *Scirtothrips citri*	Groundnut, cotton and citrus
Wheat aphid	*Macrosiphum miscanthi*	Wheat, barley and oat
Rice gall midge	*Orselia oryzae*	Rice
Pink stem borer	*Sesamia inferens*	Wheat, maize and sorghum
Pyrilla	*Pyrilla perpusilla*	Sugarcane
Eucalyptus gall wasp	*Leptocybe invasa*	Eucalyptus

15.3.1 General Principles of IPM

According to the EU Framework Directive 2009/128/EC, there are eight principles of IPM (Barzman et al. 2015, Fig. 15.3).

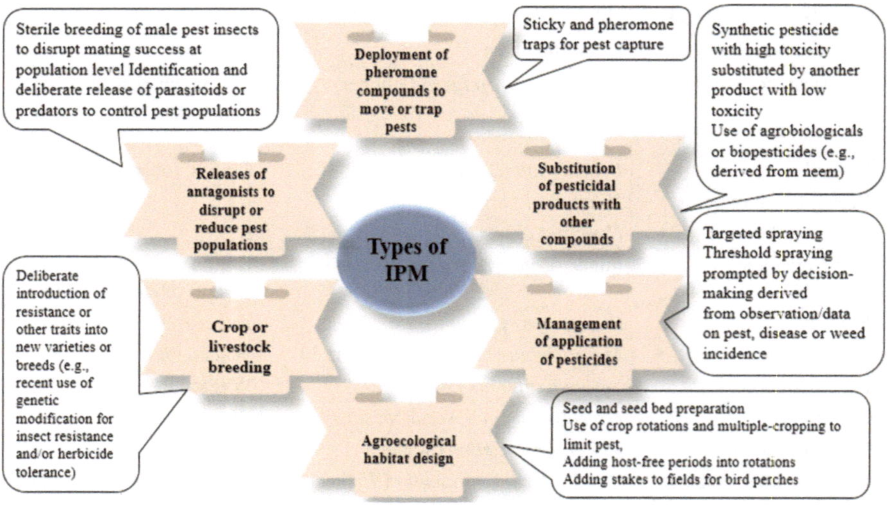

Fig. 15.2 Types of effective insect pest management

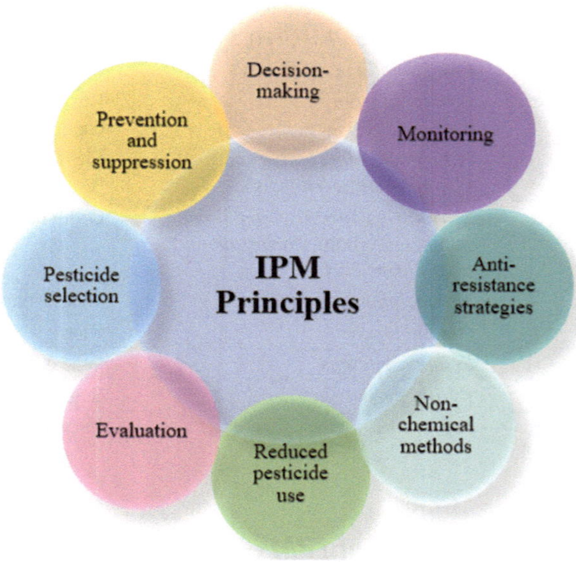

Fig. 15.3 Major principles of insect pest management

15.3.2 IPM Practices

Figure 15.4 shows four significant parts of IPM to accomplish impact results. Complete dependence on pesticides and broad utilization of synthetic concoctions for pest control is harmful to human well-being and ecological contamination.

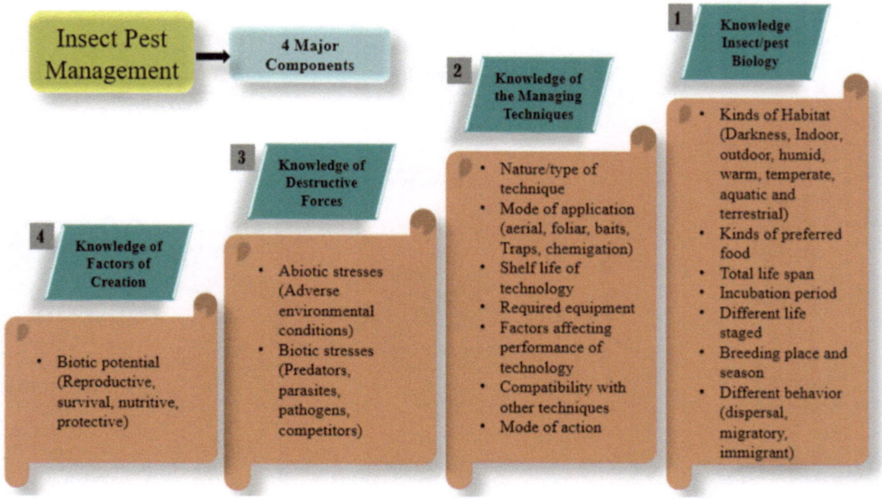

Fig. 15.4 Major components of insect pest management

15.4 Implementation of IPM Program

Effective IPM relies basically upon fundamental exploration of biological system and the comprehension of collaborations among hosts, bothers, and their normal foes. The accompanying advances ought to be taken before actualizing an IPM program (Fig 15.5).

15.5 Conclusion and Recommendations

The idea of IPM is one of a kind for the occupation of agribusiness. Entomologists with extension field staff can assume noteworthy job for moving IPM as business just at home level.

Following are some recommendations for future researchers and policymakers.

- Blend of advancements and devices, remote detecting information, Geographical Information System (GIS), Automatic Weather Stations (AWS), and internet of things (IoTs) can be utilized to advance the execution of IPM (Hussain et al. 2019, 2020).
- New age of GPS, sensors-fitted farm tool, e-tablets, and versatile applications (Plantix) could be utilized for monitoring future pest and disease threshold levels.
- Researchers can also carry out studies about converting the thoughts of farmers for the version of noninsecticidal/IPM strategies at massive level.
- In-carrier training needs to be organized for entomologists and extension area staff for distribution of new technology.

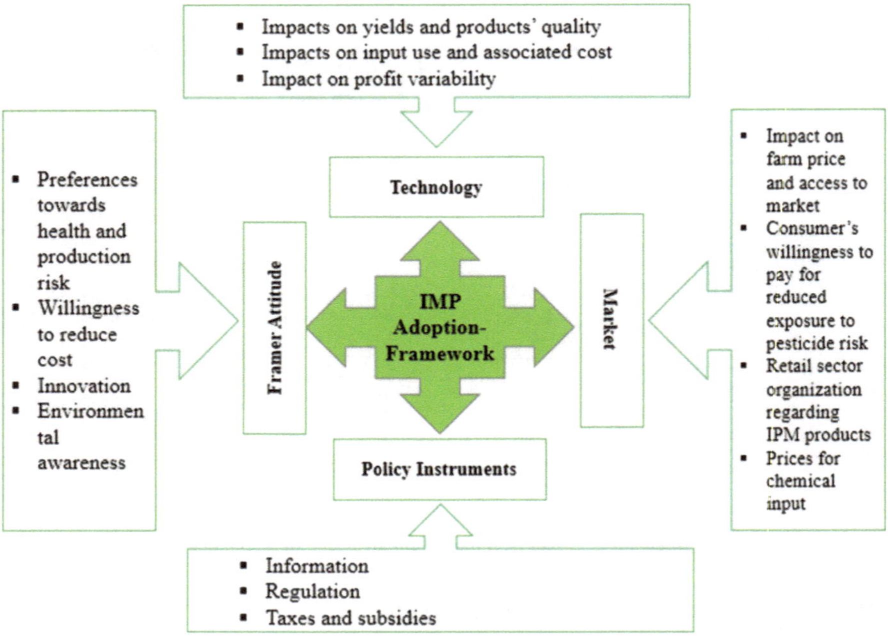

Fig. 15.5 Effective framework for insect pest management

Acknowledgment The corresponding author (Wajid Nasim Jatoi) is highly thankful to Dr Dr. Bashir Ahmad, Director CEWRI, NARC, Pakistan Agricultural Research Council, Islamabad, Pakistan. Additionally, the support and cooperation given by all colleagues especially Dr. Muhammad Aown Sammar Raza (Chairman), Dr. Muhammad Adnan Bukhari, Dr. Muhammad Aurangzaib, Dr. Muhammad Latif, Dr. Muhammad Saqib, Dr. Muhammad Usman Bashir, Dr Rashid Iqbal, Dr Abdul Rehman, Dr Farhan Khalid, Dr Muhammad Asghar Shah, Dr Muhammad Shahzad, Dr Muhammad Usman Aslam and especially Dr. Muhammad Aown Sammar Raza, Chairman, Department of Agronomy, Faculty of Agriculture and Environment, The Islamia University of Bahawalpur, Pakistan, is highly acknowledged and commendable.

References

Akram R, Turan V, Hammad HM, Ahmad S, Hussain S, Hasnain A, Maqbool MM, Rehmani MIA, Rasool A, Masood N, Mahmood F (2018b). Fate of Organic and Inorganic Pollutants in Paddy Soils. In Environmental Pollution of Paddy Soils. Springer, Cham, pp 197-214

Akram R, Turan V, Wahid A, Ijaz M, Shahid M A, Kaleem S, Hafeez A, Maqbool MM, Chaudhary HJ, Munis MFH, Mubeen M (2018a) Paddy Land Pollutants and Their Role in Climate Change. In Environmental Pollution of Paddy Soils. Springer, Cham, pp. 113–124

Akram R, Fahad S, Masood N, Rasool A, Ijaz M, Ihsan MZ, Maqbool MM, Ahmad S, Hussain S, Ahmed M, Kaleem S, Sultana SR, Mubeen M, Saud S, Kamran M, Nasim W (2019) Plant Growth and Morphological Changes in Rice Under Abiotic Stress, Editor, Nahar K, Biswas JK, Advances in Rice Research for abiotic Stress Tolerance, Woodhead publishing pp. 69–85

Ali MPDH, Nachman G, Ahmed N, Begum MA, Rabbi MF (2014a) Will climate change affect outbreak patterns of planthoppers in Bangladesh? PLoS One 9:e91678

Ali SR, Snyder J, Shehzad M, Khalid AUR, Rashad S (2014b) Associations among fungi, bacteria, and phytoplasma in trees suffering citrus decline in Punjab. J AgricTechnol 10:1343-1352

Ali M, Mubeen M, Hussain N, Wajid A, Farid HU, Awais M, Hussain S, Akram W, Amin A, Akram R, Imran M (2019) Role of ICT in Crop Management. In Agronomic Crops (pp. 637-652). Springer, Singapore. doi: https://doi.org/10.1007/978-981-32-9783-8_28

Anderson P K, et al. (2004) Emerging infectious diseases of plants: Pathogen pollution, climate change and agrotechnology drivers. Trends Ecol Evol 19:535-544

Andrew NR, Hill SJ, Binns M et al. (2013) Assessing insect responses to climate change: what are we testing for? Where should we be heading? Peer J1: e11

Ansar M, Saleem A, Iqbal A (1994) Cause and control of guava decline in the Punjab (Pakistan). Pak J Phytopathol 6:41-44

Asseng S, Ewert F, Martre P, Rötter RP, Lobell DB, Cammarano D, Kimball BA, Ottman MJ, Wall GW, White JW et al. (2015) Rising temperatures reduce global wheat production. Nat Clim Chang 5:143–147

Aziz MA, Hasan MU, Ali A, Iqbal J (2012) Comparative efficacy of different strategies for management of spotted bollworms, *Eariasspp*. on okra, *Abelmoschusesculentus*(L). Moench. Pak J Zool 44:1203-1208

Bajwa GA, Shahzad MK, Satti HK (2015) Climate change and its impacts on growth of blue pine (*Pinuswallichiana*) in Murree forest division, Pakistan. SciTechnolDevel 34:27-34

Baker MB, Venugopal PD, Lamp WO (2015) Climate change and phenology: Empoascafabae (Hemiptera: Cicadellidae) migration and severity of impact. PLoS One 10:e0124915

Bale JS et al. (2002) Herbivory in global climate change research: direct effects of rising temperature on insect herbivores. Glob Chang Biol 8:116

Barzman M, Bárberi P, Birch ANE, Boonekamp P, Dacchbrodt-Saaydeh S, Graf B, et al. (2015) Eight principles of integrated pest management. AgronSustDevel 35:1199-1215

Bolnick DI, Ingram T, Stutz WE, Snowberg LK, Lau OL, Paull JS (2010) Ecological release from interspecific competition leads to decoupled changes in population and individual niche width. Proceedings of the Royal Society of London B:Biol Sci 277:1789–1797

Chakraborty S, Newton AC (2011) Climate change, plant diseases and food security: An overview. Plant Pathol 60:214

Cooke DEL, et al. (2012) Genome analyses of an aggressive and invasive lineage of the Irish potato famine pathogen. PLoSPathog8:e1002940

Dhaliwal GS, Jindal V, Dhawan AK (2010) Insect pest problems and crop losses: Changing trends. Indian J Ecol 37:1-7

Eigenbrode SD, Davis TS, Crowder DW (2015) Climate change and biological control in agricultural systems: principles and examples from North America. In: Bjorkman, C., Niemela, P. (Eds.), Climate Change and Insect Pests pp. 119– 135

Ewald JA, Wheatley CJ, Aebischer NJ, Moreby SJ, Duffield SJ, Crick HQ, Morecroft MB (2015) Influences of extreme weather, climate and pesticide use on invertebrates in cereal fields over 42 years. Glob Chang Biol 21:3931– 3950

Fisher MC, et al. (2012) Emerging fungal threats to animal, plant and ecosystem health. Nat 484:186-194

Gadanakis Y, Bennett R, Park J, Areal FJ (2015) Evaluating the Sustainable Intensification of arable farms. J Environ Manag150:288–298

Gia MH, Andrew NR (2015) Performance of the cabbage aphid *Brevicoryne brassicae* (Hemiptera: Aphididae) on canola varieties. Gen Appl Entomol 43:1–10

Gilman SE, Urban MC, Tewksbury J, Gilchrist GW, Holt RD (2010) A framework for community interactions under climate change. Trends EcolEvol25:325–331

Gregory PJ, Johnson SN, Newton AC, Ingram JSI (2009) Integrating pests and pathogens into the climate change/food security debate. J Exp Bot 60:2827-2838

Hoekman D (2010) Turning up the heat: Temperature influences the relative importance of top-down and bottom-up effects. Ecol 91:2819–2825

Hulle M, Coeur d'Acier A, Bankhead-Dronnet S, Harrington R (2010) Aphids in the face of global changes. C R Biol 333:497–503

Hussain S, Mubeen M, Ahmad A, Akram W, Hammad H M, Ali M, Masood N, Amin A, Farid HU, Sultana SR, Fahad S (**2019**) Using GIS tools to detect the land use/land cover changes during forty years in Lodhran District of Pakistan. Environ SciPollut Res 1, 1-17. doi: https://doi.org/10.1007/s11356-019-06072-3

Hussain S, Mubeen M, Akram W, Ahmad A, Habib-ur-Rahman M, Ghaffar A, Amin A, Awais M, Farid HU, Farooq A, Nasim W (2020) Study of land cover/land use changes using RS and GIS: a case study of Multan district, Pakistan. Environ Monit Assess 192(1) 2. doi: https://doi.org/10.1007/s10661-019-7959-1

Letourneau DK (2012) Integrated pest management-Outbreaks prevented, delayed, or facilitated? In: Barbosa, P., Letourneau, D.K., Agrawal, A.A. (Eds.), Insect Outbreaks Revisited. John Wiley & Sons Ltd., Chichester, UK, pp. 371–394

Maclean IMD, Wilson RJ (2011) Recent ecological responses to climate change support predictions of high extinction risk. Proc Natl AcadSci USA 108, 12337–12342

Mauch-Mani B, Mauch F (2005) The role of abscisic acid in plant–pathogen interactions. Curr Opin Plant Biol 8:409-414

McMahon SM, Parker GG, Miller DR (2010) Evidence for a recent increase in forest growth. Proc Natl AcadSci USA 107:3611–3615

Munthali DC, Tshegofatso AB (2014) Major insect pests attacking okra; *Abelmoscusesculentos* (L) Moench, in Sebele. Botswana J AgricApplSci 9: 90-96

Nasim W, Ahmad A, Amin A, Tariq M, Awais M, Saqib M, Jabran K, Shah G M, Sultana S R, Hammad H M, Rehmani M I A (2018) Radiation Efficiency and Nitrogen Fertilizer Impacts on Sunflower Crop in Contrasting Environments of Punjab-Pakistan. Environ SciPollut Res 25:1822–1836

Nasim W, Ahmad A, Hammad H M, Chaudhary H J, Munis M F H (2012) Effect of nitrogen on growth and yield of sunflower under semiarid conditions of Pakistan. Pak J Bot 44:639–648

Nasim W, Ahmad A, Wajid, A, Akhtar J, Muhammad D (2011) Nitrogen effects on growth and development of sunflower hybrids under agro-climatic conditions of Multan. Pak J Bot 43:2083–2092

Nasim W, Belhouchette H, Tariq M, Fahad S, Hammad H M, Mubeen M, Munis M F H, Chaudhary H J, Khan I, Mahmood F, Abbas T (2016) Correlation studies on nitrogen for sunflower crop across the agroclimatic variability. Environ SciPollut Res 23:3658–3670

Nawaz A, Razaq M, Sarwar ZM, Sajjad M, Ubaid SA, Zulfiqar MA, Haider U (2019) Determination of economic threshold level (ETL) of jassid, *Amrasca bigutulla bigutulla* (Cicadellidae: Homoptera). Pak J Agric Res 32:28-32

Naz F (2017) Surveillance morpho-molecular characterization and in vitro management of *Lasiodiplodia theobromeae* causing twig dieback of loquat, 6th International Conference of Pakistan Phytopathological Society, BZU, Multan, Pakistan

Nelson WA, Bjornstad ON, Yamanaka T (2013) Recurrent insect outbreaks caused by temperature-driven changes in system stability. Sci 341:796–799

Nicholls RJ, Cazenave A (2010) Sea-level rise and its impact on coastal zones. Sci 328:1517–1520

Niziolek OK, Berenbaum MR, DeLucia EH (2012) Impact of elevated CO_2 and temperature on Japanese beetle herbivory. Insect Sci 20:513-23

Parsa S Morse S, Bonifacio A, Chancellor TCB, Condori B, Crespo-Perez V, Hobbs SLA, Kroschel J, Ba MN, Rebaudo F, et al. (2014) Obstacles to integrated pest management adoption in developing countries. Proc Natl AcadSci USA 111:3889–3894

Rosenblatt AE, Schmitz OJ (2014) Interactive effects of multiple climate change variables on trophic interactions: a meta-analysis. Clim. Chang Responses 1:1– 10

Sharma HC (2016) Climate change vis-a-vis pest management. Proceedings in conference on national priorities in plant health management. Tirupati pp 17-25

Sharp DS, Eskenazi B, Harrison R, Callas P, Smith AH (1986) Delayed health hazards of pesticide exposure, Annual Rev Pub Health 7:441-471

Shaw MW, Osborne TM (2011) Geographic distribution of plant pathogens in response to climate change. Plant Pathol 60:31-43

Singh RP, et al. (2011) The Emergence of Ug99 races of the stem rust fungus is a threat to world wheat production. Annu Rev Phytopathol 49:465-481

Smith SJ, Edmonds J, Hartin CA, Mundra A, Calvin K (2015) Near-term acceleration in the rate of temperature change. Nat Clim Chang 5:333–336

Strange RN, Scott PR (2005) Plant disease: A threat to global food security. Annu Rev Phytopathol 43:83-116

Sutherst RW, Constable F, Finlay KJ, Harrington R, Luck J, Zalucki MP (2011) Adapting to crop pest and pathogen risks under a changing climate. Wiley Interdiscip Rev 2:220–237

Woods A (2011) Is the health of British Columbia's forests being influenced by climate change? If so, was this predictable? Can. J. Plant Pathol 33: 117-126

Zahoor SA, Ahmad S, Ahmad A, Wajid A, Khaliq T, Mubeen M, Hussain S, Din MSU, Amin A, Awais M, Nasim W (2019) Improving Water Use Efficiency in Agronomic Crop Production. In *Agronomic Crops* (pp. 13-29). Springer, Singapore. doi: https://doi.org/10.1007/978-981-32-9783-8_2

Ziska LH (2014) Increasing minimum daily temperatures are associated with enhanced pesticide use in cultivated soybean along a latitudinal gradient in the mid-western United States. PLoS ONE 9:e98516

Part III
Socio-Economic and Biophysical Research

Chapter 16
Effects of Climate Change on the Socioeconomic Conditions of Farmers: A Case Study

Khuda Bakhsh, Syed Asif Ali Naqvi, and Wajid Nasim

Abstract Climate change directly affects agricultural production and thus agricultural income is very sensitive to climate change. Rural population heavily depends on earnings from agriculture and allied activities. The effects on agriculture income resulting from climate change are thus transmitted to socioeconomic conditions of farm households in particular and landless farm households in general. The objective of overcoming the effects of climate change on agriculture income can be achieved through adaptation to climate change. Farmers find difficulties in adaptation to climate change due to low level of knowledge and very little access to information. Having access to information on perception and adaptation to climate change is essential for improving socioeconomic conditions of rural population. Rural households may be adopting different strategies to cope with climate change. Some of them can be type of occupations, housing structure, preservation of food during extreme climate change events, and pattern of consumption. Similarly, farmers may be adapting to climate change through changing farm operations/practices, cropping pattern, and resistant crop varieties. All such practices are directly linked with socioeconomic conditions because adaptation strategies require financial resources and farmers/rural households may be reallocating resources from health, education, and consumption toward adaptation to climate change.

Keywords Extreme events · Rural vulnerability · Agricultural income · Access to information · Cropping pattern change · Income diversification

---

K. Bakhsh (✉)
Department of Management Sciences, COMSATS University Islamabad (CUI), Vehari, Pakistan
e-mail: kbakhsh@ciitvehari.edu.pk

S. A. A. Naqvi
Department of Economics, Government College University Faisalabad, Faisalabad, Punjab, Pakistan

W. Nasim
Department of Agronomy, Faculty of Agriculture and Environment, The Islamia University of Bahawalpur (IUB), Bahawalpur, Punjab, Pakistan

W. N. Jatoi et al. (eds.), *Building Climate Resilience in Agriculture*,
https://doi.org/10.1007/978-3-030-79408-8_16

16.1 Introduction

Climate change has become the reality in many developing countries. Extreme events related to climate change are happening in almost all developing countries with varied intensity and frequency (Ali et al. 2019). The impacts resulting from climate change related extreme events pose major threats to survival of humanity. The most crucial damaging consequences of climate change include rapidly rising temperature, fluctuating rainfall, persistent droughts, floods, and earthquakes (Hammad et al. 2017). Although all countries are exposed to adverse impacts of climate change, vulnerability of the developing countries is higher as livelihood of majority of people living in these countries highly depends on agriculture and forestry, being the climate sensitive sectors. Population is heterogeneous in terms of income level and social well-being. It is thus expected that climate change can disproportionately increase the sufferings of the poorest in the society particularly living in rural area and urban slums. Climate change exacerbates miseries of the poorest people by influencing access to education, health, food, and water (Sabagh et al. 2020).

Although all spheres of life are expected to be impacted by climate change, the most vulnerable sector is agriculture (Nasim et al. 2010). Since food security of any nation in developing countries heavily relies on production of food crops, any setbacks to food crops can jeopardize food security situation of the concerned country. Similarly, economic growth significantly depends on the performance of agriculture sector (Box 16.1). Highly dependency of the economy on agriculture also highlights the need to industrialize the economy. Otherwise, the country will continue to suffer from natural disasters because agriculture sector is more prone to vagaries of disasters.

In order to avoid and or minimize the impacts of climate change, farmers consider adaptation measures at their farms. Agriculture sector is the main pillar for improving socioeconomic conditions of rural population including small and landless farmers. Importance of this sector in providing employment opportunities can be evident from the fact that 38.5% labor force is employed in agriculture and related activities. Food crops including wheat and rice have a significant contribution to the food security of developing countries including Pakistan (Rasool et al. 2016). Agriculture sector also induces acceleration of growth in other sectors of the economy (Box 16.2). Disasters mostly related to climate change have devastated the economy of Pakistan over different points of time like prolonged drought during 1998–2001, earthquake of 2005, and floods of 2010, 2011, and 2012. These disasters badly affected livestock, crops, and human lives leaving the economy in the worst conditions.

Box 16.1: Relationship Between and Table 16.1 shows the growth of agriculture and economy. During the 1960s, growth of agriculture sector (5.1%) led to higher performance of manufacturing and services sectors resulting in overall higher growth of economy. During the decades of low performance of agriculture sector, growth of other sectors and overall economy remained very low.

Literature from Pakistan provides evidence of impacts of climate change on the productivity of rice and wheat (Mahmood et al. 2012; Siddiqui et al. 2012; Tariq et al. 2014; Bokhari et al. 2017; Khan et al. 2020). Climate change has drastic impacts on rain-fed agriculture compared to canal irrigated agriculture because rain-fed agriculture significantly depends on precipitation and small landholders dominate in this region with limited financial capacity of adaptation to climate change (Bakhsh and Kamran, 2019; Hussain et al. 2019, 2020). Naqvi et al. (2019) argue that climate change negatively affects farm production, farm returns, per capita income, and poverty in rural areas of Pakistan.

Box 16.2: Agriculture sector is among the major sectors of the economy of Pakistan. It ensures food security, generates different types of employment for labor force including unskilled, semiskilled, and skilled individuals, provides raw material to industrial sector, and contributes to foreign exchange earnings. Salient contributions are as under:

- Around 60% population depending on agriculture rely directly and indirectly on the income derived from agriculture-related activities
- Wheat and rice are two important crops having a significant role in food security of the country
- Performance of agriculture highly influences the income of rural population
- Any setback in agricultural production can have an impact on the performance of other sectors and livelihood of rural population
- Poverty is very high in rural population compared to urban population. A reduction in poverty can be achieved by increasing the productivity of agriculture as it would lead to an increase in job creation Agriculture sector in Pakistan has witnessed different climate change related extreme events resulting in huge losses to farming community and national income. In Pakistan, around 80% of disasters are related to climate change. These disasters have significantly affected the of Pakistan. Ahmad (2015) showed massive cumulative effects of disasters on the economy of Pakistan. He found that prolonged drought during 1998–2001 drastically reduced GDP (by 50%) while the flood of 2010 caused an economic loss of 5.7% of GDP. In both time periods, the major loss occurred in agriculture sector. Climate change induced disasters are also expected to impact the national economy in future as well. Khan et al. (2020) forecast that climate change would cause a loss of $19.5 billion in the form of reduction in production of wheat and rice by 2050. This reduction in production would lead to increasing prices of commodities and a reduction in consumption in the country. Rising prices and a decline in consumption would exacerbate malnutrition in the country.

Table 16.1 Relationship between agriculture and economy of Pakistan (%)

Years	GDP	Agriculture	Manufacturing	Services
1960s	6.8	5.1	9.9	6.7
1970s	4.8	2.4	5.5	6.3
1980s	6.5	5.4	8.2	6.7
1990s	4.6	4.4	4.8	4.6
2000s	4.8	3.2	7.0	5.3
2011–18	3.7	3.1	3.6	4.0

Source: Pakistan Economic Survey different issues

Dependence of livelihood of the farming community is exposed to adverse effects of climate change. Extreme events relating to climate change cause variation in sowing and harvesting time, change in input requirements, access and availability of natural resources like irrigation water, and technological changes. Farmers can face substantial productivity losses if they do not go for adaptation to climate at their farms. These losses can have drastic effects on their livelihood. On the other hand, they have to incur additional cost of adaptation for sustaining crop productivity. This chapter is designed to see the role of farmers' perception in adaptation to climate change and the consequent impact of adaptation on socioeconomic conditions of farmers.

This chapter has six sections. The first section introduces the chapter, the second part is allocated to describe the type of data and analytical methods used. Third section describes vulnerability and threats to livelihood of households. Fourth section elaborates perception and adaptation. We discuss consistencies of the results from the present study with prior expectations in fifth section. The last section of this chapter is devoted to conclusion and future research along with limitations of the study.

16.2 Data Sources

Two data sets are used in this chapter. Data on heat waves is taken from Faisalabad district. Faisalabad district is facing threats of heat waves as evident from the previous studies as number of cases of deaths and morbidity are reported. Further, being the third largest populous city of Pakistan, heat waves would be severe in the coming years in Faisalabad (Shakeel et al. 2014). Majority of population of this district lives in rural areas because out of total of 289 union councils, 166 union councils are rural. Moreover, rural union councils have little facilities of water, sanitation, education, and health thus making the population living in these union councils more vulnerable. The analysis is done based on income level. The respondents are categorized into three income groups. They included low-income, middle-income, and high-income groups to understand the severity of the problem in different income groups. Heat waves can have heterogeneous impacts on population with different levels of income due to varied level of information, perception, and adaptation.

The second type of data came from Worldwide Fund for Nature Pakistan (WWF-Pakistan). This type of data focused on adaptation in agriculture. This data was taken from rain-fed districts of Punjab province. These districts included Chakwal and Rawalpindi. This data was collected during a survey conducted under the project, "The Determinants, Impact and Cost Effectiveness of Climate Change Adaptation in the Indus Ecoregion," funded by the International Development and Research Centre (IDRC), by the project team of WWF-Pakistan.[1]

We analyzed the data by employing graphical and descriptive approaches. Bar charts were used to see the relation between perception, severity, adaptation, and socioeconomic conditions. Descriptive statistics included percentage, mean and standard deviation.

16.3 Vulnerability and Threats to Livelihood of Rural Population

Climate change affects different spheres of human life directly and indirectly in the forms of extreme events namely temperature, rainfall, drought, hailstorms, floods, heat waves, etc. Heat waves are associated with a decline in labor productivity and rising health cost of the population. In the recent past, it has caused number of deaths and hospitalization in the south of Pakistan particularly in big cities including Karachi, Faisalabad, and Multan (Ali et al. 2018). The impact of extreme events of climate change can vary across different income groups (Akram et al. 2018). Low- and middle-income households are considered highly vulnerable to extreme events including heat waves. Based on the data collected from Faisalabad on the knowledge relating to heat waves, we can see that perception, severity of climate change, barriers, and cues to action are different among three income groups (Table 16.2). Households with high income are found with higher perception compared to other two groups. However, households with low income consider themselves to be highly vulnerable to climate change as 54% of households consider that severity of climate change related extreme events is very high. This percentage

Table 16.2 Vulnerability of population to heat waves as perceived by households

Particulars	Low income	Middle income	High income
Perception (High = 1)	0.34 (0.48)	0.35 (0.48	0.43 (0.49)
Severity of climate change (High = 1)	0.54 (0.51)	0.48 (0.50)	0.43 (0.49)
Barriers to adaptation (High = 1)	0.54 (0.51)	0.55 (0.50)	0.46 (0.50)
Cues to actions (High = 1)	0.68 (0.47)	0.70 (0.46)	0.67 (0.47)
Adaptation (High = 1)	0.37 (0.49)	0.46 (0.50)	0.56 (0.49)

[1] I am highly thankful to WWF-Pakistan for providing the access to the data.

declines with an increase in income. Barriers to adaptation to heat waves are higher for households with low income than Middle- and high-income households. Although low-income households find themselves highly exposed to extreme events of climate change (namely heat waves), percentage of high adaptation is low. These results imply that low-income households face two types of difficulties relating to climate change. First, their vulnerability to climate change related extreme events is high. Second, impact lies in the fact that limited financial resources hinder them in adaptation to climate change. This further aggravates their miseries. Economic costs associated with heat wave-induced morbidity and mortality can be reduced by adaptation strategies (see, for example, Rauf et al. 2017; Bakhsh et al. 2018). Generating awareness among the masses relating to benefits of adaptation is of the utmost importance to decrease vulnerability of population.

We consider responses of the respondents regarding frequency of extreme events over the last 15 years and we find that around 78% and 79% of respondents are not aware whether floods/flash floods and drought respectively are increasing or not. This is due to the fact that such respondents are taken from the rain-fed region where farmers rarely experience floods/flash floods. The same is the case with drought because this area already represents drought-like situation throughout the year. Wind/dust storm and hailstorm are known to majority of the respondents. We further analyze the data to see the perception of the respondents relating to frequency of extreme events. These responses include "increasing," "the same," and "decreasing." Approximately 32% and 30% farmers are of the view that respective frequency of windstorms and hailstorms have increased during the last 15 years. Those claiming that the frequency of these events have decreased are 20% and 28%, respectively. The responses relating to the no change for the respective extreme events is 19% and 22%, respectively (Fig. 16.1).

Extreme events have devastated impact on the well-being of the farming community, so they are much concerned about the occurrence of these events. Table 16.3 provides information on the severity of extreme events of climate change as perceived by the farmers. Hailstorms and wind/dust storms are considered important in terms of severity since a small percentage report that they do not know about the severity of these extreme events. Among those having knowledge of the severity of

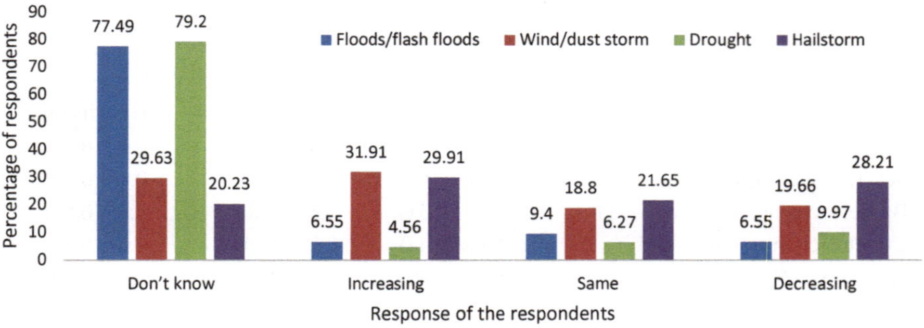

Fig. 16.1 Frequency of extreme events over the last 15 years reported by the respondents

Table 16.3 Severity of extreme events over the last 15 years (%)

Particulars	Do not know	Increasing	Same	Decreasing
Floods/flash floods	78.06	6.55	9.69	5.70
Wind/dust storm	30.20	32.48	25.36	11.97
Drought	80.34	5.98	3.99	9.69
Hailstorm	20.23	32.76	21.65	25.36

Table 16.4 Perception of farmers regarding climate change during the last 15 years

Climatic variables	Mean	Standard deviation
Change in rainfall	0.03	0.17
Less rainfall	0.36	0.52
More rainfall	0.58	0.49
Change in onset of rainfall	0.02	0.14
Change in temperature	0.04	0.19
More hot days	0.81	0.40
Less hot days	0.08	0.28
Change in onset of hot season	0.05	0.23
Change in night temperature	0.46	0.50
Change in cold spells	0.38	0.48

these events, 33% consider that severity of wind/dust storms has increased while 12% are of the view that the severity has declined and 25% of farmers' response is in the favor of the same option. Rising severity of hailstorm is reported by 33% whereas 25% provide the response in the favor of decreasing option and 22% of farmers' response shows that the severity has not changed during the last 15 years. Floods and drought are considered with little severity because these selected districts are less vulnerable to floods and droughts.

16.4 Perception and Adaptation to Climate Change Among Farming Community

It is very important to understand what farmers perceive about climate change and how perception and knowledge play the role in the decision relating to adaptation to climate (Qasim et al. 2020). In turn, this understanding helps to design programs and policies for fruitful implementation of adaptation in the agriculture sector. However, there is very limited evidence on information and knowledge relating to relation between perception and adaptation in Pakistan. Table 16.4 shows perception of farming community to climate change related extreme events. Farmers perceived that they observed overall change in rainfall during the last 15 years however this percentage is very small, i.e., 3% only. Less rainfall was reported by 36% whereas 58% of respondents were of the view that they find an increase in rainfall over a period of 15 years in their location. Temperature was another important climatic variable influencing the decisions of the farming community relating to crop choices, adaptation practices, and the cost associated with adaptation (Adnan et al. 2017). When they were asked to provide their perception whether they observed any change in temperature over a period of 15 years, only 4% agreed to this statement. For further information, the respondents were requested to provide information whether hot days have increased during the last 15 years, we find that 81% of respondents perceived that they experienced more hot days compared to the less hot days. This indicates that farmers perceived and experienced more hot days. Further, they agreed to more hot days instead when they were asked about the general statement, i.e., the change in temperature. In connection to temperature, 46% of respondents perceived that there occurred a change in night temperature during the mentioned time period whereas perception of the change in cold spell was reported by 26% of respondents. Higher perception level relating to more number of hot days and change in night time temperature and cold spells influenced farmers to make decision in the light of these climatic extreme events in order to cope with climate change impacts on their farming practices and crop production.

Farmers in developing countries make decisions to cope with short-term shock events (Sajjad et al. 2012) and these responses are mostly made with an aim of decreasing the impacts of climate change and increasing income in order to ensure sustainable livelihood. Adaptive capacity of farmers varies depending on infrastructure development. It is also the case that the farmers with better infrastructure development are more probable to have a higher potential of adaptation. Farmers consider various adaptation practices and these practices include crop diversification, mixed crop, livestock farming, a change in sowing and harvesting time, using different varieties, resistant varieties, and high yielding crop varieties. Bhushal et al. (2009) argue that farmers employ coping activities using local knowledge and innovation and they are usually not aware of the impacts of climate change.

Hasnain et al. (2017) found that almost all the world is facing climate change but the impacts of climate in the developing countries are more apparent due to higher level of vulnerability. Further, the highly exposed sector to impacts of climate change in the developing countries is agriculture. Farmers have the only option of adaptation to climate change. They are making changes in their farming practices, choice of crops, seed varieties, sowing and harvesting time, infrastructure development, and many more. Each and every adaptation strategy requires expenditures, i.e., adaptation cost. Here we present an overview on the added cost associated with a particular adaptation practice (Table 16.5). Cropping pattern is among the most used adaptation practices to climate change adopted by the farmers (Sorhang and Kristiansen 2011). Changing cropping pattern also incurs some additional costs to farmers. This practice adds Pak Rs 103 to farmers for making changes in the prevailing cropping pattern. Changing cropping pattern and other adaptation practices cause a change in the use of farm inputs. Crop diversification is assumed to serve as insurance against variation in rainfall because rainfall has different effects on various crops. Among the farm inputs employed at the farms, fertilizer is the most important input used to improve crop productivity. The adaptation practices cause

Table 16.5 Impact of adaptation strategies on livelihood (Pakistani Rs)

Change in input use due to climate change	Added cost of adaptation	
	Mean	Standard deviation
Cropping pattern	102.86	1148.44
Fertilizer	1799.15	4709.68
Seed	498.60	2075.48
Pesticide	820.23	2477.32
Labor	161.54	875.10

Table 16.6 Relationship between adaptation practices and income

	Change in income (%)	
Livelihood practices	No effect	Effect on income
Livestock and fishing	97.44	2.56
Off-farm job	64.96	35.04
Private business	81.20	18.8
Migration to urban area	96.01	3.99

an increase in the use of fertilizers and the higher use of fertilizer adds Pak Rs 1799 to the total cost. Another crucial farm input in terms of added cost is pesticide. Climate change is linked with more pest attacks on crops due to climate change and the farmers must employ a huge quantity of pesticides to control pest attacks on crops. Seed and labor are other inputs causing an increase in cost because of changing farm inputs due to climate change. These statistics imply that the farmers already facing rising cost of production face an additional challenge due to adaptation to climate change. The added cost due to changing farm inputs can affect the socioeconomic welfare of farmers, particularly small landholders.

Climate change poses serious threats to livelihood of farming community. To minimize threats of climate change, farmers are expected to consider different income generating activities in order to sustain livelihood. Income diversification is assumed to be the most widely used option to increase income. Table 16.6 shows the effect of various livelihood practices on income. Small farmers are particularly found employing income diversification activities as they find it difficult to meet their daily expenses. Livestock and or fishing is one such livelihood strategy to increase the income of households. The findings of Table 16.6 show that 2.56% of respondents report that this practice has some impact on their income level. Off-farm job is among the most important option for the farmers having impact on income thereby livelihood of farming community. 35% of respondents agree that off-farm job activities positively contribute to income. After off-farm job, private business at village level also causes an increase in income as reported by 18.8% of respondents. Since small farmers dominate in the rural areas of Punjab province of Pakistan, they move to urban centers with a hope for better employment opportunities. Uneducated and unskilled rural people find it tough to find a good job at the urban centers. They end up in doing odd jobs at low wages. Thus, only 3.99% of respondents are of the view that people migrating to urban areas have very little contribution in income. This calls for promoting agro-based industries in rural areas in order to discourage influx of rural people migrating to urban areas.

16.5 Consistencies of the Findings with Prior Expectations

We discuss the results of the study in relation to the prior expectations. Heat waves being one of the extreme events of climate change is considered to have a substantial impact on human health. Although all types of population are vulnerable to heat waves, we assume that perception, barriers of adaptation to heat waves, cue to actions, and finally adaptation can be commonly observed among high-income groups. The results of the study confirm these priori expectations.

Regarding perception of farmers to climate change related extreme events namely rainfall and temperature, priori expectations imply that farmers would provide responses in favor of rising temperature and rainfall. Bakhsh et al. (2020) argue that temperature and rainfall are significantly related to human conflicts and rising economic costs. They also maintain that farmers are aware of climate change. More rainfall and higher number of hot days are reported by 58% and 81% of farmers respectively in the present study showing that a larger portion of farming community in the study area is aware of rising temperature and an increase in rainfall. These statistics are in line with priori expectations. Further, it is assumed that adaptation to climate change in agriculture brings changes in farm practices, input uses. These changes happen at the expense of added cost to farmers. We see in the present study that farmers adapting to climate change face an addition to the existing cost of farm inputs. The most added cost is found in the use of fertilizers followed by pesticides and seeds. These added costs can have two types of impacts on farmers. One is relating to an additional financial burden to farming community possessing small landholdings and the other one is positive in the form of avoided damages.

Farmers are expected to diversify income sources in order to face the challenges of climate change. Income diversification can be found in the forms of livestock and fishing activities, off-farm jobs, private business, and migration of one or more members of household to urban center for better livelihood strategies. These diversified sources of income are expected to have a positive impact on farmers' income. Farmers reporting that off-farm job causes an increase in income by around 35%, confirming prior expectation of the role of off-farm job in enhancing income of farmers. Contribution of private business is 18.8% whereas contribution of migration to urban area is not according to prior expectations as it is reported that its impact on income is approximately 4% only. This may be due to the reason that those migrating to urban areas are mostly unskilled and such persons mostly do not find a reasonably good job and therefore the impact on households' income is very small.

16.6 Conclusion and Future Research

It is concluded that low-income population is aware of heat waves, but the population is constrained to adapt to climate change mainly due to financial constraints. Similarly, farmers in the rain-fed region of Punjab province perceive more hot days as a result of rising temperature. Perception on flood and drought is very low among the farmers due to the geographical location of the selected districts. Farmers are highly worried about adverse impacts of climate change on their crop production. However, they face challenges of not adapting to climate change due to one and other reasons. The major reason lies in the fact that adaptation practices add cost to production and the farmers are already facing rising cost of production. Adaptation practices cause a change in farm inputs and resultantly expenditures on fertilizer, pesticide, seed and labor increase.

This study can have applications to other regions of Pakistan and other developing world with similar characteristics of farmers and the environment discussed in this chapter. Considering Pakistan, farming community is characterized by small landholding, low level of education, large family size, and very little access to information particularly relating to climate change and adaptation. For such conditions of the farmers, the findings of the study can be generalized.

The shortcomings of the study include no use of statistical model, small sample size, and data taken from a few districts. Statistical model better explains to understand the precise impact of different parameters including climate change on livelihood of farming community. However, the purpose of the study was to explain the relationship between adaptation to climate change and the effects on socioeconomic conditions for the general understanding of the readers including students, academia, journalists, and policymakers. Using complex statistical methods can make it difficult for the target audience to understand the implications of the study. However, it would be better to use an appropriate statistical method for finding the impact of climate change on socioeconomic conditions of farmers using the large data sets taken from different agro-ecological zones of Pakistan. Another limitation of the study is no discussion on the socioeconomic conditions in relation to impact assessment, mitigation, and resilience. Since the data sets used in the study did not have information on these aspects, future research should incorporate impact assessment, mitigation, and resilience of farmers linked to socioeconomic conditions.

This study considers only heat waves in one district and adaptation to climate change and adaptation cost in rain-fed agriculture. Geographical conditions, access to information, and infrastructure development vary across different districts and provinces. Therefore, future research should consider these aspects by organizing a study on a wider scale. This will help to understand the role of geographical conditions, access to information, and infrastructure development in adaptation to climate change. This would also help in designing appropriate policies to encourage farmers in adapting to climate change.

References

Adnan M, Shah Z, Fahad S, Arif M, Alam M, Khan IA, Mian IA, Basir A, Ullah H, Arshad M, Rahman IU (2017) Phosphate-Solubilizing Bacteria Nullify the Antagonistic Effect of Soil Calcification on Bioavailability of Phosphorus in Alkaline Soils. Scientific Reports 2018: 07-22653

Ahmad N (2015) Economic losses from disasters. Policy brief of LEAD-Pakistan, Islamabad available at http://www.lead.org.pk/lead/attachments/briefings/LPNB3.pdf

Akram, R., Turan, V., Hammad, H. M., Ahmad, S., Hussain, S., Hasnain, A., ... & Nasim, W. (2018). Fate of organic and inorganic pollutants in paddy soils. In *Environmental pollution of paddy soils* (pp. 197-214). Springer, Cham.

Ali M, Mubeen M, Hussain N, Wajid A, Farid HU, Awais M, Hussain S, Akram W, Amin A, Akram R, Imran M (2019) Role of ICT in crop management. In: Mirza H., (eds). Agronomic Crops: ISBN 978-981-32-9150-8 (https://link.springer.com/chapter/10.1007/978-981-32-9783-8_28).

Ali S, Khan N, Nouroz FA, Erum S, Nasim W (2018) Effects of sucrose and growth regulators on the microtuberization of potato germplasm. Pakistan Journal of Botany 50(2): 763-768

Bakhsh K, Abbas K, Hassan S, Yasin AM, Ali R, Ahmad N, Chattha AWM (2020) Climate change induced human conflicts and economic costs in Pakistani Punjab. Environmental Science and Pollution Research, https://doi.org/10.1007/s11356-020-08607-5

Bakhsh K, Kamran AM (2019) Adaptation to climate change in rain-fed farming system in Punjab, Pakistan. International Journal of the Commons 13(2):833-847

Bakhsh K, Rauf S, Zulfiqar F (2018) Adaptation strategies for minimizing heat wave induced morbidity and its determinants. Sustainable Cities and Society 41(1):95-103

Bhushal Y, Tiwari KR, Timilsina YP (2009) Local peoples' perceptions on climate change, its impacts and adaptation measures in Mid-Mountain Region of Nepal (A Case study from Kaski District). B.Sc. Forestry Research Thesis, Tribhuvan University, Institute of Forestry, Pokhara, Nepal

Bokhari SAA, Rasul G, Ruane AC, Hoogenboom G, Ahmad A (2017) The past and future changes in climate of the rice-wheat cropping zone in Punjab, Pakistan. Pakistan Journal of Meteorology 13(6):9-23

Hammad HM, Farhad W, Abbas F, Fahad S, Saeed S, Nasim W, Bakhat HF (2017) Maize plant nitrogen uptake dynamics at limited irrigation water and nitrogen. Environmental Science and Pollution Research 24(3):2549-2557

Hasnain A, et al. (2017) Antibiotics resistance Genes. In: Hashmi M., Strezov V., Varma A. (eds) Antibiotics and Antibiotics Resistance Genes in Soils. Soil Biology, vol 51. Springer, Cham (https://link.springer.com/chapter/10.1007/978-3-319-66260-2_2)

Hussain, S., Mubeen, M., Ahmad, A., Akram, W., Hammad, H. M., Ali, M., ... & Nasim, W. (2019). Using GIS tools to detect the land use/land cover changes during forty years in Lodhran district of Pakistan. *Environmental Science and Pollution Research*, 1-17.

Hussain, S., Mubeen, M., Akram, W., Ahmad, A., Habib-ur-Rahman, M., Ghaffar, A., ... Nasim, W. (2020). Study of land cover/land use changes using RS and GIS: a case study of Multan district, Pakistan. *Environmental monitoring and assessment*, *192*(1), 1-15.

Khan AM, Tahir A, Khurshid N, Husnain IM, Ahmad M, Boughanmi H (2020) Economic effects of climate change-induced loss of agricultural production by 2050: A case study of Pakistan. Sustainability 12(3):1216

Mahmood N, Ahmad B, Hassan S, Bakhsh K (2012). Impact of temperature and precipitation on rice productivity in rice-wheat cropping system of Punjab province. The Journal of Animal and Plant Sciences 22(4):993-997

Naqvi A.A.S, Nadeem MA, Iqbal AM, Ali S, Noreen A (2019) Assessing the vulnerabilities of current and future production systems in Punjab, Pakistan. Sustainability 11(19):5365

Nasim W, Ahmad A, WajidSA, Hussain A, Khaliq T, Usman M, Hammad HM, Sultana SR, Mubeen M, Ahmad S (2010) Simulation of different wheat cultivars under agro-ecological condition of Faisalabad-Pakistan. Crop Environment 1(1): 44-48

Qasim MZ, Hammad HM, Abbas F, Saeed S, Bakhat HF, Nasim W, Farhad W, Rabbani F, Fahad S (2020) The potential applications of picotechnology in biomedical and environmental sciences. Environmental Science and Pollution Research (Accepted)

Rasool A, Xiao T, Farooqi A, Shafeeque M, Masood S, Ali S, Fahad S, Nasim W (2016) Arsenic and heavy metal contaminations in the tube well water of Punjab, Pakistan and risk assessment: A case study. Ecological Engineering 95(1):90-100

Rauf S, Bakhsh K, Abbas A, Hassan S, Ali A, Kächele H (2017) How hard they hit? Perception, adaptation and public health implications of heat waves in urban and peri-urban Pakistan. Environmental Science and Pollution Research 24(11):10630-10639

Sabagh, A. E., Hossain, A., Islam, M. S., Iqbal, M. A., Fahad, S., Ratnasekera, D., ... & Saneoka, H. (2020). Consequences and Mitigation Strategies of Heat Stress for Sustainability of Soybean (Glycine max L. Merr.) Production under the Changing Climate. In *Plant Stress Physiology*. IntechOpen.

Sajjad M, Khan SH, Ashfaq M, Nasim W (2012) Association of seed morphology with seedling vigor in wheat (Triticum aestivum L). Research Plant Biology 2(5):07-12

Shakeel M, Akram W, Ali A, Ali MW, Nasim W (2014) Frequency of Aphid (Aphis Gossypii G.) on Brinjal (Solanum melongena L.) and farming practices in agroclimatic conditions of Faisalabad, Pakistan. International Journal of Agriculture and Innovation Research 02(5):841-845

Siddiqui R, Samad G, Nasir M, Jalil HH (2012) The impact of climate change on major agricultural crops: evidence from Punjab, Pakistan. Pakistan Development Review 4(51):261-274

Sorhang A, Kristiansen S (2011) Climate change impacts and adaptations among Ethiopian farmers. M.Sc. Thesis, Faculty of Economic and Social Sciences for Development Studies, University of Adger, Ethiopia.

Tariq A, Tabassum N, Bakhsh K, Ashfaq M, Hassan S (2014) Food security in the context of climate change in Pakistan. Pakistan Journal of Commerce and Social Sciences 8(2):540-550

Chapter 17
Research on Climate Change Issues

Rida Akram, Tasmiya Jabeen, Maham Asif Bukari, Syed Aftab Wajid, Muhammad Mubeen, Fahd Rasul, Sajjad Hussain, Muhammad Aurangzaib, Muhammad Adnan Bukhari, Hafiz Mohkum Hammad, Muhammad Zamin, Muhammad Habib ur Rahman, Javaid Iqbal, Muhammad Ishaq Asif Rehmani, Muhammad Tariq, Ghulam Abbas, Nosheen Mirza, Hussani Mubarak, Faisal Mahmood, Muhammad Sajjad, Shaukat Ali, and Wajid Nasim

Abstract Agriculture plays important role in human welfare. An extensive research has been carried out on agriculture and changes in climate over the past decades. Variation in climate is anticipated to influence livestock and crop production, hydrologic balances, input supplies, and other constituents of agricultural systems. It is likely to have significant effect on food security in the upcoming decades, by decreasing food production and enhancing food prices. Food may become more exclusive as climate variation mitigation struggles increase energy prices. Water essential for food production may become rarer due to enhanced crop water use and dry period. Struggle for land may enhance as definite regions become climatically unsuitable for production. Though stable increases in carbon dioxide and temperature may outcome in more satisfactory conditions that could enhance the production of some crops, but in some areas, these latent production increases are likely to be limited by severe events, particularly extreme drought and heat, during flowering phase. Crop production is probably to reduce in many regions during the twenty first century due to climatic variations.

R. Akram · T. Jabeen · M. A. Bukari · M. Mubeen · S. Hussain
Department of Environmental Sciences, COMSATS University Islamabad, Vehari, Punjab, Pakistan

S. A. Wajid · F. Rasul
Agro-Climatology Lab., Department of Agronomy, University of Agriculture, Faisalabad, Punjab, Pakistan

M. Aurangzaib · M. A. Bukhari · W. Nasim (✉)
Department of Agronomy, Faculty of Agriculture and Environment, The Islamia University Bahawalpur, Bahawalpur, Punjab, Pakistan
e-mail: wajid.nasim@iub.edu.pk

W. N. Jatoi et al. (eds.), *Building Climate Resilience in Agriculture*,
https://doi.org/10.1007/978-3-030-79408-8_17

Keywords CO_2 · Temperature · Drought · Major crops · Food security

17.1 Introduction

In agricultural productivity, climate is a significant factor. Agriculture plays important role in human welfare, federal agencies have specified some concerns on the possible effects of climate variation on crop production (Holmes 2020). Due to this

H. M. Hammad
Department of Environmental Sciences, COMSATS University Islamabad (CUI), Vehari, Punjab, Pakistan

Department of Agronomy, MNS-University of Agriculture, Multan, Punjab, Pakistan

M. Zamin
Department of Agriculture, University of Swabi, Swabi, Khyber Pakhtunkhwa, Pakistan

M. H. ur Rahman
Institute of Crop Science and Resource Conservation (INRES), Crop Science Group, University of Bonn, Bonn, Germany

Department of Agronomy, MNS-University of Agriculture, Multan, Punjab, Pakistan

J. Iqbal · M. I. A. Rehmani
Department of Agronomy, Ghazi University, Dera Ghazi Khan, Punjab, Pakistan

M. Tariq
Agronomy Section, Central Cotton Research Institute (CCRI), Multan, Punjab, Pakistan

G. Abbas
Department of Agronomy, Bahauddin Zakariya University, Multan, Punjab, Pakistan

N. Mirza
Department of Environmental Sciences, COMSATS University Islamabad (CUI), Abbottabad, Khyber Pakhtunkhwa, Pakistan

Department of Soil and Environmental Sciences, Ghazi University, Dera Ghazi Khan, Punjab, Pakistan
e-mail: nmirza@gudgk.edu.pk

H. Mubarak
Department of Soil and Environmental Sciences, Ghazi University, Dera Ghazi Khan, Punjab, Pakistan

F. Mahmood
Department of Environmental Sciences & Engineering, Government College University, Faisalabad, Punjab, Pakistan

M. Sajjad
Department of Bio-Sciences, COMSATS University Islamabad (CUI), Islamabad, Pakistan

S. Ali
Global Change Impact Studies Centre (GCISC), Ministry of Climate Change, Islamabad, Pakistan

issue, extensive research has raised on agriculture and changes in climate over the past decades. Variation in climate is anticipated to effect on livestock and crop production, hydrologic balances, input supplies, and other constituents of agricultural systems (Zahoor et al. 2019). Furthermore, CO_2 is crucial for plant growth and development; rising concentrations have the tendency to enhance the yield of agriculture systems. Climate changes may also change the frequencies, types, and intensities of livestock and crop pests; the timing and availability of irrigation water; and the intensity of soil destruction (Ijaz et al. 2019). Therefore, the human reaction is complex to consider and approximate the possible impact of climatic changes on food supply and crop production. It is challenging but essential to accurately measure the impacts of climate change. Climate variations are estimating the effect on yield and cumulative demand for production such as labor, water, equipment, energy, and essential supplies (Ahmad et al. 2017; Sabagh et al. 2020).

Climate change impact on agriculture, in any case, has developed to be a more theoretical issue as compared to the hypothetical one producing voluminous literature (Amin et al. 2018). Adverse rainfall and high temperature have occurred as the key element of agriculture region produce across the agronomic and globe models of climate change manuscript a range of unfavorable climatic effect on agriculture of developing countries. Research shows that high temperature causes shorter periods of seed development, high respiration rates, and low production of biomass. Most likely, consequences of high temperatures include shorter grain filling duration, lighter and smaller grains in size, and possibly lower grain quality and lower crop production. Climate variation is corresponding to scientific modification in agricultural system which can affect the overall efficiency and can decrease or increase the yield of one factor as compared to another (Awais et al. 2018; Stefanski et al. 2020).

17.2 Climate Change and Agriculture

Agriculture is a complex part and one of the most susceptible segments to the effects of climate variation. Climate change is a crucial contribution to agricultural productivity and climate variation will inevitably have an effect on output, agricultural yield, prices, and farm incomes. History shows that most studies of agriculture productivity overlooked changes in climate (Akram et al. 2017a). Variations in climate will also have a significant effect on crop growth and production. Greater temperatures tend to lead to rapid crop development, an abbreviated grain filling phase, and reduced production. This is lessened to certain grade through enhanced stomatal conductance, due to high cons. of CO_2, which leads to improved water used ability and enhanced optimum temperatures for C3 plants. Though increase temperatures can damage plant cells, and during the flowering stage extreme heat increases barrenness. Other than that, aggressive wild plant tends to be superior adapted to a varying climate, with little immature phase, long distance seed spread, and eminent response to increased cons. of CO_2. In conclusion, there are many different paths

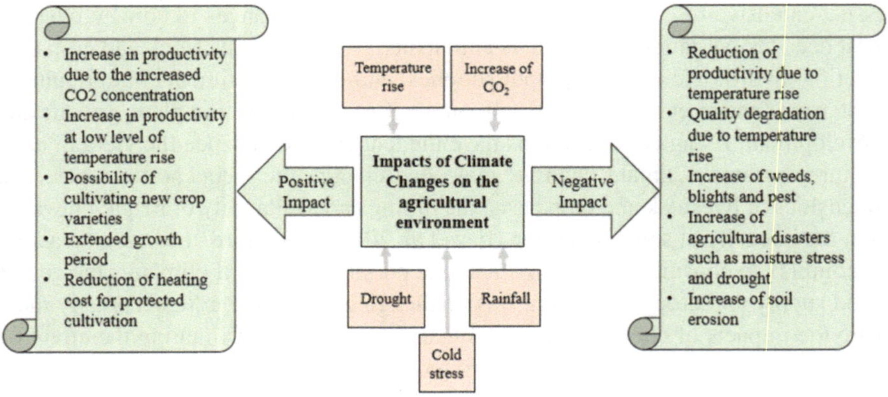

Fig. 17.1 Impacts of climate change on agriculture system

through which climate variation can affect crop growth and production (Dawson et al. 2016). Plant yields are effected by several ecological features like temperature and humidity, which may act also antagonistically or synergistically with other aspects in decisive production. Agriculture is integrally delicate to climate changeability and alteration, due to either human activities or natural causes (Akram et al. 2017b). Changes in climate triggered by GHG emissions are assessed to directly affect crop production systems for feed, fodder, or food; to affect livestock health; and to change the balance and pattern of food products and trade of food. Agricultural production has already been effected by climate change and these effects will fluctuate with the grade of warming and accompanying changes in precipitation patterns, in addition to, from one location to another (Corlett and Tomlinson 2020) (Fig. 17.1).

17.3 Impacts of Climate Change on Agriculture of Different Regions

Crop production is estimated to reduce in upcoming climate conditions, and current research proposes that production has already been compressed. The impacts of climate change on agricultural systems of different regions are given as follows:

17.3.1 Europe

All central crop production in southern and western European countries declined to 6.3–21.2% due to change in climate. It describes the yields less productivity in European countries (Brisson et al. 2010). It was observed that, a reduction in the major crops production such as barley, maize, wheat, and rapeseed for parts of the cultivated area in Russian regions and in the cereal strap of Western Siberia (Sobolev 2014). Since 1970s, in Russian Federation, the annual temperature has enhanced at approximately 0.4 °C in a decade (Sobolev 2014). The antagonistic impacts of climate change have been observed on maize, barley, and sorghum in Ukraine (Müller et al. 2016). Yearly crop production losses in southern and western Europe are enhanced though exceptions thrive as in Andalucıa in southern Spain where wheat crop production gained from mean climate variations. As well, in northern and eastern European regions crop production decrease for maize, wheat, and barley is −24.5%, −2.1%, and −9.1% respectively (Blanc et al. 2019).

17.3.2 Asia

In Asia, the average increase in climate change enhanced the overall crop production cereal crops by approximately 2% however decreased the yield production of wheat and rice. It is also observed that wheat and maize production also increased due to climate change in Huang-Huai-Hai and Heilongjiang province, respectively (Meng et al. 2014; Tao et al. 2017). In India (UP and Haryana), it has been noticed that few conditions with a determined pattern of production undergo losses in major cereal crops, like in wheat and rice −0.7% and −2.1%, respectively. The province of Laguna, Philippines, also face reduction in rice production at −0.2%, while generally rice productivity improved in the Philippines. In Turkey, wheat yield also decreased approximately at −0.8 MT. Climate change has lower consumable meal calories in lots of Asian international locations both food secured (Israel and Iran) and insecured (Bangladesh, Pakistan, Nepal, and India). The effect on crop production only due to change in temperature and rainfall is mapped. The effects of temperature are more potent in East Asia and Europe while rainfall best results are similarly strong in South Asia, sub-Saharan Africa, and Australia (Fig. 17.2).

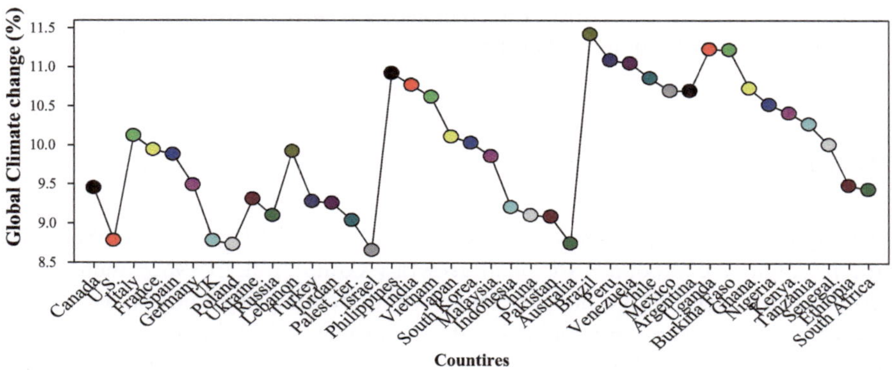

Fig. 17.2 Global climate change in different regions

17.4 Current Issues in Climate Change and Agriculure-Related Research

The research demonstrates that impacts of climate variation are decearsing the ability of natural resources (soil, biodiversity, and water) to cope with the need for food of the world's increasing population (Ahmad et al. 2017). Climate variation and food security are interlinked issues that need to be addressed immediately. Enhancing resource effectiveness in building resilience and agriculture to climate dangers are the main actions for undertaking these issues (Fahad et al. 2018). Although there are various features of climate change to which Grassland agriculture is vulnerable, dry period can impose the most widespread damage. A long period of unusually low precipitation, especially one that causes adverse effects on living and growing conditions is called drought (Nasim et al. 2018; Mubeen et al. 2020, 2021). At the start of the growing season, agriculture has always been dependent on the climate variability and state of land for the growing season. The significant adjustment for crop generation to climate change is the reliability of the conditions (Akram et al. 2019a, b). It is required for the understanding of the effect of changing climate on water, land, and temperature. High temperature extends the farmable area and growing season, it gives rise to prior development of grain and opens up for the new growing crops. However, the high temperature is favorable for the crops, the extra heat also kills the weeds. Weeds, insects, and pests tend to improve living conditions under higher temperatures. The combination of the pests, weeds, and poor herbicide performance reduce the possible crop production. At the initial stage, water is essential for production but not throughout the final stages during the growth cycle of plant. Low precipitation has a harmful impact on the germination of the seeds. Drought frequency, conditions, and intensity of dust storms all result in reduced production of main grains. In general, food production is expected to deteriorate in many dangerous areas (subtropical and tropical areas). In developing countries agriculture area is acutally beneficial where technology is more accessible where proper adaptive modifications are employed (Ali et al. 2019a, b; Akram et al. 2018a, b). Climate variation has slight impacts on overall food production, but these impacts are unfairly spread in nature. Most of the damages are suffered in lower income states, such as those in sub-humid and arid South Asia and Africa.

17.4.1 Increase in CO_2

Through the process of photosynthesis, plants use water, CO_2, and sunlight to synthesize organic compounds for plant growth (Akram et al. 2019b; Amanet et al. 2019). The effects of CO_2 increase on growth of cotton. This study showed that if the CO_2 concentrations are doubled in the atmosphere which increased the process of photosynthesis by about 40% which lead to increased yield and growth in well-watered environments. The increasing conc. of CO_2, increased the water use. From studies it was concluded that increase in yield and growth would occur as a result of higher conc. of CO_2 even in nutrient scarce or dry situations. By using this work, we could believe that increases in the conc. of CO_2 to the levels that will be forecast is 406–415 ppm and 473–555 ppm for 2020 and 2050 respectively and photosynthetic process would increase approximately to 23% and 29%, correspondingly. In cotton field, on increased level of CO_2 using free air CO_2 (FACE) facilities. They came to know that the use of radiation effectiveness was developed on average from 1.56 g MJ^{-1} to 1.97 g MJ^{-1} which results in improved biomass when the conc. of CO_2 was increased upto 550 ppm. The use of irrigation management, radiation use efficiency method was increased in CO_2 level. This recommended that an increase in CO_2 concentrations in atmosphere may partly recompense for stress in plant which is due to water scarcity. This result average lint yield was increased by 43%—as a result of a longer flowering period and increased early leaf area.

17.4.2 Increase in Temperature

There are two major influences on development and growth of cotton. First, it finds out the crop growth and rates of morphological development. Second, it also helps to determine the start and end of a growth season. Climate change results in the increase of temperatures.

17.4.3 Rainfall

The total annual and extreme rainfall patterns are expected to change with raising GHGs conc. in the US. The global precipitation increases from 5% to 25%, which enhanced the evapotranspiration that results in earth's warming, especially unreasonable rise of temperature in winter can increase hydrological cycle. Timing and the quantity of rain is a main concern for crop productivity in the US. Precipitation pattern throughout the season is estimated to alter which leads to dry summers and wet winters. The disproportion concerning less rain and high rate of evapotranspiration in summer, leading to extra dehydrating of soil at significant growing periods of crop. Due to climate change decline in rainfall was predicted by CSIRO and RegCM models were 20% and 30%, respectively in the US during summer and rise in rainfall predicted by both models was 35% and 25% during spring season (Rahman et al. 2018). On the other hand, wetted winters, which occur due to increased rainfall and temperatures, possibly will cause flooding during winter.

17.4.4 Drought

The unfavorable effects of water inadequacy on rice cultivation would be, to an enormous quantity, kept up under raised encompassing CO_2 intensities, which shows CO_2 and dry season cooperation. As indicated by an investigation, humidity stress influences rice at morphological (diminished germination, plant tallness, plant biomass, number of tillers, different root, and leaves characteristics), physiological (decreased photosynthetic rate, transpiration rate, and stomata opening and closing), biochemical, and subatomic levels and along these lines influences its production (Tariq et al. 2018). The regenerative stage is influenced by diminished grain arrangement, thwarted pollen growth during meiosis period, and panicle exertion, which can ordinarily represent 70–75% spikelet sterility due to drought (Fig. 17.3).

17.4.5 Cold Stress

Low temperature is a main constrain for rice development, and seedling is considered as most foundational stage at which affectability to chilling compression is greater. Cold stress prompts a progression of changes in physiological and subatomic procedures and results in the amassing of reactive oxygen species in plant cells.

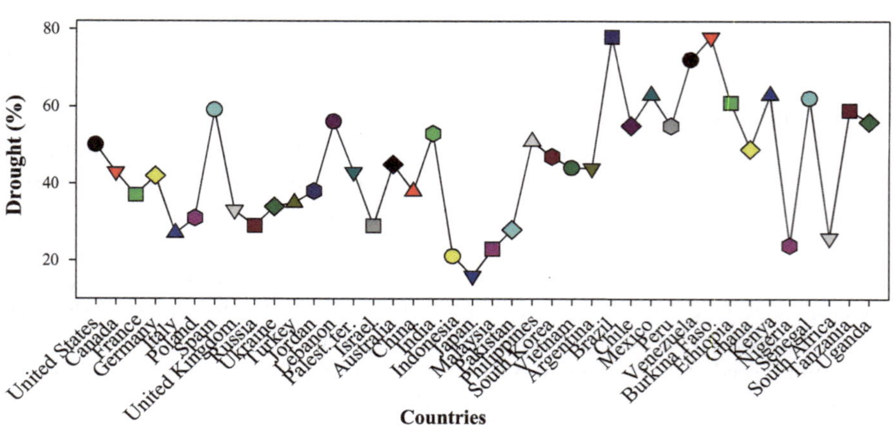

Fig. 17.3 Effects of drought percentage in different countries due to climate change

17.5 Impacts of Climate Change on Major Crops

17.5.1 Cotton

Cotton is persistent with an uncertain pattern of growth. Modern crops have these hereditary characteristics, which makes the cotton crop well modified to irregular supply of water that happens with irrigated production and rain-fed. Its development and growth is quite complex as compared with other crops. The reproductive and vegetative growth occurs concurrently making the analysis of the response of crop to climate and sometimes managing becomes hard. Climatic change has effects on phonological stages of cotton that control fiber quality and yield due to: Increases in concentration of CO_2; increased atmospheric evaporative demand and reduced water accessibility due to less rain and moisture; and high temperature (Hussain et al. 2020a).

17.5.2 Wheat

The atmospheric CO_2 conc. has raised (280–390μmol) since 1800. Future researches specify that the concentration of CO_2 will probably reach 550 and 700μmol by 2050–2100, respectively in North America. Further, nitrous oxide (N_2O) concentration and methane (CH_4) concentration have reached limits beyond the values which can be found out by past experiments. Agreeing to the US Global Change Research Program, the US agriculture system represents the emission of greenhouse gasses nationally is 8.6%, including the nitrous oxide emissions is 80% and emission of methane is 31%. Globally, 13.5% of greenhouse gasses are released due to human activities in agricultural sector. The IPCC SRES predictions show a global increase in temperature from 3–7 °F during the next century and in North America increased 5–7 °F next to 2100. The last 5 years are considered as warmest in prior century and mean surface air temperature increased by 1.08 °F globally. The period and rate of strong heat waves is possibly to rise all over the US if conc. of GHGs continues to increase. The CSIRO1 model expects that the temperatures will rise from 5 to 9 °F in which higher temperatures dominate in spring and winter months. The RegCM2 model predicted that the mean temperature will increase from 1.8 to 12.6 °F in the US, but the performance of both models is contradictory.

17.5.3 Maize

Maize yields are comparatively low when it is related to developed countries. The nation's average maize production is expected to be 1.6 ton/ha. This is primarily because of the limited input of irrigation facilities and fertilizers in the growth of maize. In the previous years, there is more decrease in yields in several countries. Even though further fundamental factors present, indiscretion in rainfall and increasing temperature has been referred to as the main reason for the constant decline of crop production (Mubeen et al. 2016). The production of maize in the coastal savannah region decreased above the previous 16 years. The decrease in maize production is also observed in the Mfantseman region. Decline in the precipitation rate and the rising of temperatures have been recommended as likely reasons of the decrease in the yield. The same decrease in the maize production in agricultural system. In 2020, according to the report of EPA it is estimated that the yield of maize decreased by 6.9% in the alteration area in comparison with standard yield production. For these estimations CERES maize model is used, to generate different climate scenarios using three GCMs (UK Meteorological Office High Resolution Model (UKHI), the Hadley Centre Model 2 (HadCM2), and the UK Meteorological Office Transient Model (UKTR)). It simulates the yield and growth rate of maize in specified environmental, weather conditions, and agricultural managing methods (Abbas et al. 2017). Another crop simulation model APSIM is used to estimate the impacts of climate change on maize and projections showed 5.0–13.4% decrease in yield production from 2046 to 2065 (Tachie-Obeng et al. 2013).

17.5.4 Rice

The changes in climatic conditions will diminish rice production by 4.5–9% by 2039 in India. Worldwide climatic expectations showed extended rate of temperature pierces and warmer evenings, applying further difficulties to accomplishing higher harvest rates. The storm precipitation is not the main climate variable upsetting the kharif rice production. Paddy yield and its reaction to environmental change can be assessed by using simulation models, for example, GIS-based-GEPIC model (Hussain et al. 2020b, c). According to the exploration the climatic boundaries, for example, temperature, precipitation, CO_2, and sunlight radiation are essential to rice cultivation. An expansion in temperature changeability and precipitation fluctuation were seen as supportive and unsafe, individually to fall and winter rice production yet these factors were sure just as unessential for summer rice cultivation (Akram et al. 2017a, b). Expanding inclination of day by day most extreme temperature may diminish the rice spikelet ripeness, which influences for decrease of the development while the expanding propensity of CO_2 conc. can enhance rice production (Din et al. 2019). The general impacts of different atmosphere factors on yield rely upon both the affectability with respect to the atmosphere factors and the degree of progress in the atmosphere factors, where temperature and solar radiation span assume a basic part in disturbing rice development and production. The increase in night temperature associated with worldwide warming declines rice production. Rice production leans to be decreased by means of high min temperature and low solar radiation, especially at some stage during final developing period. The elevated stage of CO_2 from 340 to 680 ppm may want to raise the production of foremost crops by using 10–15% particularly in C3 plants like rice.

17.6 Mitigation and Adaptation to Deal with Climate Change Issues

There are the following ways to mitigate the impacts of climate change:

- Reducing climate change involves reducing the flow of heat-trapping greenhouse gasses into the atmosphere, either by reducing sources of these gasses (for example, the burning of fossil fuels for electricity, heat, or transport) or enhancing the "sinks" that accumulate and store these gasses (such as the oceans, forests, and soil). The aim of mitigation is to minimize the major anthropological activities and interaction of human beings with the climate system.
- Environmental change influences for all intents and purposes all distinctive and monetary frameworks. This communication concerning climate change and biodiversity, land deprivation, woodlands, synthetic compounds, and worldwide waters focuses on the significance of perceiving climate change suggestions in all that we do. The GEF has the unmistakable capacity to continue normal arrangements created with frameworks believing that exploits cooperative energies to look for numerous worldwide ecological advantages across Conventions while decreasing exchange offs and replication. Inside the developing scene of climate funding, the procedure is intended to be as corresponding as conceivable to different wellsprings of atmosphere account, for example, the Green Climate Fund (Fig. 17.4).
- Despite the fact that environmental change is a widespread issue, it is felt on a nearby scale. Regions and Cities are in this manner at the front of adjustment. Without global or national atmosphere strategy bearing, neighborhood networks and urban communities around the globe have been in regards to solving their own atmosphere issues. They are attempting to fabricate flood barriers, plan for higher temperatures and heatwaves, introduce water-penetrable asphalts for stormwater and floods control, and improve water use and capacity.
- As per the 2014 report on Climate Change Impacts, Vulnerability and Adaptation from the United Nations Intergovernmental Panel on Climate Change, governments at different stages are likewise showing signs of improvement at adjustment. Atmosphere change has distinctive significant designs to be begun and plan for decline water accessibility, how to create adaptable yield assortments and how to keep vitality and open infrastructure.

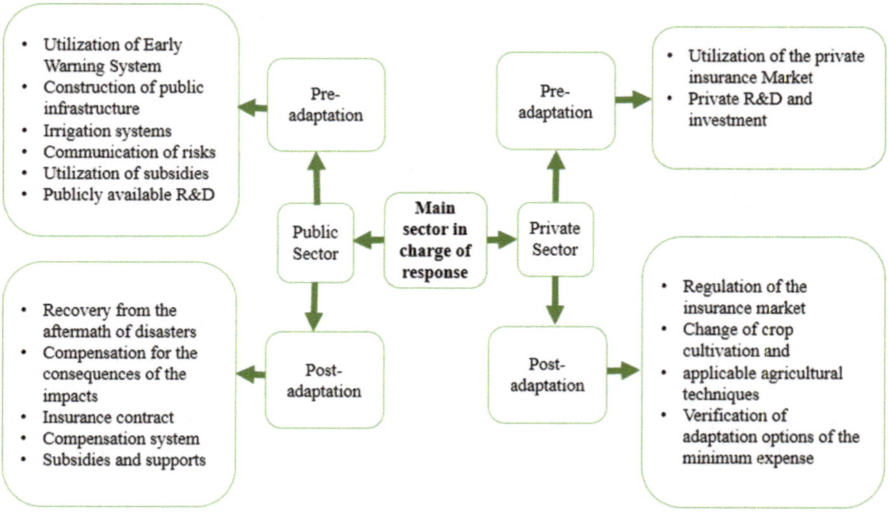

Fig. 17.4 Adaptive measure to reduce the impacts of climate change

References

Abbas G. et al. (2017) Quantification the impacts of climate change and crop management on phenology of maize-based cropping system in Punjab, Pakistan. Agricultural Forest Meteorology, 247: 42-55

Ahmad S, et al. (2017) Quantification of climate warming and crop management impacts on cotton phenology. Plants, 6 (7): 1-16

Akram R, Ahmad A, Noreen S, Hashmi MZ, Sultana SR, Wahid A, Mubeen M, Zakir A, Farooq, A, Abbas M, Shahzad K, (2019a) Global Trends of E-waste Pollution and Its Impact on Environment. In Electronic Waste Pollution (pp. 55-74). Springer, Cham.

Akram R, Amin A, Hashmi MZ, Wahid A, Mubeen M, Hammad HM, Fahad S, Nasim W (2017a) Fate of antibiotics in soil. In Antibiotics and Antibiotics Resistance Genes in Soils (pp. 207-220). Springer, Cham

Akram R, Fahad S, Hashmi MZ, Wahid A, Adnan M, Mubeen M, Khan N, Rehmani MIA, Awais M, Abbas M, Shahzad K (2019b) Trends of electronic waste pollution and its impact on the global environment and ecosystem. Environmental Science and Pollution Research, pp. 116

Akram R, Hashmi MZ, Nasim W (2017b) Role of antibiotics in climate change. In Antibiotics and antibiotics resistance genes in soils (pp. 97-104). Springer, Cham

Akram R, Turan V, Hammad HM, Ahmad S, Hussain S, Hasnain A, Maqbool MM, Rehmani MIA, Rasool A, Masood N, Mahmood F (2018a) Fate of organic and inorganic pollutants in paddy soils. In Environmental Pollution of Paddy Soils (pp. 197-214). Springer, Cham

Akram R, Turan V, Wahid A, Ijaz M, Shahid MA, Kaleem S, Hafeez A, Maqbool MM, Chaudhary HJ, Munis MFH, Mubeen M (2018b) Paddy land pollutants and their role in climate change. In Environmental Pollution of Paddy Soils (pp. 113-124). Springer, Cham

Ali M, Mubeen M, Hussain N, Wajid A, Farid HU, Awais M, Hussain S, Akram W, Amin A, Akram R, Imran M (2019a) Role of ICT in Crop Management. In Agronomic Crops (pp. 637-652). Springer, Singapore

Ali, S., et al. 2019b. Assessment of climate extremes in future projections downscaled b multiple statistical downscaling methods over Pakistan, Atmospheric Research, 222: 114-133

Amanet K, Chiamaka EO, Willie G, Quansah MM, Farid HU, Akram R, Nasim W(2019) Cotton Production in Africa. Cotton Production, p. 359

Amin, et al. (2018) Evaluation and analysis of temperature for historical (1996–2015) and projected (2030–2060) climates in Pakistan using SimCLIM climate model: Ensemble application. Atmospheric Research, 213: 422-436

Awais, M, et al. (2018) Potential impacts of climate change Climate change impact assessment and adaptation strategies for sunflower in Pakistan. Environmental Science and Pollution Research, 25 (14): 13719-13730

Blanc S, Gasol CM, Blanco JM., Muñoz P, Coello J, Casals P, Brun F (2019) Economic profitability of agroforestry in nitrate vulnerable zones in Catalonia (NE Spain). Spanish journal of agricultural research, 17(1), 1

Brisson N, Gate P, Gouache D, Charmet G, Oury FX, Huard FJFCR (2010) Why are wheat yields stagnating in Europe? A comprehensive data analysis for France. 119(1), 201-212

Corlett RT and Tomlinson KW (2020) Climate change and edaphic specialists: irresistible force meets immovable object?. Trends in Ecology & Evolution, 35(4), 367-376

Dawson N, Martin A, Sikor T (2016) Green revolution in sub-Saharan Africa: implications of imposed innovation for the wellbeing of rural smallholders. World Development, 78, 204-218

Din MSU, Ahmad I, Hussain N, Ahmad A, Wajid A, Khaliq T, Mubeen M, Imran M, Ali A, Akram R, Amanet K(2019) Agronomic Cropping Systems in Relation to Climatic Variability. In Agronomic Crops (pp. 67-82). Springer, Singapore

Fahad S et al. (2018) Consequences of high temperature under changing climate optima for rice pollen characteristics concept and perspectives. Archives of Agronomy and Soil Science, 64 (11): 1473-1488

Holmes TP (2020) Opportunities for Systematically Valuing Ecosystem Service Benefits Produced by Federal Conservation Programs. Agricultural and Resource Economics Review, 49(1), 178-191

Hussain S, Ahmad A, Wajid A, Khaliq T, Hussain N, Mubeen M, Farid HU, Imran M, Hammad HM, Awais M, Ali A, Aslam M, Amin A, Akram R, Amanet K, Nasim W (2020a) Irrigation Scheduling for Cotton Cultivation. In Cotton Production and Uses (pp. 59-80). Springer, Singapore.

Hussain S, Mubeen M, Ahmad A, Akram W, Hammad H M, Ali M, Masood N, Amin A, Farid HU, Sultana SR, Fahad S (2020b) Using GIS tools to detect the land use/land cover changes during forty years in Lodhran District of Pakistan. Environmental Science and Pollution Research **27,** 39676–39692.

Hussain S, Mubeen M, Akram W, Ahmad A, Habib-ur-Rahman M, Ghaffar A, Amin A, Awais M, Farid HU, Farooq A, Nasim W (2020c) Study of land cover/land use changes using RS and GIS: a case study of Multan district, Pakistan. Environment Monitoring and Assessment 192(1): p.2.

Ijaz M, Rehman A, Mazhar K, Fatima A, Ul-Allah S, Ali Q, Ahmad S (2019) Crop Production Under Changing Climate: Past, Present, and Future. In Agronomic Crops (pp. 149-173). Springer, Singapore

Meng Q, Hou P, Lobell DB, Wang H, Cui Z, Zhang F, Chen XJCC (2014) The benefits of recent warming for maize production in high latitude China. 122(1-2), 341-349

Mubeen M, Ahmad A, Hammad H M, Awais M, Farid H U, Saleem M, Nasim W (2020) Evaluating the climate change impact on water use efficiency of cotton-wheat in semi-arid conditions using DSSAT model. *Journal of Water and Climate Change*, *11*(4), 1661-1675

Mubeen M, Ahmad A, Wajid A, Khaliq T, Hammad H M, Sultana S R, Nasim W (2016) Application of CSM-CERES-Maize model in optimizing irrigated conditions. *Outlook on Agriculture*, *45*(3), 173-184

Mubeen M, Bano A, Ali B, Islam ZU, Ahmad A, Hussain S, Fahad S, Nasim W (2021) Effect of plant growth promoting bacteria and drought on spring maize (Zea mays L.). Pakistan Journal of Botany, 53(2): 731-739. DOI: https://doi.org/10.30848/PJB2021-2(38)

Müller D, Jungandreas A, Koch F, Schierhorn FJKIFER, Consulting P (2016) Impact of Climate Change on Wheat Production in Ukraine. 41

Nasim W, et al. (2018) Future risk assessment by estimating historical heat wave trends with projected heat accumulation using SimCLIM climate model in Pakistan. Atmospheric Research, 205-118-133

Rahman MHR et al. (2018) Multi-model projections of future climate and climate change impacts uncertainty assessment for cotton production in Pakistan. Agricultural and Forest Meteorology, 253-254: 94-113

Sabagh A E, Hossain A, Islam M S, Iqbal M A, Fahad S, Ratnasekera D, Llanes A (2020) Consequences and Mitigation Strategies of Heat Stress for Sustainability of Soybean (Glycine max L. Merr.) Production under the Changing Climate. In *Plant Stress Physiology*. IntechOpen

Sobolev V (2014) Second Roshydromet assessment report on climate change and its consequences in Russian Federation: Roshydromet.

Stefanski A, Bermudez R, Sendall KM, Montgomery RA, Reich PB (2020) Surprising lack of sensitivity of biochemical limitation of photosynthesis of nine tree species to open-air experimental warming and reduced rainfall in a southern boreal forest. Global change biology, 26(2), 746-759

Tachie-Obeng E, Akponikpe PBI, Adiku S (2013) Considering effective adaptation options to impacts of climate change for maize production in Ghana. Environmental Development, 5, 131-145

Tao, F., Xiao, D., Zhang, S., Zhang, Z., Rötter, R. P. J. A., & Meteorology, F. (2017). Wheat yield benefited from increases in minimum temperature in the Huang-Huai-Hai Plain of China in the past three decades. 239, 1-14

Tariq M, et al. (2018) The impact of climate warming and crop management on phenology of sunflower-based cropping systems in Punjab, Pakistan. Agricultural and Forest Meteorology, 256-257: 270-282

Zahoor SA, Ahmad S, Ahmad A, Wajid A, Khaliq T, Mubeen M, Hussain S, Din MSU, Amin A, Awais M, Nasim W (2019) Improving Water Use Efficiency in Agronomic Crop Production. In Agronomic Crops (pp. 13-29). Springer, Singapore

Chapter 18
Role of Modeling in Assessing Climate Change

Fahd Rasul, Ashfaq Ahmad, Syed Aftab Wajid, Hassan Munir, Ramsha Razaq, Shoaib Nadeem, M. Akhlaq Muddasir, M. Imran Khan, Sobia Shahzad, Hassan Javed Chaudhary, M. Farooq Hussain Munis, Wang Xuechun, Musaddiq Ali, and Wajid Nasim

Abstract Climatic changes are associated with fluctuations spanning over a period of three decades as a classic period of computing weather trends all around the world which, by studies till now, was proved to be harmful for life on earth. Natural processes going on in this earth were observed to be impacted significantly by these variations in our climate that are the result of anthropogenic activities. Rapid growth

F. Rasul (✉) · S. A. Wajid · H. Munir · R. Razaq · S. Nadeem · M. A. Muddasir
Agro-Climatology Lab., Department of Agronomy, University of Agriculture (UAF), Faisalabad, Punjab, Pakistan

A. Ahmad
Agriculture Sector Risk Assessment Specialist, Asian Disaster Preparedness Center (ADPC), Bangkok, Thailand

M. I. Khan
Department of Mathematics and Statistics, University of Agriculture, Faisalabad, Punjab, Pakistan

S. Shahzad
Department of Botany, Islamia University of Bahawalpur (IUB), Bahawalnagar Campus, Bahawalnagar, Punjab, Pakistan

H. J. Chaudhary · M. F. H. Munis
Department of Plant Sciences, Quaid-e-Azam University, Islamabad, Pakistan

W. Xuechun
School of Life Science and Technology, South West University of Science and Technology, Mianyang, Sichuan, China

M. Ali
Department of Environmental Sciences, COMSATS University Islamabad, Vehari, Punjab, Pakistan

W. Nasim
Department of Agronomy, Faculty of Agriculture and Environment, The Islamia University of Bahawalpur (IUB), Bahawalpur, Punjab, Pakistan

W. N. Jatoi et al. (eds.), *Building Climate Resilience in Agriculture*,
https://doi.org/10.1007/978-3-030-79408-8_18

in population demands more resources for their survival that includes the basic amenities of livelihood, i.e., nutrition, energy, and housing. Limited resources in combination with the risk of climatic changes are in fact a big problem that must be solved before it results in nonreversible damage. Modeling is the advanced approach to study climate change. Right after the Second World War, predominantly in the USA, by the end of the 1960s, representatives were being presented with the model's findings, which strongly supported the concept that the persistent intensification in greenhouse gas (GHG) emissions caused by human activities have completely changed the overall impact of global climate. With the passage of time, more advancement in modeling was observed; first of all, conceptual models were formed; those were replaced by analog models and then energy balance models were introduced by researchers. In agricultural systems, modeling as an essential tool is accomplished by scientists from different disciplines that has contributed for six decades in this field. Models have been used in ecosystem studies, hydrology, climate, crops, livestock and Hadley Climate model version 3 (HadCM3) is recently commonly used and several other Global climate models (GCMs) are in practice apart from statistical models like Statistical Downscaling Model (SDSM) are prominent among others for analytical climatic data studies. In order to study the climate changes; different climate projection scenarios have been made on the basis of previously provided data, i.e., rainfall, temperature, carbon dioxide and GHG emissions, and other components. On the basis of these scenarios, future predictions are likely to be more realistic and hopefully helpful for addressing the changing climatic situations across the globe and proactively devising mitigation practices to save the masses.

Keywords Food security · Climate model · Statistical model · Assessment · Mitigation · Adaptation · Climate change scenario · Hydrological models

18.1 Introduction

Climate science is based on computer models. To answer the questions of common people who are not much familiar with climate modeling and its science may need to understand some basic phenomenon of nature and its trends that make the climate shift pattern a realistic theme to study and ponder for well-thought prediction systems. Climate change is the additional one to existing factors that will lead to several changes as we already know it. As we know that these factors and climate hazards will lead to many disasters in this world (Wheeler and von Braun 2013; Amin et al. 2018a). When we say the climate is changing it means that there is a gradual increase in average temperatures, shifts in weather patterns, and slowly but surely rise in sea levels. The concepts that increased greenhouse gasses in the atmosphere are causing the current climate change is not accepted universally and effected different sectors like agriculture, tourism, forestry, energy consumptions, etc. Climate variability is the main environmental hazard of the twenty first century.

Sequential reports of Intergovernmental Panel on Climate Change and other different studies (e.g., Schlenker and Lobell 2010; Thornton et al. 2011) indicated that climatic change is applying multidimensional consequences on the human civilizations and their surroundings. It is evidenced by scientific studies that the prevailing global climate change is also mainly contributed by anthropogenic activities. The concentrations of greenhouse gasses (GHGs) in atmosphere like methane, carbon dioxide (CO_2), and nitrous oxide (N_2O) are noted to be considerably increased with the passage of time. For instance, the CO_2 concentrations have been raised from 280 to 394 ppm (preindustrial level) in 2012; showing a growth of 41% (www.epa.gov/climatechange/indicators) because of anthropological activities. In the preceding century, the mean temperature has been increased by 0.74 °C worldwide and now by the end of this century 1.1–5.8 °C increase is estimated as well as significant changes in rainfall patterns with an increased frequency will also be observed (IPCC 2012; Amin et al. 2018b). Climate change has always had direct or indirect effects and variability of these effects depends on different trade and industry sectors but out of all these sectors agriculture is most susceptible and intrinsically at risk due to climate change (Müller et al. 2011; Wheeler and Braun 2013; Thornton et al. 2010). It is expected that agricultural production lands will be reduced and further pressure on marginal lands would be increased due to changing patterns of rainfall and rising temperature because of global warming. Numerous evidences (Müller et al. 2011) determined Sub-Saharan Africa as the most prone territory to climate change's adverse effects in terms of economic outputs of agriculture which indicates that it is a huge challenge to manage the deleterious effects of climate change and mostly food apprehensive areas. As it is already recognized that agricultural operations are based on climatic conditions and the yield of the crops differs each year due to climate unpredictability, the agriculture zone is more in danger due to change in climate. In Europe, climate change studies indicate that the present climate change conditions in the northern parts of the continent may possibly effect the crop yield and productivity positively, whereas in the southern parts of the continent like the Mediterranean basin, the shortcomings will be in the majority with less economic yield, increased variations in yields of crops and decreased area for major crops production. The most critical problem caused by climate change is food security, providing food to the growing population and also sustaining the stressed environment. Nowadays the major concern is the effect of variability in climate change on food and agricultural production. Secondly, potential damages and significances are mainly concerned by the countries that will arise over the coming years in their territories, because of prevailing climate change conditions and it is a natural thing. Besides these global effects are also under consideration as they will change the trading patterns, international policies, domestic regional planning, resource employment, and ultimately the human welfare. Modern research on the change in climate authorizes that under the absence of climate variation some crops will respond significantly to raised CO_2 concentrations (supplementary effects due to change in precipitation frequency and intensity, increased temperatures, and probably enhanced the occurrence of life-threatening occasions, i.e., floods and droughts, will most likely cause the reduces yields and increased production threats in several global domains, increasing the gap among poor and rich nations. According to an estimate, susceptibility for climate change is more in developing countries rather than the developed nations because the developing countries are mostly agriculture-based countries and their economies predominate on agriculture, the lack of investment for adaptive techniques, their high temperature standard micro and macro environments along with sharp coverage of dangerous actions (Amin et al. 2018c). Conclusively it is proven by several research findings that climate change, shifting, and variability are serious threats to human life on earth ultimately, so the question is how to manage, adapt, or overcome these changes for which monitoring the present events and predicting our future in keen concern. Modeling is the subject that helps greatly to study the prevailing climate change events (Tariq et al. 2018).

18.1.1 Climate Change and Food Security

A lot of challenges will be faced by future food security due to increases in demand, consumption pattern changes, and change in climate. Doubled demand for agricultural food products is already predicted in future because of rapid population growth, by 2050 (Tilman et al. 2011; Kastner et al. 2012). As wheat, maize, rice, and barley are the main staple crops, their demands will ultimately be increased in the future. But at the same time during the last decade alarming situations are developed by the yield stagnation of these crops across the world, for example, the rice yield stagnation is reported in China (Ray et al. 2012) along with Triticum aestivum in some countries of Europe (Brisson et al. 2010; Ray et al. 2012), the reason is climate change in combination with bad agricultural management (Brisson et al. 2010; Ray et al. 2012). Furthermore, resource competition is also a limitation, for example, allocation of land either for biofuel or food production, reduced cropland area, and reduction in water resources are also constraints to agricultural crop production (Carberry et al. 2013). Therefore, climate changes should be tackled properly (Tester and Langridge 2010; Trnka et al. 2014). Additionally, due to chances of errors in climate projection challenges are posed for crop scientists. (Semenov et al. 2014). To feed the robustly increasing human population on this planet the food security is of key concern while climate change is also a non-neglectable phenomena. Climate smart agriculture is the term that is famous these days that can help combat future food security challenges.

18.1.2 Evidences for Global Change in Climate

Climate change along with the development of climatic scientific evidence shows that networks started exactly with industrial revolution although instrument-aided climate change observations were already inaugurated in some areas of Europe in the seventeenth century. In northern Europe piped water infrastructure was started for the sake of human health considerations. For controlled water reservoir management measured data for temperature and rainfall was required. So different approaches and instruments were developed to monitor the climatic parameters, that became standardized with time and till the mid of the nineteenth century Europe and North American regions were completed with their skeletal climate networks. In 1973, the International Meteorological Organization was found and the World Meteorological Organization in 1953 played a key role in climate change studies. By time-integrated approaches and satellite observations became of major concern and temperature time series was formulated confidently and processed to provide the estimates that climate is really changing significantly (Figs. 18.1 and 18.2).

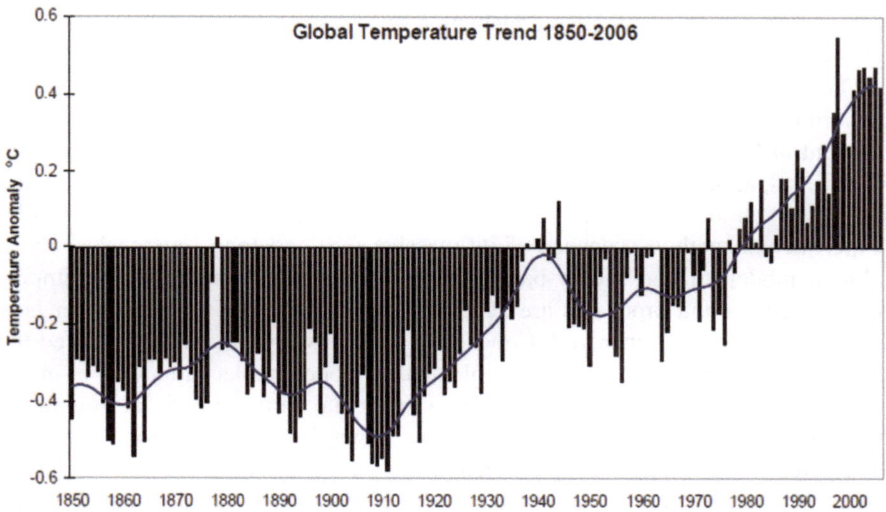

Fig. 18.1 Annual global temperature trend (difference from 1961–1990 baseline)

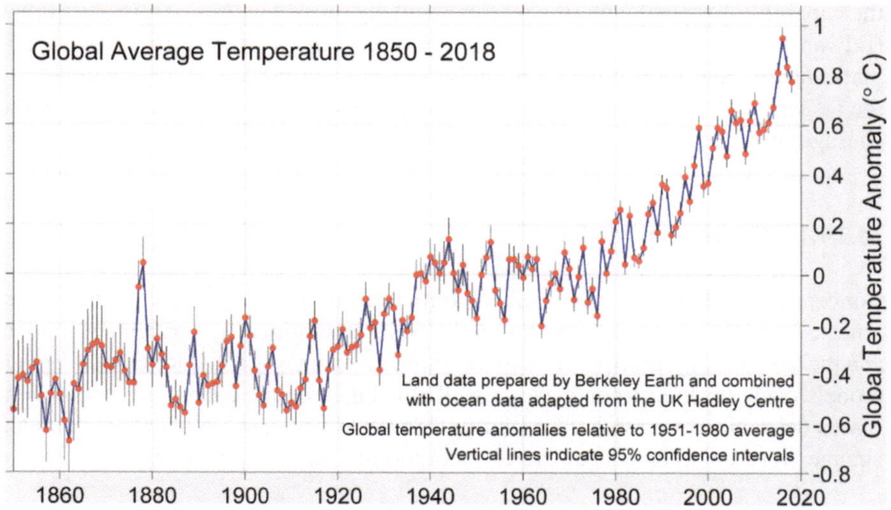

Fig. 18.2 Global Temperature Trend (Source: Berkeley Earth)

18.2 History of Modeling

18.2.1 Climate Modeling

Less than a century ago, climate model was just bigger than a thought because fundamental equations were roughly outlined on papers. Right after the Second World War, dramatic development in modeling was observed, predominantly in the US. By the end of the 1960s, representatives were being presented with the model's findings, which strongly supported the concept that the persistent intensification in greenhouse gas emissions caused by human activities would completely change the global climate in thoughtful ways.

The global climate system works as the function of transporting the heat energy toward the poles from the equator. Thus, climatology's key queries concern what amount of heat is retained by the Earth system wherever the energy inherent (in the oceans, atmosphere, land surfaces, etc.), the process of heat dissemination and its movement all over the globe. It is not possible to study the climate system by experimental methods because of its massive size and long time scales. Thus, climate models that are theory-based representations that depict or pretend important structures and processes are developed by some scientists to discover how Earth's climate is mechanized. Conceptual models were the first that started the history of climate modeling and simple analog models replaced models of mathematical systems, radiative and energy systems and shifted in the latest conceptual models. Meanwhile, in the 1950s, the general circulation models that are computer-based simulations have been primary tools of climate science globally. After 1990s till date, the field of the entire climate system is dominated by the increasing trend toward comprehensive coupled models. The intercomparison between different models and climate model evaluation are progressively changing the concept of modeling into a more reliable, modular process, giving away the capability for amalgamating research and operative features of climate studies (Edwards 2011). To monitor the energy flow in earth systems and different climatic factors and their impact on livelihood, climate models were developed by the researchers that became modified, comprehensive, and more efficient with the passage of time.

18.2.1.1 Conceptual Models

Conceptual models were the result of Initial efforts to recognize climatic occurrences. Prevalent winds around the globe were determined by conceptual models and the basic energy-transport role of the climate system was considered. Conceptual models made for the carbon cycle as well aided the climate change understanding. It was determined by researchers during 1861 that key causing agents for climate change were the geophysical cycles incorporating heat trapping gasses. With the passage of time, it was claimed that high volcanic activities during some period released huge amounts of CO_2 that warmer the earth more. During the process of slow weathering calcium carbonate is formed by combination of carbon dioxide from atmosphere and calcium from igneous rocks. The other medium for carbon absorption is organic matter. More CO_2 is absorbed and released during low volcanic activity that causes the global temperature to cool (Brooks 1951). Though, the theory of CO_2 was overruled by the early twentieth century by saying that the earth's temperature is more effected by water vapors than that of CO_2. That is why researchers assumed that changes in the concentrations of CO_2 have no influence on the temperature of the earth. Just initially there were different concepts and theories about the factors instigating the rise in earth's temperature and causing global warming but obviously evolution helped the most to better understand the thing.

18.2.1.2 Analog Models

Bowls or globes occupied with hazy and viscous solutions also called the physical models were developed not less than from the early twentieth century. Regardless of their capabilities being much narrow, these analog models validated basic principles of fluid motion on earth and motivated the leading age band of general circulation Modelers. Analog models utilization for climate change progressively helped in further studies of climate change.

18.2.1.3 Energy Balance and Radiative–Convective Models

Far ahead "energy balance" models (EBMs) were started to use to measure and calculate the values for the factors such as albedo (reflectance), solar radiation, and atmospheric absorption to figure out the worldwide radiative temperature. These models can be characteristically multidimensional having both regional as well as longitudinal (meridional) flows of energy. The radiative–convective model that is another type of climate mathematical model, emphases on upright atmospheric transmissions of energy. Atmospheric temperature profile simulation may be two dimensions vertical or one dimension, or meridional and vertical, respectively is the distinguishing character of these models. The above described three models are of keen importance in climate science studies.

Till the arrival of digital computers, conceptual, mathematical, and analog models dominated and were the subject of interest in the 1940s. Since the 1960s, GCMs (General Circulation Models) that simulates the functions and processes of atmosphere for a long time with the aid of computer have been the talk of the town in climate science even though simple models are still important both in their own function and as checks on sub-models incorporated in General Circulation Models. Later in 1975, after the introduction of coupled modeling, climate models progressively combined atmospheric and ocean GCMs. In the 1980s with the coupled AOGCMs (Atmosphere-Ocean General Circulation Models), modeling in climate has been motivated in the direction of progressively Comprehensive models. Coupling of Atmosphere–ocean GCMs occurred with other climate-related system models, i.e., vegetation, land surface, hydrology (rivers, evaporation, lakes, and rainfall), glaciers, sea ice, and snow cover by Earth system models (ESMs). Since the 1990s till now, models for substantial climatic processes have been combined progressively with AOGCMs. Many disciplines are brought together by these inclinations to seek accurate, near to reality and potentially extrapolative models for climate change. Subsequently from 2000, working authorities, as well as research agencies, started to do work in collaboration extra faithfully, looking for an integrated modeling structure for climate forecasting and predictions (Edwards 2011). The outcome is a progressively standardized modular infrastructure for simulation modeling globally, which provides significant information on the World's climate.

18.3 Agriculture Production System Modeling (APSIM)

Agricultural science dealing with the crop and livestock production systems is multidisciplinary research area which elaborates complicated agricultural mechanisms. Though the study of agricultural systems is beneficial by use of collected information which depicts how an individual system performs within a particular environment, in many situations it is impossible or unfeasible to do this. Agricultural system science provides the knowledge that enables the researchers to understand the complex problems and consider informed agricultural decisions. Modeling as an essential tool is accomplished by the scientists from different disciplines in the agricultural system that has contributed for six decades in this field (Jones et al. 2016). Models for agricultural systems play progressively significant parts in sustainable land management development through various and socioeconomic and agro-ecological circumstances as large amounts of resources are required by farm or field experiments and still enough data may not be provided to recognize applicable and efficient management strategies. The history of modeling of the agricultural system is regarded with several important happenings and events that motivated the researchers belonging to diverse disciplines for the development and utilization of models for different determinations (Fig. 18.3). Firstly, farming system models were developed including the components of biology and economics. Right after the introduction of farm system models by agricultural economists, the International Biological Program (IBP) was produced. It directed that the ecological model's development comprising grassland models the duration of 1960s and early 1970s that were also being used for studying livestock grazing. The International Biological Program was encouraged and geared up by advanced ecological scientists to generate tools for research that will help out to learn and understand the difficult ecosystem would permit them to learn the complex performance of ecosystems being affected by numerous climatic factors. In agricultural production systems concept for models came in 1960s initially. C. T. de Wit who is a physicist by profession from Wageningen

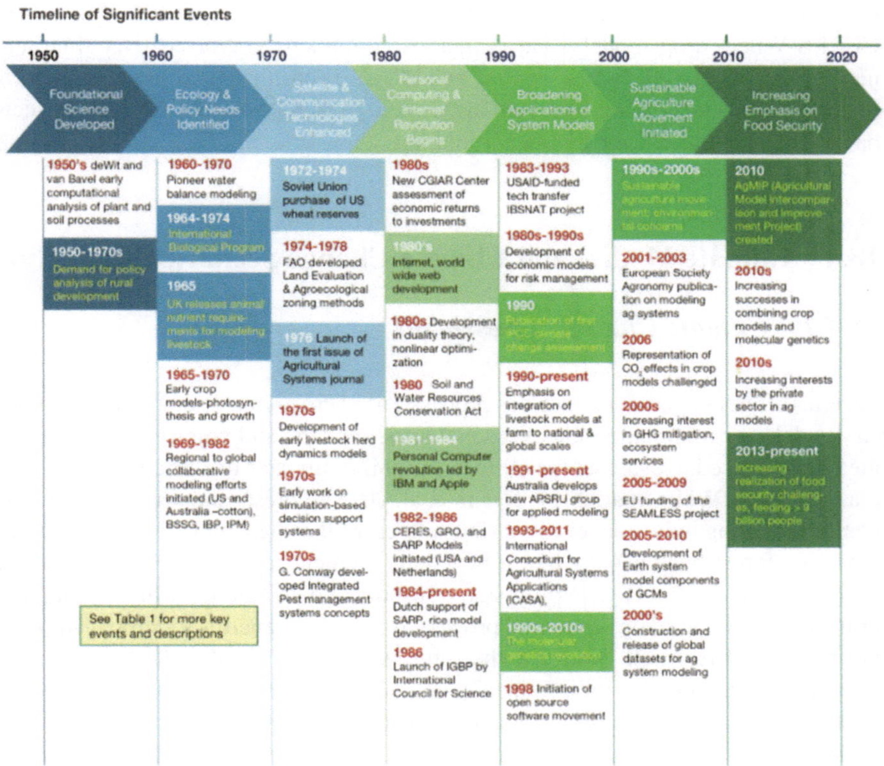

Fig. 18.3 Key events summary and factors that impacted the agricultural production system modeling development

University was the pioneer of modeling of the agricultural system who in the middle of 1960s, assumed the possibility of agricultural systems to be modeled by combined integration of physics and biological principles. Meanwhile G. Duncan was a chemical engineer, who worked in a fertilizer industry at age of 58 and then rejoined the graduate school and got Ph.D. degree in agronomy. He worked in the field of plant photosynthesis and wrote a research article on the canopy photosynthesis model; his research was a persistent way forward that his paper got citations several times since its publication by different crop modelers. He started to develop crop-specific simulation models for the very first time, i.e., for maize, cotton crop, and peanut crop), after his Ph.D. degree. The development by de Wit (1958) and his work fascinated many researchers and engineers who were aimed to start and develop crop models. Thus, the gradual progressive development in this field of study was geared up.

In 1972 the work on crop model development was boosted when the US government was amazed by the enormous wheat purchase by the Soviet Union which led to highly increase in prices and global wheat demands. After that funding for crop modeling, projects were increased and new research programs were started in combination with remote sensing for the prediction of major food crops by the US government. All these activities resulted in the production of CERES-Maize and CERES-Wheat crop models. These models continuously evolved over time and now a part of DSSAT (Jones et al. 2003, 2012). From that time till now there is remarkable progress that is observed in this field of study different research projects have been completed with significant findings.

18.4 Climate Models and Climate Change Scenarios

18.4.1 Climate Change Assessment

Several different approaches such as process-based crop simulation models, statistical models, agro-climatic indices, econometric models, and empirical field survey methods are used to evaluate the climate variability impacts on agriculture and ultimate changes (White et al. 2011; Nasim et al. 2018). Forests and specifically forest health are strong indicators of the various effects of the change in climate and monitoring of forest health looks like a beneficial element to observe climate change evolution. Conventionally, step-by-step transport of data among different scientific branches was focused on sequential process to develop scenarios for climate models in this field of research. Not only the future prediction but also the better understanding of uncertainties to become able to make decisions that will defiantly work in possible future is the ultimate goal of climate scenarios.

To understand the complicated relationships going on between the ecosystem, anthropogenic activities, climate system, and conditions, scenarios are developed and utilized by researchers. The climate scenarios describe confidently the plausible possibilities for future that several key areas (i.e., GHG emissions, environment, climate, socioeconomic and technological conditions) may be unfolded like this. Specifically, agro-ecological zones combined with four different socioeconomic future scenarios developed by IPCC are utilized for climate change studies as modeling database and framework. Analog variables, synthetic increasingly changing climate variables, and general circulation models are the three basic approaches for the development of climate scenarios. Out of all these approaches use of general circulation models is the most valid tool for climate change scenarios production (Ceglar and Kajfež-Bogataj 2012; Rötter et al. 2012; Cairns et al. 2013).

Climate scenarios are mainly sensible explanation for climatic conditions in future which are based on the assumptions of radioactive forcing and a number of climatic interactions. It could be envisioned by different regional climate models and different global environmental models, which have complex structure and are 3D mathematical depictions to represent the processes of relations among the ocean, sea ice, atmospheric system, and dearth top soil layer which occasioned from weather conditions over a longer period of time. Using the projections of climate under given specific emission scenarios is assumed to be a well-organized method for the determination of future climatic conditions instead of a forecasting tool. Although General Circulation Models have been used as beneficial tools for the simulation of significant features of present and future climates but chances for uncertainties are always present (Amin et al. 2017). There is a need to integrate different modeling approaches with the help of climate models to forecast climatic susceptibility and other climatic parameters, for example, CO_2, precipitation, and temperature. To measure the climate vulnerability Advanced Terrestrial Ecosystem Analysis and Modeling (ATEAM) platform was developed using the GCM (General Circulation Models) and to determine the climate change projections in 2080 $HadCM_3$ (Hadley center Climate Model 3) was developed, studied, and used a statistics methodology for selection of 15 ideal working models to develop the seasonal and annual average projection of temperature and rainfall in Australia. In Australia, by using historical climate data, climate variability and droughts were studied and some suggestions were made to deal with possible threats with future climate change that includes the adjustment of water allocation of ground and surface water by using different prediction models, improvement in water assimilation through enhancing using efficiency in Agroecosystem and buildup of a legal framework to manage water according to anticipated climate change. On the basis of Robust Decision support system, a novel analytical methodology for quantification of the SRES scenarios (Special Report on Emissions Scenarios) crafted employing the scenario-axes methodology for verdict analyzers. Elaborated NSRP (Neyman–Scott Rectangular Pulse) stochastic simulation method is used in precipitation scenarios generation model to originate rainfall climate information. GCM data is typically low in resolution of several degrees, due to which there is a lack in spatio-temporal precision which was complimentary for comprehensive analysis at regional scale and in many instances show errors in simulation in current climate event. In general circulation models (GCMs)' chances of errors, are there to predict the future climate conditions, but they have the ability to deliver accurate information about variation and large-scale features because of climate forcing. Since always uncertainties are present in models so more than one climate model should be dealt with for more precise results and reduced uncertainties.

18.4.2 Climate Change Scenarios

Climate projections are based on climate model simulations that predict the possible changes in future climate definitely with uncertainty chances. For the previous set of models, Special Report on Emissions Scenarios (SRES) were employed according to IPCC's 4th Assessment Report. But then according to IPCC's 5th Assessment Report, the emission scenarios of Representative Concentration Pathways (RCPs) have been used.

Special Report on Emissions Scenarios (SRES) worked with almost 40 different scenarios for greenhouse gas emissions but specifically in experiments of CMIP3 only three scenarios were used that are B1 (low), A1B1 (medium), and A2 (high). Here B1 describes a convergent world with reduced population advances but prompt variation in economic structure and providing with resource-efficient technologies.

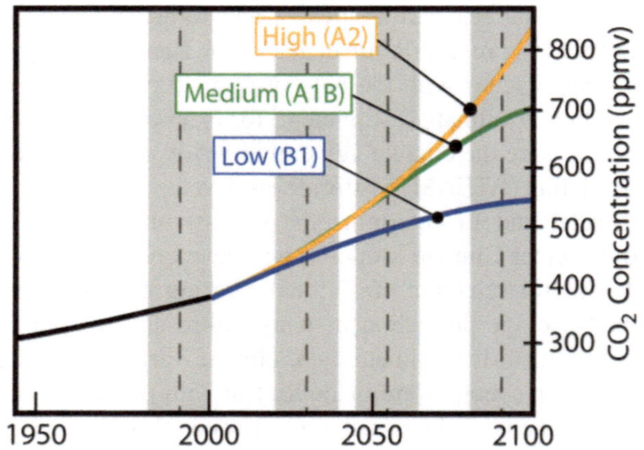

Fig. 18.4 540 ppm CO_2 concentration at 2100 with low (B1) scenario, 703 CO_2 conc. with medium (A1B) scenario, and 836 ppm CO_2 conc. with high (A1) scenario while the base period for predictions was for 20 years 1981–2000

A1B shows all the above storyline with a balanced use of energy resources. A2 defines a diversified world with less technological improvements with increased population growth (Figs. 18.4 and 18.5).

Representative Concentration Pathways (RCPs) are four and characterized according to radiative forcing (W/m^2) in relevance to the specific pathways till the end of the twenty first century. RCP8.5, RCP2.6, RCP4.5, and 6.0 depicting a high emission scenario, low emission scenario, and intermediate emission scenarios.

18.5 Crop Modeling and Climate Change Impact Assessment

Food production will be significantly affected globally as well as at regional level by different climate variables. Crop growth simulation models and experimental data are ways to check possible effects of change in climate on global food and crop production.

To estimate the effects of future climatic systems on yield, a lot of valuable approaches are provided by crop models. A variety of models are used for the simulation to check impacts of the probable climatic changes on productivity of crop, mostly to evaluate climate sensitivity to crop yield by varying climatic projection scenarios that include Info Crop, CERES-Wheat, SWAP (soil–water–atmosphere–plant), and CERES-Maize (Crop Environment Resource Synthesis) (Mubeen et al. 2016). The upcoming CO_2 concentration's impacts on the production of wheat were studied by a number of scientists these days. By 2050s, wheat can be adaptable to changing climate in Indo-Gangetic Plains and global warming would be the

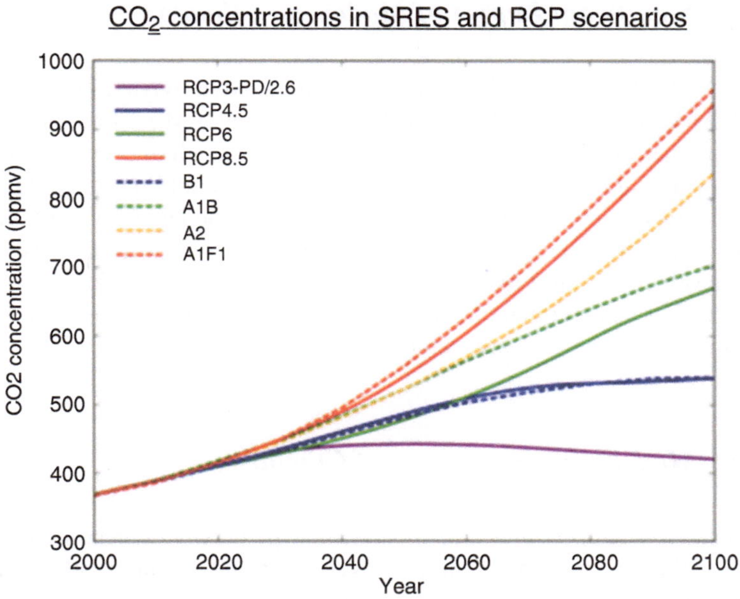

Fig. 18.5 Carbon dioxide (CO_2) concentrations (ppm) comparison for SRES scenarios B1, A1B, A2 and A1FI, and RCP2.6, RCP4.5, RCP6, and RCP8.5 (Collier et al. 2011). Base period of 20 years from 1981 to 2000 were taken for these estimations/predictions. (Source RCP database version 2.0.5)

advantage for the production of wheat that might be the cause of reduction in yield in other parts with critical temperature. Now it is the need of time to develop the heat-tolerant germplasm for different major food crops to provide the demanding food for the increasing human population. Also, introduce the climate resilient crops to different areas for efficient use of changing climate.

18.6 Models for Cereals and Their Interpretations

CERES-Wheat model has a great influence on agronomic practices along with breeding strategies, climatic, and edaphic means. CERES model is strongly capable to stimulate the different growth stages of wheat, i.e., stem growth, leaf, and grain. From different researches, it is also concluded that biomass availability directly depends on light interference and stresses which may result from changing climate and its impact on plants especially cereals.

18.6.1 Significant Impact of Models on Climate

Different wheat cultivars of wheat were calibrated through different models, CROPSIM-Wheat APSIM-Wheat, CERES-Wheat, and Nwheat for optimum sown wheat varieties. On the other hand, these models are unwell to convey the image of yield in relation to steep temperature throughout the grain filling stage. In association with temperature ranges, the simulation of these models with respect to yield for utmost planting dates is needed to improve the veracity of these models. APSIM-wheat shows accuracy just for maturity days for those cultivars which were sown on extreme earlier and late planting date. Overall, all these models need elevation in their function to respond to high temperatures.

18.6.2 Model and Uncertainty of Climate

At high and low temperatures crop models show perplexity in their simulation just because of their incorporated assimilated structures and functions belonging to the equations fitted behind the interface. By decision in crop models, the estimation of the impact of climate change through quantities are major concerns of crop models. For good response of simulation with respect to higher temperatures and increased concentration of carbon dioxide needed an elevation so that models results could be more precise for observed and simulated values, for example, in reproductive stage rather than vegetative stage, the synchronization and impact of estimation and uncertainty may vary widely (Mubeen et al. 2019). Improvements by intercomparison of models for better simulation options and cross check of final outputs of different models can give a better understanding of heat stress tolerance and precision in computations at both reproductive and vegetative stages in the crop life cycle leading to robust assessment while integrating climate change impacts in model's use.

18.6.3 Models for Rice Production and Climate Assessment

School of De wit' proposed a model regarding the modeling of products and it is known as ORYZA2000. It is specifically known as a model tailored-made for the simulation of rice growth and development. Water limitation and nitrogen limitation are also dealt by this model.

PDT was used for the development of Rice-Grow and other parameters which are covered by this PDT are genotypic parameters and management practices.

- Rice-Grow is divided into 7 sub-models
  These models generally can be understood by their simulations on

1. Phonology
2. Organ functionality
3. Biomass production
4. Photosynthesis
5. Quality
6. Yield formation
7. Nutrient balance

Three databases are used for the calibration of this model which are cultivars, snowing dates, and Nitrogen rate.

Analysis and integration of Rice-Grow were developed by using rice growth and developmental and environmental factors. Some key factors were stimulated which include organ function, biomass accumulation, photosynthesis, and organ formation.

18.6.3.1 Effect of Various Rice Models on Climate

For predication of growth and productivity of rice, this model performs as a systematic and quantitative tool. There are some climatic uncertainties which are dealt by this model such as growth as well as yield parameter. Due to this, crop adjustment can be made according to the maximum productivity of rice or yield.

18.6.4 Climatic Models for Maize

- For maize under high yielding conditions, some of the models were analyzed and evaluated which are DSSAT, CSM-IXIM, and CSM-CERES.
- Few parameters are examined which are yield of grain and Nitrogen uptake.
- CERES performed in a much better way as compared to other models and therefore that model is most suitable and appropriate as far as biomass and yield of grain is concerned.
- IXIM was most suitable for Uptake as well as grain of Nitrogen.
- Improvement has been made in IXIM due to an alternative process which can estimate Nitrogen demand at will.

18.7 Models for Non-cereal Crops

18.7.1 Sugarcane Models Interpretation

In southern Brazil, the DSSAT or CANEGRO model was simulated to conclude the significant effect on Brazilian sugarcane (Marin et al. 2011; Nasim et al. 2012). The calculation of algorithm for photosynthesis in DSSAT/CANEGRO (version 4.5.0.047) through radiation use efficiency in addition of total biomass on daily basis, and to check concentration of CO_2 in the process of fertilization were reported. In this literature, estimate the impacts of change in climate on the yield of sugarcane, in southern Brazil, water use efficiency, and irrigation requirements; moreover, it depends upon two models PRECIS and CSIRO and a sugarcane growth model.

18.7.2 Models Used for Cotton

Cotton producers in the USA since 1989 have used successfully the production model of cotton: GOSSYM-COMAX production model. At the commute of this model, production strategies have emerged, due to which model is difficult to use in other production environments. The light interception sub-model of GOSSYM does not distinguish between row spacing and plant canopy structure to consider light interception. In addition, GOSSYM affects an average plant not all populations of plants.

Thus, the progress shows that it is not suitable for crop variability, which otherwise defines the correlation between the genetic composition of the crop and its growing conditions. Farmers accommodate their agronomic practices on the basis of views derived from field observations, and crop variability is main factor which is get.

18.8 Horticultural Crops and Models

18.8.1 Model Simulations for Potato

For all inclusive study of the development and growth of potato, LINTUL-POTATO model is used. The mechanistic model is designed for the simulations of crop processes like emergence and the expansion of leaf and are used for the interception of light through leaf layers till the elimination of leaves. The old crop growth model-SUCROS was dependent on temperature and simulated for light use efficiency.

This model was performed by allowing endurance of leaf on daily basis and reducing daily dry matter accumulation to the tubers and then evaluated leaf senescence at initial crop stages. For diminishing all the possible errors, certain modifications were taken in the LINTUL-POTATO and derived LINTUL-POTATO-DSS with classical equations. These equations were used to evaluate environmental conditions throughout production, agro-ecological zone distribution, climatic threats, climate change, and analysis of the gap between potential and actual yield. These modifications were helpful to reduce the input parameters and unreliable data which were collected for unknown reasons.

The model worked for the development of specific temperatures from ground cover 0–100% at emergence. Radiation use efficiency is an activity which depends upon temperature. To assess crop end, fixed harvest index linearly increases and at tuber initiation dry matter distribution to the tubers occurs. This end of the crop is used as model input; it was stated that the crop cycle found out by ripening synchronized the duration of frost and heat free availability of the growing season.

Hence, LINTUL-POTATO-DSS consists of novel calculations to find out potato-tuber quality attributes like concentration of dry matter, tuber size distribution, which depends on crop environment during growth and development, management, and potential yields and initial crop development.

18.9 Hydrological Modeling and Climate Assessment

Water is the basis of life and the importance of water resources for human prosperity and crop yield is well understood. The world's water supplies, hydroelectric power, and water agriculture productions depend mainly on various key parts of the hydrological system, that also included the normal replacement of groundwater as well as surface water reservoirs. Nowadays' water scarcity and availability are an alarming issue that includes how much water could be diverted, utilized efficiently, and stored for ground and surface water reservoirs. To assess the water availability is not of major concern for sustainable human life, environment, and biodiversity but it is also helpful for farmers and water authorities for successful water management. Alongside growing population, increasing land use changes, and pollution, climate change is also among the extreme pressures on hydrological cycles.

Possible precipitation decrease due to climate change in some parts of the earth is threatening the water resources. The increase in water demand, uncertainties in climate variability, and socioeconomic environmental effects are insisting today to formulate the rules and regulations for the efficient use of water and produce more water resources. Water resources reduction in the event of more snow cover buildup as well as glacier melting process hydrological structure would be highly vulnerable to climate change. All around the world, several researchers studied the influence of climate change on water availability and streamflow (Hussain et al. 2020). The effect of changing climate on water resources was studied with help of GIS (geographic information system) and GCMs (general circulation models) in People's Republic of China (PRC) and it was concluded that variation in rainfall patterns is more threatening to water runoff as compared to rise in temperature. Integrated water management was the solution provided for this problem. In the Shiyang River northwest of China, climate variability influences on yearly streamflow were observed that resulted in the mean annual streamflow which was reduced by 64% due to less precipitation in meantime and the catchment streamflow was more prone to precipitation than potential evapotranspiration. To investigate water reservoirs in the Seyhan River under the present and future climatic conditions dynamic downscaling of GCMs and the linked river basin hydrological models were used which concluded that increased water demands will cause the projected water shortage that will probably be due to frequent irrigation practices water shortage will occur only with the increased water demands, e.g., owing to increased irrigation practices thus to tackle the future water sufferings managed and judicial use of water is necessary. A water availability modeling (WAM) system was provided by some scientists to judge the water supply proficiencies and investigate the impacts of climate on water availability and hydrological system that assessed possible reduction in mean streamflow due to climate. In the San Joaquin Basin, California, a system integration approach for evaluation of climate variability impact on water resources was developed that can provide a reference to manage the climate vulnerability effectively. In the Georgia Basin, Canada, a multicriteria decision expert system was developed to examine the availability of water under stressful climatic conditions that concluded it is critical to assess the socioeconomic features and environment under changing climate. Food production in Russia is under threat due to extreme climate events, high runoff, and increase in average water availability will be the feature of future climate change. Frequency of floods and drought would be increased due to climate variation in South Africa. Bayesian approach is a beneficial and easily applicable approach to determine the climate change impact's uncertainties and improbability assessment on water resources. In Romania, VIDRA rainfall runoff model, a tool for estimation of climate change impact on hydrological resources in 2075 was used that concluded water will increase in future in relevant observed area. In Churchill–Nelson River Basin, central Canada, Mann-Kendall trend test and in Missouri River Basin, an ANOVA test were applied to determine the variation in water resources due to climate change. Different river basins possess climate change impact differently. The droughts and floods frequency will be probably increased under prevailing climate conditions. Evapotranspiration is less sensitive to rainfall than to runoff and streamflow.

18.9.1 Lake Modeling

In 2010, Jørgensen introduced a number of models for interpretation of all the significant processes of the lake ecosystems and their diversified interaction. The model was published during the last four decades. Major task performed by these models is the point analysis for the interrelated ecosystem management. The ecological quality of lakes and their ecological qualities are mainly in danger by a variety of polygenic factors that causes stress, such as eutrophication, over consumption, and invasive species. Moreover, changes in land utilization, pollution causing factors, and hydrological system in catchment and the most important climate change, all are worst disturbing the ecological quality (Mooij et al. 2010). A variety of models are examined and published at the same time. Two major claims were identified: one defines with the number of models and the second one with the diversification of models. If the number of models are considered, results show that newly established models more of the time resemble the characters of the previously found models. This situation moves toward the adaptation of already existing models instead of newly created models. Improvement in the existing model could be a better option than striving for new. The restrictions related to varieties of models are that they are working with limited concerns and do not observe at wide range. This matter deprives the knowledge for a vast range and lack predictions for significant observations.

Different models that are used for lake modeling are static type models such as regression, steady and state models, complex dynamic models that include CAEDYM, LakeWeb, MyLake, PCLake, CE-QUAL-W2, SALMO, LakeMab, Delft 3D-ECO, PROTECH, minimal dynamic models, and structurally dynamic models (Mooij et al. 2010). CE-QUAL-W2 that is 2 dimensional hydrodynamic and model for analysis of water quality that predicts the steep and end long changes in ecosystem characters. CAEDYM also called Computational Aquatic Ecosystem Dynamics Model works with the process that includes the sub-models for water quality, geochemical and biological properties of the lake system.

This model was run with the Dynamic Reservoir simulation Model (DYRESM: 1D Lagrangian vertical stratification model) or Estuary and Lake Computer model (ELCOM: 3-D-structured grid hydrodynamics model) for examining water shifting, flow, and mix up. All these models are widely and successfully used for inflow and outflow dynamics of water bodies, stratification in dams, reservoirs for drinking water, and lakes. The CAEDYM can monitor oxygen, nutrients, i.e., Carbon, Nitrogen, Phosphorus, and Silicon, suspended solids, zooplankton, and fish. Geochemistry (ion redox, Ph, etc.), sediment's nutrient, metal, and oxygen flux.

For reducing the effect of changing climate on resources of water, water scarcity, and drought, efficient use of water and management integration would be progressively significant. Even though to alleviate climate change impacts, several approaches to manage water were practiced, but still there is a need to find out solutions locally. It is the demand of time to know the efficient water use techniques for irrigated regions and basins of river, to keep an eye on the available water, storability of water for water scarcity duration, and quantification of water resource dynamics on long term basis.

18.9.2 Types of Hydrological Models

Hydrological model comprises of following kinds of models.

18.9.2.1 Empirical Models (Metric Model)

Empirical models are distinct from others by considering information only from the actual data rather than focused on the hydrological systems, features, and processes. This is the reason that they were represented as data-driven models. The model used synchronal time series data for the derivation of different mathematical equations. To explore the relationship on function basis among all inputs and outputs, the model used statistical approaches of regression and correlation. Machine learning techniques for the hydro informatics method also included artificial neural network (ANN) and fuzzy regression. It clearly shows the model validation depends on the given boundaries and Hydrograph is one of the examples of this method.

18.9.2.2 Conceptual Methods (Parametric Models)

Hydrological processes and their different components are defined in parametric models. It consists of several interconnected reservoirs. Catchment areas were recharged by precipitation, infiltration and percolation were imputed by drainage, evaporation, runoff, etc. Moreover, this method also used semi-empirical equations. Model factors are examined through calibration and field data. The calibration involves curve representation of the data that made difficult interpretation and thus it effects on land use land change (LULC) and weakens the prediction with much confidence.

For best calibrations, huge amount of meteorological and hydrological records are prescribed for calibrations. Many conceptual models have been worked out with varying the degree of difficulty. Stanford by Crawford and Linsley formed introduced Watershed Model IV which is the first major conceptual model (SWM) in 1966 having parameters 16–20.

18.9.2.3 Idealized Physical Model

Representation of the actual phenomenon in an ideal way leads to principles of the physical processes simulated by these mechanistic models. State variables of time and space are measured and considered as functioning tools in this model. Water movement in hydrological processes is defined by finite difference equations. The model considered the evaluation of great number of physical attributes of catchment instead of the comprehensive data of hydrological and meteorological processes. The present methodology required big data like topography, soil moisture content (%), initial water depth (cm), topology and river flow systems, etc..

Modelers define the benefit of this physical model as that it overcomes many faults of the other two models because it used those traits which had physical reasoning. Even outside the boundary it can grant a large amount of information and can be used for a broad range of situations. SHE/MIKE SHE model is one of the examples.

18.10 Rainfall Pattern Models

18.10.1 GLMMs (Generalized Linear Mixed Models)

For the single site measurements, daily time series of rainfall are used to introduce the use of GLMs. Coe and Stern in 1982 and 1984 have introduced the GLMs (Generalized Linear Models). These models are an example of the stochastic weather models which are widely used. Actually, GLMs are parametric models and used for properly quantifying the variation associated with the contributing factors and/or covariates in the variability of the output variable (say daily rainfall). The effect of observed covariates like account for seasonality, climatological variables like regional forecasts, surface temperature, and sea-level rise will be effectively incorporated by using GLMs.

Usually, model's data for the insignificant distribution of the daily rainfall considered has a point mass at zero, also known as dry days. This feature splits the model into two functional parts.

1. Model based on binary data, for example, two-state Markov chain model which works on the simulation of the rainfall at a specific day and tentative analysis from the previous rainfall occurrence.
2. A right-skewed distribution works for the simulation of rainfall on wet days, such as the exponential, gamma, or mixed exponential distribution.

For the systematic modeling, functions of sine and cosine of the various periods can be added or deleted with respect to the results of the standard tests. It may be a likelihood ratio test for the given hypotheses which studied different analytical sources of seasonality. Seasonality further depends on the rotation of the earth around the sun every year, lower and higher frequency of season occurrence due to different rainfall patterns, such as state of ENSO (El Niño/Southern Oscillation), which usually occurs within 3–7 years and ranges of atmospheric situations over the period of 1 month.

The most prominent advantage of GLMs is that they encompass the climate change effect with different variables over a long period of time. Wherever likelihood and Bayesian methods are implemented, here GLMs are the most fitted models.

18.10.2 HMM Models

The Hidden Markov models (HMM) have been considered since 1960s for studying different aspects related to climate change (Mares et al. 2014). In HMMs, an invisible chain of random states has been generated out of specific visible finite number of observations while having transition probabilities (a kind of conditional probabilities) for each state. HMM invisibly changes throughout particular time of the first order Markov chain and thus interprets the distributions independently with respect to invisible changes. Nonhomogeneous HMMs or NHMMs are the models run on the nonstationary processes and by multivariable.

The NHMMs for rainfall data were introduced by Hughes and Guttorp and allocated and allowed for seasonality in the process of rainfall occurrence, on the same time scientists' more recent work has included the seasonality in rainfall amounts process. NHMMs can use the EM algorithm or methods of Bayesian; in some cases, they are given a predetermined value of K, the number of hidden states. With regard to application, the use of K depends on the predictive error measured for out of sample through cross-validation method and then interpretations done with scientific databases of hidden values. HMMs design for autocorrelation can be improved more by providing high dependency between unopened states or by losing the strict independence among the values.

Thus, NHMM shows that the stability of NHMMs for measurement of the data of rainfall lies in their capability to give real scientific phenomena, such as atmospheric conditions of location, studied under the invisible values.

There is a gradual increase in comparing different statistical models (HMM, NHMM, and KNN) for studying rainfall at multi-sites (Ghamghami et al. 2019).

18.10.3 Nonparametric Models

Contrasting feature of this model to parametric models is that they do not describe the data regarding rainfall processes and that causes the alteration in data flexibility against other models. This prominent characteristic of Nonparametric Models makes them an alternative to GLMs and HMMs. Resampling algorithm technique for simulation of daily time series of rainfall, resampled all observed data and give auto synchronize and interrelate the rainfall and weather parameters.

Thus, the distinct character of nonparametric models represents the robust-typed relationships between variables and often vaguely describe the effect of climate change with respect to rainfall processes. So, this feature restricted the model from working only on already observed data to regenerate the values.

18.10.4 Semi-parametric Models

The objective of semi-parametric model is to achieve state of model's parsimonious (figuring out best prediction using fewer predictors) while having the both parametric and nonparametric components. Semi-parametric models are more suitable in case of extreme value data, usually happened in case of rainfall (Tencaliec et al. 2020). The advent of advanced computing facilities led researchers to consider more sophisticated modeling approaches: semi-parametric models. There has been an increasing number of publications studying climate changes through semi-parametric models since the late 2000s. Different approaches can be considered through semi-parametric models for effective rainfall forecasting. These are not limited to the examples such as hybrid semi-parametric regression used for rainfall prediction (Wu 2013), Max-stable processes have been used for modeling highest amount of rainfall at various durations (Tyralis and Langousis 2018), Kriging has been used for calibrating rainfall (Nikahd et al. 2015), and a special type of quantile regression used for forecasting rainfall (Nguyen-Huy et al. 2020). Bayesian semi-parametric approaches have been also used for modeling the rainfall precipitation (Kottas and Fellingham 2012) and to study daily precipitation for estimating flood risk to all near about of "Richelieu Valley Basin in Quebec, Canada" (Jalbert et al. 2019).

18.10.5 Mechanistic Models

The fourth type of model Mechanistic Models work with the radar data at high resolution in time and space that provides the physical process of rainfall.

These models are especially designed to record the data regarding numbers, locations, and frequencies of rainfall within the given limit of area over 1000 km^2. Thus, it divides storm into rain cells at different times at different locations. Rain fallen within the cells across space with random velocities. Resolution for the analysis of data in models is set according to both space within the range of 2 km^2 and time of 5-min intervals. Classic feature of mechanistic model is that they are stationarily fitted with the objective of prediction of floods in any catchment area.

Meanwhile, with respect to this model applications, it has major drawbacks that these models are not well-designed to incorporating covariates that explains the variation in rainfall over long periods of time and they required very high-resolution data of radar and this data is not available in most of the developing countries in which index insurance is being used.

18.11 Changing Climate Raises Earthquake Risk

When temperature increases due to change in climate its imbalanced tectonic plates show the Gangetic delta in South Asia is unsafe. Kolkata-based geographer, "SujibKar" concluded that whenever an earthquake is built up under the vast delta it interlinked with increase in temperature causing climate change. According to Kar, if you want to comprehend the dynamics of earthquakes, you must take the Richter scale from 2001 to 2015 and look for five or more earthquakes on prone areas. Thus, the increase in global temperature was phenomenal. He said that the relationship between the amount of temperature and number of earthquakes is noticeable from data adjusted for the two sets over a given period. Moreover, he also said that the risk of earthquake is increasing due to global warming. The recent threat is observed in a subduction zone where earth crust or tectonic plates slowly pushing one another. This zone basically falls under the land instead of being under the ocean bed due to which risk of earthquakes are multiplied. Through conclusion of Kar's studied Richter scale that shown in 2001 across the Globe, the earthquake observed was 157 whereas after a decade and half it increases ten times in 2015.

According to global data that was recorded by the National Centers for Environmental Information of the US government from 2001 to 2015 these were the warmest top 16 years where warmest ever was 2015. Now models are developed which can predict the hazard by calculating the size of earthquake through data.

18.11.1 Climate Change Connection with Earthquake

According to Kar's study with the passage of time sea water level is raising, which is basically misbalancing the tectonic plates which increased the number of earthquakes. Moreover, size of winter is getting prolong especially in the northern hemisphere of the world due to climate change. This lengthening of the duration of winter leads toward the increase of heat evolution from the soil surface rather than from water bodies.

Kar said that earthquake is mainly affected by three factors movement of the sun, due to increase in atmospheric temperature by the globalization of thermal fluctuations at plate margins, primarily at mounting sites and subduction regions. Kar's study of last 200 years has shown that majority of earthquakes occurred between the month of November and March in the Northern hemisphere while in the southern hemisphere majority of earthquakes occurred between May and July.

Thus, many researchers following Kar across the world concluded that whenever sea level increases load related to the movement of the crust along the boundaries of the ocean basins might be in time enough to unclasp the coastal restrictions and the irregular patterns of quake show that more variation in the temperature among crust and mantle activates some parts of plates that will cause powerful divergence.

18.12 Conclusion

Climate changes are proved to be harmful to life existing on earth. Anthropogenic activities led to disaster on earth by moving away from nature. Modeling as an advanced technique had been used to study climate change to avoid or minimize its harmful effects. To monitor the energy flow in earth systems and different climatic factors and their impact on livelihood, climate models were developed by the researchers that became modified, comprehensive, and more efficient with the passage of time. Different conceptual and analog models were developed to understand the basic energy-transport role of the climate system and the basic principles of fluid motion on earth. Modeling in agriculture as an essential tool has been contributed for six decades in the agricultural system that play an important role in the sustainable land management and development through various socioeconomic and agro-ecological circumstances as the large amounts of resources are required by farm or field experiments and still enough data may not be provided to recognize applicable and efficient management strategies. Different crop models like CERES-Maize, CERES-Wheat, DSSAT, etc., continuously developed by the times helped in obtaining maximum yield and sustainable agriculture. To feed the robustly increasing human population on this planet the food security is of key concern while climate change is also a non-neglectable phenomenon. Different models for cereals and non-cereal crops also have been developed. Different crop simulation models are used to study the climatic variability, its impact on agriculture, and the ultimate changes that occur due to climate change; these are the model's statistical model, economic models, agro-climatic indices, econometric models, and empirical field survey methods.

Effect of changing climate on water resources was studied with help of GIS (geographic information system) and GCMs (general circulation models) and it was concluded that variation in rainfall patterns is more threatening to water runoff as compared to rise in temperature. Different models that are used for lake modeling are static type models such as regression, steady and state models, complex dynamic models that include CAEDYM, LakeWeb, MyLake, PCLake, CE-QUAL-W2, SALMO, LakeMab, Delft 3D-ECO, PROTECH, minimal dynamic models, and structurally dynamic models. CE-QUAL-W2 that is 2 dimensional hydrodynamic and model for analysis of water quality that predicts the steep and end long changes in ecosystem characters. CAEDYM also called Computational Aquatic Ecosystem Dynamics Model works with the process that includes the sub-models for water quality, geochemical and biological properties of the lake system.

To study rainfall pattern, its location, quality, duration, and climate and rainfall interrelation, different models have been developed like GLMs, HMM, Nonparametric, semi-parametric models, and mechanistic models. How climate change raises earthquake risk is studied by different models. So, models are helpful in every field of life because with the help of models we can estimate the climatic variability, its effects on natural resources or other human beings, and it also gives future predictions.

References

Amin A, et al. 2018a. Regional climate assessment of precipitation and temperature in Southern Punjab (Pakistan) using SimCLIM climate model for different temporal scales. Theoretical and Applied Climatology. January 2018, Volume 131, Issue 1–2, pp 121–131

Amin, A. et al. 2017. Comparison of future and base precipitation anomalies by SimCLIM statistical projection through ensemble approach in Pakistan. Atmospheric Research 294:214-225.

Amin, A. et al. 2018b. Evaluation and analysis of temperature for historical (1996-2015) and projected (2030-2060) climates in Pakistan using SimCLIM climate model: Ensemble application. Atmospheric Research, 213: 422-436

Amin, A. et al. 2018c. Simulated CSM-CROPGRO-cotton yield under projected future climate by SimCLIM for southern Punjab, Pakistan. Agricultural Systems 167: 213–222

Brisson N, Gate P, Gouache D, Charmet G, Oury F-X, Huard F (2010) Why are wheat yields stagnating in Europe? A comprehensive data analysis for France. Field Crops Research 119, 201–212.

Brooks CEP (1951) Geological and historical aspects of climatic change. In: Malone TF, ed. Compendium of Meteorology. Boston: American Meteorological Society; 1951, 1004–1018.

Cairns JE, Hellin J, Sonder K, Araus JL, MacRobert JF, Thierfelder C, Prasanna B (2013) Adapting maize production to climate change in Sub-Saharan Africa. Food Security 5, 345-360

Carberry PS, Liang W, Twomlow S, Holzworth DP, Dimes JP, McClelland T, Huth NI, Chen F, Hochman Z, Keating BA. 2013. Scope for improved eco-efficiency varies among diverse cropping

Ceglar A, Kajfež-Bogataj L (2012) Simulation of maize yield in current and changed climatic conditions: addressing modelling uncertainties and the importance of bias correction in climate model simulations. European Journal of Agronomy 37, 83-95.

Collier, M.A., Jeffrey, S.J., Rotstayn, L.D., Wong, K.K., Dravitzki, S.M., Moseneder, C., Hamalainen, C., Syktus, J.I., Suppiah, R., Antony, J., El Zein, A., 2011. The CSIROMk3.6.0Atmosphere-Ocean GCM: participation in CMIP5 and data publication. In: 19th International Congress on Modelling and Simulation. Perth, Australia, 12–16 December 2011. http://mssanz.org.au/modsim2011

de Wit, C.T. (1958). Transpiration and crop yields. Volume 64 of Agricultural research report/ Netherlands Volume 59 of Mededeling (Instituut voor Biologisch en Scheikundig Onderzoek va Landbouwgewasses) Verslagen van landbouwkundige onderzoekingen. Institute of Biological and Chemical Research on Field Crops and Herbage (1958).

Edwards PN (2011) History of climate modeling. Climate change vol 2, 128-139.

Ghamghami M, Ghahreman N, Olya H (2019) Comparison of three multi-site models in stochastic reconstruction of winter daily rainfall over Iran. Model. Earth Syst. Environ. 5, 1319–1332 doi: https://doi.org/10.1007/s40808-019-00599-7

Hussain, S. et al. (2020). Study of land use/land cover changes using RS and GIS: A case study of Multan district, Pakistan. Environmental Monitoring and Assessment (2020) 192: 2

IPCC (2012). Managing the Risks of Extreme Events and Disasters to Advance Climate Change Adaptation. A Special Report of Working Groups I and II of the Intergovernmental Panel on Climate Change [Field, C.B., V. Barros, T.F. Stocker, D. Qin, D.J. Dokken, K.L. Ebi, M.D. Mastrandrea, K.J. Mach, G.-K. Plattner, S.K. Allen, M. Tignor, and P.M. Midgley (eds.)]. Cambridge University Press, Cambridge, UK, and New York, NY, USA, pp 582

Jalbert J, Murphy OA., Genset C, Neˇslehov´a JG. 2019. Modelling extreme rain accumulation with an application to the 2011 Lake Champlain flood. Appl. Statist 68:831-856

Jones JW, Hoogenboom G, Porter CH, Boote KJ, Batchelor WD, Hunt LA, Wilkens PW, Singh U, Gijsman AJ, Ritchie JT (2003) The DSSAT cropping system model. European Journal of Agronomy, 18(3–4), 235–265.

Jones JW, Naab J, Fatondji D, Dzotsi K, Adiku S, He J (2012) Uncertainties in simulating crop performance in degraded soils and low input production systems. In Improving soil fertility recommendations in Africa using the Decision Support System for Agrotechnology Transfer (DSSAT) (pp. 43–59). Springer, Dordrecht.

Jones JW, Antle JM, Basso BO, Boote KJ, Conant RT, Foster I, Godfray HCJ, Herrero M, Howitt RE, Janssen S, Keating BA, Munoz-Carpena R, Porter CH, Rosenzweig C, Wheeler TR, (2016) Towards a new generation of agricultural system models, data, and knowledge products: state of agricultural systems science. Agric.Syst. 155:269–288 (in this issue).

Jørgensen SE (2010) A review of recent developments in lake modelling. Ecol Modell 221:689–692

Kastner T, Rivas MJI, Koch W, Nonhebel S (2012) Global changes in diets and the consequences for land requirements for food. Proceedings of the National Academy of Sciences U S A 109:6868–6872.a

Kottas A, Fellingham GW (2012) Bayesian semiparametric modeling and inference with mixtures of symmetric distributions. Stat Comput 22:93–106

Mares C, Mares I, Huebener H, Mihailescu M, Cubasch U, Stanciu P (2014) A Hidden Markov Model Applied to the Daily Spring Precipitation over the Danube Basin. Advances in Meteorology Volume 2014, Article ID 237247, 11 pages

Marin, F.R., Jones, J.W., Royce, F., Suguitani, C., Donzeli, J.L., Filho, W.J.P. and Nassif, D.S., 2011. Parameterization and evaluation of predictions of DSSAT/CANEGRO for Brazilian sugarcane. Agronomy Journal, 103(2):304–315.

Mooij, W.M., Trolle, D., Jeppesen, E., Arhonditsis, G., Belolipetsky, P.V., Chitamwebwa, D.B., Degermendzhy, A.G., DeAngelis, D.L., Domis, L.N.D.S., Downing, A.S. and Elliott, J.A., 2010. Challenges and opportunities for integrating lake ecosystem modelling approaches. Aquatic Ecology, 44(3), pp. 633–667.

Mubeen M. et al. 2016. Application of CSM-CERES-Maize Model in Optimizing Irrigated conditions. Outlook on Agriculture. 45(3) 173–184

Mubeen, M. et al. 2019. Evaluating the climate change impact on crop water requirement of cotton- wheat in semi-arid conditions using DSSAT model. Accepted in Journal of Water and Climate Change. doi: https://doi.org/10.2166/wcc.2019.179

Müller, C., Cramer, W., Hare, W.L., Lotze-Campen, H., 2011. Climate change risks for African agriculture. Proceedings of the National Academy of Sciences 108, 4313-4315.

Nassif D.S.P., Marin F.R., Pallone Filho W.J., Resende R.S., Pellegrino G.Q., 2012. Parametrização e avaliação do modelo DSSAT/Canegro para variedades brasileiras de cana-de-açúcar. Pesquisa Agropecuária Brasileira, 47, 311–318.

Nasim, W., et al. 2018. Future risk assessment by estimating historical heat wave trends with projected heat accumulation using SimCLIM climate model in Pakistan. Atmospheric Research 205 (2018) 118–133.

Nguyen-Huy T., Deo R.C., Mushtaq S., Khan S., 2020. Probabilistic seasonal rainfall forecasts using semiparametric d-vine copula-based quantile regression. Handbook of Probabilistic Models 203-227https://doi.org/10.1016/B978-0-12-816514-0.00008-4

Nikahd A, Hashim M, Mirzaie AA, Ghosiraie ZN (2015) Advanced of Mathematics-Statistics Methods to Radar Calibration for Rainfall Estimation; A Review. International Journal on Recent and Innovation Trends in Computing and Communication 3(1):96-105

Ray DK, Ramankutty N, Mueller ND, West PC, Foley JA (2012) Recent patterns of crop yield growth and stagnation. Nature Communications 3, 1293.

Rötter RP, Carter TR, Olesen JE, Porter JR (2012) Crop-climate models need an overhaul. Nature Climate Change 1, 175-177.

Schlenker W, Lobell DB (2010) Robust negative impacts of climate change on African agriculture. Environmental Research Letters 5: 014010

Semenov MA, Stratonovitch P, Alghabari F, Gooding MJ (2014) Adapting wheat in Europe for climate change. Journal of Cereal Science 59:245–256.

Tariq M et al.2018. The impact of climate warming and crop management on phenology of sunflower-based cropping systems in Punjab, Pakistan. Agricultural and Forest Meteorology 256–257 (2018) 270–282

Tencaliec P, Favre A-C, Naveau P, Prieur C, Nicolet G(2020) Flexible semiparametric generalized Pareto modeling of the entire range of rainfall amount Environ metrics 31(2) https://doi.org/10.1002/env.2582

Tester M, Langridge P (2010) Breeding technologies to increase crop production in a changing world. Science 327:818–822.

Thornton PK, Jones PG, Alagarswamy G, Andresen J, Herrero M (2010) Adapting to climate change: agricultural system and household impacts in East Africa. Agricultural Systems 103:73-82.

Thornton PK, Jones PG, Ericksen PJ, Challinor AJ (2011) Agriculture and food systems in Sub-Saharan Africa in a 4°C+ world. Philosophical Transactions of the Royal Society A: Mathematical, Physical and Engineering Sciences 369:117-136

Tilman D, Balzer C, Hill J, Befort BL (2011) Global food demand and the sustainable intensification of agriculture. Proceedings of the National Academy of Sciences U S A 108(50):20260–20264.

Trnka M, Rötter RP, Ruiz-Ramos M, Kersebaum KC, Olesen JE, Zalud Z, Semenov MA (2014) Adverse weather conditions for European Wheat production will become more frequent with climate change. Nature Climate Change 4, 637–643.

Tyralis H, Langousis A (2018) Modelling of rainfall maxima at different durations using max-stable processes. European Geosciences Union General Assembly 2018 Geophysical Research Abstracts Vol. 20 https://www.researchgate.net/publication/325908779

Wheeler T, von Braun J (2013) Climate change impacts on global food security. Science 341 (6145):508–513.

White JW, Hoogenboom G, Kimball BA, Wall GW (2011) Methodologies for simulating impacts of climate change on crop production. Field Crops Research 124:357-368

Wu J (2013) An Effective Hybrid Semi-Parametric Regression Strategy for Rainfall Forecasting Combining Linear and Nonlinear Regression Book chapter accessible at https://www.igi-global.com/chapter/content/74935

Chapter 19
Nutrient Dynamics and the Role of Modeling

Mukhtar Ahmed, Muhammad Aqeel Aslam, Fayyaz-ul-Hassan, Rifat Hayat, Wajid Nasim, Muhammad Akmal, Muhammad Mubeen, Sajjad Hussain, and Shakeel Ahmad

Abstract Nutrients are required for plant growth and development, absence or shortage of this could limit crop productivity. However, the misappropriate application of nutrients could cause environmental challenges like greenhouse gas emission, global warming, and climate change. Nutrient dynamics also seem to be affected by climate change as all the processes in which nutrients are taken up,

M. Ahmed (✉)
Department of Agricultural Research for Northern Sweden, Swedish University of Agricultural Sciences, Umeå, Sweden

Department of Agronomy, Pir Mehr Ali Shah Arid Agriculture University, Rawalpindi, Punjab, Pakistan
e-mail: mukhtar.ahmed@slu.se

M. A. Aslam · Fayyaz-ul-Hassan
Department of Agronomy, Pir Mehr Ali Shah Arid Agriculture University, Rawalpindi, Punjab, Pakistan

R. Hayat · M. Akmal
Institute of Soil Science, PMAS-Arid Agriculture University Rawalpindi, Rawalpindi, Punjab, Pakistan

W. Nasim
Department of Agronomy, Faculty of Agriculture & Environmental Sciences (FA&ES), Islamia University of Bahawalpur (IUB), Bahawalpur, Punjab, Pakistan

M. Mubeen
Department of Environmental Sciences, COMSATS University Islamabad, Vehari, Punjab, Pakistan

S. Hussain
Department of Horticulture Bahauddin Zakariya University, Multan, Punjab, Pakistan
e-mail: sajjad.hussain@bzu.edu.pk

S. Ahmad
Department of Agronomy, Bahauddin Zakariya University, Multan, Punjab, Pakistan

W. N. Jatoi et al. (eds.), *Building Climate Resilience in Agriculture*,
https://doi.org/10.1007/978-3-030-79408-8_19

transferred, and cycled over time in an ecosystem are linked with climatic factors directly or indirectly. Many biogeochemical models including APSIM (Agricultural Production Systems Simulator), CropSyst, CERES-EGC, DayCent, DNDC (DeNitrification DeComposition), DSSAT (Decision Support System for Agrotechnology Transfer), EPIC (Environmental Policy Integrated Climate), PaSim, RothC (Rothamsted Carbon Model), and STICS (Simulateur mulTIdisciplinaire pour les Cultures Standard, or multidisciplinary simulator for standard crops) can be used to study the nutrients dynamics which includes uptake from the soil, assimilation, and remobilization in plants. This chapter presents the application of different biogeochemical models to simulate nutrients dynamics, mainly Carbon (C), Nitrogen (N), and Phosphorus (P). Soil organic carbon (SOC) dynamics and loss of N as nitrous oxide (N_2O) emission is also discussed by using models like APSIM and DNDC. Finally, life cycle assessment (LCA) is presented as a valuable tool to study environmental impacts associated with all steps of nutrients distribution among different systems. In conclusion, process-based biogeochemical cycles are valuable tools that can be used to study and manage nutrients in soil-crop system under changing climate.

Keywords Nutrient dynamics · Biogeochemical models · Soil organic carbon · Life cycle assessment

19.1 Introduction

Nutrients are chemicals vital to biological functions. Nutrients required at higher than 1–150 g kg^{-1} (>1000 mg kg^{-1} dry weight) of plant dry matter are called macronutrients, and these include N, P, K, Ca, Mg, and S. However, nutrients which are required at the concentration of 0.1–100 mg kg^{-1} (<100 mg kg^{-1} dry weight) of plant dry matter are called as micronutrients (Fahad et al. 2016; Ahmed et al. 2020a). These include Fe, Zn, Mn, Cu, B, Mo, and Cl. Elements like Al, Si, Co, Na, and Se are not essential according to the criteria but widely taken up by the plants to perform different metabolic functions. However, it should be kept in mind that nutrients applied in excess amount could cause environmental pollution like global warming, greenhouse gas (GHG) emission (Ahmed 2020a; Hammad et al. 2018); these may also cause above and underground water deterioration. Nutrient dynamics is the process by which nutrients are taken up by the plants. Plants take these nutrients in different forms, as shown in Table 19.1. Similarly, Fig. 19.1 elaborates the role of these nutrients on plant growth.

Different types of biogeochemical models can be used to study the dynamics of nutrients, which include uptake from soil, assimilation, and remobilization in plants. Mostly biogeochemical modeling has been focused on the dynamics of C or its integration with N (Ahmed and Hassan 2011; Ahmed 2012). RothC (Rothamsted Carbon Model) is the well-known monthly time step model to calculate total organic

Table 19.1 Macronutrients and Micronutrients with forms in which they are absorbed by plants

Elements	Symbols	Form absorbed by plants
Macronutrients		
Hydrogen	H	H_2O
Carbon	C	CO_2
Oxygen	O	H_2O
Nitrogen	N	NO_3^{-1}, NH_4^{+1}
Phosphorus	P	PO_4^{3-}, HPO_4^{2-}, $H_2PO_4^-$
Potassium	K	K^+
Magnesium	Mg	Mg^{2+}
Sulfur	S	SO_4^{2-}
Calcium	Ca	Ca^{2+}
Micronutrients		
Iron	Fe	Fe^{2+}, Fe^{3+}
Manganese	Mn	Mn^{2+}
Zinc	Zn	Zn^{2+}
Copper	Cu	Cu^{2+}
Boron	B	BO_3^{2-}, $B_4O_7^{2-}$
Molybdenum	Mo	MoO_4^{2-}
Nickel	Ni	Ni^{2+}
Chlorine	Cl	Cl^-

C (TOC) and microbial biomass C. RothC is a widely used model to predict variations in C stocks of arable soils (Hammad et al. 2020). Peltre et al. (2012) studied shifts in soil C stocks from long-term experiments in Denmark (31years), France (14 years and 11 years), and Sweden (52 years) after application of EOM. RothC provided a successful outcome after modification of the partition coefficients of TOC in EOM. Thus, suggested partitioning of EOM-TOC could allow us to predict soil C storage regardless of sites and composition of EOM. The RothC model was used by Yokozawa et al. (2010) for arable soils in Japan to estimate carbon (C) sequestration potential of organic matter application. Twenty-five years' simulation was conducted after creating a baseline simulation of soil organic carbon for 1990. They concluded that the model has the potential to simulate potential C sequestration against different treatments. The RothC model was used to develop the soil C calculation system by Shirato (2020). The model was evaluated from long-term field data in China and Thailand, and two pools of plant litter (Decomposable plant material (DPM) and resistant plant material (RPM)). The web-based tool they developed can be used to spread mitigation options widely. Similarly, RothC can be used to assess the potential of C storage under a wide range of agricultural practices (Chenu et al. 2019).

Biogeochemical models such as RothC, DAISY, DNDC, EPIC, CENTURY, and SPACSYS have a long history of simulating C and N in agricultural systems. These models can be used to estimate future changes in soil biogeochemistry as they can

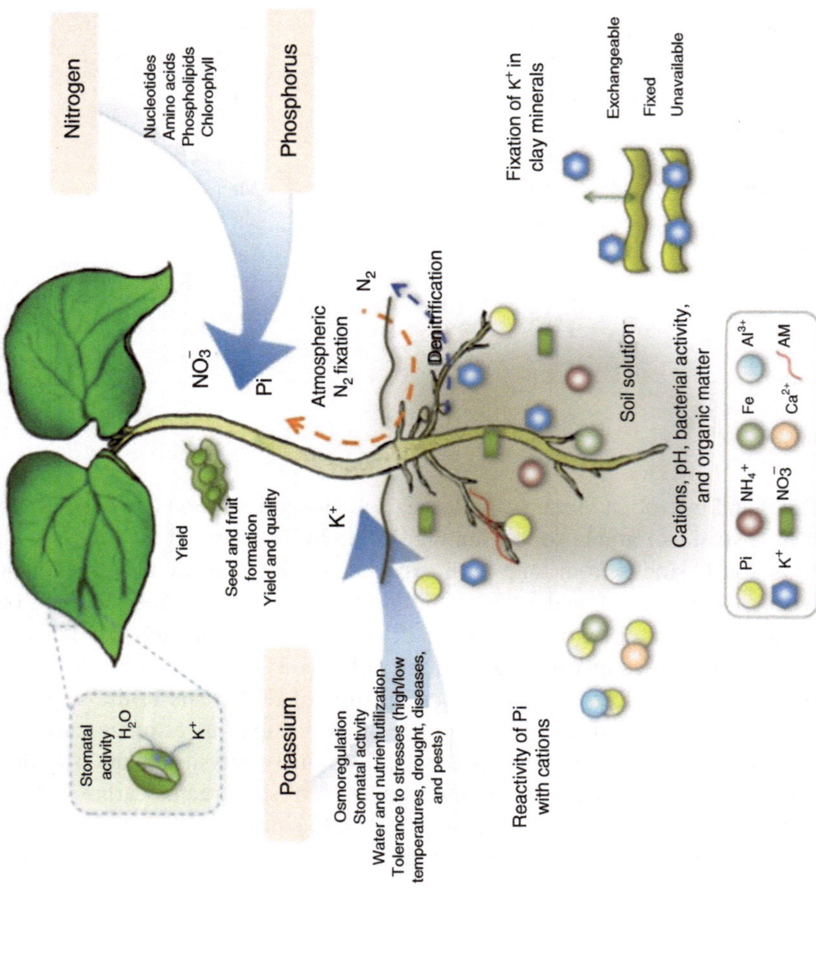

Fig. 19.1 Plant growth and nutrients role

integrate C–N–P cycling. The International Soil Modeling Consortium (ISMC) promotes the integration of soil modeling in different disciplines and facilitates to integrate and advance soil systems modeling.

19.2 Biogeochemical Models and C, N, P Dynamics

The detail of models which can be used for the soil nutrient, energy, solutes, and water dynamics studies are presented in Table 19.2.

19.2.1 Modeling C Dynamics

Soil organic carbon acts as a possible C sink for C sequestering atmospheric CO_2 as well as it improves soil quality and crop productivity (Ahmed and Ahmad, 2019). The soil organic matter (SOM) modules in DSSAT model are shown in Fig. 19.2. The SOM modules of DSSAT consist of two models (i) CENTURY and (ii) CERES-based SOM model. The components of DSSAT-CENTURY which maintain SOM pools and fresh organic matter pool are illustrated in Fig. 19.3. Furthermore, the C flow in the Roth C model has been shown in Fig. 19.4. The Rothamsted carbon model (RothC) was used by Liu et al. (2019) to simulate SOC from long-term field experiments. Jebari et al. (2018) studied climate change impacts on SOC sequestration by using four IPCC (Intergovernmental Panel on Climate Change) scenarios from 2010 to 2100 using RothC model. The model predicted a general increase in SOC under all scenarios compared to baseline. However, the greatest loss in C stocks was simulated in the highest temperature rise and rainfall drop scenario, i.e., ECHAM4. A range of biogeochemical models (APSIM, CERES-EGC, DayCent, DNDC, DSSAT, EPIC, PaSim, RothC, and STICS) was used to study GHG source/sink status and C sequestration. This work concluded that biogeochemical research in crops and grasslands ecosystems should be considered an essential step for future research (Brilli et al. 2017; Zamin et al. 2019; Ali et al. 2019). Luo et al. (2019) used APSIM to simulate SOC dynamics. The validated model was then used to simulate SOC change from 2009 to 2070, which was further used to develop surrogate models.

19.2.2 Modeling N Dynamics

Complex soil, plant, climate, and manure management interactions impact on nitrous oxide (N_2O) emissions could be simulated by using DeNitrification DeComposition (DNDC) model. The model was calibrated and evaluated using multi-year datasets of measured N_2O fluxes and other parameters in contrasting climates of Canada (He et al. 2020). The DeNitrification DeComposition (DNDC)

Table 19.2 Biogeochemical models and their applications

S. No.	Models	Applications	References
1.	MOHID-Land	Simulation of water cycle	Canuto et al. (2019)
2.	CPlantBox	Modeling of water and C flows	Zhou et al. (2019)
3.	VISIT (Vegetation integrative Simulator for trace gasses)	C, Water, and N-Cycle Simulation	Inatomi et al. (2010)
4.	APEX (Agricultural policy/ Environmental eXtender)	Whole farm model	Ford et al. (2015)
5.	Answer Application	Water and salt balance modeling	Ben-Gal et al. (2008)
6.	AgroC (Extension of Soil CO_2/ RothC model with the addition of growth module SUCROS)	Modeling fluxes of soil heat, water, and C in agricultural systems	Herbst et al. (2008)
7.	Cop-soil	Modeling dynamics of organic pollutant during organic matter decomposition	Zhang et al. (2014), Lashermes et al. (2013)
8.	CNMM (Catchment nutrient management model)	Modeling of energy balance, hydrology, plant/crop growth, biogeochemistry of C, N, and P	Li et al. (2005)
9.	CANDY (Carbon and N Dynamics)	Modeling C and N dynamics	Franko et al. (1995)
10.	BASFOR (Basic forest model)	Modeling of forest biogeochemistry	Cameron et al. (2013)
11.	DEMENT	Prediction of OM decomposition	Allison (2012)
12.	DAISY	Water, energy, C, and N simulation	
13.	Criteria	Soil water and crop modeling	Bittelli et al. (2010)
14.	CoupModel	Soil water and heat process modeling	Wu et al. (2012)
15.	ECOSSE (Estimation of C in Organic soils-Sequestration and Emissions)	Simulation of soil C and N dynamics	Abdalla et al. (2016), Dondini et al. (2015)
16.	EPIC (Environmental Policy Integrated Climate)	Modeling of the physiochemical process in agricultural systems	Izaurralde et al. (2006)
17.	DNDC (DeNitrification DeComposition)	Modeling of C and N biogeochemistry in agroecosystems	Han et al. (2014)
18.	Hydrus 2D-3D	Water flow and solute transport	Autovino et al. (2018)
19.	Expert-N	Simulation of N cycle	Biernath et al. (2013)
20.	ORCHIDEE	Modeling C, water, and energy fluxes	Krinner et al. (2005)
21.	RootBox	Water, nutrient, and root modeling	Leitner et al. (2014)

(continued)

Table 19.2 (continued)

S. No.	Models	Applications	References
22.	R-SWMS	Solute transport and water flow modeling	Javaux et al. (2013)
23.	OpenSimRoot	Root modeling with solute transport	Chen et al. (2013)
24.	RUSLE (Revised Universal loss equation) 2015	Erosion model	Panagos et al. (2015)
25.	SiSPAT (Simple soil-plant atmosphere transfer)-Isotope	Heat and water modeling	Braud et al. (2005)
26.	Saltirsoil-M	Simulation of inorganic ions (Na^+, K^+, Ca^{2+}, Mg^{2+}, Cl^-, SO_4^{2-}, NO_3^{1-}, and alkalinity)	Visconti et al. (2014)
27.	SPOTPY (Statistical parameter optimization tool)	Calibrate, optimize, and analyze parameters	Houska et al. (2015)
28.	SoilGen	Pedogenesis study	Finke and Hutson (2008)
29.	SWAP	Simulation of water, solutes, and heat transport	van Dam et al. (2008)

model was used to simulate alfalfa (*Medicago sativa*L.) biomass and yield by considering RCP4.5 and 8.5 scenarios. This study suggests the use of winter hardy cultivars could mitigate the ill effects of changing climate (He et al. 2019). Jiang et al. (2019) compared two process-based models (DSSAT and DNDC) to simulate dynamics of C and N. They suggested that fertilizer use efficiency and crop production can be improved by using multiple models. Nitrogen transformation through different processes in response to different environmental conditions and nitrous oxide (N_2O) emissions were simulated by Vogeler et al. (2013) using APSIM and DNDC. The results showed that in APSIM, nitrification was affected by temperature while in DNDC, water played a significant role. However, denitrification in DNDC demonstrated a response to temperature and organic C while in APSIM, it was linked with water content at drain upper limit. These models depicted a differential response to N load as increasing rainfall intensity decreased APSIM-simulated N_2O emissions while opposite was observed in the case of DNDC (Vogeler et al. 2013). The impact of leguminous green manure treatments on SOC stocks and total nitrogen stocks were simulated by using the RothC model (Yao et al. 2019).

Soil hydrology, which strongly influences biogeochemical processes, was studied using the Denitrification Decomposition model (DNDC) in comparison with RZWQM2 (Root Zone Water Quality Model) by Smith et al. (2020). DNDC was able to capture water and N between conventional and controlled drainage management with sub-irrigation losses (Hussain et al. 2020). Soil organic carbon dynamics was studied through CENTURY model across France in response to different agricultural practices. Different initialization scenarios were used, and results depicted that the SOC stock can be simulated with reasonable accuracy (Dimassi et al. 2018).

Nitrogen (N) is one of the limiting elements in the production of crops. Nitrogen diminishes due to continuous crop production, and it is required in large quantities;

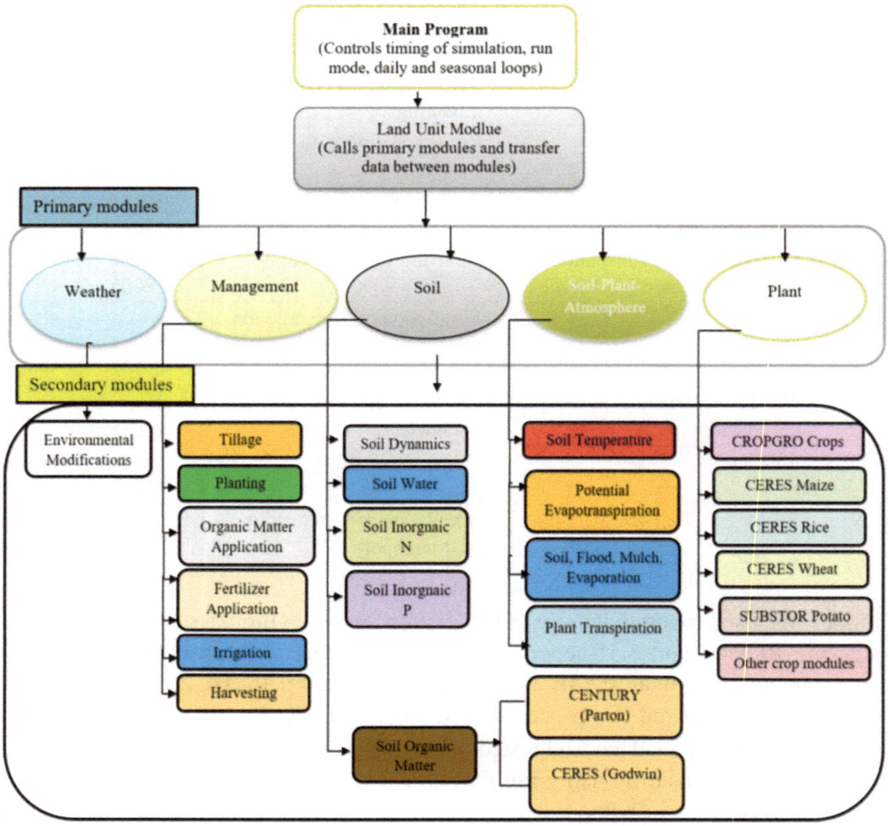

Fig. 19.2 Components of DSSAT-CSM showing two SOM modules (The Century and CERES-Based Modules)

therefore, to improve production, the addition of N fertilizer is necessary (Srivastava et al. 2018; Fatima et al. 2018; Abbas et al. 2020a, b; Ahmad and Hasanuzzaman 2020). Wheat produces 68 kg of grain per kg of N while for rice it is 44 kg of grain per kg of N and for maize 49 kg of grain per kg of N is produced (Naz et al. 2013; Danish et al. 2019). Nitrogen occurs in the environment in the elemental form. In the presence of extreme temperature and high-pressure N, the gas reacts with oxygen to produce nitrogenous oxides (NO and N_2O) by oxidation reactions. If there is rain, nitrogen dioxide in reaction with water is converted into nitric acid (HNO_3). Thus, nitrates produced by combining N and oxygen are used by plants as their food. Ammonia can also be produced synthetically by the Haber process. In this process, ammonia is produced by reacting hydrogen with N in the presence of a catalyst at high temperature and pressure. This synthetic NH_3 may be applied directly to the plants as fertilizer, or it can be converted into nitric acid by reacting with oxygen. Ammonium nitrate (NH_4NO_3) is manufactured with the reaction of NH_3 and nitric acid that can be directly used as N fertilizer. Another source of N for

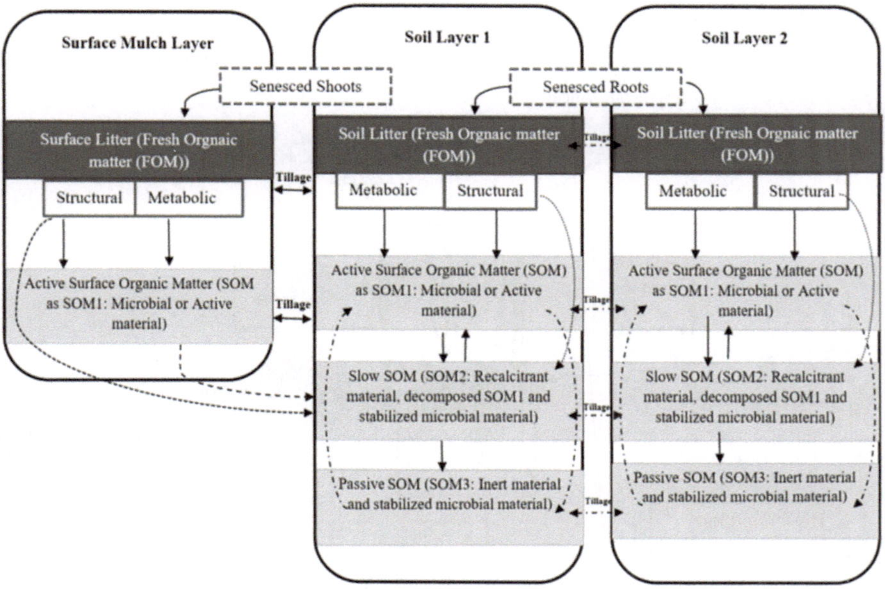

Fig. 19.3 The flow structure of DSSAT-CENTURY organic matter module with carbon pool

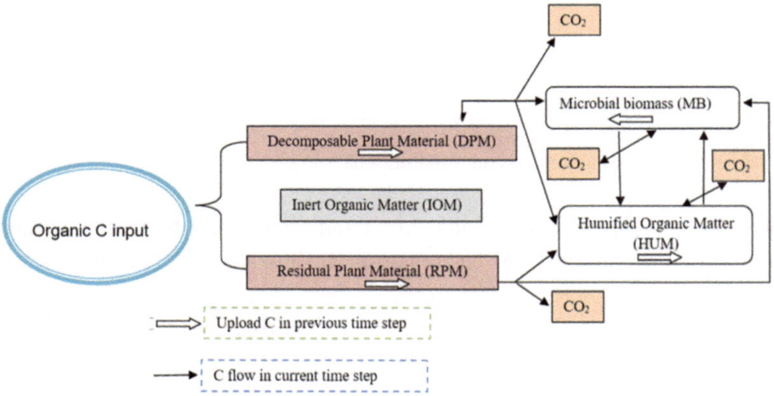

Fig. 19.4 Carbon (C) flow in RothC model

plants is animal wastes; after decomposition of these wastes, nitrates are released into the soil which are directly used by crop plants (Nasim and Akram 2017). Nitrogen availability and unavailability pathway are shown in Fig.19.5.

Nitrogen fixing microorganisms have an enzyme called nitrogenase that is sensitive to O_2. It interacts with Fe of the protein, deactivated when exposed to O_2. Aerobic spp. suffer due to this problem, while anaerobic does not have such problems of denaturing of enzymes. Rhizobium spp. has an association with legumes, while Brady rhizobium is present in the root nodules of chickpea and soybean. By

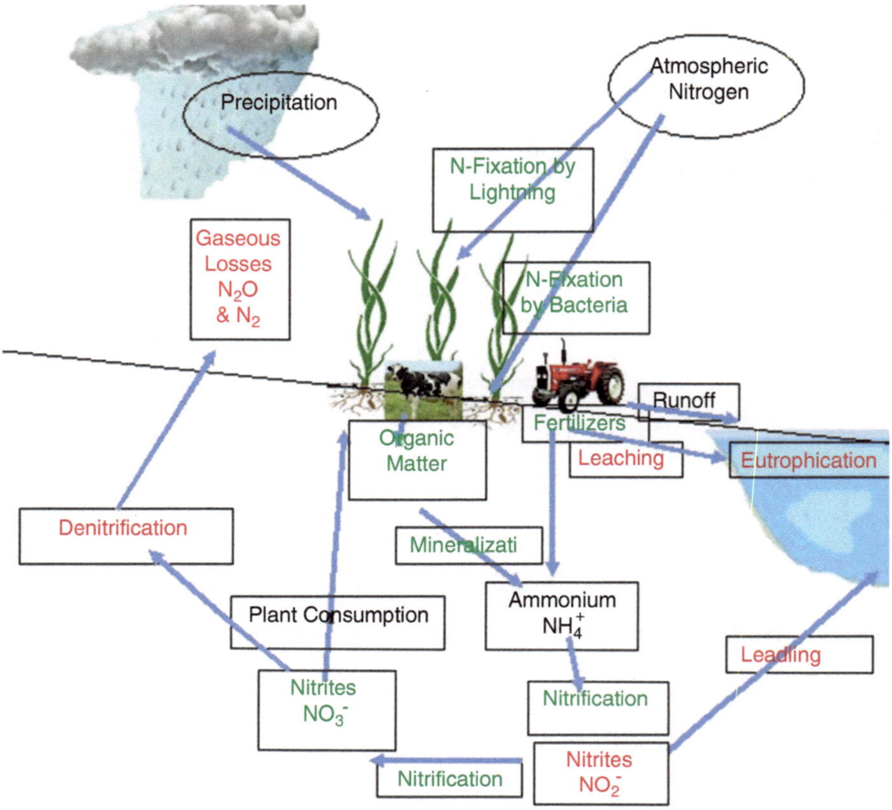

Fig. 19.5 Nitrogen availability and unavailability pathway

all these means, N is converted into nitrate and ammonium form. Plants take them through roots in solution form. The equation to show this mechanism:

$$N_2 + 8\,\text{flavodoxin}^- + 16\,\text{MgATP}^{2-} + 18H_2O \xrightarrow{\textit{nitrogenase}} 2NH_4^+ + 2OH^- + 8\,\text{flavodoxin} + 16\,\text{MgATP}^- + 16H_2PO_4^- + H_2$$

Meanwhile, within the plant, N participates in RNA, DNA, amino acids, and proteins biosynthesis and it could be elaborated by the following equation.

$$NH_4^+ + \alpha - \text{ketoglutarate} \xrightarrow{\textbf{GDH}} \text{glutamine} \rightarrow \text{Amino acids / protein}$$

$$\text{DNA} \rightarrow \text{RNA} \rightarrow \text{Protein}(\text{Central Dogma of cellular Biology})$$

After death and decay, these nitrogenous compounds are denatured, and N is released into the atmosphere.

Nutrient N is exceptionally vulnerable to environmental hazards and has less than 50% recovery. It was found that as the plant progresses, amount in the plant body declines due to a rise in cell wall and a decrease in cytoplasm. Likewise, Harper et al. (1987) highlighted diminished N concentration as long as the wheat plant reaches maturity (Ahmad et al. 2017). The earth's atmospheric temperature is increasing by global warming. The earth's temperature has risen by 0.8 °C from the start of the last century. This increase in temperature is contributing to changes in global wheat production (Liu et al. 2019). Plant growth and development may increase by global warming on a short-term basis, but in the long-run rise in temperature would disturb N-cycle and may cause deterioration of quality (Asseng et al. 2019; Zahoor et al. 2019).

Nitrogen losses have been accredited to the mutual effects of denitrification, volatilization, and percolation. A large number of N losses to the atmosphere could cause serious climatic hazards like toxification of underground moisture. N leaching might be minimized by applying N at an adjusted level. Urea-N, undergoes transformation processes, i.e., rapid hydrolysis to NH_4^+, followed by ammonia volatilization. N could be lost by the leaching of N from higher plants, e.g., by rainfall. Many researchers have found that plant roots also release N into the soil, and these losses range from 6% to 33% total N of the plant. In northwestern Europe, experiments on wheat for fertilizer uptake confirmed that N uptake was considerably lower at early stages (germination and tillering) than application at later stages (shooting).

19.2.3 Modeling P Dynamics

Phosphorus (P) being major macronutrients is needed to be part of the model so that model's performance remains perfect under P-deficient conditions. Dzotsi et al. (2010) in their work reported that DSSAT-P module could simulate *P* transformations between different pools. The soil-plant *P* model in DSSAT consists of two soil and one plant module (Fig. 19.6) and *P* exists as labile, active, and stable form.

$$P_i\left(Labiletoactive\right) = K_{LA} \times P_{iLabile}$$

$$P_i\left(Activetolabile\right) = K_{AL} \times P_{iActive}$$

$$P_i\left(Activetostable\right) = K_{AS} \times P_{iActive}$$

$$P_i\left(Stabletoactive\right) = K_{SA} \times P_{iStable}$$

Here P_{iLabile}: Inorganic labile P pools (mg kg^{-1}), P_{iActive}: Inorganic active P pools (mg kg^{-1}) and P_{iStable}: Inorganic stable *P* pools (mg kg^{-1}), and *P* transformation has units of mg kg^{-1} day^{-1}. The coefficients K_{LA}, K_{AL}, K_{AS}, and K_{SA} are the respective transformation rate constants. The validity of DSSAT-CSM-CROPGRO-Cotton-P model was tested by Amin et al. (2018) to optimize P use in cotton. Their results showed that DSSAT-P model can be effectively used to manage P in semiarid climates

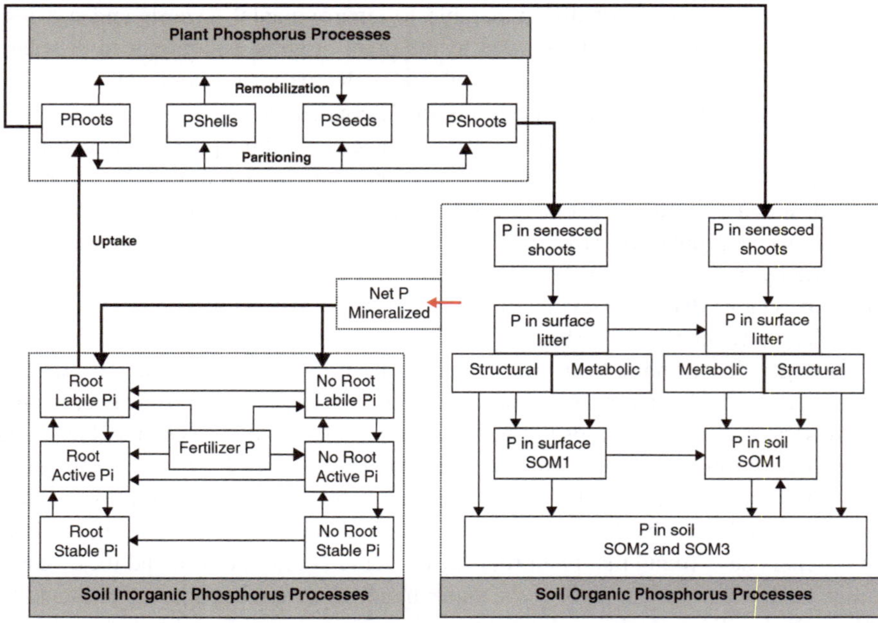

Fig. 19.6 Soil-Plant phosphorus model in DSSAT (Source with permission from Elsevier: Dzotsi et al. 2010)

(Mubeen et al. 2021). APSIM-SoilP module was tested by Ahmed et al. (2018) for rain-fed environments to manage P for the wheat crop. The results depicted that APSIM-SoilP has the potential to be used for the management of P under P-deficient environments.

19.3 Models as Decision Support Tools

Simulation modeling is a procedure of generating and investigating a numerical pattern of a corporeal model to forecast its application in the real world (Ahmed et al. 2013; Ahmed, 2020a, b; Ahmed et al. 2020a, b, c, d; Ahmed & Ahmad, 2020; Amin et al. 2018; Fahad et al. 2017; Nasim et al. 2012; Nielsen & Halvorson, 1990; Wang et al. 2020; Wallach et al. 2018). Simulation modeling is helpful to forecast weather and climate. A model can be defined as the purest demonstration of a system or the mathematical representation of a living system. Models are of different types. The most common examples are dynamic models and statistical models. Dynamic models are based upon time (Rosenberg 2010). There are varieties of models used by the researchers for the study of nutrients dynamics in agricultural systems. These include DNDC (DeNitrification DeComposition), CropSyst, DAISY, MINERVA, RZWQM (Root zone water Quality Model) (Ahuja et al. 2000), ANIMO, APSIM, and MANNER. These models simulate mineralization, volatilization, leaching,

Table 19.3 Coefficients used by different models

b_n: Nitrifying population
BIO: Nitrifier-biomass
BN: Total potential N need during crop growth
C_r : Nitrogen concentration at the root surface
$(C/N)_i$: C: N ratio of pool i
$(C/N)_r$: Crop residue C-to-N ratio
f_{df}: Denitrifying factor
f_{ifre}: Nitrogen fraction in root exudates
$F_{C/N}$: Correction factor for C-to-N ratio
$F_{d(NO3)}$: Maximum gas flux for a given nitrate concentration
F_{NO3}: Correction factor for nitrate
FX_D: Crop nitrogen demand
FX_N: Soil nitrogen content effect on nitrogen demand supplied by N_2 fixation
FX_P: Crop development effect on nitrogen demand supplied by N_2 fixation
FX_W: Water stress effect on nitrogen demand supplied by N_2 fixation
k_0: Final crop nitrogen content
k_1: Crop nitrogen content at the first development step
k_{max}: Maximum crop nitrogen content
K_n: Nitrification rate under aerobic conditions
$N_{\%}$: Nitrogen content of legume dry matter
$N_{amend, amend\_i}$: Nitrogen soil amendment input
N_A: Nitrogen is taken up by the crop
N_{dmd}: Nitrogen demand for uptake
N_D: Denitrified nitrogen
N_F: Fixed nitrogen
N_H: Nitrogen leaching during the winter period
N_L: Leached nitrogen
N_m: Nitrogen content of microbial biomass
N_{max}: Maximum nitrogen uptake by crop
N_M: Mineralized nitrogen
N_N: Nitrified nitrogen
N_r: Nitrogen content of the rhizosphere
N_{RO}: Initial crop residue nitrogen content
N_{Vmax}: Maximum volatilized nitrogen

nitrification, denitrification, and uptake. Different coefficients used by different models to study the dynamic of nutrients are listed in Table 19.3. It is challenging to predict temporal and spatial variation in precipitation, soil N in the form of nutrient contents, water in the soil profile, disease, and pest status. Only in a few studies, the worth of temporal variability was evaluated. Edelfeldt and Fritzson (2008) compared two ecological models to simulate N in wetlands. Simulation language, Modelica modeling, and additional tools were used to simulate, visualize, and appliance models. The variations and resemblances among the MathModelica Model and ecological modeling tools were estimated. Their experiment findings depicted that MathModelica Model Editor can simulate and it can be modeled by modeling approaches. N declined in a fabricated treatment, so wetland should be defined and simulated using the nitrification/denitrification model as this model. Gouis et al. (2000) performed an experiment to estimate varietal differences for N use at two N levels. Twenty genotypes of wheat were selected for 2 years; the soil was deep loam, and the N rate was 0 kg and 170 kg ha^{-1} as ammonium nitrate. During both years, results were nonsignificant for genotype × year but highly significant in terms of N × genotype. Wheat genotypes behaved differently during both years for N utilization and grain production.

All dynamic models simulate N and its uptake by plants, but N-fixation is not simulated by these models. The high input of N fertilizers and N balance is not dependent upon N-fixation and the difficulty of N-fixation simulation is two reasons not to incorporate N-fixation in the final simulation of N. Volatilization and denitrification and their impacts on the climate should also be taken into account while simulating N balance of a crop.

19.4 APSIM and Nutrient Dynamics

APSIM SOILN is a dynamic model that shares physical, biochemical, and physiological processes taking place in the soil to precisely predict soil N. More information is needed to predict N precisely. APSIM SOILN module is inherited from CERES-Maize on experience basis. There were restrictions to simulate crop and soil managing effects on N supply to the crop. APSIM SOILN is differentiated from the CERES model by distributing SOM into biom and hum pools. The first pool predominantly represents biotic factors like soil microbial biomass and their products while the second represents the rest of soil organic matter. Probert et al. (1998) predicted soil N with 82% accuracy, i.e., $R^2 = 0.82$. However, for the wheat-based cropping system, there are four types of modules NWHEAT, SOILWAT, SOILN, and RESIDUE are considered in APSIM to simulate wheat crop, soil water, soil N, and crop residue, respectively. These NWHEAT, SOILWAT, SOILN, and RESIDUE were developed from CERES, soil modules, and the PERFECT model based on experiments. Wang et al. (2003) used APSIM-Wheat module to simulate LAI, biomass, grain yield, grain N, and N uptake for the wheat crop with good accuracy.

19.5 Life Cycle Assessment (LCA)

Life cycle assessment is a valuable tool to limit the environmental impact of nutrients from different fertilizers. It can help to have appropriate use of fertilizer for sustainable crop production. Life Cycle Assessment was conducted for flax cultivation using N emissions derived from the DeNitrification DeComposition (DNDC) method and the genericIPCC method. The DNDC model was able to give more reliable estimations from the agricultural life cycle phase (Deng et al. 2017). Hence, LCA is an excellent tool to see the impact of different agricultural activities on nutrient dynamics in the changing climate.

19.6 Conclusion

This study elaborated that different biogeochemical models can be used to study the dynamics of nutrients (C, N, and P). These nutrients help in the different metabolic and physiological processes in plants. Thus, it is essential to study the dynamics of these nutrients under a set of different management and climate change scenarios. There are numerous types of models for simulating the nutrients processes in the plant. However, they have some limitations such as spatial heterogeneity, climate dependency, data limitations, over-parameterization of models and simulations under drier, cold, and wet soils. On the basis of the studies mentioned above, it can be concluded that nutrients dynamics modeling can be used to design adaptation strategies for improvement of the agricultural systems under climate change conditions.

References

Abbas G, Ahmad S, Hussain M, Fatima Z, Hussain S, Iqbal P, Ahmed M, Farooq M (2020b). Sowing date and hybrid choice matters production of maize-maize system. International Journal of Plant Production, (https://link.springer.com/article/10.1007/s42106-020-00104-6)

Abbas G, Fatima Z, Hussain M, Hussain S, Atique-ur-Rehman, Sarwar N, Ahmed M, Ahmad S. (2020a) Nitrogen rate and hybrid selection matters productivity of maize-maize cropping system under irrigated arid environment of southern Punjab, Pakistan. International Journal of Plant Production, (https://link.springer.com/article/10.1007/s42106-020-00086-5)

Abdalla M, Richards M, Pogson M, Smith JU, Smith P (2016) Estimating the effect of nitrogen fertilizer on the greenhouse gas balance of soils in Wales under current and future climate. Regional Environmental Change 16(8):2357-2368

Ahmad S, Hasanuzzaman A (2020) Cotton production and uses. Springer Nature Singapore Pte Ltd. https://doi.org/10.1007/978-981-15-1472-2

Ahmad, S., Q. Abbas, G. Abbas, Z. Fatima, Atique-ur-Rehman, S. Naz, H. Younis, R.J. Khan, W. Nasim, M Habib Ur Rehman, A. Ahmad, G. Rasul, M.A. Khan, M. Hasanuzzaman, 2017. Quantification of Climate Warming and Crop Management Impacts on Cotton Phenology. Plants, 6 (7): 1-16

Ahmed M, Hassan FU (2011) APSIM and DSSAT models as decision support tools. 19th International Congress on Modelling and Simulation, Perth, Australia, 12–16 December 2011,http://mssanz.org.au/modsim2011

Ahmed M (2012) Improving Soil Fertility Recommendations in Africa Using the Decision Support System for Agrotechnology Transfer (DSSAT); A Book Review. Exp Agri. 48 (4): 602-603

Ahmed M, Asif M, Hirani AH, Akram MN, Goyal A (2013) Modeling for Agricultural Sustainability: A Review. In Gurbir S. Bhullar GS, Bhullar NK (ed) Agricultural Sustainability Progress and Prospects in Crop Research. Elsevier, 32 Jamestown Road, London NW1 7BY, UK

Ahmed M, Ijaz W, Ahmad S (2018) Adapting and evaluating APSIM-SoilP-Wheat model for response to phosphorus under rainfed conditions of Pakistan. Journal of Plant Nutrition 41, 2069-2084.

Ahmed M, Ahmad S (2019) Carbon Dioxide Enrichment and Crop Productivity. In: Hasanuzzaman M (ed) Agronomic Crops: Volume 2: Management Practices. Springer Singapore, Singapore, pp 31-46. doi:https://doi.org/10.1007/978-981-32-9783-8_3

Ahmed M (2020a) Introduction to Modern Climate Change. Andrew E. Dessler: Cambridge University Press, 2011, 252 pp, ISBN-10: 0521173159. Sci Total Environ 734, 139397. https://doi.org/10.1016/j.scitotenv.2020.139397

Ahmed M (2020b) Systems Modeling, Springer Nature Singapore Pvt. Ltd., pp. 409. https://doi.org/10.1007/978-981-15-4728-7

Ahmed M, Hasanuzzaman M, Raza MA, Malik A, Ahmad S (2020a) Plant Nutrients for Crop Growth, Development and Stress Tolerance. R. Roychowdhury et al. (eds.), Sustainable Agriculture in the Era of Climate Change, https://doi.org/10.1007/978-3-030-45669-6_3

Ahmed K, Shabbir G, Ahmed M, Shah KN (2020b) Phenotyping for drought resistance in bread wheat using physiological and biochemical traits. Sci Total Environ 729, 139082. https://doi.org/10.1016/j.scitotenv.2020.139082

Ahmed M, Ahmad S (2020). Systems Modeling. In: Ahmed M (ed.), Systems Modeling, Springer Nature Singapore Pvt. Ltd., pp. 1-44. https://doi.org/10.1007/978-981-15-4728-7_1

Ahmed M, Raza MA, Hussain T (2020c) Dynamic Modeling. In: Ahmed M (ed.), Systems Modeling, Springer Nature Singapore Pvt. Ltd., pp. 111-148. https://doi.org/10.1007/978-981-15-4728-7_4

Ahmed M, Ahmad S, Raza MA, Kumar U, Ansar M, Shah GA, Parsons D, Hoogenboom G, Palosuo T, Seidel S (2020d) Models Calibration and Evaluation. In: Ahmed M (ed.), Systems Modeling, Springer Nature Singapore Pvt. Ltd., pp. 149-176. https://doi.org/10.1007/978-981-15-4728-7_5

Ahuja LR, Rojas KW, Hanson JD, Shaffer MJ, Ma L (2000) Root Zone Water Quality Model: Modeling management effects on water quality and crop production. p. 372. Water Resources Publications, LLC, Highland Ranch, CO

Allison SD (2012) A trait-based approach for modelling microbial litter decomposition. Ecology Letters 15(9):1058-1070

Ali M, Mubeen M, Hussain N, Wajid A, Farid HU, Awais M, Hussain S, Akram W, Amin A, Akram R, Imran M (2019) Role of ICT in Crop Management. In Agronomic Crops (pp. 637-652). Springer, Singapore

Amin A, Nasim W, Mubeen M, Ahmad A, Nadeem M, Urich P, Fahad S, Ahmad S, Wajid A, Tabassum F, Hammad HM, Sultana SR, Anwar S, Baloch SK, Wahid A, Wilkerson CJ, Hoogenboom G. 2018. Simulated CSM-CROPGRO-Cotton yield under projected future climate by SimCLIM for southern Punjab, Pakistan. Agricultural Systems 167:213–222.

Asseng S, Martre P, Maiorano A, Rötter RP, O'Leary GJ, Fitzgerald GJ, Girousse C, et al. (2019) Climate change impact and adaptation for wheat protein. Global Change Biology 25 (1):155-173. doi:https://doi.org/10.1111/gcb.14481

Autovino D, Rallo G, Provenzano G (2018) Predicting soil and plant water status dynamic in olive orchards under different irrigation systems with Hydrus-2D: Model performance and scenario analysis. Agricultural Water Management 203:225-235

Ben-Gal A, Ityel E, Dudley L, Cohen S, Yermiyahu U, Presnov E, Zigmond L, Shani U (2008) Effect of irrigation water salinity on transpiration and on leaching requirements: A case study for bell peppers. Agricultural Water Management 95(5):587-597

Biernath C, Bittner S, Klein C, Gayler S, Hentschel R, Hoffmann P, Högy P, Fangmeier A, Priesack E (2013) Modeling acclimation of leaf photosynthesis to atmospheric CO_2 enrichment. European Journal of Agronomy 48:74-87

Bittelli M, Tomei F, Pistocchi A, Flury M, Boll J, Brooks ES, Antolini G (2010) Development and testing of a physically based, three-dimensional model of surface and subsurface hydrology. Advances in Water Resources 33(1):106-122

Braud I, Bariac T, Gaudet JP, Vauclin M (2005) SiSPAT-Isotope, a coupled heat, water and stable isotope (HDO and H218O) transport model for bare soil. Part I. Model description and first verifications. Journal of Hydrology 309(1):277-300

Brilli L, Bechini L, Bindi M, Carozzi M, Cavalli D, Conant R, Dorich CD, Doro L, Ehrhardt F, Farina R, Ferrise R, Fitton N, Francaviglia R, Grace P, Iocola I, Klumpp K, Léonard J, Martin R, Massad RS, Recous S, Seddaiu G, Sharp J, Smith P, Smith WN, Soussana J-F, Bellocchi G (2017) Review and analysis of strengths and weaknesses of agro-ecosystem models for simulating C and N fluxes. Science of The Total Environment 598:445-470

Cameron DR, Van Oijen M, Werner C, Butterbach-Bahl K, Grote R, Haas E, Heuvelink GBM, Kiese R, Kros J, Kuhnert M, Leip A, Reinds GJ, Reuter HI, Schelhaas MJ, De Vries W, Yeluripati J (2013) Environmental change impacts on the C- and N-cycle of European forests: a model comparison study. Biogeosciences 10 (3):1751-1773

Canuto N, Ramos TB, Oliveira AR, Simionesei L, Basso M, Neves R (2019) Influence of reservoir management on Guadiana streamflow regime. Journal of Hydrology: Regional Studies 25:100628

Chen YL, Dunbabin VM, Postma JA, Diggle AJ, Siddique KHM, Rengel Z (2013) Modelling root plasticity and response of narrow-leafed lupin to heterogeneous phosphorus supply. Plant and Soil 372 (1):319-337

Chenu C, Angers DA, Barré P, Derrien D, Arrouays D, Balesdent J (2019) Increasing organic stocks in agricultural soils: Knowledge gaps and potential innovations. Soil and Tillage Research 188:41-52

Danish S., et al., 2019. Alleviation of Cr toxicity in Maize by Fe fortification and Cr tolerant ACC deaminase producing PGPR. Ecotoxicology and Environmental Safety, 185:109706

Deng Y, Paraskevas D, Cao S-J (2017) Incorporating denitrification-decomposition method to estimate field emissions for Life Cycle Assessment. Science of The Total Environment 593-594:65-74

Dimassi B, Guenet B, Saby NPA, Munoz F, Bardy M, Millet F, Martin MP (2018) The impacts of CENTURY model initialization scenarios on soil organic carbon dynamics simulation in French long-term experiments. Geoderma 311:25-36

Dondini M, Jones EO, Richards M, Pogson M, Rowe RL, Keith AM, Perks MP, McNamara NP, Smith JU, Smith P (2015) Evaluation of the ECOSSE model for simulating soil carbon under short rotation forestry energy crops in Britain. GCB Bioenergy 7(3):527-540

Dzotsi, K.A., Jones, J.W., Adiku, S.G.K., Naab, J.B., Singh, U., Porter, C.H., Gijsman, A.J., 2010. Modeling soil and plant phosphorus within DSSAT. Ecological Modelling 221, 2839-2849.

Edelfeldt S, Fritzson P (2008) Evaluation and comparison of models and modeling tools simulating nitrogen processes in treatment wetlands. Simulation Modelling Practice and Theory 16:26–49

Fahad, S., A.A. Bajwa, U. Nazir, S.A. Anjum, A. Farooq, A. Zohaib, S. Sadia, W. Nasim, S. Adkins, S. Saud, M.Z. Ihsan, H. Alharby, C. Wu, D. Wang, J. Huang, 2017. Crop Production under drought and heat stress: Plant responses and management options. Frontier in Plant Sciences. 08: 1147: 01-16

Fahad, S., S. Hussain, S. Saud, S. Hassan, M. Tanveer, M.Z. Ihsan, A.N. Shah, A. Ullah, Nasrullah, F. Khan, S. Ullah, H. Alharby, W. Nasim, C. Wu, J. Huang, 2016. A combined application of biochar and phosphorus alleviates heat-induced adversities on physiological, agronomical and quality attributes of rice. Plant Physiology and Biochemistry, 103: 191-198

Fatima Z., Abbas Q., Khan A., Hussain S., Ali MA, Abbas G, Younis H, Naz S, Ismael M, Shahzad MI, Nadeem M, Farooq U., Khan SU, Javed K, Khan AA, Ahmed M, Khan MA,

Ahmad S. (2018). Resource use efficiencies of C_3 and C_4 cereals under split nitrogen regimes. Agronomy, 8, 69.

Finke PA, Hutson JL (2008) Modelling soil genesis in calcareous loess. Geoderma 145(3):462-479

Ford W, King K, Williams M, Williams J, Fausey N (2015) Sensitivity analysis of the agricultural policy/environmental eXtender (APEX) for phosphorus loads in tile-drained landscapes. Journal of Environmental Quality 44(4):1099-1110

Franko U, Oelschlägel B, Schenk S (1995) Simulation of temperature-, water- and nitrogen dynamics using the model CANDY. Ecological Modelling 81(1):213-222

Gouis LJ, Beghin D, Heumez E, Pluchard P (2000) Genetic differences for nitrogen uptake and nitrogen utilisation efficiencies in winter wheat. European Journal of Agronomy 12:163–173

Hammad HM, Khaliq A, Abbas F, Farhad W, Fahad S, Aslam M, Shah, GM, Nasim W, Mubeen M, Bakhat HF (2020) Comparative effects of organic and inorganic fertilizers on soil organic carbon and wheat productivity under Arid Region. Communications in Soil Science and Plant Analysis https://doi.org/10.1080/00103624.2020.1763385

Hammad, HM. et al. 2018. Offsetting Land Degradation through Nitrogen and Water Management during Maize Cultivation under Arid Conditions. Land Degradation and Development, 29 (5): 1366-1375

Han J, Jia Z, Wu W, Li C, Han Q, Zhang J (2014) Modeling impacts of film mulching on rainfed crop yield in Northern China with DNDC. Field Crops Research 155:202-212

Harper LA, Sharpe RR, Langdale GW, Evans JE (1987) Nitrogen cycling in a wheat crop: soil, plant, and aerial nitrogen transport. Agronomy Journal 79:965-973

He W, Dutta B, Grant BB, Chantigny MH, Hunt D, Bittman S, Tenuta M, Worth D, VanderZaag A, Desjardins RL, Smith WN (2020) Assessing the effects of manure application rate and timing on nitrous oxide emissions from managed grasslands under contrasting climate in Canada. Science of The Total Environment 716:135374

He W, Grant BB, Smith WN, VanderZaag AC, Piquette S, Qian B, Jing Q, Rennie TJ, Bélanger G, Jégo G, Deen B (2019) Assessing alfalfa production under historical and future climate in eastern Canada: DNDC model development and application. Environmental Modelling & Software 122:104540

Herbst M, Hellebrand HJ, Bauer J, Huisman JA, Šimůnek J, Weihermüller L, Graf A, Vanderborght J, Vereecken H (2008) Multiyear heterotrophic soil respiration: Evaluation of a coupled CO_2 transport and carbon turnover model. Ecological Modelling 214 (2):271-283

Houska T, Kraft P, Chamorro-Chavez A, Breuer L (2015) SPOTting model parameters using a ready-made python package. PLOS ONE 10 (12):e0145180

Hussain S, Ahmad A, Wajid A, Khaliq T, Hussain N, Mubeen M, Farid HU, Imran M, Hammad HM, Awais M, Ali A, Aslam M, Amin A, Akram R, Amanet K, Nasim W (2020) Irrigation Scheduling for Cotton Cultivation. In Cotton Production and Uses (pp. 59-80). Springer, Singapore

Inatomi M, Ito A, Ishijima K, Murayama S (2010) Greenhouse gas budget of a cool-temperate deciduous broad-leaved forest in Japan estimated using a process-based model. Ecosystems 13 (3):472-483

Izaurralde RC, Williams JR, McGill WB, Rosenberg NJ, Jakas MCQ (2006) Simulating soil C dynamics with EPIC: Model description and testing against long-term data. Ecological Modelling 192 (3):362-384

Javaux M, Couvreur V, Vanderborght J, Vereecken H (2013) Root water uptake: From three-dimensional biophysical processes to macroscopic modeling approaches. Vadose Zone Journal (2013) 12(4): vzj2013.02.0042

Jebari A, del Prado A, Pardo G, Rodríguez Martín JA, Álvaro-Fuentes J (2018) Modeling Regional Effects of Climate Change on Soil Organic Carbon in Spain. Journal of Environmental Quality 47(4):644-653

Jiang R, He W, Zhou W, Hou Y, Yang JY, He P (2019) Exploring management strategies to improve maize yield and nitrogen use efficiency in northeast China using the DNDC and DSSAT models. Computers and Electronics in Agriculture 166:104988

Krinner G, Viovy N, de Noblet-Ducoudré N, Ogée J, Polcher J, Friedlingstein P, Ciais P, Sitch S, Prentice IC (2005) A dynamic global vegetation model for studies of the coupled atmosphere-biosphere system. Global Biogeochemical Cycles 19 (1)

Lashermes G, Zhang Y, Houot S, Steyer JP, Patureau D, Barriuso E, Garnier P (2013) Simulation of organic matter and pollutant evolution during composting: The COP-Compost Model. Journal of Environmental Quality 42 (2):361-372

Leitner D, Felderer B, Vontobel P, Schnepf A (2014) Recovering root system traits using image analysis exemplified by two-dimensional neutron radiography images of Lupine. Plant Physiology 164 (1):24-35

Li Y, Chen D, Zhang Y, Edis R, Ding H (2005) Comparison of three modeling approaches for simulating denitrification and nitrous oxide emissions from loam-textured arable soils. Global Biogeochemical Cycles 19(3):1-15

Liu B, Martre P, Ewert F, Porter JR, Challinor AJ, et al. 2019 Global wheat production with 1.5 and 2.0°C above pre-industrial warming. Global Change Biology 25 (4):1428-1444. doi:https://doi.org/10.1111/gcb.14542

Luo Z, Eady S, Sharma B, Grant T, Liu DL, Cowie A, Farquharson R, Simmons A, Crawford D, Searle R, Moore A (2019) Mapping future soil carbon change and its uncertainty in croplands using simple surrogates of a complex farming system model. Geoderma 337:311-321

Mubeen M, Bano A, Ali B, Islam ZU, Ahmad A, Hussain S, Fahad **S,** Nasim W (2021) Effect of plant growth promoting bacteria and drought on spring maize (Zea mays L.). Pakistan Journal of Botany, 53(2): 731-739

Nasim W and Akram R, 2017. Electronic Wastes published by Technology Times Publishers: (http://www.technologytimes.pk/electronic-waste/)

Nasim, W., A. Ahmed, M. Tariq and S.A. Wajid, 2012. Studying the comparative performance of wheat cultivars for growth and grains production. International Journal of Agronomy and Plant Production, 3 (9), 306-312

Naz I., Hussain MH, Kamran MA, Mufti R, Mukhtar T, Rasul F, Nasim W, Chaudhary HJ, 2013. Effect of different Fungicides on the incidence of Maize pathogen Helminthosporium maydis. Jokull Journal, 63. 196-207

Nielsen DC, Halvorson AD (1990) Nitrogen fertility influence on water stress and yield of winter wheat83(6)*:1065-1070*

Panagos P, Borrelli P, Poesen J, Ballabio C, Lugato E, Meusburger K, Montanarella L, Alewell C (2015) The new assessment of soil loss by water erosion in Europe. Environmental Science & Policy 54:438-447

Peltre C, Christensen BT, Dragon S, Icard C, Kätterer T, Houot S (2012) RothC simulation of carbon accumulation in soil after repeated application of widely different organic amendments. Soil Biology and Biochemistry 52:49-60

Probert ME, Dimes JP, Keating BA, Dalal RC, Strong WM (1998) APSIM's water and nitrogen modules and simulation of the dynamics of water and nitrogen in fallow systems. Agricultural Systems 56(1):l-28

Rosenberg NJ (2010) Climate change, agriculture, water resources: what do we tell those that need to know? Climate Change 100:113-117

Shirato Y (2020) Use of models to evaluate carbon sequestration in agricultural soils. Soil Science and Plant Nutrition 66 (1):21-27

Smith W, Grant B, Qi Z, He W, VanderZaag A, Drury CF, Helmers M (2020) Development of the DNDC model to improve soil hydrology and incorporate mechanistic tile drainage: A comparative analysis with RZWQM2. Environmental Modelling & Software 123:104577

Srivastava RK, Panda RK, Chakraborty A, Halder D (2018) Enhancing grain yield, biomass and nitrogen use efficiency of maize by varying sowing dates and nitrogen rate under rainfed and irrigated conditions. Field Crops Research 221:339–349.

van Dam JC, Groenendijk P, Hendriks RFA, Kroes JG (2008) Advances of Modeling Water Flow in Variably Saturated Soils with SWAP. Vadose Zone Journal 7 (2):640-653

Visconti F, de Paz JM, Martínez D, Molina MJ (2014) Irrigation recommendation in a semi-arid drip-irrigated artichoke orchard using a one-dimensional monthly transient-state model. Agricultural Water Management 138:26-36

Vogeler I, Giltrap D, Cichota R (2013) Comparison of APSIM and DNDC simulations of nitrogen transformations and N_2O emissions. Science of the Total Environment 465:147-155

Wallach D, Martre P, Liu B, Asseng S, Ewert F, et al. (2018) Multimodel ensembles improve predictions of crop–environment–management interactions. Global Change Biology 24 (11):5072-5083. doi:https://doi.org/10.1111/gcb.14411

Wang E, Van Oosterom E, Meinke H, Asseng S, Robertson M, Huth N, Keating B, Probert M The new APSIM-Wheat Model—performance and future improvements. In: Proceedings of the 11th Australian Agronomy Conference, 2003. vol 6. Australian Society of Agronomy Geelong, Victoria, Australia

Wang S, Zhao Y, Wang J, Gao J, Zhu P, Cui Xa, Xu M, Zhou B, Lu C (2020) Estimation of soil organic carbon losses and counter approaches from organic materials in black soils of north-eastern China. Journal of Soils and Sediments 20 (3):1241-1252

Wu SH, Jansson P-E, Kolari P (2012) The role of air and soil temperature in the seasonality of photosynthesis and transpiration in a boreal Scots pine ecosystem. Agricultural and Forest Meteorology 156:85-103

Yao Z, Zhang D, Liu N, Yao P, Zhao N, Li Y, Zhang S, Zhai B, Huang D, Wang Z, Cao W, Adl S, Gao Y (2019) Dynamics and sequestration potential of soil organic carbon and total nitrogen stocks of leguminous green manure-based cropping systems on the Loess Plateau of China. Soil and Tillage Research 191:108-116

Yokozawa M, Shirato Y, Sakamoto T, Yonemura S, Nakai M, Ohkura T (2010) Use of the RothC model to estimate the carbon sequestration potential of organic matter application in Japanese arable soils. Soil Science and Plant Nutrition 56(1):168-176

Zahoor S.A. et al. (2019) Improving Water Use Efficiency in Agronomic Crop Production. In: Hasanuzzaman M. (eds) Agronomic Crops. Springer, Singapore (https://link.springer.com/chapter/10.1007/978-981-32-9783-8_2)

Zamin M., S. Fahad, et al., 2019. Developing the first halophytic turfgrasses for the urban landscape from native Arabian desert grass. Environmental Science and Pollution Research (Accepted)

Zhang Y, Lashermes G, Houot S, Zhu YG, Barriuso E, Garnier P (2014) COP-compost: a software to study the degradation of organic pollutants in composts. Environmental Science and Pollution Research 21(4):2761-2776

Zhou X-R, Schnepf A, Vanderborght J, Leitner D, Lacointe A, Vereecken H, Lobet G (2019) CPlantBox, a whole plant modelling framework for the simulation of water and carbon related processes. bioRxiv:810507

Part IV
Innovative Approaches to Achieve Climate Resilience in Agriculture

Chapter 20
Climate Smart Agriculture (CSA) Technologies

Sajjad Hussain, Asad Amin, Muhammad Mubeen, Tasneem Khaliq, Muhammad Shahid, Hafiz Mohkum Hammad, Syeda Refat Sultana, Muhammad Awais, Behzad Murtaza, Muhammad Amjad, Shah Fahad, Khizer Amanet, Amjed Ali, Mazhar Ali, Naveed Ahmad, and Wajid Nasim

Abstract Agricultural production is low in many parts of the world which is subjected to limited adaptability toward adverse events. Climate change (CC) is estimated to decrease productivity on even lower levels and has made production more inconsistent. Many countries all over the world have planned to accept climate smart agriculture (CSA) approach to make improvements in agriculture. The CSA refers to a combined set of technologies and practices that simultaneously improve farm productivity as well as monetary return, enhance adaptability to CC, and minimize the emissions of greenhouse gas (GHG). The concept of CSA is gaining sig-

S. Hussain · M. Mubeen (✉) · M. Shahid · H. M. Hammad · S. R. Sultana · B. Murtaza · M. Amjad · K. Amanet · M. Ali
Department of Environmental Sciences, COMSATS University Islamabad, Vehari, Punjab, Pakistan
e-mail: muhammadmubeen@cuivehari.edu.pk

A. Amin
Queensland Alliance for Agriculture and Food Innovation (QAAFI), The University of Queensland,
Brisbane, QLD, Australia

T. Khaliq
Department of Agronomy, University of Agriculture, Faisalabad, Punjab, Pakistan

M. Awais · W. Nasim
Department of Agronomy, The Islamia University of Bahawalpur (IUB), Bahawalpur, Punjab, Pakistan

S. Fahad
Department of Agronomy, The University of Haripur, Haripur, Pakistan

A. Ali
Department of Agronomy, University College of Agriculture, University of Sargodha (UoS), Sargodha, Punjab, Pakistan

N. Ahmad
Department of Zoology, University of Education, Vehari, Punjab, Pakistan

W. N. Jatoi et al. (eds.), *Building Climate Resilience in Agriculture*,
https://doi.org/10.1007/978-3-030-79408-8_20

nificance at national and international levels to cope with future challenges of the agricultural-related plannings. The CSA is a notion that calls for the combination of the need of adaptability to CC and strategies for mitigation in agriculture to ensure food security. For instance, strategies like the use of renewable energy for agriculture, i.e., pyrolysis units, solar panels, windmills, as well as water pumps are very vital for food production. The concept of CSA comprehends three major stakes (productivity, adaptation, and mitigation), but the literature has not fully addressed them in an integrated way. Adaptation and productivity were of prime importance for poor and under developing countries, while mitigation was mainly addressed in developed countries. The utmost challenge for policymakers and stakeholders to operationalize CSA is the identification, assessment (cost-benefit ratio), as well as subsequent ordering of CSA options and portfolios for investment. The aims of CSA are to sustainably increase farmer's resilience, improve agricultural productivity, achieve food security and sustainable development goals in addition to reduce emissions of GHG.

Keywords Climate smart agriculture (CSA) · Climate change (CC) · Agricultural production · Food security

20.1 Background and Introduction

> Climate smart agriculture (CSA) was first introduced in an international conference on climate change (CC), agriculture and food security at Hague in the Netherlands by the food and agriculture organization (FAO) in 2010 (FAO 2013). The FAO describes CSA as "agriculture that sustainably raises production and incomes, in addition to adapting and promoting resilience to CC and decreasing greenhouse gasses (GHG) emissions (Amin et al. 2015). The CSA is a new approach to upgrade technical, policy as well as investment measures for environment to improve agricultural growth for food production under CC."

There are three important trade-offs of CSA according to FAO 2015 :

1. Increasing agricultural incomes and productivity on sustained basis
2. Building resilience to CC
3. Decreasing GHG emissions according to IPCC standards

The CSA approach aims to support food production of farmers having small landholdings by implementing suitable technologies and methods for the production, processing as well as marketing of agriculture-related commodities, and also improve the agricultural income (Di Falco et al. 2011; Awais et al. 2018). Nowadays, the CSA approach has been developed as an agenda to apprehend the idea that agriculture production systems can be applied to increase the food production and rural livelihoods, at the same time to provide mitigation benefits and facilitate CC

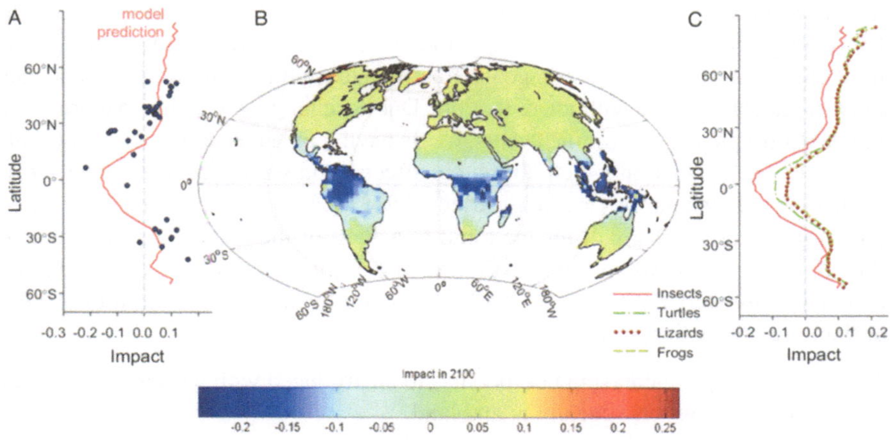

Fig. 20.1 Pattern in CC is projected to be greatest poisonous for insects in hot regions Source: Deutsch et al. (2008)

adaptation. The CSA includes farm-based land conservation practices like crop residue management, increasing livelihood, agro-forestry, and efficient water management (Scherr et al. 2012; Fellmann 2012; Saud et al. 2017). The CSA is a novel technique which can guide the required modifications of Agro-based-systems and provided the need to equally address the CC and food production (FAO 2013).

The adoption of CSA methods has a positive effect on the agricultural profits of both rich as well as poor households. Conservation agriculture (CA), a mixture of soil management practices that includes crop rotation, improved soil cover, and reduced soil disturbance, is frequently labeled CSA (Jat et al. 2020). The CSA can also help to overcome hunger while supporting urban populations to adapt CC by managing natural resources sustainably and curbing rising temperatures (Rahman et al. 2020) (Fig. 20.1).

20.2 Why Is Climate Smart Agriculture (CSA) Needed?

The expected population in 2050 will be 9 billion but our current crop production systems will not be able to feed so many humans. Out of these, two billion people will habitat in developing countries while most people out of this surplus population will be living in urban areas. According to FAO prediction, agricultural production will have to rise by 60% by 2050 to feed that big population. But current crop genetic gain is much lower than expected. Agriculture must be transforming itself if it wants to feed an increasing worldwide population as well as give some good foundation to decrease poverty and increase economic development (Gommes et al. 2010; Nasim et al. 2017a, b). Under a business-as-usual scenario, CC will make this task very difficult due to bad effects on agriculture, requiring spiraling adaptation. Lower production intensities and adaptation to CC will be essential to attain agricultural growth and food production goals. Based on the natural resource these changes must be accomplished without reduction. Due to changing weather patterns and increased frequency of extreme events, CC is having an adverse impact on agriculture and food production which ultimately affects food security. Smallholder farmers in developing countries are being affected due to these variations. By FAO, there are various reasons that call for the rapid transition of the current agriculture production system to CSA in the backdrop of increasing risks from CC and climate associated disasters (Scherr et al. 2012). There are five main points to improve the agriculture production system in CSA.

1. Food demand is increasing but food has to be produced with the same amount of resources like water, vegetation, and land.
2. Farmers are exposed to the effects of CC and there is urgency for sustainable adaptation to CC.
3. There is a total degradation and depletion of natural resources that sustain agriculture production.
4. In the expression of CC in agricultural production systems have to be other productive, strong to risks, efficient, better constancy in their outputs in long-term CC.
5. There is an essential need for increasing food production along with mitigating CC as well as protecting the natural resource (IPCC 2007).

20.3 Strategies for Climate Smart Agriculture (CSA)

20.3.1 Efficient Resource Management

Resource management is the most important feature of CSA and also future climate. In all stages of food security, food waste is found till food application. Almost, one-third part of produced food is lost all over the world. The energy consumed in yearly world food losses is nearly 38% of the total energy used in the food chain (Gustavsson et al. 2011). In food chains, agriculture, processing, cooking, consumption, and management are likely regions for improving the use of energy efficiently. In Africa, 90% of removed wood is utilized for cooking and industrial household. Enhanced energy saving, as well as resource management in cooking stoves, can support to reduction of deforestation (FAO 2012).

20.3.2 Integrated Renewable Energy Technologies for Farming Perceptions

The proper energy tools, technologies as well as various services in agriculture fields are significant to make the stable variation to capable for food and energy smart systems. New CSA approaches can be most significant for integrated food and energy systems include (Khoi and Murayama 2010):

- bioenergy-operated water pumps,
- cooking stoves,
- distillation and fermentation processes,
- hydrothermal conversion exchange tools,
- information and communication technologies (ICT),
- monitoring observing systems,
- photovoltaic lights,
- power generators,
- pyrolysis units,
- renewable energy-powered vehicles,
- solar panels,
- solar-wind electricity production by solar panels,
- windmills.

These recently developed methods increase the value of production and the accessibility of raw properties. These approaches can be interrelated on a farm in terms of food and energy smart systems (Fig. 20.2).

20.3.3 Soil Management in CSA

Agriculture is extremely dependent on climate, up to now, evidence of perceived variations related to CC and especially to water, has been difficult to find. Agriculture is intensely changing due to climate factors, agricultural practices, market prices, technological advances, and management practices. Policymaker in agriculture will be required to revise the way in which water, land, and soil nutrients are achieved to confirm that these resources are utilized most efficiently to attain the three main objectives of CSA. The CSA is an extension of earlier maintainable efforts which includes conservation , pest management, crop improvements, horticulture, soil management, aquaculture, fodder crops, agroforestry, and new policies. Soils are made up of different elements like decayed plant, variety of dead animals, and plants. Due to these changes in local climate, properties of soil (topography, geology, vegetation, and variable landscapes in depth) are also changing. The abundance of nutrients that are available in the soil bring more diversity. A handful of soil that can hold various organisms can also play an important role in preserving soil health (Corsi et al. 2012).

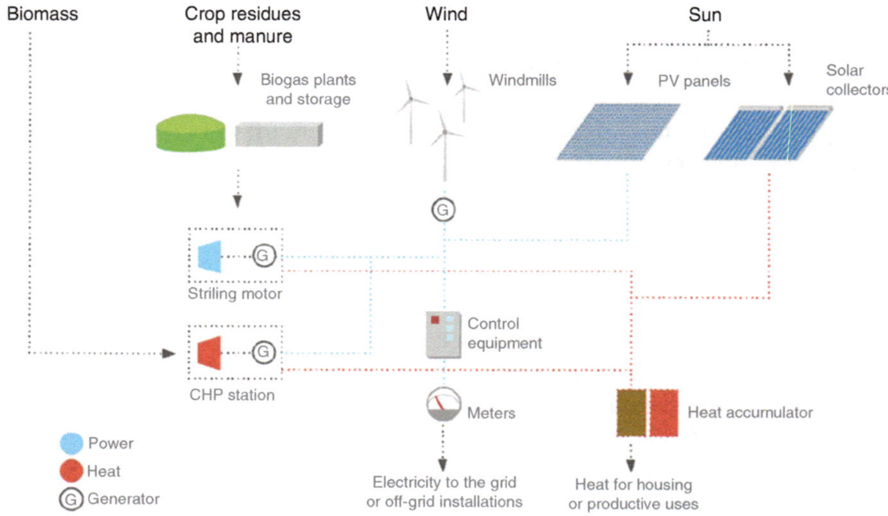

Fig. 20.2 Integrated renewable energy technologies (Source: Amin et al. 2015)

20.3.4 Water Management in CSA

Agriculture is most exposed to vulnerable impacts of CC than other economic areas, it consumes 80% of freshwater in the whole world (Mundial 2012). According to the UN report (2014) population is estimated to reach nine billion people by 2050. On other hand, weather patterns have turned out to be irregular, which has led to low food production in different countries around the world. The CSA technology has been considered as a friendly method that applies a political aspect to attain sustainable development goals (Mundial 2012). It combines three ways of sustainable progress (community-based, financial, and biological) to tackle CC issues and food security (Zahoor et al. 2019). Water demand has been increasing in industries and cities as an outcome of fast economic development in developing countries (Mubeen et al. 2020). Pollution has been increased by industries, cities, and agriculture activities which have affected the ecosystems. Water demand was projected to increase in the whole world in next coming years as population will spread nine billion upto 2050.

20.3.5 Land Use (LU) Management

Practices of land use (LU) management like reduced tillage and integration of soil biodiversity can play a significant role in accretion of carbon in the soil. Therefore, the issues of CC can be resolved by CSA like effective resource management and conservation management practices, improved weather predicting, better pest

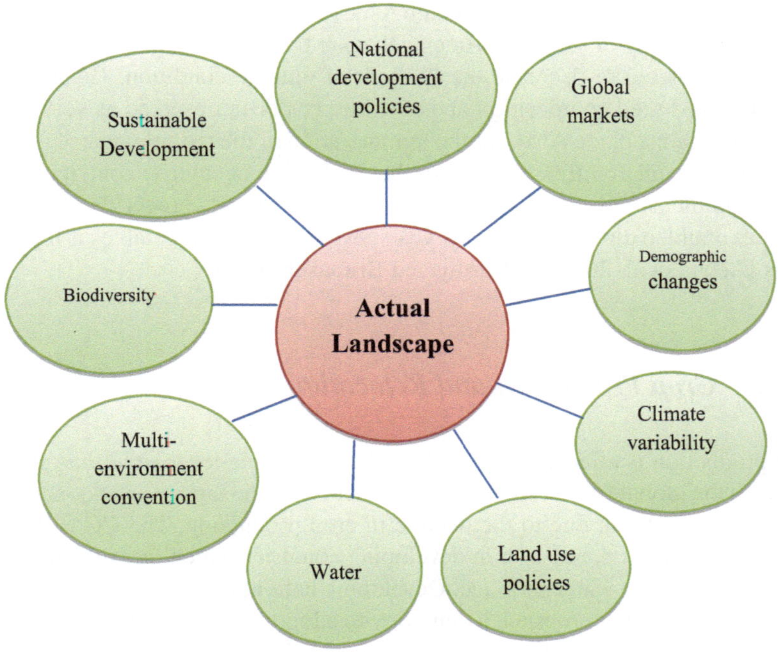

Fig. 20.3 Ten principles of LA to the integration of conservation agriculture and competing LU (Moussa et al. 2016)

control, crop variation, and crop modeling. Future LU policy procedures and involvement of national values of farmers will be crucial for food security. This method also contributes to the solution of the environmental problems as well as the projected sustainable growth initiatives (on LA, FAO 2015).

Several advantages of the landscape approach (LA) include the following:

- Economies of scale and scope: water and land users in a LA share their services as well as cost advantages subsequent from included food production;
- It needs that LU policy-making and planning are measured contrarily with respect to scale taking into account spatial components;
- It is not a "one-size-fits-all" method but rather includes a few areas, cultural, stakeholders as well as further geographic boundaries that generally make natural sense (Fig. 20.3).

20.3.6 Crops Genetic Modification

Environmental stress generally influences the decay of organic matter in soil, and accessibility of water and nutrients to plant (Akram et al. 2018). Crop production is projected to reduced by 2.5–10% after 2020 and by 5–30% upto 2050, with maximum reduction in central and south Asia. Adjustment methods can be improved by the accessibility of yield assortments that are tolerant to salinity, drought as well as heat and therefore decrease the dangers of climate condition. Genetic variety with improved seed composition and structure has been observed as very effective protection against plant. Also, for the remuneration of the rising issues of crop cycle reductions as well as different ideas for the production of environment, it is essential to build up the assortments. Because of the genetic change in worldwide vegetable yield has doubled during the last few years, overall trade in vegetables is higher than grains (Aslam et al. 2013; McCarthy and Brubaker 2014).

20.3.7 Crop Production and Relocation

Crop production is affected by the CC at the worldwide level and it has great concern with the global food security. On the other hand, the farmer and poor communities are highly at risk due to the decline in crop production. The CC is the global issue for maximum population in developing countries which faces the health risk, food insecurity, and water shortage. Developing nations are highly at risk due to low income, absence of appropriate technology to adopt the CC. The few crops may get benefit in different areas all over the world under the CC. For example, the precipitation will be affected by the change in extreme weather events such as flood and drought (Mubeen et al. 2013a). The increase in the temperature will lead to the rapid spread of the pest in crops and also favors the growth of weeds. The rapid increase in the sea level will limit crop production and also the long time crop failure all over the world. Currently, the whole world is facing great challenges like volatility, loss of soil fertility, increase in urbanization, CC, and water scarcity. Approximately 86% of the rural communities are dependent on agribusiness for their livelihood. High-level adaptation plan will decrease the risk of CC and protects the livelihood of the rural communities as well as also decrease food insecurity (FAO 2012). In practical terms, we need first to maintain the production capacity of the agricultural system under the CC; it will only be possible when we apply the implementation actions at the national level according to country problems as well as preference of action to need and also quick response from vulnerable communities (McCarthy et al. 2011; Fahad et al. 2015, 2016).

20.3.8 Efficient Pest Management (EPM)

Differences in rainfall and temperature suddenly affect pest, disease occurrence as well as extremeness on main crops. It is due to the influence of CC that will essentially impact the association of the host population as well as pest/weed host relations. Various potential adaptation practices include (Lipper et al. 2014):

1. Growth of cultivars impervious to pests as well as diseases;
2. The EPM selection having more reliance on organic control and change in social practices;
3. Implementation of additional security of crops systems, and places that are impervious to pests as well as different dangers.

20.3.9 GIS Mapping

Geographical information system (GIS) and remote sensing (RS) is mostly used in mapping as well as analysis which helped in the computation and estimation of LU management, land surface temperature (LST), urbanization, and food security. The study integrated infrastructure, urban sprawl, population allocation as well as other resources (Hussain et al. 2020a, b, c). These images and photographs were utilized in the study of the seashore because of hot cyclones as well as rising ocean levels (Hussain 2018). Hazard and danger maps can be made at various potential scales to show the risk designation crosswise over various land areas. These areas can be site explicit, include local governmental regions as well as small national landscapes, like coastlines, river basins, and lakes. Current progressions in the field of administration in databases, as well as the computer information, made that GIS is a perfect instrument for investigating satellite maps in land management by utilizing a multi-situational approach. Numerous topographic factors of elevation, coastal proximity, slope, water canals, and vegetation cover are integrated by the analytical hierarchy process (AHP) to create a weighting scheme for the geospatial variables (Suppasri et al. 2013).

20.3.10 Forecasting

The forecast of climate is very supportive to decrease the dangers of climate misfortunes. The ICT (information and communications technology) can proficiently support researchers as well as administrators in arranging possibility programs. In forecasting, a novel method MCA (multi-criteria analysis) tool has been used to assess CC strategy alert on mitigation and adaptation strategies (Ali et al. 2019). The best suitable considerations of policymakers and stakeholders are to execute in their policy-making, that is a different MCA actualized to discover potential variational practices.

20.3.11 Crop Modeling

Crop modeling is a new and creative advanced tool for dealing with hazards in agriculture (Abbas et al. 2020). Computer-aided simulation models can also play a significant role to assess crop yield and crop management responses. Crop modeling can also use to monitor the potential threat of CC on future crop yields, CSA mitigation, and development procedures. However, two crop management systems are Decision Support System for Agro Technology Transfer (DSSAT) as well as Agricultural Production System Simulator (APSIM) frequently used all over the world (Mubeen et al. 2013b).

Despite the fact that several uncertainties owing to less information and crop production models can obviously assess the effect of explicit water stress situations (possible due to CC) on crop efficiency if these are well validated and calibrated in field experiments (Sabagh et al. 2020). The crop modeling permits the difference of climate factors like temperature, water regime, and simulates the crop response via numerous determined growth parameters. For the unpredictability of the issue, research improvements continue to be regularly being made to models, such as assessment of drought effects. Mubeen et al. (2016) associated the different models like Crop Environment Resource Synthesis (CERES)-Wheat and SWHEAT with measurements from wheat grown under drought (Mubeen et al. 2013c). CERES-Wheat has shown to be helpful for estimating the drought impacts on different crops in explicit areas. CERES-Wheat is well-referenced and has been effectively verified in several research, particularly for CC belongings on crop growth (Ahmad et al. 2014; Wajid et al. 2014; Fahad et al. 2017, 2018).

20.3.12 Climate Smart Agriculture (CSA) in Livestock

There are many effects of climate change on livestock production, some may be direct and others may be indirect. Thes are given in Table 20.1. An effective use of natural resources is an important approach for decoupling development in livestock region from adverse ecological effects. Proficiency in the use of natural resources is estimated by the proportion between uses of natural resources as contribution to security as well as the output from security. Examples of opportunities that fall within this technique are greater feed productivity, water efficiency, greater yields per hectare, better management of fertilizers and manure as well as diminished damages along the food chain (Davis et al. 2012).

The actual prices of inputs, for example, feed, water, and land, utilized in livestock production frequently do not reflect true shortages. Any future strategies to secure the climate will have to present sufficient market pricing for natural resources. Confirming best management rules, under communal as well as private ownership of natural resources, is a further key strategy component for improving the

Table 20.1 Indirect and direct effects of CC on livestock production (Derpsch et al. 2010)

Sr.	Direct effects	Indirect effects
1	Frequency of dangerous climate events increased	Change in fodder quantity and quality
2	Increased magnitude and frequency of floods and droughts	Variation in host-pathogen relation ensuing in an improved incidence of emerging diseases
3	Productivity losses due to increase in temperature	Disease epidemics
4	Variation in water availability	Improved prices of different resource s

utilization of resources. Livestock makes an important involvement in food production, particularly in marginal areas where it signifies a single source of micronutrients, protein, and energy. The influence in livestock regions on food production could be supported, especially in regions where the present status of consumption of livestock yields are very low (Dawar et al. 2021).

20.3.13 Resource Conserving Technologies (RCTs)

The RCTs are able to apply the inputs in a sustainable way and improve the efficiency of resource management, also some remarkable economic advantages like improvement in the crop yield through timely crop sowing, reduce the input cost like water, labor, and fuel. The example of the RCT is zero tillage (ZT), growing of crop seeds in uncultivated fields. The main benefit of the ZT is to reduce cost of production. The farmer sown their crops like wheat after harvesting the cotton, the wheat heads appear, and the grain filled before the pre-monsoon weather. Early sowing is most important for the wheat crop in those regions where the average temperature rises rapidly (Pathak et al. 2012).

Approximately 5 million hectares of areas in Nepal, India, Bangladesh as well as Pakistan have practiced ZT due to the rotation of rice and wheat (Saeed et al. 2013; Jabran et al. 2016). The ZT reduces the emissions of CO_2 because it enhances the retention duration of organic matter and this store in the form of carbon into the soil. Therefore the ZT is a good way to sequester the carbon into the soil and reduce GHG emissions (Zingore 2010). But in developing countries (like Pakistan) various studies were conducted about variations in the soil carbon under different tillage practices (World Bank 2015). The ZT is considered as a good carbon sequestration strategy because it reduces carbon from 367 to 3667 kg per hectare in one year as compared to the conventional tillage practices (Tebrügge and Epperlein 2011).

20.3.14 Fisheries and Aquaculture

Aquaculture and the fisheries sectors have great importance in the development at national level and also provide livelihood supports. Furthermore, the sector is facing important challenges in sustaining its vital involvement to these parts (Merino et al. 2012).

- The CC leads toward the increase in sea level and acidification but aquatic food demands are increasing such as fishes with the passage of time. So we need to establish the climate smart aquaculture and fisheries sectors which will require efficient inputs to achieve sustainable fish production, which fulfills the fish demand among the communities and also contribute to national development.
- There is no absence of direction for the sector. Responsible fisheries as well as the environment approach to aquaculture and fisheries outline the values that are integral to guaranteeing the maintainability of the sector.
- The change to CSA in aquaculture and fisheries will be essential to take place at total levels (national, regional, community, individual, and business) as well as time scales. All stakeholders from the public and private areas will be associated with the improvement of explicit options to confirm the aquaculture as well as fisheries sector is CSA.
- To make the change to CSA in aquaculture and fisheries, it will be compulsory to confirm that the maximum production structures, susceptible states, stakeholders, and communities have the possibility to develop as well as apply CSA methods (NHB 2011) (Fig. 20.4).

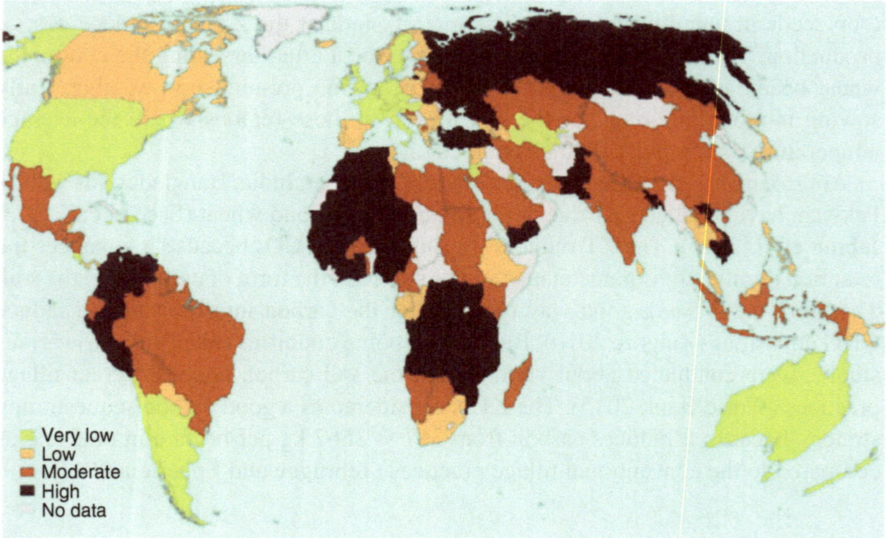

Fig. 20.4 Worldwide mapping of national economies' susceptibility due to CC effects by fisheries (Sources: Allison et al. 2009)

20.4 Improving Climate Smart Agriculture (CSA)

20.4.1 Climate Risk Management Through Water and Land Management

Strengthening the resilience of smallholding farmers is becoming an increasingly important political agenda as the magnitude and frequency of shocks, as well as tensions, increase due to important changes in socioeconomic and biophysical factors. Energy and food price unpredictability, economic downturn, CC as well as soil degradation is the latest main variations that have augmented the susceptibility of smallholders to shocks and stress (Barrett and Constas 2014; Torero 2015). The influence of these shocks on food production and the well-being of small farmers is usually huge. Maize, the area's maximum significant crop, accounting for 13% of the region, is expected to decline by 22% production by 2050, with the greatest effect on the most important crops in the CSA (Nasim et al. 2012). Similarly, it is estimated that sorghum and millet production are reduced by 17% (Iqbal et al. 2019). Land degradation not only reduces agricultural productivity but also increases production risk particularly for small farmers who do not make much effort to cover the negative effects of land degradation.

Sustainable land and water management (SLWM) which contains a mixture of inorganic and organic soil fertility as well as water management more than compensates for the impact of CC on crop yields under existing management methods. In addition, SLWM is also profitable and increases family income to combat against poverty (Sandhu et al. 2010). Training for agricultural advisory services providers on sustainable land management (SLM) and CC needs to be strengthened both are moderately new to much older agricultural advisory services. In addition, advisory services for the management and development of irrigation systems remained weak (Mubeen et al. 2021). This applies in particular to recommendations on irrigation technology, which are largely limited to large-scale irrigation systems (Nkonya et al. 2016). Due to these factors, water loss in irrigation systems in Africa is over 50%. Short-term training with an operative focus on these significant topics will be more operative and practical than long-term trainings. Furthermore, the gender of advisory service sources has a significant effect on the type of advisory services offered to the recipients (Hussain et al. 2020a).

20.4.2 Improving the Resilience of Agriculture

Agriculture is a vital industry and a key source of income for poor rural population of central Asia. Production of agriculture is sensitive to climate and weather changeability if adaptation measures are not taken. In general, CSA has three objectives:

1. Increase agricultural production in a sustainable and supportive in terms of fair increase in agricultural development, food production, and income.
2. Establish the flexibility of food production and agricultural systems toward CC.
3. Reduction in GHG emission in agriculture (including plants, fisheries, and livestock).

In their comprehensive assessment of environmentally sound and SLM technologies in different areas, Pender et al. (2009) also found the cost-benefit ratio (BCR) of many SLM technologies, like being positive in the region. On the other hand, the acceptance rates of these methods stay comparatively low. These low acceptance rates are frequently due to numerous barriers deliberated above, like lack of knowledge, information services, and to input and output markets. Financial institutions are often reluctant to lend to small households with unknown risk characteristics and a lack of collateral to guarantee a loan. In different cases, these agricultural households rent their area only from the state without the legal right to use their land as security for a loan (Takeshima and Edeh 2013).

20.4.3 Managing Environmental Risk in Presence of Climate Change (CC)

At first stage of analysis, it was emphasized that the risk related to the environmental characteristics of the company, like soil fertility and access to information, are important determining factors for adaptation. These results are constant on adaptation to CC, food production, and adoption of irrigation technology in the event of production uncertainty. Mubeen et al. (2016) highlight that informed farmers value the opportunities which are less and therefore will use new technologies more often than other farmers. This means that the expectation of new and improved information can be positive, as well as the provision of information on CC can decrease the value of the quasi-option related to adaptation (Nasim et al. 2017a, b). Development policies aimed at raising the level of education can have a positive impact on the adaptation and adoption of technologies (Hussain et al. 2020b).

Secondly, modification would be more beneficial for agricultural households that would not have formerly been adjusted. This group would reduce the risk of reduction than the group of adapters. This leads us to the third conclusion that there are some significant causes of heterogeneity and variances among adapters and non-adapters so that, regardless of CC, non-adapters are at a lower risk of reduction than adapters. These changes signify sources of difference between the two groups, which may or may not be taken into account when evaluating with different models, counting a dummy variable for adaptation to CC. Lastly, CC adaptation is an effective risk management approach that makes adapters additional resistant to climate situations. Non-adapters are considered precious by the precipitation in both the long and short raining periods while the adapters are less precious by CC (Gbetibouo et al. 2010; Westhoek et al. 2011).

20.4.4 Role of Institutions for CSA Improvement

Institutions are important which needed to create useful information and its sharing, and also to help people interpret, understand, and work on new technologies like CSA Institutions, e.g, farmer schools that educate and support farmers in practicing new techniques. Agricultural radio programs provide information to rural people related to weather and agriculture; exhibitions of agricultural parcels for the community; and discussion of ideas among farmers. Profits made using management methods that are sustainable for the country should first become visible. During this time, the farmers must bear the general costs, including labor, cash, and land. Then poor farmers do not have the resources to approach credit and markets and cannot adapt to these new methods to make CSA successful. So, there is a vital need that strong institutions must support agricultural markets and financing mechanisms. Using the most effective technologies, private sector investors, policymakers, stakeholders, community members, and researchers can jointly describe the problems that are expected to be addressed and resolved.

Programs and policies are also effective if their application is supported by reliable institutions. Therefore, it is important to increase the institutional capacity to apply and replicate CSA strategies. Institutions are also crucial for the development of agriculture and the creation of sustainable livelihoods. They are not only a tool for decision-makers and farmers but also the most important way to improve and maintain climate-friendly growing methods (FAO 2013). Other significant formal institutions may include national and regional institutions in various perspectives. Nontraditional actors may also contain private sector and market-like insurance companies as well as agro-consulting companies.

20.5 Benefits of CSA Technologies

- By using CSA method, Improving food security by moderate CC, sustainable use of all products and natural resource, have greater constancy as well as less in consistency in their outputs.
- Computer-aided crop simulation models can monitor to find out the imaginable risk of CC on future crop yields, CSA mitigation, and development procedures. The crop models permit different ecological factors like temperature and water system as well as simulate the crop response through various projected growth parameters, for example, crop yield.
- Therefore CSA is a practice to increase the scientific rule and investment setting to achieve the maintainable agriculture progress to confirm the food production under CC (FAO 2012). The CSA is a way to get green economy objectives and sustainable developments (Braimoh and Osaki 2010).
- One of the most significant uses of GIS is the presentation as well as analysis of information for ecological decision-making. The GIS-based transforms and

combines spatial information (input) into a resultant decision (output) (Akinci et al. 2013).

- The various models for example AHP consists of criteria, sub-criteria, and objectives. To assess the criteria, scoring was created with preference scale, and a PWCM (pair wise comparison matrix) was made, for which consistency should be below 0.10.
- The MCE (multicriteria evaluation) model discovers solutions to decision-making issues characterized by various factors, and WLC (weighted linear combination) cumulative them into a final land suitability index (Schlenker and Lobell 2010).

20.6 Conclusion

FAO estimated that agricultural and food production should increase 60% by 2050 if the current consumption and production rates continue, to meet the requirements of food of world's population. Food security is highly dependent on the CSA because this system provides the opportunity to mitigate and adopt the CC. Crop modeling can also use to monitor the potential threat of CC on future crop yields, CSA mitigation, and development procedures. To attain agricultural development and food security goals with adaptation to CC will be necessary to lower emission intensities per output. Further flexible as well as fruitful agriculture needs a maximum significant variation in the way of use in resources management, land, water, and soil nutrients by using CSA practices. The CSA technologies can help to increase the resilience of agricultural producers in different areas and increase climate variability lead by CC. The CSA is a new method to transforming agricultural systems and supporting food production in CC, offering concrete and flexible solutions. Under a business-as-usual scenario, CC will make this task very difficult due to bad effects on agriculture, requiring spiraling adaptation. Lower production intensities and adaptation to CC will be essential to attain agricultural growth and food production goals.

References

Abbas G, Fatima Z, Tariq M, Ahmed M,ur Rahman MH, Nasim W, Ahmad S (2020) Applications of Crop Modeling in Cotton Production. In *Cotton Production and Uses* (pp. 429-445). Springer, Singapore

Ahmad MI, Ali A, Ali MA, Khan SR, Hassan, SW, Javed MM (2014) Use of Crop Growth Models in Agriculture: A Review. Sci Inter pp: 26

Akinci H, Ozalp AY, Turgut B (2013) Agricultural land use suitability analysis using GIS and AHP technique. Computers Electronics in Agriculture 97: 71-82

Akram R, Turan V, Hammad HM, Ahmad S, Hussain S, Hasnain A, Maqbool MM, Rehmani MIA, Rasool A, Masood N, Mahmood F (2018) Fate of organic and inorganic pollutants in paddy soils. In Environmental Pollution of Paddy Soils (pp. 197-214). Springer, Cham. https://doi.org/10.1007/978-3-319-93671-0_13

Ali M, Mubeen M, Hussain N, Wajid A, Farid HU, Awais M, Hussain S, Akram W, Amin A, Akram R, Imran M (2019) Role of ICT in Crop Management. In Agronomic Crops (pp. 637-652). Springer, Singapore. https://doi.org/10.1007/978-981-32-9783-8_28

Allison EH, Perry AL, Badjeck MC, Adger WN, Brown K, Conway D, Halls S, Pilling GM, Reynolds JD, Andrew NL, Dulvy NK (2009) Vulnerability of national economies to the impacts of climate. Fish and Fisheries, 24 pp

Amin A, Mubeen M, Hammad HM, Nasim W (2015) Climate Smart Agriculture: an approach for sustainable food security. Agric Res Commun 2(3): 13-21

Aslam M, Zamir M S I, Afzal I, Yaseen M, Mubeen M, Shoaib A (2013) Drought stress, its effect on maize production and development of drought tolerance through potassium application. *Cercetări Agronomice în Moldova*, *46*(2), 99-114

Awais M et al. (2018). Potential impacts of climate change Climate change impact assessment and adaptation strategies for sunflower in Pakistan. Environmental Science and Pollution Research, 25 (14): 13719-13730

Barrett C B, Constas M A (2014) Toward a theory of resilience for international development applications. Proceedings of the National Academy of Sciences, 111(40): 14625–14630

Braimoh AK, Osaki M. (2010) Land use change and environmental sustainability. Sustain Sci 5: 5-8

Corsi S, Friedrich T, Kassam A, Pisante M, Sa JM (2012) Soil organic carbon accumulation and carbon budget in conservation agriculture: a review of evidence. Vol. 16. FAO Integrated Crop Management

Davis K, Nkonya E, Kato E, Mekonnen D A, Odendo M, Miiro R, Nkuba J (2012) Impact of farmer field schools on agricultural productivity and poverty in East Africa. World Develop 40(2): 402–413

Dawar K, Rahman U, Alam S S, Tariq M, Khan A, Fahad S, Noor M (2021) Nitrification Inhibitor and Plant Growth Regulators Improve Wheat Yield and Nitrogen Use Efficiency. *Journal of Plant Growth Regulation*, 1-11

Derpsch R, Friedrich T, Kassam A, Li H (2010) Current status of adoption of no-till farming in the world and some of its main benefits, Inter J Agri Bio Engineer 3: 1-25

Deutsch CA, Tewksbury JJ, Huey RB, Sheldon KS, Ghalambor CK, Haak DC, Martin PR (2008) Impacts of climate warming on terrestrial ecotherms across latitude. Proc. Natl Acad Sci USA 105: 6668–6672

Di Falco S, Veronesi M, Yesuf M (2011) Does adaptation to climate change provide food security? A micro-perspective from Ethiopia. American J Agri Economics 93(3): 829–846

Fahad S, Bajwa A A, Nazir U, Anjum S A, Farooq A, Zohaib A, Huang J (2017) Crop production under drought and heat stress: plant responses and management options. *Frontiers in plant science*, *8*, 1147

Fahad S, Hussain S, Saud S, Khan F, Hassan S, Nasim W, Huang J (2016) Exogenously applied plant growth regulators affect heat-stressed rice pollens. *Journal of agronomy and crop science*, *202*(2), 139-150

Fahad S, Hussain S, Saud S, Tanveer M, Bajwa A A, Hassan S, Shah F (2015) A biochar application protects rice pollen from high-temperature stress. *Plant Physiology and Biochemistry*, *96*, 281-287

Fahad S, Ihsan M Z, Khaliq A, Daur I, Saud S, Alzamanan S, Wang D (2018) Consequences of high temperature under changing climate optima for rice pollen characteristics-concepts and perspectives. *Archives of Agronomy and Soil Science*, *64*(11), 1473-1488

FAO (2012) Mainstreaming climate-smart agriculture into broader landscape approach: Second Global Conference on Agriculture, Food Security and Climate Change. Hanoi, Vietnam. 2012

FAO (2013) Climate-Smart Agriculture Source Book. The Food and Agriculture Organization of the United Nations. 2013. http://www.fao.org/docrep/018/i3325e/i3325e.pdf. Accessed 12th June, 2015

FAO (2015) Regional overview of food insecurity: African food insecurity prospects brighter than ever. Accra

Fellmann T (2012) The assessment of Climate Change-related vulnerability in the agricultural sector: reviewing conceptual frameworks. FAO/OECD Workshop. Building Resilience for Adaptation to Climate change in the Agriculture sector. Red Room, FAO

Gbetibouo G, Hassan R, Ringler C (2010) Modelling farmers' adaptations strategies to climate change and variability: The case of the Limpopo Basin, South Africa. Agrekon 49(2): 217–234

Gommes R, Acunzo M, Baas S, Bernardi M, Jost S, Mukhala E, Ramasamy S (2010) Communication approaches in applied agrometeorology, In K. Stigter, ed. Applied Agrometeorology, Heidelberg, Springer, pp. 263–287

Gustavsson J, Cederberg C, Sonesson U, van Otterdijk R, Meybeck A (2011) Global food losses and food waste: extent, causes and prevention. Rome, FAO

Hussain S (2018) Land Use/Land Cover Classification by Using Satellite NDVI Tool for Sustainable Water and Climate Change in Southern Punjab. COMSATS University Islamabad. DOI: https://doi.org/https://doi.org/10.13140/RG.2.2.32363.69923

Hussain S, Ahmad A, Wajid A, Khaliq T, Hussain N, Mubeen M, Farid HU, Imran M, Hammad HM, Awais M, Ali A, Aslam M, Amin A, Akram R, Amanet K, Nasim W (2020a) Irrigation Scheduling for Cotton Cultivation. In Cotton Production and Uses (pp. 59-80). Springer, Singapore. https://doi.org/10.1007/978-981-15-1472-2_5

Hussain S, Mubeen M, Ahmad A, Akram W, Hammad H M, Ali M, Masood N, Amin A, Farid HU, Sultana SR, Fahad S (2020b) Using GIS tools to detect the land use/land cover changes during forty years in Lodhran District of Pakistan. Environmental Science and Pollution Research **27**, 39676–39692. https://doi.org/10.1007/s11356-019-06072-3

Hussain S, Mubeen M, Akram W, Ahmad A, Habib-ur-Rahman M, Ghaffar A, Amin A, Awais M, Farid HU, Farooq A, Nasim W (2020c) Study of land cover/land use changes using RS and GIS: a case study of Multan district, Pakistan. Environment Monitoring and Assessment 192(1): p. 2. https://doi.org/10.1007/s10661-019-7959-1

Intergovernmental Panel on Climate Change (IPCC) (2007) Technical Summary. In Climate change 2007: mitigation. Contribution of Working Group III to the Fourth Assessment Report of the IPCC. Cambridge, UK, and New York, USA, Cambridge University Press

Iqbal J, DitommasoA, RehmaniMIA, JabranK, Hussain S, Nasim W, Fahad S, Shehzad MA, Ali A (2019) Purple nutsedge (Cyperus rotundus) control through interference by summer crops. International Journal of Agriculture Biology, 21: 1083–1088

Jabran K, HussainM, Fahad S, Farooq M, Bajwa AA, Alharrby H, Nasim W (2016) Economic assessment of different mulches in conventional and water-saving rice production systems. Environmental Science and Pollution Research, 23 (9): 9156-9163

Jat HS, Choudhary M, Datta A, Yadav A K, Meena MD, Devi R, Sharma PC (2020) Temporal changes in soil microbial properties and nutrient dynamics under climate smart agriculture practices. Soil and Tillage Research, 199, 104595. https://doi.org/10.1016/j.still.2020.104595

Khoi DD, Murayama Y (2010). Delineation of suitable cropland areas using a GIS based multi-criteria evaluation approach in the Tam Dao National Park Region, Vietnam. Sustain 2: 2024-2043

Lipper L, Thornton P, Campbell BM, Baedeker T, Braimoh A, Bwalya M, Caron P, Cattaneo A, Garrity D (2014) Climate-smart agriculture for food security. In: Nature Climate Change pp. 1068–1072. doi: https://doi.org/10.1038/nclimate2437

McCarthy N, Brubaker J (2014) Climate-Smart Agriculture and Resource Tenure in Sub-Saharan Africa: A Conceptual Framework. Rome, FAO

McCarthy N, Lipper L, Branca G (2011) Climate-smart agriculture: smallholder adoption and implications for climate change adaptation and mitigation. FAO, MICCA Series No. 4. Rome, FAO

Merino G, Barange M, Blanchard JL, Harle J, Holmes R, Allen I, Allison EH, Badjeck MC, Dulvy NK, Holt J, Jennings S, Mullon C, Rodwell LD (2012) Can marine fisheries and aquaculture meet fish demand from a growing human population in a changing climate? Global

Environ Change 22(4): 795—806. available at http://www.sciencedirect.com/science/article/pii/S0959378012000271

Moussa B, Nkonya E, Meyer S, Kato E, Johnson T, Hawkins J (2016) Economics of land degradation and improvement in Niger. In Economics of Land Degradation and Improvement–A Global Assessment for Sustainable Development, Springer International Publishing pp. 499–539

Mubeen M, Ahmad A, Hammad H M, Awais M, Farid H U, Saleem M, Nasim W (2020) Evaluating the climate change impact on water use efficiency of cotton-wheat in semi-arid conditions using DSSAT model. *Journal of Water and Climate Change*, *11*(4), 1661-1675

Mubeen M, Ahmad A, Khaliq T, Sultana S R, Hussain S, Ali A, Nasim W (2013a) Effect of growth stage-based irrigation schedules on biomass accumulation and resource use efficiency of wheat cultivars

Mubeen M, Ahmad A, Wajid A, Bakhsh A (2013b) Evaluating different irrigation scheduling criteria for autumn-sown maize under semi-arid environment. *Pak. J. Bot*, *45*(4), 1293-1298

Mubeen M, Ahmad A, Wajid A, Khaliq T, Bakhsh A (2013c) Evaluating CSM-CERES-Maize Model for Irrigation Scheduling in Semi-arid Conditions of Punjab, Pakistan. *International Journal of Agriculture & Biology*, *15*(1)

Mubeen M, Ahmad A, Wajid A, Khaliq T, Hammad H M, Sultana S R, Nasim W (2016) Application of CSM-CERES-Maize model in optimizing irrigated conditions. *Outlook on Agriculture*, *45*(3), 173-184

Mubeen M, Bano A, Ali B, Islam ZU, Ahmad A, Hussain S, Fahad S, Nasim W (2021) Effect of plant growth promoting bacteria and drought on spring maize (Zea mays L.). Pakistan Journal of Botany, 53(2): 731-739. DOI: https://doi.org/10.30848/PJB2021-2(38)

Mundial B (2012) Agricultural Innovation Systems: An Investment Sourcebook. World Bank, Washington, DC

Nasim W, Ahmad A, Ahmad S, Nadeem M, Masood N, Shahid M, Mubeen M, Hoogenboom G (2017a) Response of sunflower hybrids to nitrogen application grown under different Agro-environments. Journal of Plant Nutrition, 40 (1) : 82-92

Nasim W, Ahmad A, Khaliq T, Wajid A, Hussain A, Hammad M, Sultana SR, Mubeen M (2012) Effect of organic and inorganic fertilizer on maize hybrids under agro-ecological conditions of Faisalabad-Pakistan. African Journal of Agricultural Research, 07 (15): 2713-2719

Nasim W, Akram R, Mubeen M (2017b) Ozone layer protection-greening the blue published by Technology Times Publishers: (http://www.technologytimes.pk/ozone-layer-protection-greening-the-blue/)

NHB (2011) National Horticulture Database, 2011, Ministry of Agriculture, Govt. of India, Gurgaon, pp: 278

Nkonya E, Johnson T, Kwon HY, Kato E (2016) Economics of land degradation in sub-Saharan Africa In: E. Nkonya, A. Mirzabaev and J. von Braun (eds). Economics of Land Degradation and Improvement – A Global Assessment for Sustainable Development. Springer, New York: 215–260

Pathak H, Aggarwal PK, Singh SD (2012) Climate Change Impact, Adaptation and Mitigation in Agriculture: Methodology for Assessment and Applications. Indian Agricultural Research Institute, New Delhi. pp: 302

Pender J, Mirzabaev A, Kato E (2009) Economic Analysis of Sustainable Land Management Options in Central Asia. Final report for the ADB. IFPRI/ICARDA, 168

Rahman MH, Ahmad I, Ghaffar A, Haider G, Ahmad A, Ahmad B, Ahmad S (2020) Climate Resilient Cotton Production System: A Case Study in Pakistan. In Cotton Production and Uses (pp. 447-484). Springer, Singapore. https://doi.org/10.1007/978-981-15-1472-2_22

Sabagh A E, Hossain A, Islam M S, Iqbal M A, Fahad S, Ratnasekera D, Llanes A (2020) Consequences and Mitigation Strategies of Heat Stress for Sustainability of Soybean (Glycine max L. Merr.) Production under the Changing Climate. In *Plant Stress Physiology*. IntechOpen. https://doi.org/10.5772/intechopen.92098

Saeed HS, RasulF, Sarfaraz M, Mubeen M, Nasim W (2013) Allelopathic Potential Assessment of Jaman (Syzygium cumini L.) on Wheat. Int. Poster J. Sci. Tech. 03 (1) : 09-14

Sandhu et al. (2010) Organic agriculture and ecosystem services. Environ Sci Policy 13: 1–7

Saud S, Fahad S, Yajun C, Ihsan MZ, Hammad HM, Nasim W, Arif M, Alharby H (2017) Effects of Nitrogen Supply on Water Stress and Recovery Mechanisms in Kentucky Bluegrass Plants. Frontier in Plant Sci 8:983: 01-18

Scherr SJ, Shames S, Friedman R (2012) From climate-smart agriculture to climate-smart landscapes. Agriculture & Food Security 2012, 1:12, Agriculture and Food Security. http://www.agricultureandfoodsecurity.com/content/1/1/12

Schlenker W, Lobell DB (2010) Robust negative impacts of climate change on African agriculture. Environ Res Letters 5(1): 014010

Suppasri A, Mas E, Charvet I, Gunasekera R, Imai K, Fukutani Y, Abe Y, Imamura F (2013) Building Damage Characteristics Based on Surveyed Data and Fragility Curves of the 2011 Great East Japan Tsunami. Natural Hazards, **66**: 319-341. https://doi.org/10.1007/s11069-012-0487-8

Takeshima H, Edeh H (2013) Typology of Farm Households and Irrigation Systems: Some Evidence from Nigeria, IFPRI Discussion Paper 01267. International Food Policy Research Institute, Washington D.C

Tebrügge F, Epperlein J (2011) ECAF Position Paper: The Importance of Conservation Agriculture within the Framework of the Climate Discussion In, ECAF, European Conservation Agriculture Federation, http://www.ecaf.org/docs/ecaf/positionpaperco2ecaf.pdf

Torero M (2015) Consistency between Theory and Practice in Policy Recommendation by International Organizations for Extreme Price and Extreme Volatility Situations

Wajid A, Ahmad A, Hussain M, ur Rahman MH, Khaliq T, Mubeen M, Sultana SR (2014) Modeling growth, development and seed-cotton yield for varying nitrogen increments and planting dates using DSSAT. Pak J Agric Sci 51: 641-650

Westhoek H, Rood T, Van den Berg M, Janse J, Nijdam D, Reudink M, Stehfest E (2011) The protein puzzle: the consumption and production of meat, dairy and fish in the European Union. The Hague, PBL Netherlands Environmental Assessment Agency. (available at http://www.pbl.nl/sites/default/files/cms/publicaties/Protein_Puzzle_web_1.pdf)

World Bank (2015) Policy and Legislation Agbiz leads on Climate Smart Agriculture (CSA) in LCTPi. Policy

Zahoor SA, Ahmad S, Ahmad A, Wajid A, Khaliq T, Mubeen M, Hussain S, Din MSU, Amin A, Awais M, Nasim W (2019) Improving Water Use Efficiency in Agronomic Crop Production. In Agronomic Crops (pp. 13-29). Springer, Singapore. https://doi.org/10.1007/978-981-32-9783-8_2

Zingore S (2010) SOC sequestration in farming systems in Africa: Potential, opportunities and challenges, In: 2nd Meeting of the Round Table on Organic Agriculture and Climate Change, RTOACC, 2010

Chapter 21
Internet of Things (IoT) and Sensors Technologies in Smart Agriculture: Applications, Opportunities, and Current Trends

Muhammad Zeeshan Mehmood, Mukhtar Ahmed, Obaid Afzal, Muhammad Aqeel Aslam, Raja Zoq-ul-Arfeen, Ghulam Qadir, Saida Komal, Muhammad Adnan Shahid, Adeem Arshad Awan, Mohamed Ali Awale, Aashir Sameen, Tahira Kalsoom, Wajid Nasim, Fayyaz-ul-Hassan, and Shakeel Ahmad

M. Z. Mehmood · O. Afzal · M. A. Aslam · G. Qadir · A. A. Awan · M. A. Awale · A. Sameen · Fayyaz-ul-Hassan
Department of Agronomy, Pir Mehr Ali Shah Arid Agriculture University, Rawalpindi, Punjab, Pakistan

M. Ahmed (✉)
Department of Agronomy, Pir Mehr Ali Shah Arid Agriculture University, Rawalpindi, Punjab, Pakistan

Department of Agricultural Research for Northern Sweden, Swedish University of Agricultural Sciences, Umeå, Sweden
e-mail: ahmadmukhtar@uaar.edu.pk

R. Zoq-ul-Arfeen
School of Food and Agricultural Sciences, University of Management and Technology, Lahore, Punjab, Pakistan

S. Komal
Department of Agronomy, University of Poonch, Rawalakot, Azad Kashmir, Pakistan

M. A. Shahid
Department of Agriculture, Nutrition, and Food Systems, College of Life Sciences and Agriculture, University of New Hampshire, Durham, NH, USA

T. Kalsoom
Department of Horticulture, Pir Mehr Ali Shah Arid Agriculture University, Rawalpindi, Punjab, Pakistan

W. Nasim
Department of Agronomy, Islamia University, Bahawalpur, Punjab, Pakistan

S. Ahmad
Department of Agronomy, Bahauddin Zakariya University, Multan, Punjab, Pakistan

W. N. Jatoi et al. (eds.), *Building Climate Resilience in Agriculture*,
https://doi.org/10.1007/978-3-030-79408-8_21

Abstract For sustainable agricultural production and timely preparations to mitigate the climate change impacts, innovative modern technologies can be used. These technologies have great potential for monitoring agricultural systems and valuable solutions to combat climate change in order to offset the adverse impacts on agricultural production. Farmers need continuous information throughout the crop life cycle for implementing profitable farming decisions. Internet of things (IoT) is one of the advanced technologies in smart agriculture. IoT is the network of Internet connected devices to obtain and transfer real-time data. Now, the manual and conventional procedures are being replaced with automated technologies globally. IoT is becoming popular in agriculture sector as compared to conventional agriculture due to its distinguishing features such as less energy requirement, good global connectivity, and real-time data collection. On the other hand, device compatibility is the major limitation in IoT but now the solutions are being developed with technological advancements. This chapter focuses on the role of information communication technology (ICT) and IoT in agriculture domain and proposes the benefits of these wireless technologies. Use of IoT technology in smart farming can serve as a solution for several management and decision-making for building climate resilience in agriculture.

Keywords Climate change · Smart agriculture · Internet of things · Sensors · Precision agriculture · E-Agriculture

21.1 Introduction

Nowadays, climate change is becoming one of the most important barriers for agriculture production and sustainability globally. Climate variability has an impact on the number of natural events involved in agriculture such as increased erratic rainfall, temperature rise, more invasive pathogens, heat waves, and floods. (Puranik et al. 2019; Amin et al. 2018; Ashraf et al. 2017; van Ogtrop et al. 2014). Hence, the changing patterns in the trends of environment are compromising the overall success of agriculture along with its sustainability. This global climatic change is predicted to affect global food security by disrupting food production, availability, and quality. For instance, temperature extremes and more erratic rainfalls are dominantly minimizing agricultural production (Malavade and Akulwar 2016; Ali et al. 2019; Ahmed et al. 2014). Increase in the severity and frequency of these climatic events is predicted in the upcoming future. Hence, it indicates more vulnerability to sustainable agriculture (Fig. 21.1). These scenarios lead to the evolution of climate smart agriculture based on smart climate monitoring methods and technologies to increase preparation and sustainability in agriculture (Lipper et al. 2014; Rahman

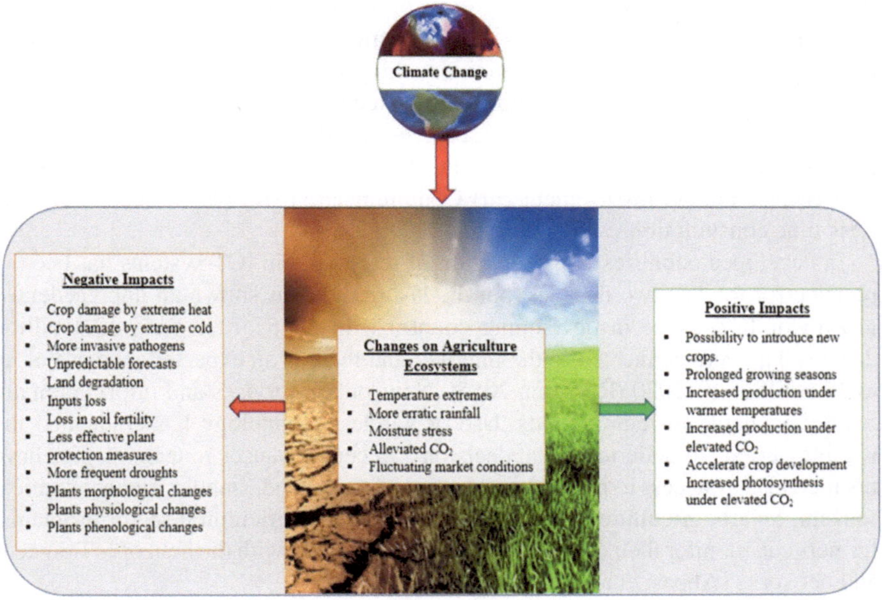

Fig. 21.1 Impacts of climate change on agriculture

et al. 2020; Jabeen et al. 2017; Ijaz et al. 2017). Therefore, to boost agricultural productivity and sustainability, more innovative techniques and technologies are needed to be utilized in agriculture (Ahmed, 2012, 2020a, b, Ahmed and Ahmad, 2019, 2020; Ahmad and Hasanuzzaman, 2020; Ahmed et al. 2013, 2018, 2020a, b, c, d, e; Ahmed and Stockle, 2016; Ahmed and Hassan, 2011). One of these innovative technologies, the Internet of things (IoT) is rapidly developing as an applied technology in wireless environments. Internet of things technologies have great application potential in the climate, food, and agriculture domain especially in the context of climatic challenges faced in these sectors (Patil and Kale 2016). The IoT technologies can transform the agriculture sector by contributing to food security, monitoring farm environments, increasing preparation, and optimizing the agricultural inputs to reduce wastes. However, the success of these technologies is linked with the remarkable change in the culture (Brewster et al. 2017).

Agriculture informatics is referred to as the use of innovative techniques, ideas, and scientific knowledge to expand the use of computer science in agriculture. Information communication technology (ICT) and IoT are used for management, land use, and analysis of agricultural data. Use of IoT and ICT in agriculture is also known as E-agriculture (Gakuru et al. 2009). E-agriculture focuses on enhancing agricultural development by using advanced information and communication technology. It involves development, application, and evaluation of innovative means to use ICT in agriculture (Dlodlo and Kalezhi 2015; Aslam et al. 2017b). IoT is the framework for connecting physical things (like sensor, devices, etc.) to the Internet that enables the monitoring and controlling of the physical world from remote locations (Kopetz 2011). IoT provides an ICT infrastructure to facilitate exchange of things and the main function is to minimize the gap among things in physical world and their representation in information system technology (Weber and Weber 2010).

IoT consists of software, networks, devices, and different types of sensors. There are many reasons that IoT is an efficient technology including (i) Global connectivity (ii) Less human involvement (iii) Communication (iv) Quick access, and (v) Less time consumption.

In developed countries, representation of real-world in ICT systems has been in practice for the last two decades, but the last decade has shown an unprecedented growth of ICT usage in developing countries. In 2015, there were 13.4 billion devices that were connected to the Internet, and there is an expected increase of up to 38.5 billion by 2020 (Research 2015). Now, public services and information are readily available in remote areas. Use of wireless technology has eradicated the waiting periods to undertake vital decisions. Recent advances in technologies have resulted in easy access to networks and more sophisticated, smaller, and economical sensors. Smart agriculture can serve as a solution for agricultural sector problems; farmers can monitor their agriculture sector individually with the help of IoT devices and networks (Abbasi et al. 2014).

The ICT and IoT drivers in agriculture have the advantage of (i) Connectivity and low cost, (ii) Adaptability and affordability of tools, (iii) Data exchange and storage advances, (iv) Innovative models for business, and (v) Demand of information services in agriculture (Nlerum and Onowu 2014). However, in rural areas, there are some barriers that should be addressed by ICTs broadband. These barriers include (i) Distance barriers, (ii) Economic barriers, and (iii) Social barriers (Stratigea 2011).

IoT in agriculture can be used in various scenarios and can enhance agricultural processes. Cloud enabled systems can be used for agriculture data and its uses in simulated systems. By using IoT technology, farmers can get timely knowledge about the recent trends in agriculture. IoT devices can monitor the soil properties, weather variables, plant characteristics, etc. Therefore, it can play a vital role in timely management of agricultural systems under swiftly occurring climate changing scenarios and boosting agriculture production.

21.2 IoT System

This section delineates several components of the IoT system, classification, and development trends (Fig. 21.2).

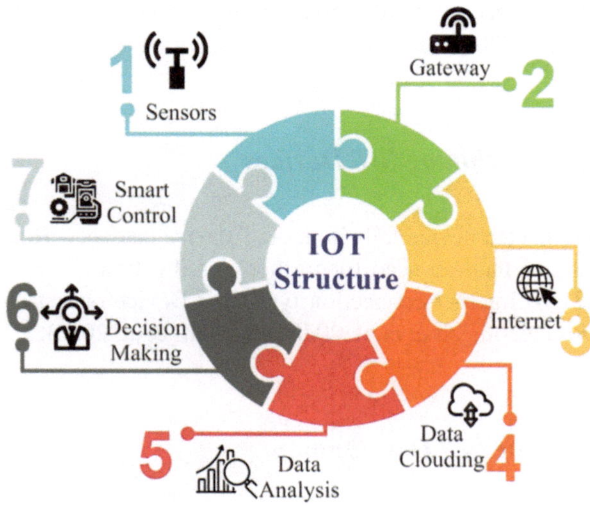

Fig. 21.2 IoT structure

21.2.1 *IoT Platforms, Standards, and Protocols*

IoT has the potential to grow in the future and many companies (The Yield, Agrosmart, Fieldin, Cowlar, HerdDogg, etc.) are focusing on this technology that is leading to the emergence of solutions and new platforms. Device compatibility is the main challenge that needs to develop solutions for general use. Single board computer device is the main concept for most IoT devices that have sufficient computing power, use open source solution, and have low energy needs. Advanced Reduced instruction set computing Machine (ARM) processor is commonly used in IoT devices. These devices have operating system for both MS Windows and Linux platforms (Dusadeerungsikul et al. 2020; Bacenetti et al. 2020).

21.2.2 *IoT Networks*

For proper operation of IoT devices, Internet connection is a basic need; mostly this connection is wireless. These connection technologies have different standards and parameters. The common characteristics of wireless connections depend on (i) Energy consumption, (ii) Downlink and uplink data rate, (iii) Size of package, (iv) Range, and (v) Frequency. IoT comprises of many technologies and networks primarily designed for IoT (such as SigFox and LoRaWAN), while on the other hand, it also uses technologies that are developed for other purposes (such as Wi-Fi, GSM, and LTE). Lower energy use is the main feature of IoT designed networks. It is assumed that with the advancement in technology, IoT devices will be enabled to operate for years or decades with only a simple battery (Dusadeerungsikul et al. 2020; Bacenetti et al. 2020).

21.2.3 Classification of IoT Devices

Any Internet connected device falls in the IoT device category. However, there are some other features that are used for further classification of devices, such as (i) Usage purpose, (ii) Internet connection type, (iii) Device or sensor type, and (iv) Energy use (Dusadeerungsikul et al. 2020; Bacenetti et al. 2020).

21.2.4 Trends in IoT Development

IoT has the potential to be used in each area of human activity. Current trends in IoT development are (i) Specific IoT network development, (ii) Reliable security, (iii) Minimal energy use, (iv) Miniature devices, and (v) User-friendly controls, solutions, and settings (Dusadeerungsikul et al. 2020; Bacenetti et al. 2020).

21.3 Wireless Technologies in Smart Agriculture

In smart agriculture, several kinds of IoT devices are used (such as wireless sensors) for the purpose of data collection of environmental and physical attributes (Fig. 21.3). In the agriculture domain, sensors are used for the following reasons:

(i) Weather, crop, and soil monitoring and data collection.
(ii) Fertilizer and irrigation management.
(iii) Increased preparedness to respond to abrupt climatic changes.

However, these sensors or IoT devices utilize some network protocols to transmit the data remotely. In this section, common wireless technologies used for data acquisition and transmission in smart agriculture are presented and compared for distinctive features. Weather monitoring and prediction is becoming more challenging under the current climate changing scenarios. Moreover, these extreme climatic events are severely impacting the agricultural environments and making farming practices harder and vulnerable. Therefore, there is a need to put in place the more effective strategies to combat the increasing food demands and climate variability (Ahmed et al. 2017). Several cellular and wireless technologies have been developed to play a vital role in smart agriculture (Andreev et al. 2015). A new IoT system has been developed on the basis of Long-Term Evolution (LTE) features and is known as Narrowband IoT (NB-IoT). The main functions of this system are to

Fig. 21.3 Data chain in IoT

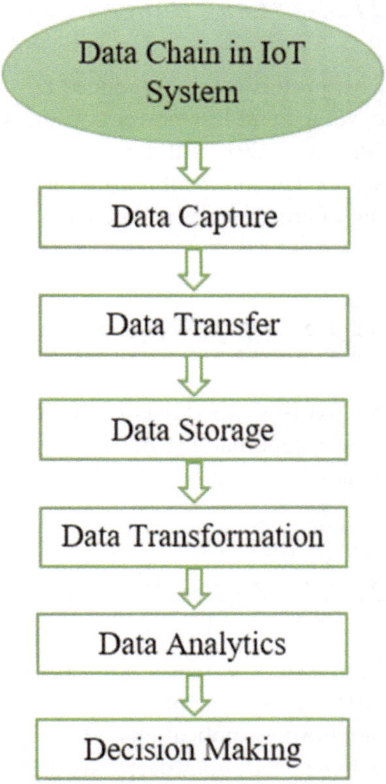

Table 21.1 Comparison of all the wireless technologies

	Range	Cost	Power Consumption
LoRa	5 km	Low	Low
SigFox	10 km	Low	Low
ZigBee	100 m	Low	Low
4G/3G/GPRS	10 km	Medium	Medium
Wi-Fi	100 m	High	High
GPS and BT	10 m	Low	Medium

increase the coverage area and reduce power consumption (Ratasuk et al. 2016). In the future, NB-IoT technologies like Long Range Radio (LoRa) will be used in agriculture on a larger scale due to its lower energy use and enhanced coverage (Table 21.1).

21.3.1 Long Range Radio (LoRa)

This protocol was introduced by the LoRa Alliance. It is a system with wide area coverage and less power consumption as compared to other wireless technologies (Pitì et al. 2017). It consists of LoRa end device, gateway device, and a network server. This protocol is used for moisture, humidity, temperature, and light monitoring in greenhouses and fields as well (Ilie-Ablachim et al. 2016).

21.3.2 SigFox Protocol

SigFox is a narrow band cellular network (Pitì et al. 2017). This protocol was used in a system that was developed for locating animals in pastures and grazing lands. This system is also used for animal tracking in mountains and pastures.

21.3.3 ZigBee Protocol

ZigBee is a wireless, low-cost IoT network technology (Sarode and Chaudhari 2018). Low cost of this technology allows it to be widely adopted in wireless monitoring applications. As an advanced technology, it has gradually become popular and considered as the best choice for agriculture. This technology identifies and obtains real-time data for crops pests, moisture and drought, and transfers data to remote monitoring centers. With real-time information of field data, automated devices can be used for controlling irrigation, fertilization, and pesticide (Cancela et al. 2015). ZigBee can be used up to 100 m effectively (Jawad et al. 2017).

21.3.4 4G/3G/GPRS

These are packet data services used for cellular phones. These modules can be coupled with sensors to control the irrigation to crops depending on weather and soil monitoring (Gutiérrez et al. 2014). It is a cost-effective method for monitoring agricultural information.

21.3.5 Wi-Fi

It is one of the most commonly used wireless technologies in portable devices. Mobile phone applications are connected and used by utilizing Wi-Fi and 3G technologies to monitor and regulate agricultural operations (Chung et al. 2015). However, it requires high cost and power (Mohapatra and Lenka 2016). It can be used up to 20–100 m. This technology can be used in agriculture for weather and soil monitoring in fields and greenhouses.

21.3.6 GPS and Bluetooth (BT)

BT is used for connection and communication between devices like mobiles, laptops, etc. It is used to satisfy various agricultural operations (Ojha et al. 2015). A system was developed to monitor weather, moisture, and temperature by utilizing Global Positioning System (GPS) and BT technologies. The purpose of this system is to conserve water and increase productivity by controlling irrigation (Kim and Evans 2009; Mubeen et al. 2013). BT can be used up to 10 m only. It has been used in agriculture for its lower energy requirements and ease of use (Vellidis et al. 2016). Using this technology, fertilizer and pesticide use and weather and soil conditions can be monitored while irrigation can be controlled.

21.3.7 Crop Simulation Modeling

Modeling is the representation of real systems using equations or sets of equations. Crop simulation models mimic the plant growth and development (Oteng-Darko et al. 2013; Mahmood et al. 2017; Aslam et al. 2017a; Mehmood et al. 2017, 2020). These models simulate plants and environment relationship for yield prediction, decision-making, crop management, and studying the climate change impacts on global food security (Kasampalis et al. 2018; Asseng et al. 2019; Liu et al. 2019). The basic purpose of agricultural model development is to increase the scientific understanding of underlying process in crop production and to improve the decision-making based on scientific understanding (Ahmed, 2012, 2020a, b; Ahmed et al. 2013, 2018, 2020a, b, c, d; Ahmed and Stockle, 2016; Ahmed and Ahmad, 2019, 2020). Crop simulation models predict on the basis of weather, soil, fertilizer, genetic, and environment information as well as their interactions (Wallach et al. 2018). Recent advances in crop simulation models have enabled the insects, pests, and diseases prediction as well. Decision Support System for Agrotechnology Transfer (DSSAT) (Hoogenboom et al. 2019; Jones et al. 2003), Agricultural Production Systems Simulator (APSIM) (Keating et al. 2003) and Environmental

Policy Integrated Climate (EPIC) are very well-known crop simulation models worldwide.

21.3.8 Remote Sensing

It is an art as well as a science to acquire knowledge and information about the object from distance. Information is remotely sensed by sensors through several platforms like satellites, UAVs (Un-manned Aerial Vehicles), airplane, and handheld devices (Bregaglio et al. 2015). Remotely sensed satellite images have very useful application over the last few years (Kasampalis et al. 2018). Remotely sensed data is used to monitor the crops, vegetation type, vegetation indices, vegetation vigor (Silleos et al. 2006), canopy, leaf area index, absorbed photosynthetically active radiations, evapotranspiration, biomass, and yield (Franch et al. 2019; Leroux et al. 2019; Campoy et al. 2020). This data is used to develop primary productivity models and simulation of productivity (Han et al. 2020; Zhang et al. 2020; Chen and Tao 2020). Remote sensing in combination with crop simulation models can help in yield assessment. Remote sensing can provide the enhanced spatial information and actual field condition that can improve the growth and yield assessment under the anticipated climate change (Kasampalis et al. 2018; Nasim et al. 2011).

21.3.9 UAVs (Un-manned Aerial Vehicles) and Drones

UAVs and drones are aerial vehicles equipped with multispectral cameras, sensors, and microcontrollers to facilitate the daily work in the fields. In recent years precision agriculture is the main focus of community research (Hassan-Esfahani et al. 2014; Pederi and Cheporniuk 2015). Monitoring the health of crop is essential for yield increase in agriculture (Carbone et al. 2018). For agricultural monitoring use of drones and UAVs are now becoming very common. These are used to monitor the growth and development of crops as well as plant height, soil moisture, pest population, and variations occurring in the field (Potrino et al. 2018). These drones and UAVs help in remote sensing of agricultural crops by taking images of crops and fields. These images are used for analysis by using different kinds of software to improve the understanding. Drones and UAVs have the advantage of providing high-resolution images than satellite images (Carbone et al. 2018). These drones and UAVs are also used for spraying of chemicals like insecticides, pesticides, etc. (Pederi and Cheporniuk 2015).

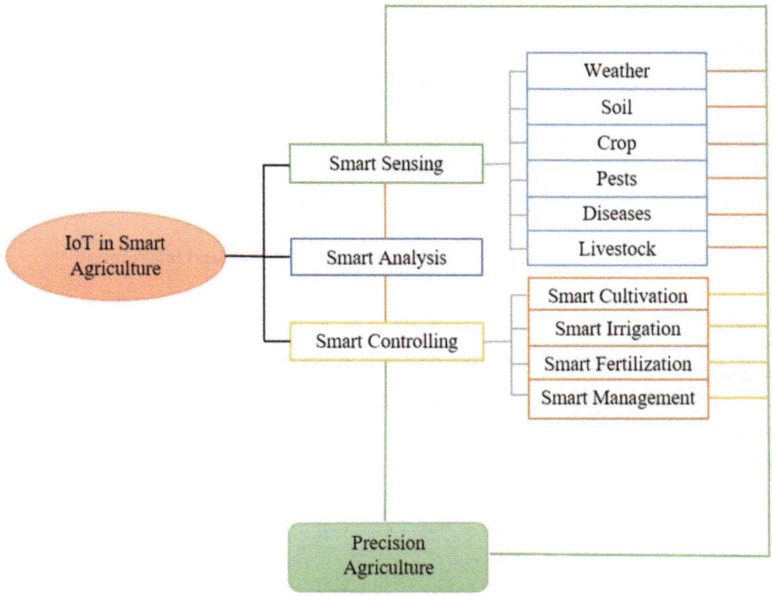

Fig. 21.4 IoT in smart agriculture

21.4 Potential Applications, Opportunities, and Current Usage Trends of IoT System in Agriculture

IoT system has numerous applications in the agriculture sector. It is used for monitoring and management of several processes with the help of modern technology (Fig. 21.4).

For decision support, raw sensor data is processed. Then, information is used for decision-making and performing different farm operations. A variety of sensors used for monitoring and data acquisition in agriculture are described in Table 21.2.

21.4.1 Weather Forecasting

To reduce agricultural risks associated with climate change, previous weather data can be used for weather forecasting through big data analysis. Environmental conditions can be monitored with the help of different environmental sensors. Depending on the collected data, timely decision-making, and farm management can be done.

Table 21.2 Sensors used in smart agriculture

Weather sensors	
Met station one (MSO)	Detection of wind speed and direction, temperature, humidity, and barometric pressure
Compact weather (CM-100)	Detection of wind speed and direction, temperature, humidity, and barometric pressure
T and R.H sensor (HMP45C)	Measurement of air relative humidity (0–100%) and temperature (−40 to +60 °C)
T and R.H sensor (SHT71)	Measurement of the relative humidity and temperature
Temperature sensor (107-L)	Measurement of air, water, and soil temperature
Soil sensors	
Moisture sensor (ECH_2O)	Detection of soil water content
Soil moisture sensor (MP406)	Monitoring of volumetric moisture content of soils
EC sensor (EC250)	Measurement of the electrical conductivity of soil
Hydra probe II soil sensor	Detection of moisture, conductivity, salinity, and temperature
Pogo portable soil sensor	Detection of soil temperature, conductivity, moisture, etc.
Plants/leaves sensors	
Photosynthesis (TPS-2)	Monitoring of leaf photosynthesis
Leaf wetness sensor (LW100)	Detection of leaf surface wetness and rainfall
Chlorophyll sensor (YSI6025)	Estimation of phytoplankton concentrations by detecting the fluorescence from chlorophyll
Temperature sensor (LT-2M)	Monitoring of absolute leaves temperature

21.4.2 *Pest Control*

Along with the weather and climate data, pest life cycle can also be monitored. It can play an important role in predicting pest outbreaks. For pest management, data of environmental variables (like humidity, precipitation, leaf wetness, etc.) are collected using different kinds of sensors. A disease known as Phytophthora was monitored and reduced with the help of sensor (TNOdes) in potato production (Baggio 2005).

21.4.3 *On-farm Water Management*

Irrigation and water management play a key role in agriculture (Zia et al. 2017). Climate change has resulted in erratic rainfall, water shortage, and health problems (Rasool et al. 2018). Remotely controlled irrigation systems such as drip irrigation and sprinkler irrigation can be opted in water-stressed areas. Linking the data of various variables (like radiation, humidity, temperature, etc.) from different types of sensors can control the amount and placement of water depending on the needs. Website real-time presentation of a river basin can be done. Sensors are used that feed the information into the websites. It enables users to monitor the river basin and react to changing patterns. As flooding is the main problem in river basins, it can be used as risk management strategy in agricultural communities.

21.4.4 Greenhouse Management

Monitoring the environment in the greenhouse is very critical (Ahmed 2017). Sensors are used in greenhouses for controlling and monitoring the moisture, temperature, humidity, etc. These sensors can be linked to a system, and Internet and can lead to smart agriculture that would help in effective management (Fig. 21.5).

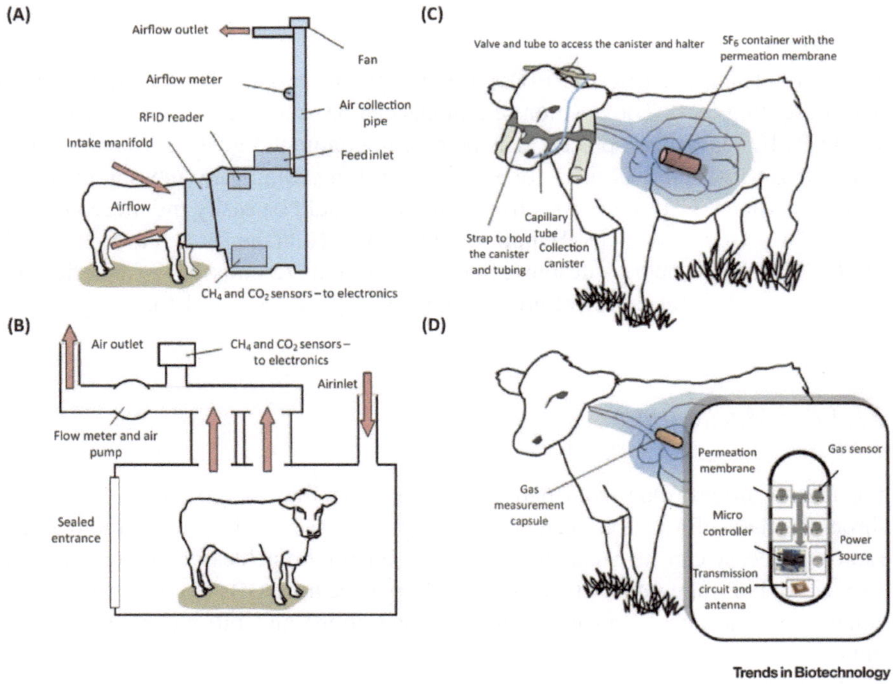

Fig. 21.5 Different sensors-based gas measurements techniques (Source: Hill et al. 2016 with permission from Elsevier)

21.4.5 Forest Management

Forest management can also be done effectively using IoT technology. Forest fires and illegal cuttings are the major issues in forest management. Fire can be detected by satellite technology using photos and heat sensors. Scannable plastic barcodes can be inserted in trees, to prevent illegal cutting. In this way, trees can be tracked from forests to the consumers.

21.4.6 Pollution Detection

IoT can play a vital role in pollution detection. In massive water bodies, satellite light radiations are used to detect water pollution. This technique would become convenient in aquaculture. By using the radiation wavelength, the type of pollutants can also be identified (Mubarak et al. 2016).

21.4.7 Livestock Monitoring

Radio frequency identifiers (RFIDs) are attached to animals for tracking animals and preventing theft. The position of animals can be seen in control or data centers where RFID readers are placed. It is helpful in communal grazing systems where animals get lost. GPS is used for location tracking. Similarly, measurements of enteric methane (CH_4) emissions from ruminants could be easily monitored by different sensors-based devices (Hammond et al. 2016; Huhtanen et al. 2015; Hill et al. 2016). This could help to design strategies to minimize emission of enteric CH_4 which is a significant source of greenhouse gas (Figs. 21.5 and 21.6).

21.4.8 Marketing

For market prices, data from national markets are filtered out and disseminated through small information centers having Internet access. In more remote areas, radio broadcast can be used. IoT can be used for branchless banking services. Especially in rural communities where farmers have no access to banks within reasonable distance, they can make money transactions and bill payments at retail outlets.

Precision agriculture and smart agriculture are aimed to maximize the net returns on investment. IoT technology can play a vital role in precision agriculture. Decision support systems running on smart devices can assist in better farming management. Mobile applications can help farmers in decision-making in farming operations

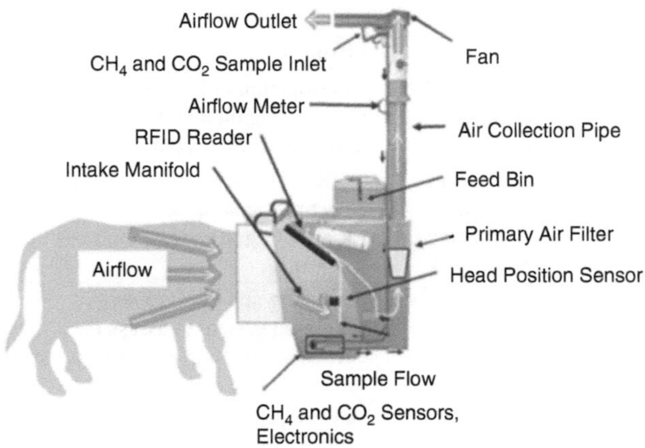

Fig. 21.6 Layout of GreenFeed system (Source: Huhtanen et al. 2015 with permission from Elsevier)

(irrigation, disease and pest management, etc.) and diagnosing the diseases. IoT technology systems track and monitor animals. Similarly, farm product delivery can also be facilitated by IoT technology. Sensors and mobile Internet cut out the role of middleman and enable farmers to directly contact with consumers. Near Field Communication (NFC) systems can facilitate the purchasing and buying of farm inputs and farm products without using cash. Electronic transactions are replaced with cash and enable branchless banking in remote areas far from banks (Dlodlo and Kalezhi 2015).

Precision agriculture has become a prime focus worldwide (Jiang 2014) and use of modern IoT technologies in smart agriculture has gained importance in recent years (Table 21.3).

21.5 Scope and Future of IoT

By using smart agriculture technologies, more area can be brought under cultivation to meet the increased demands. Livestock management can be done more effectively with the help of IoT and other modern technologies to improve the quality and cleanliness. However, on the other hand, affordability, sustainability, and scalability are the major obstacles for adoption of these technologies. Farmer literacy and technical education are vital to use modern advanced technologies.

Developing countries are lagging behind the developed countries in the agriculture sector. Therefore, it is a prime necessity to take serious initiatives to promote modern technology to make agriculture more efficient and profitable. Hence, smart agriculture will result in optimizing resources, improving product quality, reducing production losses as well as improving the farmers' and country's economic conditions.

Table 21.3 Applications of IoT and ICT in smart agriculture

Region	Crop	Purpose	Findings	Reference
USA	Floriculture crops	Plant tissue nitrogen detection	Development of low-cost image analysis technique for determination of tissue nitrogen content	Adhikari et al. (2020)
USA	Soybean	Monitoring of nitrogen fertilization responses	OptRx™ sensor has the potential to detect nitrogen responses under varied nitrogen application rates in soybean	Sivarajan et al. (2020)
Malaysia	Rice	Monitoring soil Electrical Conductivity (EC).	Designing and development of IoT-based system to assess the nutrients availability by measuring soil EC in rice	Othaman et al. (2020)
India	Banana	Temperature responses detection	Development of SQLite local web database to monitor and track the banana growth and generate alerts on the basis of weather detection	Geethanjali and Muralidhara (2020)
China	Grasses	Monitoring of grasslands	Implementation of dynamic threshold method to construct a time series of Satellite Pour l'Observation de la Terre Vegetation (SPOT Vegetation) and Normalized Difference Vegetation Index (NDVI) data from 1999 to 2012 to monitor the green-up date	Guo et al. (2019)
USA	Cotton	Plants phenotyping	Development of terrestrial LiDAR-based high throughput phenotyping system for cotton	Sun et al. (2018)
Japan	Tomato	Leaf area detection	3D depth sensor (Microsoft Kinect) can be used to measure the light interception characteristics and leaf area nondestructively	Umeda et al. (2017)
Malaysia	Multiple crops	Plants nutrients stress	Spectral reflection pattern measurements in visible and infrared ranges can be used to detect plant stresses	Me et al. (2017)
UK	Wheat	Plants phenotyping and growth detection	Results indicate that multi-temporal, very high spatial resolution, 3D digital surface models via Structure from Motion (SfM) photogrammetry can be used successfully to monitor crop height and growth rate	Holman et al. (2016)
China	Potato	Shape and size detection	Ellipse axis method can be used to detect the shape of potato, with 98.8% accuracy level	Liao et al. (2015)
Israel	Apple	Mature apple detection	Apple detection can be done with shape analysis of using by implying the convexity test	Kelman and Linker (2014)

(continued)

Table 21.3 (continued)

Region	Crop	Purpose	Findings	Reference
China	Cotton	Foreign fiber detection	Image analysis segmentation method can be used with great accuracy and speed than the other methods to remove the foreign fibers	Wu et al. (2014)
Spain	Maize	Crop stand detection	Development and verification of automatic expert image system to assess the maize cropping geometry and aid in different treatment applications in field	Guerrero et al. (2013)
Germany	Multiple crops	Leaf detection	Ellipse approximation method in combination with active shape models can be used for individual plant identification under overlapping conditions	Pastrana and Rath (2013)
USA	Apple	Apple grading	Automatic adjustable algorithm method for segmentation of color images, using linear support vector machine (SVM) and Otsu's thresholding method can be robustly used for apple sorting and grading	Mizushima and Lu (2013)
Portugal	Grape	Grapes color detection	The proposed image analysis method can be used for the detection and location of crops in fields	Reis et al. (2012)

21.6 Benefits of IoT Technology in Agriculture

IoT technology can improve living standards and alleviate the poverty of farmers. For example, a wide range of crops can be grown in organic greenhouses that results in extra income and contribute to country's gross domestic product (GDP) (Akram et al. 2018). The role of the middleman is lessened with mobile Internet because farmers can directly contact consumers.

Agriculture surveillance programs can be run with IoT that enable farmers to take preventive measures before the economic threshold level is reached. Therefore, even in droughts, bumper harvests can be obtained through precision agriculture. Smart health cards for crops and livestock can facilitate effective and efficient diagnosis and treatment with historic information of affected crops and livestock (Alonso et al. 2020).

IoT facilitates farm product tracking. Buyer can know where the product is and when it will be delivered. It can also empower transporters by providing information about farmers who require transport. Near Field Communication (NFC) system use facilitates the seller and buyer in paperless transactions and minimizes theft and fraud. It is also beneficial for rural area people who do not have bank access. Satellite transmissions can serve as means to obtain information in remote areas about markets, prices, government services, facilities, etc. The major advantages of IoT in agriculture are (i) Efficient water management without water wastage (Kumari and

Iqbal 2020), (ii) Continuous monitoring and timely management (Shafi et al. 2020), (iii) Minimal labor and time requirements (Panda and Bhatnagar 2020), (iv) Soil and plant management (Othaman et al. 2020), (v) Disease diagnosis in plants and animals (Nayak et al. 2020), and (vi) Global marketing access (Panda and Bhatnagar 2020).

21.7 Challenges in Wireless Technologies

There are a number of challenges in observing different agricultural climates. Challenges and recommendations are listed below:

21.7.1 Communication Range

Communication range is a major challenge in wireless technologies. This range is short in agricultural applications. In ZigBee technology, it can be extended up to 100 m. In a farm field, this range can be extended using UAVs and drones.

21.7.2 Cost

Cost of sensors and monitoring equipment is important. The equipment should have low cost and robust performance. Cost reduction can make these technologies available to be adopted in poor countries.

21.7.3 Power Consumption

Power consumption of these systems should be minimal. On the other hand, this challenge can be overcome by introducing alternative energy harvesting ways like solar, wind, etc.

21.7.4 Reliability

Information collected from sensors are directed to farmers and organizations therefore information reliability is essential to deal immediately with dangerous information.

21.7.5 *Real-Time Data*

In precision agriculture real-time data is critical. Crops are vulnerable to changing climatic conditions. Therefore, real-time data is required to avoid crop failures and disasters.

21.7.6 *Fault Tolerance*

Sensors and systems should be fault tolerant. Communication or sensor fault can lead to serious crop damage. For example, if the sensor controlling irrigation is not working correctly or communicating, it can cause water stress damage to plants.

21.8 Conclusion

This chapter has reviewed the IoT applications in agriculture for sustainable developments and management of agricultural systems. IoT can benefit agriculture in several domains under the changing climate like weather forecasting, drought, crop, livestock, water and disease management, etc. Agricultural and rural development policies influence IoT technology adoption. Regions and sites specific IoT technologies should be developed to target poverty alleviation and economic uplift of farmers. Large number of IoT devices can be classified on the basis of type, usage, connection, place, etc. Instead of developing greater amount and variety of IoT structures, focus must be directed towards the development of reliable, more secure, minimal energy using, and more user-friendly devices. Agriculture is a more suitable area for IoT implementation, and there is room for development. IoT standards, platforms, and protocols should be open source software and evolving solutions for device compatibility issues. These solutions can decrease implementation costs and broaden the implementation of IoT as well.

Markets grow and collapse, disruptive business ideas emerge and die, but the people on the planet always need food for living. Therefore, developments and advancements in food and agriculture will always be a top priority, particularly under the current observed and predicted dynamics of climate extremes on the earth. Hence, application of IoT in agriculture has a promising future in terms of enhanced sustainability, efficiency, and scalability of agricultural systems.

References

Abbasi AZ, Islam N, Shaikh ZA (2014) A review of wireless sensors and networks' applications in agriculture. Computer Standards & Interfaces 36 (2):263-270

Adhikari R, Li C, Kalbaugh K, Nemali K (2020) A low-cost smartphone controlled sensor based on image analysis for estimating whole-plant tissue nitrogen (N) content in floriculture crops. Comput Electron Agric 169:105173

Ahmad S, Hasanuzzaman A (2020) Cotton production and uses. Springer Nature Singapore Pte Ltd. https://doi.org/10.1007/978-981-15-1472-2

Ahmed M, Hassan FU (2011) APSIM and DSSAT models as decision support tools. 19th International Congress on Modelling and Simulation, Perth, Australia, 12–16 December 2011,http://mssanz.org.au/modsim2011

Ahmed M (2012) Improving Soil Fertility Recommendations in Africa Using the Decision Support System for Agrotechnology Transfer (DSSAT); A Book Review. Exp Agri. 48 (4): 602-603

Ahmed M, Asif M, Hirani AH, Akram MN, Goyal A (2013) Modeling for Agricultural Sustainability: A Review. In Gurbir S. Bhullar GS, Bhullar NK (ed) Agricultural Sustainability Progress and Prospects in Crop Research. Elsevier, 32 Jamestown Road, London NW1 7BY, UK

Ahmed, M., Stockle, C.O. (2016). Quantification of climate variability, adaptation, and mitigation for agricultural sustainability. Springer Nature Switzerland AG.part of Springer Nature.

Ahmed M (2017) Greenhouse Gas Emissions and Climate Variability: An Overview. In: Ahmed M, Stockle CO (eds) Quantification of Climate Variability, Adaptation and Mitigation for Agricultural Sustainability. Springer International Publishing, Cham, pp 1-26. doi:https://doi.org/10.1007/978-3-319-32059-5_1

Ahmed M, Fayyaz-ul-Hassan, Ahmad S (2017) Climate Variability Impact on Rice Production: Adaptation and Mitigation Strategies. In: Ahmed M, Stockle CO (eds) Quantification of Climate Variability, Adaptation and Mitigation for Agricultural Sustainability. Springer International Publishing, Cham, pp 91-111. doi:https://doi.org/10.1007/978-3-319-32059-5_5

Ahmed M, Ijaz W, Ahmad S (2018) Adapting and evaluating APSIM-SoilP-Wheat model for response to phosphorus under rainfed conditions of Pakistan. Journal of Plant Nutrition 41, 2069-2084.

Ahmed M, Ahmad S (2019) Carbon Dioxide Enrichment and Crop Productivity. In: Hasanuzzaman M (ed) Agronomic Crops: Volume 2: Management Practices. Springer Singapore, Singapore, pp 31-46. doi:https://doi.org/10.1007/978-981-32-9783-8_3

Ahmed M (2020a) Introduction to Modern Climate Change. Andrew E. Dessler: Cambridge University Press, 2011, 252 pp, ISBN-10: 0521173159. Sci Total Environ 734, 139397. https://doi.org/10.1016/j.scitotenv.2020.139397

Ahmed M (2020b) Systems Modeling, Springer Nature Singapore Pvt. Ltd., pp. 409. https://doi.org/10.1007/978-981-15-4728-7

Ahmed M, Hasanuzzaman M, Raza MA, Malik A, Ahmad S (2020a) Plant Nutrients for Crop Growth, Development and Stress Tolerance. R. Roychowdhury et al. (eds.), Sustainable Agriculture in the Era of Climate Change, https://doi.org/10.1007/978-3-030-45669-6_3

Ahmed K, Shabbir G, Ahmed M, Shah KN (2020b) Phenotyping for drought resistance in bread wheat using physiological and biochemical traits. Sci Total Environ 729, 139082. https://doi.org/10.1016/j.scitotenv.2020.139082

Ahmed M, Ahmad S (2020). Systems Modeling. In: Ahmed M (ed.), Systems Modeling, Springer Nature Singapore Pvt. Ltd.. pp. 1-44. https://doi.org/10.1007/978-981-15-4728-7_1

Ahmed M, Raza MA, Hussain T (2020c) Dynamic Modeling. In: Ahmed M (ed.), Systems Modeling, Springer Nature Singapore Pvt. Ltd., pp. 111-148. https://doi.org/10.1007/978-981-15-4728-7_4

Ahmed M, Ahmad S, Raza MA, Kumar U, Ansar M, Shah GA, Parsons D, Hoogenboom G, Palosuo T, Seidel S (2020d) Models Calibration and Evaluation. In: Ahmed M (ed.), Systems Modeling, Springer Nature Singapore Pvt. Ltd.. pp. 149-176. https://doi.org/10.1007/978-981-15-4728-7_5

Ahmed M, Ahmad S, Waldrip HM, Ramin M, Raza MA (2020e). Whole Farm Modeling: A Systems Approach to Understanding and Managing Livestock for Greenhouse Gas Mitigation,

Economic Viability and Environmental Quality. In Animal Manure (eds H. Waldrip, P. Pagliari and Z. He). doi:https://doi.org/10.2134/asaspecpub67.c25

Ahmed M, Fayyaz-ul-Hassan, Van Ogtrop FF (2014) Can models help to forecast rainwater dynamics for rainfed ecosystem? Weather and Climate Extremes 5–6 (0):48-55. doi: https://doi.org/10.1016/j.wace.2014.07.001

Akram R, Turan V, Hammad HM, Ahmad S, Hussain S, Hasnain A, Maqbool MM, Rehmani MIA, Rasool A, Masood N, Mahmood F, Mubeen M, Sultana SR, Fahad S, Amanet K, Saleem M, Abbas Y, Akhtar HM, Hussain S, Waseem F, Murtaza R, Amin A, Zahoor SA, Sami ul Din M, Nasim W (2018) Fate of Organic and Inorganic Pollutants in Paddy Soils. In: Hashmi MZ, Varma A (eds) Environmental Pollution of Paddy Soils. Springer International Publishing, Cham, pp 197-214. doi: https://doi.org/10.1007/978-3-319-93671-0_13

Ali S, Eum H-I, Cho J, Dan L, Khan F, Dairaku K, Shrestha ML, Hwang S, Nasim W, Khan IA, Fahad S (2019) Assessment of climate extremes in future projections downscaled by multiple statistical downscaling methods over Pakistan. Atmospheric Research 222:114-133. doi: https://doi.org/10.1016/j.atmosres.2019.02.009

Alonso RS, Sittón-Candanedo I, García Ó, Prieto J, Rodríguez-González S (2020) An intelligent Edge-IoT platform for monitoring livestock and crops in a dairy farming scenario. Ad Hoc Networks 98:102047

Amin A, Nasim W, Fahad S, Ali S, Ahmad S, Rasool A, Saleem N, Hammad HM, Sultana SR, Mubeen M, Bakhat HF, Ahmad N, Shah GM, Adnan M, Noor M, Basir A, Saud S, Habib ur Rahman M, Paz JO (2018) Evaluation and analysis of temperature for historical (1996–2015) and projected (2030–2060) climates in Pakistan using SimCLIM climate model: Ensemble application. Atmospheric Research 213:422-436. doi: https://doi.org/10.1016/j.atmosres.2018.06.021

Andreev S, Galinina O, Pyattaev A, Gerasimenko M, Tirronen T, Torsner J, Sachs J, Dohler M, Koucheryavy Y (2015) Understanding the IoT connectivity landscape: a contemporary M2M radio technology roadmap. IEEE Communications Magazine 53 (9):32-40

Ashraf R, Fayyaz-ul-Hassan, Ahmed M, Shabbir G (2017) Wheat Physiological Response Under Drought. In: Ahmed M, Stockle CO (eds) Quantification of Climate Variability, Adaptation and Mitigation for Agricultural Sustainability. Springer International Publishing, Cham, pp 211-231. doi:https://doi.org/10.1007/978-3-319-32059-5_10

Aslam MU, Shehzad A, Ahmed M, Iqbal M, Asim M, Aslam M (2017a) QTL Modelling: An Adaptation Option in Spring Wheat for Drought Stress. In: Ahmed M, Stockle CO (eds) Quantification of Climate Variability, Adaptation and Mitigation for Agricultural Sustainability. Springer International Publishing, Cham, pp 113-136. doi:https://doi.org/10.1007/978-3-319-32059-5_6

Aslam Z, Khattak JZK, Ahmed M, Asif M (2017b) A Role of Bioinformatics in Agriculture. In: Ahmed M, Stockle CO (eds) Quantification of Climate Variability, Adaptation and Mitigation for Agricultural Sustainability. Springer International Publishing, Cham, pp 413-434. doi:https://doi.org/10.1007/978-3-319-32059-5_17

Asseng S, Martre P, Maiorano A, Rötter RP, O'Leary GJ, Fitzgerald GJ, Girousse C, Motzo R, Giunta F, Babar MA, Reynolds MP, Kheir AMS, Thorburn PJ, Waha K, Ruane AC, Aggarwal PK, Ahmed M, Balkovič J, Basso B, Biernath C, Bindi M, Cammarano D, Challinor AJ, De Sanctis G, Dumont B, Eyshi Rezaei E, Fereres E, Ferrise R, Garcia-Vila M, Gayler S, Gao Y, Horan H, Hoogenboom G, Izaurralde RC, Jabloun M, Jones CD, Kassie BT, Kersebaum K-C, Klein C, Koehler A-K, Liu B, Minoli S, Montesino San Martin M, Müller C, Naresh Kumar S, Nendel C, Olesen JE, Palosuo T, Porter JR, Priesack E, Ripoche D, Semenov MA, Stöckle C, Stratonovitch P, Streck T, Supit I, Tao F, Van der Velde M, Wallach D, Wang E, Webber H, Wolf J, Xiao L, Zhang Z, Zhao Z, Zhu Y, Ewert F (2019) Climate change impact and adaptation for wheat protein. Global Change Biology 25 (1):155-173. doi:https://doi.org/10.1111/gcb.14481

Bacenetti J, Paleari L, Tartarini S, Vesely FM, Foi M, Movedi E, Ravasi RA, Bellopede V, Durello S, Ceravolo C, Amicizia F, Confalonieri R (2020) May smart technologies reduce the environmental impact of nitrogen fertilization? A case study for paddy rice. Science of The Total Environment 715:136956. doi:https://doi.org/10.1016/j.scitotenv.2020.136956

Baggio A Wireless sensor networks in precision agriculture. In: ACM Workshop on Real-World Wireless Sensor Networks (REALWSN 2005), Stockholm, Sweden, 2005. Citeseer,

Bregaglio S, Frasso N, Pagani V, Stella T, Francone C, Cappelli G, Acutis M, Balaghi R, Ouabbou H, Paleari L (2015) New multi-model approach gives good estimations of wheat yield under semi-arid climate in Morocco. Agronomy for sustainable development 35 (1):157-167

Brewster C, Roussaki I, Kalatzis N, Doolin K, Ellis K (2017) IoT in agriculture: Designing a Europe-wide large-scale pilot. IEEE communications magazine 55 (9):26-33

Campoy J, Campos I, Plaza C, Calera M, Bodas V, Calera A (2020) Estimation of harvest index in wheat crops using a remote sensing-based approach. Field Crops Res 256:107910

Cancela J, Fandiño M, Rey B, Martínez E (2015) Automatic irrigation system based on dual crop coefficient, soil and plant water status for Vitis vinifera (cv Godello and cv Mencía). Agric Water Manage 151:52-63

Carbone C, Garibaldi O, Kurt Z (2018) Swarm robotics as a solution to crops inspection for precision agriculture. KnE Engineering:552-562

Chen Y, Tao F (2020) Improving the practicability of remote sensing data-assimilation-based crop yield estimations over a large area using a spatial assimilation algorithm and ensemble assimilation strategies. Agricultural and Forest Meteorology 291:108082

Chung S-O, Kang S-W, Bae K-S, Ryu M-J, Kim Y-J (2015) The potential of remote monitoring and control of protected crop production environment using mobile phone under 3G and Wi-Fi communication conditions. Engineering in Agriculture, Environment and Food 8 (4):251-256

Dlodlo N, Kalezhi J The internet of things in agriculture for sustainable rural development. In: 2015 international conference on emerging trends in networks and computer communications (ETNCC), 2015. IEEE, pp 13-18

Dusadeerungsikul PO, Liakos V, Morari F, Nof SY, Bechar A (2020) Chapter 5 - Smart action. In: Castrignanò A, Buttafuoco G, Khosla R, Mouazen AM, Moshou D, Naud O (eds) Agricultural Internet of Things and Decision Support for Precision Smart Farming. Academic Press, pp 225–277. doi:https://doi.org/10.1016/B978-0-12-818373-1.00005-6

Franch B, Vermote EF, Skakun S, Roger J-C, Becker-Reshef I, Murphy E, Justice C (2019) Remote sensing based yield monitoring: Application to winter wheat in United States and Ukraine. International Journal of Applied Earth Observation and Geoinformation 76:112-127

Gakuru M, Winters K, Stepman F Inventory of innovative farmer advisory services using ICTs. In 2009. Forum for Agricultural Research in Africa (FARA), Accra, GH

Geethanjali B, Muralidhara B (2020) A Wireless Sensor System to Monitor Banana Growth Based on the Temperature. In: Information and Communication Technology for Sustainable Development. Springer, pp 271-278

Guerrero JM, Guijarro M, Montalvo M, Romeo J, Emmi L, Ribeiro A, Pajares G (2013) Automatic expert system based on images for accuracy crop row detection in maize fields. Expert Systems with Applications 40 (2):656-664

Guo J, Yang X, Niu J, Jin Y, Xu B, Shen G, Zhang W, Zhao F, Zhang Y (2019) Remote sensing monitoring of green-up dates in the Xilingol grasslands of northern China and their correlations with meteorological factors. Int J Remote Sens 40 (5-6):2190-2211

Gutiérrez J, Villa-Medina JF, Nieto-Garibay A, Porta-Gándara MÁ (2014) Automated irrigation system using a wireless sensor network and GPRS module. IEEE transactions on instrumentation and measurement 63 (1):166-176

Hammond KJ, Crompton LA, Bannink A, Dijkstra J, Yáñez-Ruiz DR, O'Kiely P, Kebreab E, Eugène MA, Yu Z, Shingfield KJ, Schwarm A, Hristov AN, Reynolds CK (2016) Review of current in vivo measurement techniques for quantifying enteric methane emission from ruminants. Animal Feed Science and Technology 219:13-30. doi: https://doi.org/10.1016/j.anifeedsci.2016.05.018

Han C, Zhang B, Chen H, Liu Y, Wei Z (2020) Novel approach of upscaling the FAO AquaCrop model into regional scale by using distributed crop parameters derived from remote sensing data. Agric Water Manage 240:106288

Hassan-Esfahani L, Torres-Rua A, Ticlavilca AM, Jensen A, McKee M Topsoil moisture estimation for precision agriculture using unmanned aerial vehicle multispectral imagery. In: 2014 IEEE Geoscience and Remote Sensing Symposium, 2014. IEEE, pp 3263-3266

Hill J, McSweeney C, Wright A-DG, Bishop-Hurley G, Kalantar-zadeh K (2016) Measuring Methane Production from Ruminants. Trends in Biotechnology 34 (1):26-35. doi: https://doi.org/10.1016/j.tibtech.2015.10.004

Holman F, Riche A, Michalski A, Castle M, Wooster M, Hawkesford M (2016) High throughput field phenotyping of wheat plant height and growth rate in field plot trials using UAV based remote sensing. Remote Sensing 8 (12):1031

Hoogenboom G, Porter C, Boote K, Shelia V, Wilkens PW. (2019) The DSSAT crop modeling ecosystem. Burleigh dodds Science Publishing. UK

Huhtanen P, Cabezas-Garcia EH, Utsumi S, Zimmerman S (2015) Comparison of methods to determine methane emissions from dairy cows in farm conditions. Journal of Dairy Science 98 (5):3394-3409. doi: https://doi.org/10.3168/jds.2014-9118

Ijaz W, Ahmed M, Fayyaz-ul-Hassan, Asim M, Aslam M (2017) Models to Study Phosphorous Dynamics Under Changing Climate. In: Ahmed M, Stockle CO (eds) Quantification of Climate Variability, Adaptation and Mitigation for Agricultural Sustainability. Springer International Publishing, Cham, pp 371-386. doi:https://doi.org/10.1007/978-3-319-32059-5_15

Ilie-Ablachim D, Pătru GC, Florea I-M, Rosner D Monitoring device for culture substrate growth parameters for precision agriculture: Acronym: MoniSen. In: RoEduNet Conference: Networking in Education and Research, 2016 15th, 2016. IEEE, pp 1-7

Jabeen M, Gabriel HF, Ahmed M, Mahboob MA, Iqbal J (2017) Studying Impact of Climate Change on Wheat Yield by Using DSSAT and GIS: A Case Study of Pothwar Region. In: Ahmed M, Stockle CO (eds) Quantification of Climate Variability, Adaptation and Mitigation for Agricultural Sustainability. Springer International Publishing, Cham, pp 387-411. doi:https://doi.org/10.1007/978-3-319-32059-5_16

Jawad H, Nordin R, Gharghan S, Jawad A, Ismail M (2017) Energy-efficient wireless sensor networks for precision agriculture: A review. Sensors 17 (8):1781

Jiang J-A Becoming technologically advanced-IOT applications in smart agriculture. In: 38th meeting of, 2014.

Jones JW, Hoogenboom G, Porter CH, Boote KJ, Batchelor WD, Hunt LA, Wilkens PW, Singh U, Gijsman AJ, Ritchie JT (2003) The DSSAT cropping system model. European Journal of Agronomy 18 (3):235-265. doi:https://doi.org/10.1016/S1161-0301(02)00107-7.

Kasampalis D, Alexandridis T, Deva C, Challinor A, Moshou D, Zalidis G (2018) Contribution of remote sensing on crop models: a review. Journal of Imaging 4 (4):52

Keating BA, Carberry PS, Hammer GL, Probert ME, Robertson MJ, Holzworth D, Huth NI, Hargreaves JNG, Meinke H, Hochman Z, McLean G, Verburg K, Snow V, Dimes JP, Silburn M, Wang E, Brown S, Bristow KL, Asseng S, Chapman S, McCown RL, Freebairn DM, Smith CJ (2003) An overview of APSIM, a model designed for farming systems simulation. European Journal of Agronomy 18 (3–4):267–288. doi:http://dx.doi.org/10.1016/S1161-0301(02)00108-9

Kelman EE, Linker R (2014) Vision-based localisation of mature apples in tree images using convexity. Biosys Eng 118:174-185

Kim Y, Evans R (2009) Software design for wireless sensor-based site-specific irrigation. Comput Electron Agric 66 (2):159-165

Kopetz H (2011) Internet of things. In: Real-time systems. Springer, pp 307-323

Kumari V, Iqbal M (2020) Development of Model for Sustainable Development in Agriculture Using IoT-Based Smart Farming. In: New Paradigm in Decision Science and Management. Springer, pp 303-310

Leroux L, Castets M, Baron C, Escorihuela M-J, Bégué A, Seen DL (2019) Maize yield estimation in West Africa from crop process-induced combinations of multi-domain remote sensing indices. European Journal of Agronomy 108:11-26

Liao G, Wang X, Jin J, Li J Potato size and shape detection using machine vision. In: MATEC Web of Conferences, 2015. EDP Sciences, p 15003

Lipper L, Thornton P, Campbell BM, Baedeker T, Braimoh A, Bwalya M, Caron P, Cattaneo A, Garrity D, Henry K (2014) Climate-smart agriculture for food security. Nature climate change 4 (12):1068-1072

Liu B, Martre P, Ewert F, Porter JR, Challinor AJ, Müller C, Ruane AC, Waha K, Thorburn PJ, Aggarwal PK, Ahmed M, Balkovič J, Basso B, Biernath C, Bindi M, Cammarano D, De Sanctis G, Dumont B, Espadafor M, Eyshi Rezaei E, Ferrise R, Garcia-Vila M, Gayler S, Gao Y, Horan H, Hoogenboom G, Izaurralde RC, Jones CD, Kassie BT, Kersebaum KC, Klein C, Koehler A-K, Maiorano A, Minoli S, Montesino San Martin M, Naresh Kumar S, Nendel C, O'Leary GJ, Palosuo T, Priesack E, Ripoche D, Rötter RP, Semenov MA, Stöckle C, Streck T, Supit I, Tao F, Van der Velde M, Wallach D, Wang E, Webber H, Wolf J, Xiao L, Zhang Z, Zhao Z, Zhu Y, Asseng S (2019) Global wheat production with 1.5 and 2.0 °C above pre-industrial warming. Global Change Biology 25 (4):1428-1444. doi: https://doi.org/10.1111/gcb.14542

Mahmood FH, Belhouchette, W., Nasim, T., Shazad, S., Hussain, O., Therond, S., Fahad, Wery J. (2017) Economic and environmental impacts of introducing grain legumes in farming systems of Midi-Pyrenees region (France): a simulation approach. International Journal of Plant Production 11 (1):65-87. doi: https://doi.org/10.22069/ijpp.2017.3310

Malavade VN, Akulwar PK (2016) Role of IoT in agriculture. IOSR Journal of Computer Engineering (IOSR-JCE):56-57

Me C, Balasundram SK, Hanif AHM (2017) Detecting and monitoring plant nutrient stress using remote sensing approaches: A review. Asian J Plant Sci 16:1-8

Mehmood A, Ahmed M, Fayyaz-ul-Hassan, Akmal M, ur Rehman O (2017) Soil and Water Assessment Tool (SWAT) for Rainfed Wheat Water Productivity. In: Ahmed M, Stockle CO (eds) Quantification of Climate Variability, Adaptation and Mitigation for Agricultural Sustainability. Springer International Publishing, Cham, pp 137-163. doi:https://doi.org/10.1007/978-3-319-32059-5_7

Mehmood MZ, Afzal O, Aslam MA, Riaz H, Raza MA, Ahmed S, Qadir G, Ahmad M, Shaheen FA, Shah ZH (2020) Disease Modeling as a Tool to Assess the Impacts of Climate Variability on Plant Diseases and Health. In: Systems Modeling. Springer, pp 327–351

Mizushima A, Lu R (2013) An image segmentation method for apple sorting and grading using support vector machine and Otsu's method. Comput Electron Agric 94:29-37

Mohapatra AG, Lenka SK (2016) Neural network pattern classification and weather dependent fuzzy logic model for irrigation control in WSN based precision agriculture. Procedia Computer Science 78:499-506

Mubarak H, Mirza N, Chai L-Y, Yang Z-H, Yong W, Tang C-J, Mahmood Q, Pervez A, Farooq U, Fahad S, Nasim W, Siddique KHM (2016) Biochemical and Metabolic Changes in Arsenic Contaminated *Boehmeria nivea* L. BioMed Research International 2016:1423828. doi:https://doi.org/10.1155/2016/1423828

Mubeen M, Ahmad A, Khaliq T, Sultana SR, Hussain S, Ali A, Ali H, Nasim W (2013) Effect of Growth Stage-Based Irrigation Schedules on Biomass Accumulation and Resource Use Efficiency of Wheat Cultivars. American Journal of Plant Sciences Vol. 04 No. 07:8. doi:https://doi.org/10.4236/ajps.2013.47175

Nasim W, Ahmad A, Wajid A, Akhtar J, Muhammad D (2011) Nitrogen effects on growth and development of sunflower hybrids under agro-climatic conditions of Multan. Pak J Bot 43 (4):2083-2092

Nayak P, Kavitha K, Rao CM (2020) IoT-Enabled Agricultural System Applications, Challenges and Security Issues. In: IoT and Analytics for Agriculture. Springer, pp 139-163

Nlerum F, Onowu E (2014) Information Communication Technologies in Agricultural Extension Delivery of Agricultural Transformation Agenda. International Journal of Agricultural Science, Research and Technology in Extension and Education Systems 4 (4):221-228

Ojha T, Misra S, Raghuwanshi NS (2015) Wireless sensor networks for agriculture: The state-of-the-art in practice and future challenges. Comput Electron Agric 118:66-84

Oteng-Darko P, Yeboah S, Addy S, Amponsah S, Danquah EO (2013) Crop modeling: A tool for agricultural research–A. J Agricultural Res Develop 2 (1):001-006
Othaman NC, Isa MM, Murad S, Harun A, Mohyar S Electrical conductivity (EC) sensing system for paddy plant using the internet of things (IoT) connectivity. In: AIP Conference Proceedings, 2020. vol 1. AIP Publishing LLC, p 020005
Panda CK, Bhatnagar R (2020) Social Internet of Things in Agriculture: An Overview and Future Scope. In: Toward Social Internet of Things (SIoT): Enabling Technologies, Architectures and Applications. Springer, pp 317-334
Pastrana JC, Rath T (2013) Novel image processing approach for solving the overlapping problem in agriculture. Biosys Eng 115 (1):106-115
Patil K, Kale N A model for smart agriculture using IoT. In: 2016 International Conference on Global Trends in Signal Processing, Information Computing and Communication (ICGTSPICC), 2016. IEEE, pp 543-545
Pederi Y, Cheporniuk H Unmanned aerial vehicles and new technological methods of monitoring and crop protection in precision agriculture. In: 2015 IEEE International Conference Actual Problems of Unmanned Aerial Vehicles Developments (APUAVD), 2015. IEEE, pp 298-301
Pitì A, Verticale G, Rottondi C, Capone A, Lo Schiavo L (2017) The role of smart meters in enabling real-time energy services for households: The Italian case. Energies 10 (2):199
Potrino G, Palmieri N, Antonello V, Serianni A Drones Support in Precision Agriculture for Fighting Against Parasites. In: 2018 26th Telecommunications Forum (TELFOR), 2018. IEEE, pp 1-4
Puranik V, Ranjan A, Kumari A Automation in Agriculture and IoT. In: 2019 4th International Conference on Internet of Things: Smart Innovation and Usages (IoT-SIU), 2019. IEEE, pp 1-6
Rahman MHu, Ahmad I, Ghaffar A, Haider G, Ahmad A, Ahmad B, Tariq M, Nasim W, Rasul G, Fahad S, Ahmad S, Hoogenboom G (2020) Climate Resilient Cotton Production System: A Case Study in Pakistan. In: Ahmad S, Hasanuzzaman M (eds) Cotton Production and Uses: Agronomy, Crop Protection, and Postharvest Technologies. Springer Singapore, Singapore, pp 447-484. doi:https://doi.org/10.1007/978-981-15-1472-2_22
Rasool A, Farooqi A, Xiao T, Ali W, Noor S, Abiola O, Ali S, Nasim W (2018) A review of global outlook on fluoride contamination in groundwater with prominence on the Pakistan current situation. Environmental Geochemistry and Health 40 (4):1265-1281. doi:https://doi.org/10.1007/s10653-017-0054-z
Ratasuk R, Vejlgaard B, Mangalvedhe N, Ghosh A NB-IoT system for M2M communication. In: Wireless Communications and Networking Conference (WCNC), 2016 IEEE, 2016. IEEE, pp 1-5
Reis MJ, Morais R, Peres E, Pereira C, Contente O, Soares S, Valente A, Baptista J, Ferreira PJS, Cruz JB (2012) Automatic detection of bunches of grapes in natural environment from color images. Journal of Applied Logic 10 (4):285-290
Research J (2015) Internet of Things Connected Devices to Almost Triple to Over 38 Billion Units by 2020", Juniper Research.
Sarode K, Chaudhari P (2018) Zigbee based Agricultural Monitoring and Controlling System. International Journal of Engineering Science 15907
Shafi U, Mumtaz R, Hassan SA, Zaidi SAR, Akhtar A, Malik MM (2020) Crop Health Monitoring Using IoT-Enabled Precision Agriculture. In: IoT Architectures, Models, and Platforms for Smart City Applications. IGI Global, pp 134-154
Silleos NG, Alexandridis TK, Gitas IZ, Perakis K (2006) Vegetation indices: advances made in biomass estimation and vegetation monitoring in the last 30 years. Geocarto International 21 (4):21-28
Sivarajan S, Maharlooei M, Kandel H, Buetow RR, Nowatzki J, Bajwa SG (2020) Evaluation of OptRx™ active optical sensor to monitor soybean response to nitrogen inputs. J Sci Food Agric 100 (1):154-160
Stratigea A (2011) ICTs for rural development: potential applications and barriers involved. Netcom Réseaux, communication et territoires (25-3/4):179-204

Sun S, Li C, Paterson AH, Jiang Y, Xu R, Robertson JS, Snider JL, Chee PW (2018) In-field high throughput phenotyping and cotton plant growth analysis using LiDAR. Frontiers in Plant Science 9:16

Umeda H, Mochizuki Y, Saito T, Higashide T, Iwasaki Y Diagnosing method for plant growth using a 3D depth sensor. In: International Symposium on New Technologies for Environment Control, Energy-Saving and Crop Production in Greenhouse and Plant 1227, 2017. pp 631-636

van Ogtrop F, Ahmad M, Moeller C (2014) Principal components of sea surface temperatures as predictors of seasonal rainfall in rainfed wheat growing areas of Pakistan. Meteorological Applications 21 (2):431-443. doi:https://doi.org/10.1002/met.1429

Vellidis G, Liakos V, Andreis J, Perry C, Porter W, Barnes E, Morgan K, Fraisse C, Migliaccio K (2016) Development and assessment of a smartphone application for irrigation scheduling in cotton. Comput Electron Agric 127:249-259

Wallach D, Martre P, Liu B, Asseng S, Ewert F, Thorburn PJ, van Ittersum M, Aggarwal PK, Ahmed M, Basso B, Biernath C, Cammarano D, Challinor AJ, De Sanctis G, Dumont B, Eyshi Rezaei E, Fereres E, Fitzgerald GJ, Gao Y, Garcia-Vila M, Gayler S, Girousse C, Hoogenboom G, Horan H, Izaurralde RC, Jones CD, Kassie BT, Kersebaum KC, Klein C, Koehler A-K, Maiorano A, Minoli S, Müller C, Naresh Kumar S, Nendel C, O'Leary GJ, Palosuo T, Priesack E, Ripoche D, Rötter RP, Semenov MA, Stöckle C, Stratonovitch P, Streck T, Supit I, Tao F, Wolf J, Zhang Z (2018) Multimodel ensembles improve predictions of crop–environment–management interactions. Global Change Biology 24 (11):5072-5083. doi:https://doi.org/10.1111/gcb.14411

Weber R, Weber R (2010) Internet of Things: Legal Perspectives, vol. 49. Xia, F, Yang, LT, Wang, L, &Vinel, A (2012) Internet of things International Journal of Communication Systems 25 (9):1101

Wu Y, Li D, Li Z, Yang W (2014) Fast processing of foreign fiber images by image blocking. Information Processing in Agriculture 1 (1):2-13

Zhang J, Chen Y, Zhang Z (2020) A remote sensing-based scheme to improve regional crop model calibration at sub-model component level. Agricultural Systems 181:102814

Zia Z, Bakhat HF, Saqib ZA, Shah GM, Fahad S, Ashraf MR, Hammad HM, Naseem W, Shahid M (2017) Effect of water management and silicon on germination, growth, phosphorus and arsenic uptake in rice. Ecotoxicology and Environmental Safety 144:11-18. doi: https://doi.org/10.1016/j.ecoenv.2017.06.004

Chapter 22
World Nations Priorities on Climate Change and Food Security

Muhammad Sami Ul Din, Muhammad Mubeen, Sajjad Hussain, Ashfaq Ahmad, Nazim Hussain, Muhammad Anjum Ali, Ayman El Sabagh, Mabrouk Elsabagh, Ghulam Mustafa Shah, Saeed Ahmad Qaisrani, Muhammad Tahir, Hafiz Muhammad Rashad Javeed, Muhammad Anwar-ul-Haq, Musaddiq Ali, and Wajid Nasim

---

M. S. U. Din · M. Mubeen (✉) · S. Hussain · G. M. Shah · S. A. Qaisrani · M. Tahir · H. M. R. Javeed · M. Ali
Department of Environmental Sciences, COMSATS University Islamabad, Vehari, Punjab, Pakistan
e-mail: muhammadmubeen@cuivehari.edu.pk

A. Ahmad
Program Chair, Climate Change, US.-Pakistan Centre for Advanced Studies in Agriculture and Food Security, University of Agriculture, Faisalabad, Punjab, Pakistan

N. Hussain
Department of Agronomy, Bahauddin Zakariya University, Multan, Punjab, Pakistan

M. A. Ali
Directorate General Agriculture, Lahore, Punjab, Pakistan

A. El Sabagh
Department of Field Crops, Faculty of Agriculture, Siirt University, Siirt, Turkey

Department of Agronomy, Faculty of Agriculture, Kafrelsheikh University, Kafr El-Sheikh, Egypt

M. Elsabagh
Department of Animal Production and Technology, Faculty of Agricultural Sciences and Technologies, Niğde Ömer Halisdemir University, Niğde, Turkey

Department of Nutrition and Clinical Nutrition, Faculty of Veterinary Medicine, Kafrelsheikh University, Kafr El-Sheikh, Egypt

M. Anwar-ul-Haq
Institute of Soil & Environmental Sciences, University of Agriculture, Faisalabad, Punjab, Pakistan

W. Nasim
Department of Agronomy, Faculty of Agriculture and Environment Sciences, The Islamia University of Bahawalpur (IUB), Bahawalpur, Punjab, Pakistan

W. N. Jatoi et al. (eds.), *Building Climate Resilience in Agriculture*,
https://doi.org/10.1007/978-3-030-79408-8_22

Abstract The present food system (including production, transportation, processing, packaging, storing, retail, and consumption) is a source of nutrition for the great majority of the world population in addition to supporting the livelihoods of about 200 million people. Food supply per capita has increased by more than 30% since 1961, this is accompanied by more use of nitrogen fertilizers (showing an increase of about 800%) and water resources for irrigation (with an increase of more than 100%). Global food security will continue to be an international concern for the coming 50 years and even beyond. Crop yield has fallen in many areas recently due to decreasing investments in infrastructure and research, as well as due to growing water scarcity. Climate change is a global concern irrespective of borders. The poor nations are highly vulnerable to climate change and are at high risk. Food security is directly dependant on the food chain and the associated food system process. All dimensions of food security could be affected by climate change in complex ways. Approximately, 15 countries are highly vulnerable to food insecurity due to climate change, from Asia and Africa. Most of these nations are not able to cope with or counter the impact of climate change on an urgent basis. However, some countries have developed their national strategies and adaptation plans to alleviate the negative impacts of climate change.

Keywords Climate change · Food security pillars · Nations at risk · Management of risk · Climate resilient agriculture

22.1 Introduction

Food is considered a basic element for human development and human well-being as well as to achieve food security. Millions of people do not have enough food for better survival, facing malnutrition and global food demand is increased over the last few decades at global level. In the future, the condition will be very severe due to climate change. In the next five decades, the growing population will raise the food demand at a global level (Godfray et al. 2010; Senker 2011). It is very difficult to fulfill the global food demand due to climate change. All of these changes affect partially the overall food system (Ingram 2011). Land use land cover changes (LULCC) are also a major limiting factor that affect food security at a global level (Ramaswamy and Feed 2015; Hussain et al. 2019, 2020b). Food insecurity increased due to climate change (CC) because CC affects agriculture and declines crop production due to extreme weather events such as a change in temperature, rainfall patterns, and an increase in carbon dioxide. Also, climate change negatively affects the profitability and sustainability of animal agriculture enterprises as well as the quality of animal products, which play a major role in ensuring food security (Nardone et al. 2010; Sabagh et al. 2020). Vice versa, animal agriculture operations are a source of greenhouse gas emissions (methane and nitric oxide) that contribute to global warming (Grossi et al. 2019).

In the future, food security will be a great concern for all researchers and scientists at a global level. Water shortage is a major issue and agricultural production decreases due to limited water availability for irrigation. According to Stocker (2014) at the end of this century, the temperature will be increased to 2.4–6.4 °C if we could not mitigate the active carbon. It was reported that African food and agriculture are highly at risk due to climate change and extreme weather events in the review of historical perspective (Muller et al. 2011a, b). Projected impacts relative to current production levels range from −100% to +168% in econometric, from −84% to +62% in process-based, and from −57% to +30% in statistical assessments (Muller et al. 2011a, b). South Asia is considered as most vulnerable region due to climate change in the world. A decline in agricultural yield is a major problem that is induced by climate change (Ali et al. 2019). Climate variability also influences the economics of the countries such as the price of food products increased in Pakistan, Nepal, Bangladesh, India, and Siri Lanka which was examined by a global dynamic computable general equilibrium model (Bandara et al. 2014). Therefore, climate variability threatens food prices, food production systems, and especially food security. In 2080, food demand will be increased by 300% due to high population growth; higher demands of resources for food production system will affect the food supply without any environmental changes. Food security and food prices are influenced by changing global environment due to its continuous effects on food production system (Hussain et al. 2020a). The models relevant to climate change predicted that the probability of extreme weather events such as floods send droughts will be increased in the next few decades. Global warming also decreases the average cereal production over the entire globe.

22.2 Food Security

The situation in which people have access to safe, sufficient, and nutritious food for a healthy life as well as to fulfill their dietary needs is known as food security (Barrett 2010). This definition needs several improvements like access and availability of the food to be culturally appropriate. Significantly the access to food differs across the regions but the huge difference lies between developing nations and developed ones. The major difference is due to the variation in the source and amount of income among these nations. The good health and nutrition level of humans is directly linked with food security (Havas et al. 2011).

Improvements in the agriculture and food production system are considered as main stream to ensure food security (USDA 2019). Access to good quality and quantity of food has positive impacts on societies as well as among the nations, including:

- The creation of jobs and growth of economics

- Reduction in the poverty level
- Maximum trade opportunities
- Increase in global stability and security
- Improved healthcare

22.2.1 Pillars of Food Security

The WHO (World Health Organization) described that food security has three major pillars including the availability of food, access to safe food, and utilization of the food. Recently FAO (Food and Agriculture Organization) incorporates the fourth pillar: stability of the above three pillars. Now food security has four pillars availability, access, utilization, and stability (Farsund et al. 2015).

1. **Food Availability**: Sufficient food is available for every individual all of the time from commercial imports or donors and domestic food production within the available resources.
2. **Food Access:** To maintain the nutritional level individual has sufficient resources or income to purchase the food is known as food access.
3. **Food Utilization:** The proper biological use of food and a diet that provides sufficient energy is known as food utilization. Productive food utilization is based on the processing techniques for food, the principle for basic nutrition, and knowledge for food storage at the household level.
4. **Food Stability:** The ability to achieve food all over time is known as food stability. Food security may be chronic, transitory, and seasonal. Instability of markets, natural disasters, and civil conflicts can affect food stability in different regions of the world (Fan and Brzeska 2016; Henegedara and Management 2015).

22.2.2 Food Security Challenges

- Climate change
- Poverty
- Rising population
- Rising incomes, changing diets
- Falling water tables
- Slowing irrigation
- More foodless days
- Increasing soil erosion
- Melting water reserves

22.3 Relationships Between Food Security and Climate Change

The link between food security and climate change has been studied while discussing the decline in crop productivity and the failure of the agricultural system. The experimental findings of Gregory show that 5% of rice yield is declined and the duration of the wheat is decreased due to the heat stress. The rapid increase of CO_2 in the atmosphere is considered detrimental because due to this, the temperature increased which limits agricultural productivity especially cereal crop yield. Some other reviews which assessed the further impacts of climate variability on the yield and growth of the crops (Fig. 22.1).

22.4 Food Insecurity Level Across the Regions of the World

Approximately 2 billion peoples are facing food insecurity all over the world: 9% (188 million) peoples in America, 34% (676 million) peoples in Africa, and 52% (1.04 billion) peoples in Asia. Figure 22.2 also described population food insecurity level across the different regions of the world (FAO 2019).

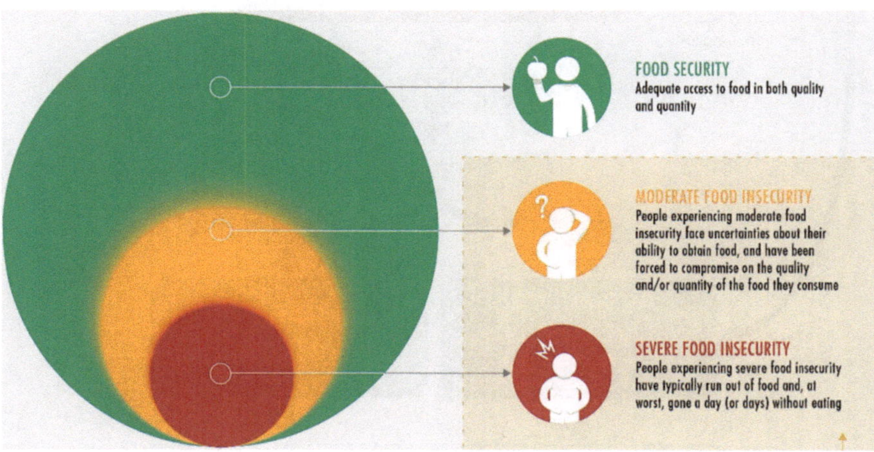

Fig. 22.1 Importance of global food security

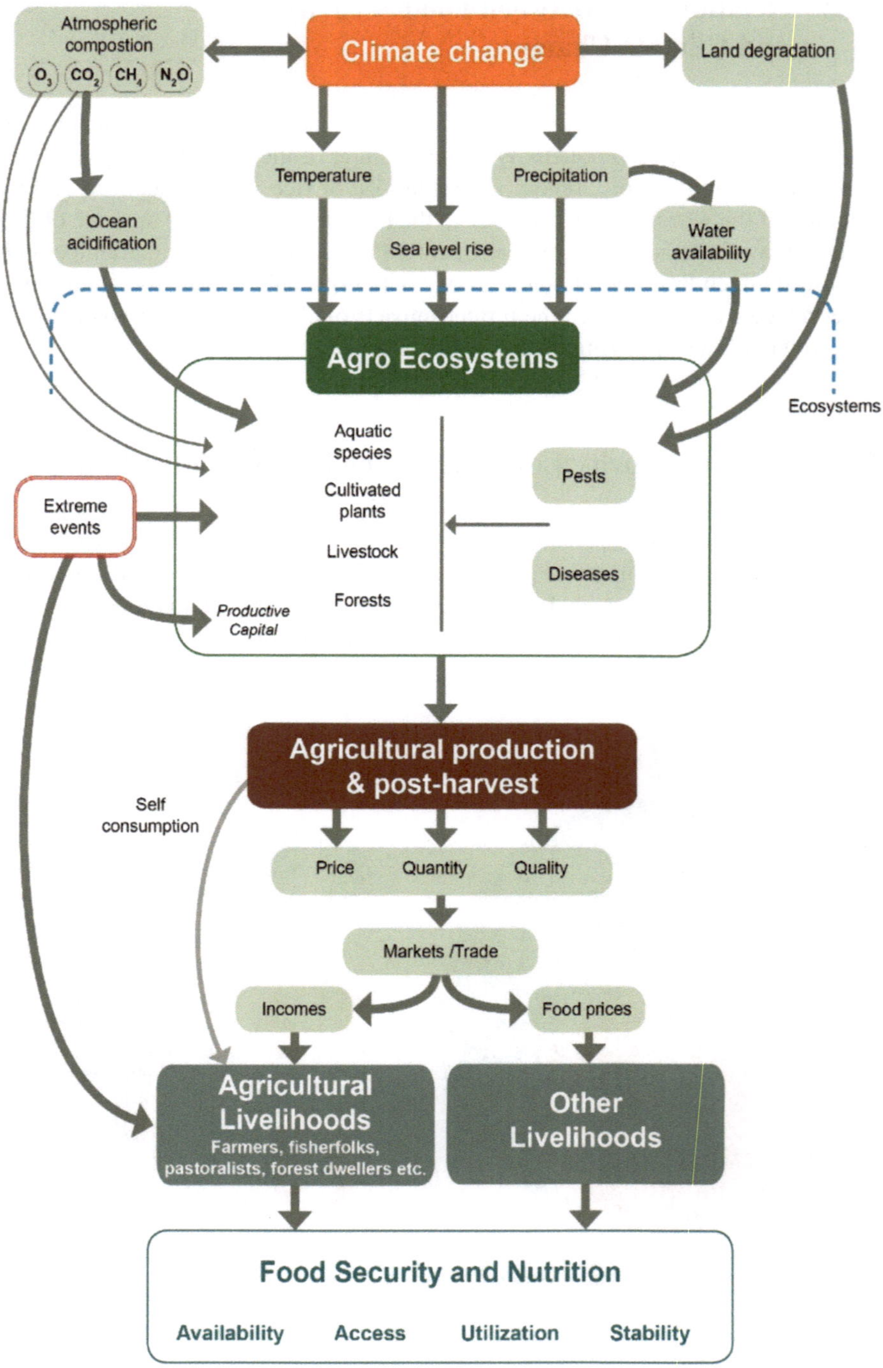

Fig. 22.2 Interlink of Climate Change and Food Security

22.5 Nations at Risk and Priorities to Manage the Risk

In 2050 worldwide food demand will be increased by 60% as compared to the current food consumption level. It is very difficult to achieve food security without climate change adaptations for reduction of poverty, hunger, and malnutrition because the rapid growth of the population increased the vulnerability in those regions of the world which are already facing undernourishment and also highly affected by climate change (Delaporte and Maurel 2018). Extreme weather events, rise in sea level, drought, and heat stress are considered as the impact of climate change. These variations in climate are very severe and have already declined agricultural production and increased food insecurity across the globe (Zahoor et al. 2019). Like food security, the livelihood of communities and households from developing tropical areas are affected by climate change. Nearly about 15 countries in the world are most vulnerable to climate change and also these developed some institutional frameworks to respond the climate change.

22.5.1 South and Central Asia

Bangladesh

Bangladesh is considered a populated and poor country in the world. Particularly Bangladesh is one of the most vulnerable countries to climate change due to maximum reliance on natural resources and minimum institutional framework as well as capacity (USDA 2019). The situation is very alarming because extreme weather events such as intense cyclones, variation or shift in rainfall, and droughts are increasing with time. Agriculture of Bangladesh is highly at risk increasing difficulties for local farmers. Tea, vegetables, rice, jute, and maize are the major crops of this country (Delaporte and Maurel 2018) (Fig. 22.3).

Priorities to Manage Risk:

- Reduced 40% carbon emissions by modern cooking stoves as compared to traditional stoves.
- To minimize the salinity problem and to save freshwater for agriculture purposes, ten dams have been built in Bangladesh.
- To reduce the food insecurity, the government have started fruit trees plantation on the roadsides which also have enhanced the household income.
- To reduce the risk of climate change in the flood, coastal, and drought areas government implements the safe drinking water plants.
- To ensure food security, improving the livestock, fisheries, and developing a cropping system which resilient to climate change (salinity, drought, and flood-tolerant variety of crops).

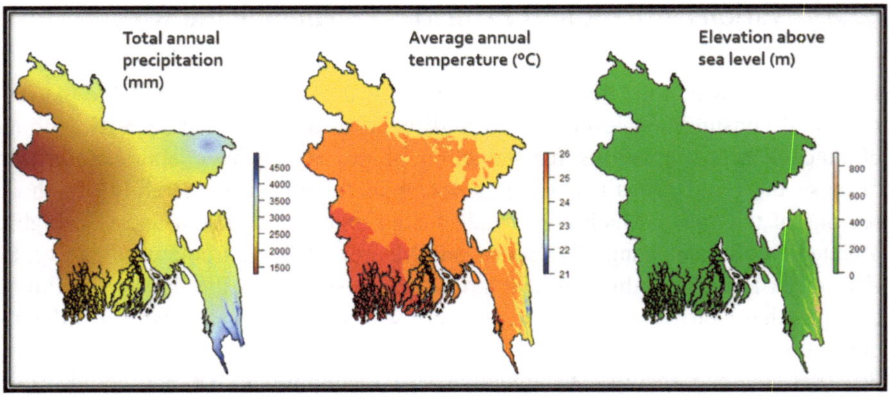

Fig. 22.3 Demographic information of Bangladesh

India

India is a big country in the world that covers an area of 3.28 Million Square kilometers. In the world, India is considered the 2nd most populous country. The economy of this country is diversified but 64% economy depends on agriculture. Forestry and agriculture are highly at risk due to climate change (Dev and Sharma 2010). Water availability is decreasing due to glacier melting, food security is at risk due to an increase in flash floods and decreases in rainfall, rural livelihood disturbed by variations in the natural ecosystem due to climate change, and coastal areas are highly affected by the rise in sea level (Fig. 22.4).

Priorities to Manage Risk:

- Government of India is introducing new plans for climate change adaptions to ensure food security and also for development.
- Government is spending more than 2% of GDP to support adaptations such as improvement of agriculture system, disaster, water, and coastal area management system.
- Some new projects have been launched in India by the Special Climate Change Fund (SCCF), (Asian Development Bank, Global Environment Facility (GEF), and Special Climate Change Fund (SCCF); the purpose of these projects is to enhance the adaptation level.

Nepal

Nepal is one of the least developing countries and highly vulnerable to climate change within small-distance micro-climates, very complex topography, and agriculture contribute to 31% of total GDP. More than 70% of people's livelihood depends on forestry and agriculture (Joshi et al. 2017). Approximately two-third of the agricultural area is rain-fed and farmers have an average landholding size of 0.8 ha (Shrestha et al. 2014). Nepal has become more vulnerable because of lack of infrastructure, lack of adaptation tools, severe weather, weak production systems, and lack of awareness (Aryal et al. 2014, 2016; Islam et al. 2016) (Fig. 22.5).

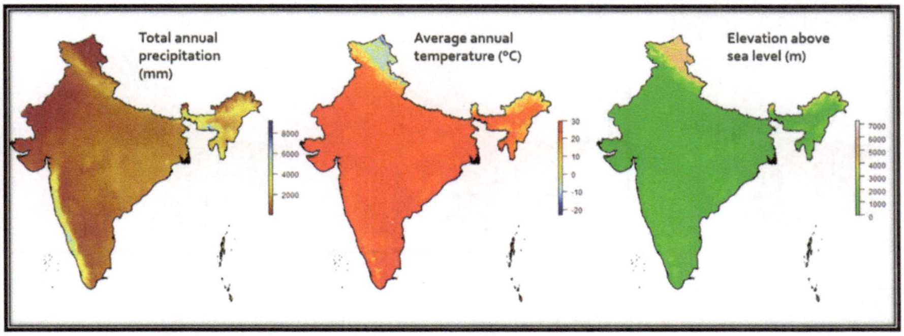

Fig. 22.4 Demographic information of India (https://gadm.org/maps/IND.html)

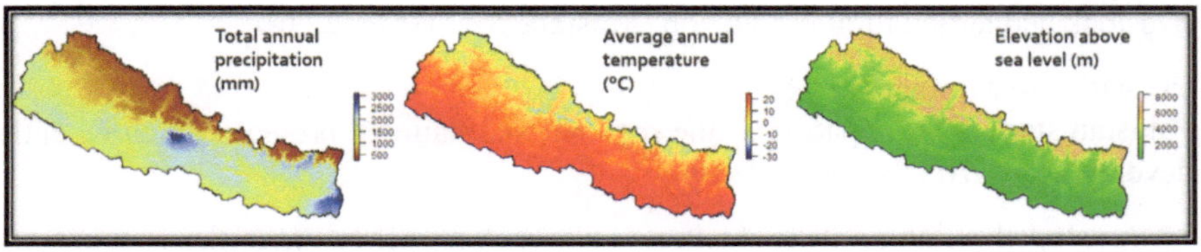

Fig. 22.5 Demographic information of Nepal (https://gadm.org/maps/NPL.html)

Priorities to Manage Risk:

- Nepal government has introduced NAPA (National Adaptation Program of Action) to address the people about climate change.
- Also, the government approved the LAPA (Local Adaptation Plans for Action) in 2011 to ensure food security and implement the village development committees and municipalities for districts that are most vulnerable to climate change.
- Recently Nepal Climate Change Support Program (NCCSP) helps to implement LAPA actions for the capacity building of peoples from climate-vulnerable communities.
- Purpose of these actions to build a climate smart agriculture and urban settlement system, and management of water resources under changing climate.
- To enhance the efficiency of the food production system now farmers are using technology and also some changes in practices related to agricultural management.

Pakistan

Pakistan is diverse in topography like coastal areas, temperate regions, subtropical, and tropical ecosystems. The economy of Pakistan depends on agriculture and most affected country by climate change. Additionally, farmer communities are highly vulnerable due to climate change. The maximum agriculture system of Pakistan is irrigated by water that comes from snowmelt from northern areas. Water availability is decreasing over time due to climate change and lack of water management system (Abid et al. 2015; Ashraf et al. 2019; Iqbal and Arif, 2010). Now the change in monsoon rainfall is a great concern of Pakistan (Abid et al. 2015). Climate change increases the risk for the nation of Pakistan such as in the form of natural ecosystem

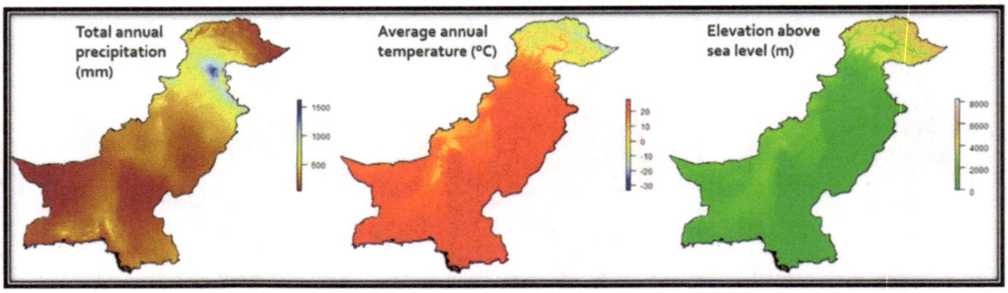

Fig. 22.6 Demographic information of Pakistan

degradation, more insect and pest attacks, food security at risk due to a decrease in livestock and agricultural production (Mustafa 2014) (Fig. 22.6).

Priorities to Manage Risk:
Pakistan started participation in the regional and national projects which recently developed, as given below:

- Developed an information-sharing system with farmer communities to cope with climate change.
- Building the climate smart agribusiness and irrigation system.
- Pakistan government developed a National Climate Change Policy in 2012 which focused on all of the problems Pakistan facing due to climate change.
- The great trend of hydropower generation and reforestation founded in the province Khyber Pakhtunkhwa and built solar energy plants in the Punjab province.
- Government ensuring food security by Implementation of National Climate Change Strategies.

Tajikistan
Tajikistan is situated in Central Asia and land lock country in the mountains (Barbone et al. 2010) which is the biggest country in cotton production and also rich in water resources (Barbone et al. 2010). Recently reported that country facing the Extreme Hydro-meteorological Hazards (Xenarios et al. 2019). The current climate of Tajikistan is semiarid, continental, and subtropical extent with few desert areas. However, the climate varies with respect to elevation drastically. The instant increase in temperature will affect the glaciers' bodies already the country has lost 2.5% (20 Billion Cubic Meter volume) of the glacier due to temperature variations throughout the twentieth century. Forests have great importance for Tajikistan because they provide raw material and also regulate the climate. The country has harsh climate conditions such as cold winter with a continental climate, expansion of desertification and erosion, scarcity of resources for livelihood due to mountain areas (Fig. 22.7).

Priorities to Manage Risk:

- Tajikistan has developed strategies and legislative documents for disaster management and risk reduction.

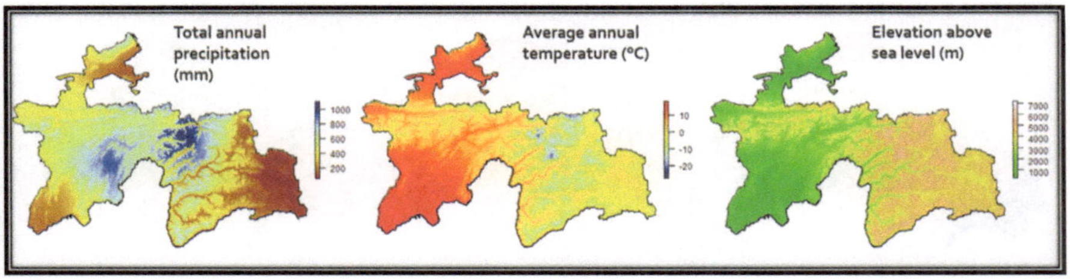

Fig. 22.7 Demographic information of Tajikistan (https://gadm.org/maps/TJK.html)

- The government established a mechanism for coordination adaptations to mitigate climate change.
- Environmental Protection State Committee plays a major role in implementing and facilitating the project related to the adaption of climate change.
- The country has set a goal to reduce greenhouse gas emissions to 65–75% until 2030 as compared to 1990.
- In Tajikistan Green Climate Fund helps to ensure the food production system, resilient agriculture, protect the forest areas, and sustainable management of water resources.

22.5.2 African countries

Burkina Faso

Burkina Faso's characterized as the long dry season, short rainy season, and dry tropical climate. The country's climate instance varies annually and seasonally because the area lies within the hinterland desert. The probability of extreme weather events (intense rainfall, floods, and droughts) is increased due to climate change in this region. The food production system of many areas in Burkina Faso already affected by climate variability in the form of a decline in crop production. The evapotranspiration rate will increase due to the increase in temperature which leads to dry soil conditions and not good for agriculture. In Burkina Faso, the sectors of forestry, agriculture, water, and pastoralism are highly at risk due to climate change (Fig. 22.8).

Priorities to Manage Risk:

- In 2007 the government of Burkina Faso introduced the NAPA. NAPA aims to access the current and future vulnerability level due to climate change
- NAPA has played a significant role in adaptation strategies to change in cropping practices, integrated water resource management, and the development of new institutes to cope with climate change.
- The government starts to give subsidies, help local farmers to dig wells, and build the water reservoir at a small scale for efficient water use in the agriculture system.
- After setting the national development strategies Burkina Faso reduced the poverty level and increased economic growth.

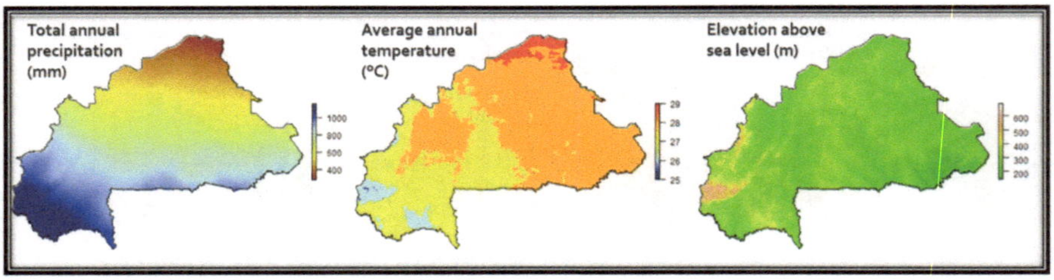

Fig. 22.8 Demographic information of Burkina Faso

- Maximum peoples of Burkina Faso have adapted to climate change by using the products of the forest as a supplement in agriculture and livelihood.

Ghana

In Ghana, climate change is expected to affect food security, crop production, water resources, and energy supplies. The country is most vulnerable to climate change due to a lack of adaptation strategies and also not have sufficient about the economic costs of climate variability. The peoples of Ghana already facing flooding in coastal areas which is a major reason behind the soil erosion, decrease in the agricultural production system due to problems induced by climate change. Ghana has five major rivers that flow into the sea, the north area of Ghana continuously facing floods and droughts (Fig. 22.9).

Priorities to Manage Risk:

- Ghana of Ghana developing the framework for national policy and National Determined Contributions (NDC) give full support and guidance to implement the strategies to respond to the climate change.
- In 2016, Ghana submitted the first NDC to UNFCCC. The NDC contains information about the Master Plan for National Climate Change (2015–2020). Ghana's priority sectors are:

  (a) Sustainable land use, including food security;
  (b) Climate-proof infrastructure and equitable social development;
  (c) Sustainable mass transportation and sustainable energy security;
  (d) Sustainable forest management and alternative urban waste management.

Mali

Mali is the West African country straddling the Sahel of the subtropical band. The country is facing variations in rainfall and frequent droughts. The south part of Mali has a tropical climate and wetter conditions as compared to the north part of Mali which has dry conditions like the Sahara Desert. Drought and rainfall variability are considered the main extreme weather event in Mali. More than 80% of the people of Mali livelihood depend on agriculture which is rain-fed and highly vulnerable to climate change (Fig. 22.10).

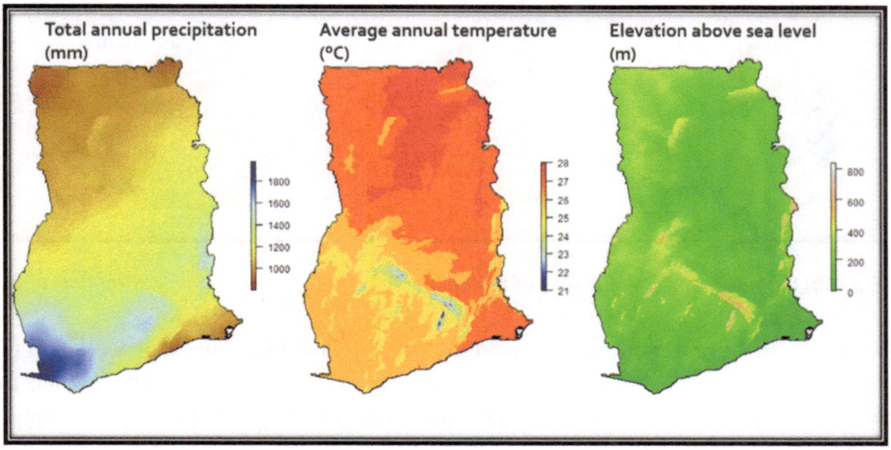

Fig. 22.9 Demographic information of Ghana

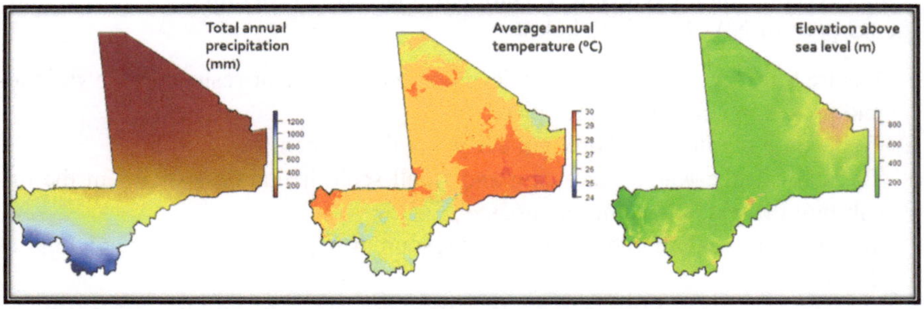

Fig. 22.10 Demographic information of Mali (https://gadm.org/maps/MLI.html)

Priorities to Manage Risk:

- There are two main bodies for implementation of the protective measures in Mali to overcome the climate risk; (a) Agency for Environment and Sustainable Development (AEDD) and (b) National Climate Change Committee (CNCC)
- With the support of USAID MCCAA (Mali Climate Change Adaptation Activity) building resilience in vulnerable farmer communities and providing sufficient information about climate change for timely decisions.
- MCCAA also helps at the local household level to adapt better solutions to cope with climate change.
- The main focus of all these projects to attain sustainability in agriculture, develop and adopt drought-resistant seeds, ensure food security, and use of advanced technologies for the capacity building of farmers.

Ethiopia

Ethiopia is situated in Africa and covered an area of 1.2 Million Square Kilometers. In the Horn of Africa, Ethiopia is considered a land lock country. Flash floods and droughts imposed heavy costs in recent decades. In the future, climate change may

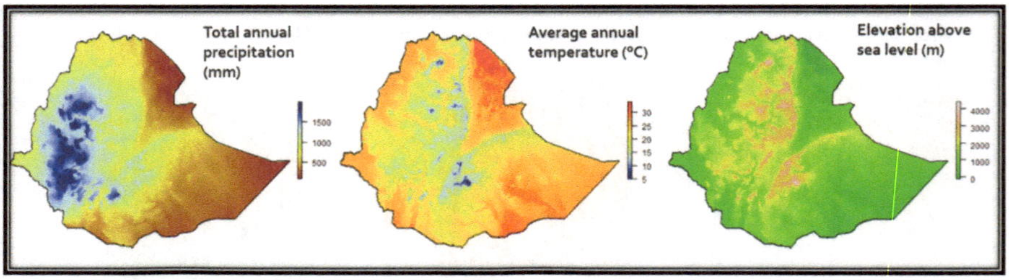

Fig. 22.11 Demographic information of Ethiopia

affect the management of water resources, increase food insecurity, as well as economic growth which limits development. The government has started the projects to access the vulnerability level and appropriate adaptations to cope the climate change (Elala 2011) (Fig. 22.11).

Priorities to Manage Risk:
The government approved NAPA (National Adaptation Program of Action) in 2007 the purpose of NAPA is given below:

- Ensure food security through the development of drought-resistant crop varieties and promoting crop insurance programs.
- To develop the flood and drought early warning system.
- To establish the water harvesting and small-scale irrigation systems in the dry sub-humid, arid, and semiarid areas.
- To ensure the sustainable utilization of resources in pastoral and wetland areas of the country.
- To establish the programs for adaptation and capacity building of people in respect of climate change.
- Management of water resources at Genale Sawa Basin to enhance the water availability for the food production system
- To launch projects in the Rift Valley for carbon sequestration among all of the communities.

Tanzania
Tanzania is situated in south of the equator on the East African coast. Like many other countries in this region, about 80% of population of Tanzania depends on agriculture for their employment, livelihoods, and income. Harsh weather events, including droughts, storms, and floods have previously imposed heavy costs in Tanzania. Recurrent droughts in most part of the country are being experienced; these have associated consequences on food production and water scarcity. The latest severe droughts have hit most parts of the country leading to severe food insecurity, food shortages, hunger,water scarcity, and acute shortage of power; this signifies the vulnerability of the country to the impacts of climate change (Fig. 22.12).

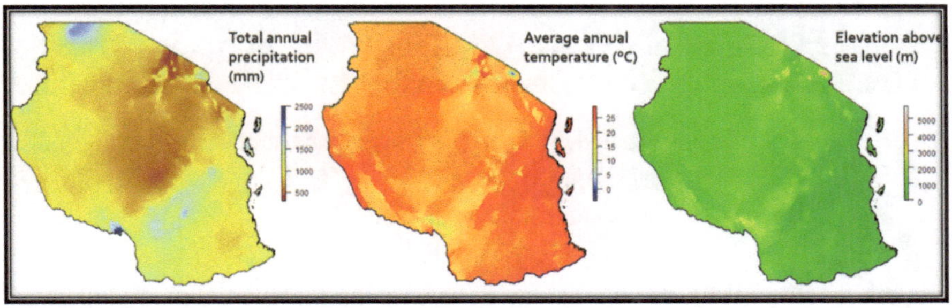

Fig. 22.12 Demographic information of Tanzania (https://gadm.org/maps/TZA.html)

Priorities to Manage Risk:
Tanzania is obliged to the UNFCCC for developing a National Adaptation Program of Action (NAPA). Tanzania fulfilled this obligation in 2007.

- Climate change-related vulnerabilities in many sectors which are important to the Tanzania economy.
- It takes measures to tackle climate change, predominantly the impacts of climate change and associated adaptation measures
- The main objective of NAPA is to recognize the most crucial measures. Majority of the projects in Tanzania are concerned about agriculture and water resource management (including irrigation, water-saving, and rainwater collection).
- Overall, 72 projects have been evaluated, 14 of which were selected to start the implementation phase. Two of these selected activities of the project are:

    (a) Irrigation efficiency for crop production to increase production as well as conserve water in all areas.
    (b) Alternative farming systems and water harvesting.

22.6 Steps to be Taken by Local Government to Ensure Food Security

The local government is actively concerned with the social, economic, and environmental needs of their communities. In addition, local government is increasingly aiming to build strong, self-reliant, and resilient communities. As illustrated above, food security can have a significant impact on the health and well-being of individuals and the community as a whole. This means that improving food security is a significant local issue for local government. Nonetheless, looking at the major contributors to food security in Fig. 22.13, it is clear that local government is active in many of these areas. Within this wide scope, local government can influence key supply and access factors to improve food security. Local government needs to take the following steps on the priority basis:

22.7 Role of Research and Technologies to Ensure Food Security

The planet is becoming food-insecure due to climate change, population growth, unpredictable food prices, ineffective supply chains, and unequal food access. In the current global scenario, it is the prime objective to apply new technology to agricultural development in order to increase food production for at least equal, if not surpass, the population growth. Enhanced agricultural production in the Third World can best be achieved through science and technology. This process comprises the growing involvement of farmers in the country's broader economic life. Advanced

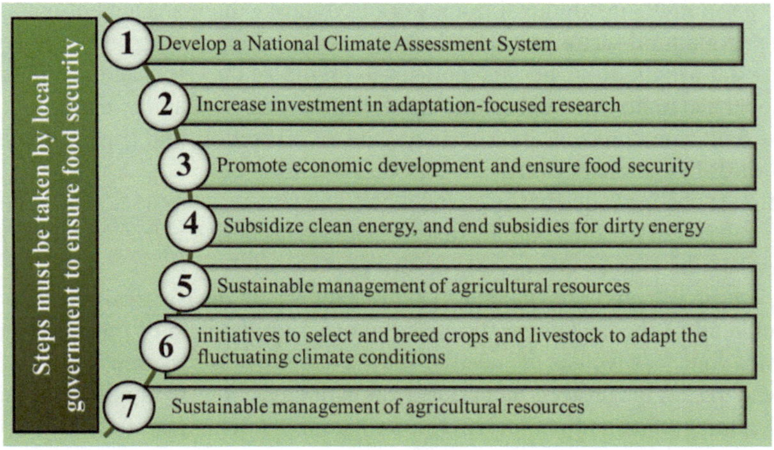

Fig. 22.13 Steps taken by local government to ensure food security

seed varieties, improved fertilizer efficiency, enhanced mechanization, more credit, and better plant protection are the main elements for the technological change in agriculture.

22.8 Recommendations and Future Needs for Meeting the Challenges

Food security is based on the agricultural production system. Climate change is closely linked with the agriculture and food sectors. Food security will be a great challenge in the future for developing countries as discussed above. The expected extreme weather events like hurricanes, floods, an increase in storm frequency, extreme drought, rise in sea level, variations in rainfall patterns, and heat waves all of these changes will decline the capacity of the agricultural production system which is the main pillar of food security. We already observed that climate change affected crop production at the global level. In 2050, the demand for agricultural products at the global level will be increased by 50% according to FAO observations. Approximately the world population will be 9.7 Billion in 2050, food demand will be high in Asia and Africa due to loss in soil fertility, high population growth, and lack of funds to adopt climate change. The demand for fruits, meat, vegetables, and milk lead to an increase in expected incomes especially in middle-income and low-income countries. So, we need

- To develop a better understanding of the challenges which limit agricultural and food production.
- To manage the resources and agricultural inputs according to the size of the field.

- To develop the institutional framework to improve the management practices in the agricultural sector.
- To establish the strategies and policies to minimize the crises and conflicts at the global and regional level because conflict affects the availability of food, disrupts the food care, threatens the social systems of protection, and leads communities towards hunger and poverty.
- Climate change is one of the global issues and global issues do not care about the boundaries of national borders that are why we need to improve the collaboration to adapt the climate change and ensure food security.
- Scientific research projects play a significant role to devise sensible solutions to tackle all of the challenges discussed above with the help of policymakers and practitioners as well as the need to learn from previous experiences.
- Need to start the projects and prioritization of the actions to improve the capacity of communities and countries to cope the climate change, after this all of the solutions will be context-specific. The direct engagement of the stakeholders will be a great achievement for better solutions. The third is to ensure participation of the focused vulnerable people to climate change.

22.9 Conclusion

Food security is a phenomenon having multiple dimensions. Food security, food production, and climate change are all interwoven. Changes in one element would have negative influences on other elements. Climate change changes agricultural production and food systems, and hence it alters the approach for transforming agricultural systems to support global food security and poverty alleviation. Climate change is a cause of greater uncertainty and risk for farmers and policymakers. An evidence-based, integrated and transformative approach for addressing food and climate security at all levels involve coordinated activities from the global to local stages, from research to investments and policies , and across private, public, and civil society sectors for achieving the rate and scale of change required. With the right policies, practices, and investments, the agriculture sector can transform to climate smart agriculture pathways, for reduced food insecurity and poverty in the short term. Food security will be a serious problem under the changing climate so we need to plan advance research and implement the actions to deal with the world's emerging issue of food security. Need collaboration at the global level because in the next decades the food system improvement will demand transformative actions. Such collaborations will make developing countries self-sufficient and less reliant on assistance from developed countries. Especially focus on the outcomes from the food production system first need to improve the research culture and involve the stakeholders. The final is in the food security context needs to address the mitigation plans and adaptation strategies at the global, national, and farm levels. The developing nations must channelize their proportion efforts and resources towards the protection of the environment as well as to cope with climate change.

References

Abid, M., Scheffran, J., Schneider, U., Ashfaq, M.J.E.S.D., et al. (2015). Farmers' perceptions of and adaptation strategies to climate change and their determinants: the case of Punjab province, Pakistan. 6, 225-243.

Ali M, Mubeen M, Hussain N, Wajid A. et al. (2019) Role of ICT in Crop Management. In Agronomic Crops (pp. 637-652). Springer, Singapore. https://doi.org/10.1007/978-981-32-9783-8_28.

Aryal, S., Cockfield, G. and Maraseni, T.N.J.C.C. (2014). Vulnerability of Himalayan transhumant communities to climate change. 125, 193-208.

Aryal, S., Cockfield, G., Maraseni, T.N.J.C. and Development (2016). Perceived changes in climatic variables and impacts on the transhumance system in the Himalayas. 8, 435-446.

Ashraf, M.Q., Khan, S.A., Khan, R., Iqbal, M.W.J.I.J.O.E. and Geology, E. (2019). Determinants of Adaptation Strategies to Climate Change by Farmers in District Sargodha, Pakistan. 16-20.

Bandara, J.S., Cai, Y.J.E.A. and Policy (2014). The impact of climate change on food crop productivity, food prices and food security in South Asia. 44, 451-465.

Barbone, L., Reva, A. and Zaidi, S. (2010). In: Tajikistan: key priorities for climate change adaptation, (The World Bank).

Barrett, C.B.J.S. (2010). Measuring food insecurity. 327, 825-828.

Delaporte, I. and Maurel, M.J.C.P. (2018). Adaptation to climate change in Bangladesh. 18, 49-62.

Dev, S.M. and Sharma, A.N. (2010). Food security in India: Performance, challenges and policies.

Elala, D. (2011). Vulnerability assessment of surface water supply systems due to climate change and other impacts in Addis Ababa, Ethiopia.

Fan, S. and Brzeska, J.J.G.F.S. (2016). Sustainable food security and nutrition: Demystifying conventional beliefs. 11, 11-16.

FAO. (2019). Climate Change and Food Security: Risks and Responses. "www.fao.org/publications ISBN 978-92-5-108998-9"

Farsund, A.A., Daugbjerg, C. and Langhelle, O.J.F.S. (2015). Food security and trade: reconciling discourses in the Food and Agriculture Organization and the World Trade Organization. 7, 383-391.

Giampiero Grossi, Pietro Goglio, Andrea Vitali, Adrian G. Williams, Livestock and climate change: impact of livestock on climate and mitigation strategies, Animal Frontiers, Volume 9, Issue 1, January 2019, Pages 69–76, https://doi.org/10.1093/af/vfy034

Godfray, H.C.J., Beddington, J.R., Crute, I.R., Haddad, L., Lawrence, D., Muir, J.F., Pretty, J., Robinson, S., Thomas, S.M. and Toulmin, C.J.S. (2010). Food security: the challenge of feeding 9 billion people. 327, 812-818.

Havas, K., Salman, M.J.I.J.O.F.S., Nutrition and Health, P. (2011). Food security: its components and challenges. 4, 4-11.

HENEGEDARA, G.J.C.I.J.O.R.I.C. and Management (2015). AGRICULTURAL INNOVATIONS AND FOOD SECURITY IN SRI LANKA. 6.

Hussain S, Ahmad A, Wajid A, Khaliq T, Hussain N. et al. (2020a) Irrigation Scheduling for Cotton Cultivation. In Cotton Production and Uses (pp. 59-80). Springer, Singapore.https://doi.org/10.1007/978-981-15-1472-2_5

Hussain S, Mubeen M, Ahmad A, Akram W. et al. (2019) Using GIS tools to detect the land use/land cover changes during forty years in Lodhran District of Pakistan. Environ Sci Pollut Res 1: 1-17.https://doi.org/10.1007/s11356-019-06072-3

Hussain S. et al. (2020b) Study of land use/land cover changes using RS and GIS: A case study of Multan district, Pakistan. Environl Monit Assess 192 (1) 2. https://doi.org/10.1007/s10661-019-7959-1

Ingram, J.J.F.S. (2011). A food systems approach to researching food security and its interactions with global environmental change. 3, 417-431.

Iqbal, M.M. and Arif, M.J.A.J.O.S.F.D. (2010). Climate-change aspersions on food security of Pakistan. 15.

Islam, M., Kotani, K., Managi, S.J.E.A. and Policy (2016). Climate perception and flood mitigation cooperation: A Bangladesh case study. 49, 117-133.

Joshi, G.R., Joshi, B.J.F.O.F.J.O.F., Agriculture and Society (2017). Household food security: Trends and determinants in mountainous districts of Nepal. 5, 42-55.

Müller, C., Cramer, W., Hare, W. L., & Lotze-Campen, H. (2011a). Climate change risks for African agriculture. Proceedings of the National Academy of Sciences, 108(11), 4313-4315.

Müller, C., Cramer, W., Hare, W.L. and Lotze-Campen, H.J.P.O.T.N.A.O.S. (2011b). Climate change risks for African agriculture. 108, 4313-4315.

Mustafa, Z. (2014). Climate change and its impact with special focus in Pakistan. Agricultural Sciences. Paper No. 243.

Nardone A., Ronchi B., Lacetera N., Ranieri M.S., Bernabucci U.Effects of climate changes on animal production and sustainability of livestock systems Livestock Sci., 130 (1–3) (2010), pp. 57-69

Ramaswamy, S.J.J.O.I.A.F. and Feed (2015). Setting the table for a hotter, flatter, more crowded earth: insects on the menu? 1, 171-178.

Sabagh AE, Hossain A, Islam M S, Iqbal M A, Fahad S, Ratnasekera D, Llanes A (2020) Consequences and Mitigation Strategies of Heat Stress for Sustainability of Soybean (Glycine max L. Merr.) Production under the Changing Climate. In Plant Stress Physiology. IntechOpen. https://doi.org/10.5772/intechopen.92098

Senker, P. (2011). Foresight: the future of food and farming, final project report. (Taylor & Francis).

Shrestha, K., Shrestha, G., Pandey, P.R.J.J.O.A. and Environment (2014). Economic analysis of commercial organic and conventional vegetable farming in Kathmandu valley. 15, 58-71.

Stocker, T. (2014). In: Climate change 2013: the physical science basis: Working Group I contribution to the Fifth assessment report of the Intergovernmental Panel on Climate Change, (Cambridge University Press).

USDA (2019). Global Food Security " https://nifa.usda.gov/topic/global-food-security".

Xenarios, S., Gafurov, A., Schmidt-Vogt, D., Sehring, J., Manandhar, S., Hergarten, C., Shigaeva, J. and Foggin, M.J.R.E.C. (2019). Climate change and adaptation of mountain societies in Central Asia: uncertainties, knowledge gaps, and data constraints. 19, 1339-1352.

Zahoor SA, Ahmad S, Ahmad A, Wajid A. et al. (2019) Improving Water Use Efficiency in Agronomic Crop Production. In Agronomic Crops (pp. 13-29). Springer, Singapore. https://doi.org/10.1007/978-981-32-9783-8_2

Chapter 23
Importance of Carbon Sequestration in the Context of Climate Change

Khurram Shahzad, Henry Sintim, Fiaz Ahmad, Muhammad Abid, and Wajid Nasim

Abstract Terrestrial carbon (C), the main reservoir of soil C, is continuously depleting due to easily decomposable C sources and improper agricultural practices. The continuous depletion of soil C has implications on climate change effects, which is a major global issue. In this chapter, we discussed the importance of C sequestration in the context of climate change, and proposed management practices to curtail the depletion of terrestrial C. Soil C sequestration is routinely defined as the persistent increase in C contents in the soil, and this can be achieved through two major pathways: (a) practices that reduce soil C depletion and (b) the addition of C sources to the soil. Therefore, stable C sources like biochar and improved agricultural practices, such as conservation tillage, could help to reduce CO_2 emission into the atmosphere. In addition to biochar and conservation tillage, increased biomass production, agroforestry, deep-rooted crops, improved pastures, residue management, optimal fertilizer application, and crop rotation could help maintain a high C balance in soils and induce negative efflux of CO_2. The accurate measurement of soil C change over time is important to determine the effectiveness of different management practices on soil C sequestration. Despite the countless benefits of C sequestration, there are some limitations, especially the economic feasibility of certain management practices, water and nutrient requirements, as well as climate and site-specific conditions. Moreover, the effectiveness of different C sequestration

---

K. Shahzad
Lasbela University of Agriculture, Water and Marine Sciences, Uthal, Balochistan, Pakistan

H. Sintim (✉)
Department of Crop and Soil Sciences, University of Georgia, Tifton, GA, USA
e-mail: hsintim@uga.edu

F. Ahmad
Central Cotton Research Institute, Multan, Punjab, Pakistan

M. Abid
Department of Soil Science, Bahauddin Zakariya University, Multan, Punjab, Pakistan

W. Nasim
Department of Agronomy, The Islamia University of Bahawalpur (IUB), Bahawalpur, Punjab, Pakistan

W. N. Jatoi et al. (eds.), *Building Climate Resilience in Agriculture*,
https://doi.org/10.1007/978-3-030-79408-8_23

practices varies with soil type and ecosystems. Carbon sequestration holds promise to mitigate global warming and prevent the adverse impacts of climate change, but the implementation of management practices must be tailored to different ecosystems and it must not be cost-prohibitive.

Keywords Climate impact on sequestration · Soil organic carbon (SOC) · Biochar · Conservation tillage · Biomass production

23.1 Introduction

It is of paramount importance to reduce the emission of greenhouse gasses to meet global climate challenges. Carbon dioxide (CO_2) level has risen to an alarming level of 400 ppm from 280 ppm in 1970, which is primarily due to increases in anthropogenic activities on earth (Hashimoto 2019; Hussain et al. 2020). Atmospheric CO_2 concentration is approximately increasing at 2.2 ppm/year, which has raised the temperature of this planet by 0.8 °C since 1980, and it is expected to increase to 3–7 °C by 2100. The increase in temperature has caused many drastic changes in the earth's environment, such as the reduction in snow cover in terrestrial and arctic regions, lower crop yields, increased frequency and severity of hurricanes in different parts of the world, reduced biodiversity, and frequent occurrence of drought events (Amin et al. 2018; Zhang and Li 2019). These undesirable changes as the result of climate change are great threats to food security. However, there is a need to ensure food security, especially for the ever-increasing population of this planet, which may reach 9.5 billion before 2050 (Lal 2015).

Soil C sequestration has implications on mitigating climate change impact and enhancing food security (Lal 2016). For instance, cultural practices to increase soil C (Fig. 23.1) will improve soil quality and mitigate climate change impact, and these conditions are critical to promoting food security (Fig. 23.2). The objective of this chapter is to discuss the importance of C sequestration in mitigating climate change impact. We also highlight the need for continuous measurement of C change over time and the adaptation of tailored strategies to increase C sequestration and achieve a net negative efflux of CO_2.

23.2 Soil C Pools

Terrestrial C is the main source of C, which is approximately 2500 billion tons to a depth of 2 m in soil (Lal 2011) (Fig. 23.3). However, this huge reservoir of C pool is continuously depleting by anthropogenic activities and soil degradation processes like erosion, acidification, nutrient depletion, salinization, deforestation, residue removal, and biomass burning (Lal 2015). Most soils have also lost their natural

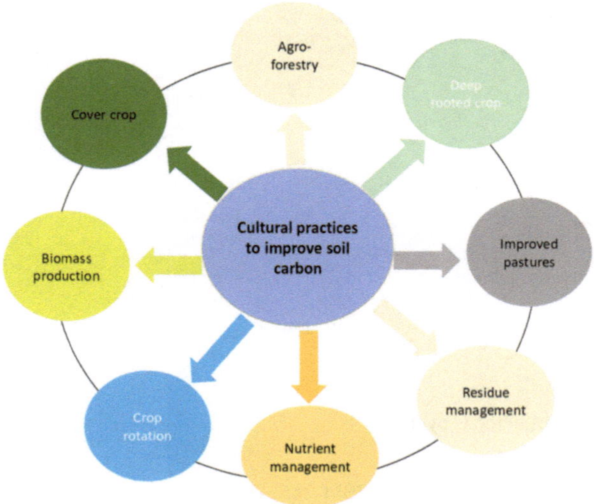

Fig. 23.1 Cultural practices that could be adopted to improve soil carbon

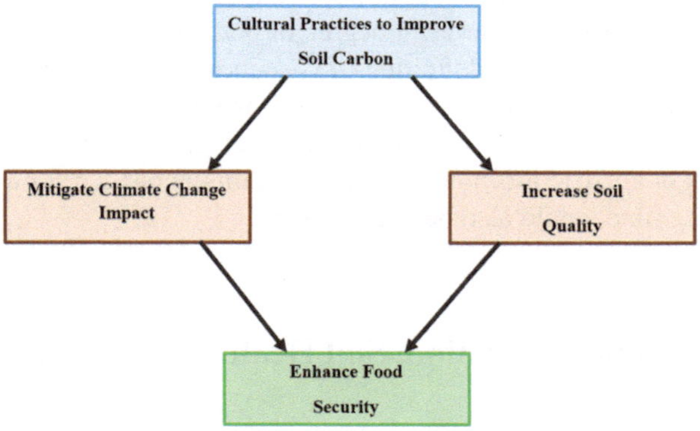

Fig. 23.2 Schematic indicating cultural practices to improve soil carbon will increase soil quality, mitigate climate change impact, and ultimately enhance food security

balance of C due to the transformation of natural ecosystems into intensive agricultural ecosystems. These anthropogenic activities have depleted the soil organic C pool by up to 78 ± 12 billion tons C due to intensive agriculture production and soil degradation processes (Lal 2011). Agricultural land use contributes approximately 20% of global CO_2 emission. Average soil organic carbon (SOC) in the top 30 cm soil layer amounts to 15 Mg ha^{-1}, but 20–30% SOC was lost after the first 20 years of cultivation (Stockmann et al. 2013). However, the depletion of soil C depends on many factors, which include the composition of organic matter, cropping history, soil type, the intensity of cultivation, moisture content, and socioeconomic conditions of a farming community.

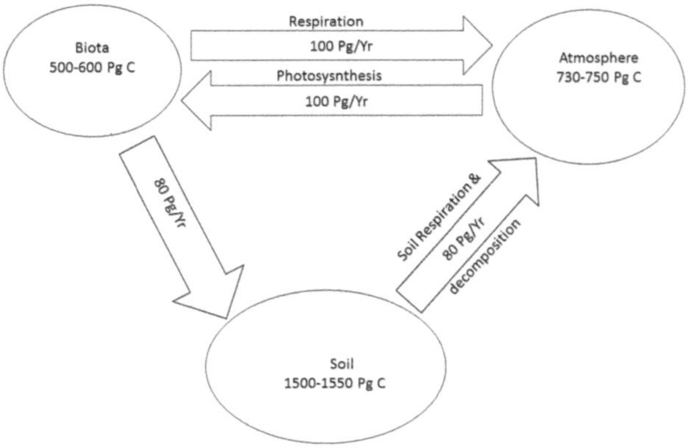

Fig. 23.3 Role of soil in the C cycle

23.3 Definition of C Sequestration

Carbon sequestration can be defined as the persistent increase in C contents in soil, plant, or sea. Some schools of thought only consider recalcitrant C as sequestered C, but soil C residence time varies in degree of performance which may extend up to 1000 years. Thus, a proper definition of C sequestration does not exist, and prevailing definitions are arbitrary, which depend upon the residence time. Management practices and climate change are important variables that determine C residence time (Schmidt et al. 2015). If both variables remain unchanged for a longer period, the C residence time would also be extended.

23.4 Carbon Sequestration, Soil Health, and Food Security

Soil organic carbon is continuously being depleted due to intensive agricultural production. The recommended SOC concentration for cropland soils of the tropics is 1.1%, and for temperate regions, SOC concentration should be ~2% (Lal 2011). Nevertheless, the SOC levels in some regions declined to <0.5%, and it even dropped down to 0.1% in some cases (Lal 2011). Owing to the low levels of SOC in these soils, the health of these soils has also declined. The main reasons for the decrease in SOC are (1) poor aggregation of soil particles, (2) low soil porosity or high bulk density, (3) low water holding capacity, (4) limited soil biodiversity, (5) low cation exchange capacity, (6) high losses of applied fertilizers, and (7) low soil buffering capacity (Lal 2015).

For sustainable crop production, improvement in the SOC of farmland is very important. Increasing the SOC status of degraded soils will improve crop productivity. It was found that an increase of SOC pool at 1 ton C/ha/year could increase grain production by 32 ± 11 million tons per year in developing countries (Lal 2011). Therefore, an increase in the SOC pool at an equivalent rate could feed large populations in areas that are prone to food security threats.

23.5 Measurement of Soil C Change Over Time

It is important to measure the actual change in soil C over time to determine the effective agricultural practices in the context of soil C sequestration. The methods of soil C change measurement are direct soil C change measurement, indirect measurement (Flux and spectral method), repeated soil surveys at national and international levels, long-term experiments on soil C changes, and the use of models. In the direct measurement of SOC stock changes, soil samples are collected from the fields, processed, and SOC contents are measured by the dry combustion method. To overcome the soil spatial variability, special study design, and sampling protocol are required (Minasny and McBratney 2016). However, on large-scale measurement, this method can be expensive in terms of the total cost of sample collection, storage, processing, and C measurement. The indirect methods for SOC change measurements are flux and spectral methods. The C flux is measured by the difference in C uptake through photosynthesis and C loss through respiration (soil, plant, and litter) by using a chamber method (Baldocchi et al. 2018). The spectral method relies on the reflectance of light on the soil in the infrared region. The organic bonds and minerals in the soil absorb light at a specific wavelength, enabling the measurement of soil C (Nayak et al. 2019). Repeated soil sampling schemes are getting popular in different countries of the world like New Zealand, England, and Wales (Schipper et al. 2014). These countries resample and measure soil C after a specific period. The main advantage of this resampling scheme is that it measures actual SOC contents over a large spatial scale and a long period. Long-term experiments on soil C change are also getting importance because soil C changes take time to occur. The long-term experiments on SOC exist in various parts of the world, some even from the nineteenthcentury. In 1990, there were efforts to bring the isolated long-term experiment in two networks: soil organic matter network (SOMNET) and EuroSOMNET. The different C models are also used to estimate the change in soil C like the full carbon accounting models (FullCAM) in Australia, the CENTURY model in Canada, Yasso07 in Finland (Palosuo et al. 2016), Soil carbon model ICBM-region in Sweden, CARBINE soil carbon accounting model in the United Kingdom (Matthews et al. 2014), and DAYCENT biogeochemical model in the USA. The calibrated model, supported by measurement, can be helpful to establish the relationship between soil C and management change in soil, climate, and land use including the estimation of uncertainty.

23.6 Impact of Climate on C Sequestration

Soil organic carbon stock generally increases with a decrease in temperature worldwide. Cold and humid climates are characterized by high soil C contents as compared to semiarid and arid climates. The climate and type of organic input to the soil are important factors that influence the efflux of CO_2 (Qi et al. 2016). High temperatures cause additional efflux of CO_2 from the soil. The global rise in temperature is causing a reduction in the SOC pool by increasing the decomposition rate and ultimately reducing the C residence time in the soil (Hammad et al. 2020). Therefore, it is considered that soils would be a major source of CO_2 emission in the future.

Soil organic matter is derived from plant and animal residues after decomposition (Akram et al. 2018). A continuous decomposition of these plant materials results in more stable forms called humus. Generally, the humus can stay in the soil for up to about 27 years. Now it is widely admitted that the decomposition of soil organic matter is ecosystem property instead of composition alone (Schmidt et al. 2015). In a given ecosystem, the decomposition of organic materials depends upon environmental and biological conditions. The decomposition process also affects the biodiversity in the soil. An increased decomposition process in soil favors the buildup of microbial communities. Variables such as moisture, temperature, material decomposability, and microbial community composition impact the residence time of added C source in the soil. Therefore, these factors must be taken into consideration during soil C management programs.

23.7 Soil C Pool Management

Increasing soil C pool is important to maintain the quality of water, soil, and vital natural resources. Efforts to increase soil C pool could include afforestation, conservation tillage, stable C sources, and improved agricultural practices. The implementation of such practices will enhance the absorption of CO_2 from the atmosphere and the long-term retention of C in the soil and thus will offset anthropogenic C emissions (Conant et al. 2017). Increased C sequestration will result in increased biodiversity and food security. The improved management practices could include shifting from conventional tillage to conservation agriculture, use of organic and inorganic mulches to lower nutrient losses by leaching and volatilization, cover crops, balanced use of macro and micronutrients, application of compost, plant growth-promoting rhizobacteria, biofertilizers, and adaption of diversified cropping system (Hammad and Elbagory 2019; Paustian et al. 2019). However, the adoption of different practices may vary, partly or wholly, across different regions, soil types, and environmental conditions. Losses of terrestrial C, as CO_2 emissions, vary among humid and dry regions. The addition of biochar and the adoption of conservation tillage and modified cultural practices are viable options to overcome the problem of continuously decreasing the SOC pool.

23.8 Biochar Role in C Sequestration

The capture of CO_2 from the atmosphere ought to be long-term with minimum chances of its leakage back to the atmosphere. One possible technology to achieve these goals is the use of biochar for C sequestration. Trees, crop residues, grass clips, leaves, and poultry waste are converted through pyrolysis into biochar under a limited supply of oxygen. Biochar contains twofold more C contents as compared to original materials. Biochar prepared under low temperatures can be used as an effective tool to increase the soil C pool. Biochar has the potential to increase the soil C sequestration at the approximate level of 1 billion tons C/year or even more. The use of long-term application of biochar could be more productive to increase C sequestration, in addition to crop residue retention in the soil. Carbon in the form of biochar is locked for a long duration with a low decomposition rate. Wang et al. (2016) estimated the retention time of biochar to be between 108 and 556 years; a clear indication of biochar's long-term residence in the soil. However, the actual residence time of biochar is under debate. It is difficult to estimate any sudden change in the loss of biochar after incorporation into the soil. Nonetheless, in comparison to fresh plant residues, biochar stays in the soil much longer. Therefore, a certain amount of C can be taken out from the C cycle by converting biomass into biochar, which would otherwise be returned to the atmosphere after plant biomass decomposition. Through this intervention we can possibly have two C cycles: a) plant biomass to atmosphere C cycle and b) biochar to atmosphere C cycle (Hammond et al. 2011). Therefore, by increasing the biochar-environment cycle, more C in the soil may be retained by making it act as a sink instead of a source of CO_2 emission. Thus, a negative C flux can be created by C sequestration (Zhao et al. 2019). An increase in soil C will cause a reduction in the atmospheric CO_2 levels from increased plant biomass produced through photosynthesis. However, the increase in soil C will depend upon the type of biochar used (Table 23.1). The wide

Table 23.1 Impact of different biochar types on soil organic carbon content

Type of biochar	Pyrolysis temperature (°C)	Study type	Rate of application	Soil carbon (g kg^{-1})	Reference
Oak wood biochar	350–600	Incubation	0.5, 1.0, and 2%	5.7, 5.3, and 4.1	Demisie et al. (2014)
Bamboo biochar	350–600	Incubation	0.5, 1.0, and 2%	5.7, 5.6, and 5.4	Demisie et al. (2014)
Prosopis wood	350	Incubation	1, 2, 3, 4, and 5%	6.3, 9.5, 11.2, 13.2, and 16.1	Shenbagavalli and Mahimairaja (2012)
Rice hull	500	Pot	1, 2, and 3%	18.9, 19.2, and 20.5%	Kim et al. (2016)
Poultry litter	450	Pot	0, 10, 25 Mg ha^{-1}	20.0, 23.8, 28.0, and 36.0	Chan et al. (2008)
Poultry litter	550	Pot	0, 10, 25 Mg ha^{-1}	19.5, 22.7, 24.8, and32.0	Chan et al. (2008)

application of biochar in agricultural systems can however be cost-prohibitive (Shahzad et al. 2019).

23.9 Role of Conservation Tillage in C Sequestration

Tillage has a direct impact on soil C contents. Tillage could affect the soil C pool in both negative and positive ways. The negative aspects of tillage include erosion, leaching, and mineralization, while the positive aspects include the humification of plant residues, aggregation of minerals, and the formation of organic compounds, as well as the deep placement of C in soil horizons. Conservation tillage is the general term, which aims to conserve moisture and reduce runoff losses. Conservation tillage may comprise of no-tillage, reduced tillage, ridge tillage, and mulch tillage. However, all these tillage practices have the same basic principles i) minimum disturbance of soil, ii) retention of crop residues on the soil surface, and iii) no or minimum traffic on the farm (Busari et al. 2015).

The retention of crop residues on the surface acts as organic mulch and provides great protection against raindrop impact on soil aggregates, which otherwise would have disintegrated and promoted soil erosion. The results of many experiments with conservation tillage showed promising results on SOC sequestration. Studies have shown that the adoption of conservation tillage caused an 8% increase in SOC in a Ultisol.

Conservation tillage also has great impact on the diversity and activities of soil fauna. The diversity and activities of soil fauna because of conservation tillage could improve SOC pool through the mixing of soil and borrowing activity that could place SOC in deep soil horizons. The borrowing activities of soil fauna develop channels in the soil, which help to translocate the C from the active degradable topsoil to lower soil depths, which are less prone to degradation (Sintim et al. 2020). The process will retain C for the long-term due to less exposure to the atmosphere (Filser et al. 2016). Further, higher activity of soil fauna produces soil mucilage, which could increase the aggregates of soil particles.

Conservation tillage is also directly helpful in the formation of soil aggregates and their stability (Blanco-Canqui and Ruis 2018). Tillage practices also determine the turnover time of soil C (Table 23.2). The direct advantage of conservation tillage is the conservation of soil water. The availability of ample soil water in the plant root zone promotes more biomass production and thus ultimately increases the SOC pool, cultural practices, and C sequestration.

The residence of SOC in soil depends upon natural ecological factors, which include rainfall, atmospheric temperature, and soil textural class (Willaarts et al. 2016). In addition to climatic and soil class, the residence of SOC depends upon soil and crop management practices (Schmidt et al. 2015). Cultural practices that could improve SOC include crop residue management, cultivation of deep-rooted crops, promotion of agroforestry system, and the inclusion of cover crops in crop cycle

Table 23.2 Impact of different tillage practices on soil organic carbon content

Location	Soil texture/ Soil type	Cropping system	Tillage system	Soil organic matter (g kg^{-1})	Location
Nebraska, USA	Loam	Winter wheat fallow	No-till	17.0	Blanco-Canqui et al. (2009)
			Reduced tillage	12.0	
Colorado, USA	Loam	Winter wheat fallow	No-till	10.5	Blanco-Canqui et al. (2009)
			Reduced tillage	9.0	
			Conventional tillage	9.0	
Kansas, USA	Silt loam	Winter wheat sorghum	No-till	17.0	Blanco-Canqui et al. (2009)
			Reduced tillage	13.0	
			Conventional tillage	14.2	
Kentucky, USA	Silt loam	Corn	No-till	21.6	Thomas et al. (1995)
			Conventional tillage	13.75	
USA	Subtropical soils	Corn	Conventional Tillage	10.1	Six et al. (2002)
			No-tillage	22.7	
Spain	Sandy loam to silty clay loam	Variable	Conventional tillage	10.6	Blanco-Moure et al. (2016)
	Silt loam soil		No-till	10.2	
USA		Wheat-pea rotation	Conventional tillage	17.2	Awale et al. (2017)
			No-till	17.9	
USA	iSlty clay loam	corn–soybean	Conventional tillage	18.5	Kibet et al. (2016)
			No-tillage	19.5	

(Fig. 23.1). These cultural practices can increase biomass production and enhance the humification of organic compounds present in the soil system.

23.9.1 Biomass Production

A system that promotes biomass production has great potential to increase the SOC pool. Biomass production can be enhanced by converting natural ecosystems to well-managed agricultural ecosystems. An increase of 21% SOC pool was observed in a study after 2 years of conversion of the natural ecosystem to an agricultural ecosystem. A model was proposed by Nye and Green land (1960) and Young (1976) to predict or estimate net change in SOC during a fallow period:

$$I = A(1 - p)$$

Where I is the change in SOC, A is the addition of SOC to the soil, and p is the SOC in a natural ecosystem.

23.9.2 Cover Crop

Fallow land causes high decomposition of residence SOC due to high exposure to climatic factors, whereas the growing of cover crops could conserve the soil from these adverse environmental factors. In addition, the residues of cover crops are a source of SOC when incorporated into the soil (Poeplau and Don 2015). Lal et al. (1978) found an increase in SOC contents after a 2 years experiment in an Alfisol when the fallow land was cultivated with grasses and leguminous cover crops. The data presented in Table 23.3 shows that cover crops could increase the SOC contents from 1.91 to 31.44%. However, the increase in SOC contents depends on the soil depth and the type of cover crop (Table 23.3).

23.9.3 Agroforestry

The promotion of an agroforestry system could increase SOC contents. A monoculture system often results in a low level of soil C. Forest trees continuously shed their leaves in autumn seasons, which are a rich source of SOC (De Stefano and Jacobson 2018). The deep roots of trees, after decomposition, also become a source of SOC in the subsoil horizons. Nevertheless, the removal of intensive biomass (cutting of trees) in this system could result in lower SOC contents in the soils. Further, the potential of C sequestration in agroforestry depends on climatic, soil, and site-specific biological and management practices. The world area under agroforestry is about 1.02 billion hectares, which is about 10% of agricultural land (Agevi et al. 2017). Agroforestry systems are believed to have more potential to sequester C as compared to field crops and pastures. Therefore, the inclusion of forest in a crop and pasture system could result in increased C sequestration in both deep and shallow soil depths. The C sequestration potential of agroforestry could range from 0.29 to 15.21 Mg ha^{-1} y^{-1}.

Table 23.3 Impact of cover crop on the soil organic matter at different depths

Country	Soil depth (cm)	Soil organic matter ($g\ kg^{-1}$) Fallow land	Cover crop	Increased over fallow land (%)	Cover crop name	Reference
USA	0–15	15.7	16.6	5.73	Rye	Kuo et al. (1997)
USA	0–15	15.7	16.0	1.91	Austrian Winter Pea	Kuo et al. (1997)
USA	0–15	15.7	16.6	5.73	Rye grass	Kuo et al. (1997)
USA	0–15	15.7	15.8	0.64	Vetch	Kuo et al. (1997)
USA	0–15	15.7	15.4	−1.91	Canola	Kuo et al. (1997)
Argentina	0–5	19.4	25.5	31.44	Wheat	Duval et al. (2016)
Argentina	0–5	19.4	24.3	25.26	Oat	Duval et al. (2016)
Argentina	0–5	19.4	24.5	26.29	Oat+ vetch	Duval et al. (2016)
Argentina	0–5	19.4	22.5	15.98	vetch	Duval et al. (2016)
Japan	0–15	39.41	44.40	12.66	Hairy Vetch	Higashi et al. (2014)
Japan	0–15	39.41	43.10	9.36	Rye	Higashi et al. (2014)

23.9.4 Deep-Rooted Crop

The production of deep-rooted crops with higher production of biomass could result in an increase in SOC in the subsoil horizon where the C contents reside for a longer duration. In deeper horizons, the processes of mineralization and decomposition are slow due to less exposure of residues to environmental effects. Deep and prolific root systems generally provide a favorable condition for C sequestration in the subsoil. The decomposition of root decreases with depth due to limited microbial activities as found in the study of Pries et al. (2018). Roots have longer turnover times and more storage times in deeper depths. In another study, it was found that soil edaphic delayed the decomposition of plant litter by a factor of 1.5 in deeper soils as compared to surface soils (Newey 2005). The delay in decomposition was mainly due to a deficiency of nitrogen in deep soils. In addition, bulk density and oxygen deficiency could be a factor of low decomposition in lower subsoils. Moisture and temperature also influence root decomposition with depth. Sala et al. (1992) reported that water potential > −1 MPa occurs mostly in soils between 5 and 15 cm which is a low frequency of plant-available water in the deepest soil layers. Therefore, by promoting deep-rooted crops, the decomposition of plant litter could be prolonged thereby reducing CO_2 emission.

23.9.5 Improvement of Pastures

Controlled grazing can increase SOC contents with minimum chances of losses. Overgrazing results in the losses of soil vegetative cover that leads to the mineralization of C present on the surface layer of soil as noted in different studies (Naeth et al. 1991; Sanjari et al. 2008; Zhou et al. 2010). Moreover, the continuous movement of livestock on grazed land increases soil bulk density and compaction (Drewry et al. 2008). Therefore, time-controlled grazing, a grazing system of short intensive grazing followed by a long period of rest, is becoming popular in the world. The time-controlled grazing system promotes ground litter accumulation and reduces soil compaction over time. Sanjari et al. (2008) found higher total organic C in time-controlled grazing as compared to continuous grazing and no grazing fields. In another study, Zhou et al. (2010) found a significant increase in SOC contents in the top 30 cm depth with 26 years of controlled grazing. The soil C stock increased from 2.05 Mg ha^{-1} in 1981 to 27.98 Mg ha^{-1} in 2006.

23.9.6 Residue Management

Burning and poor management of residues result in the loss of SOC and causes high emission of CO_2 to the atmosphere. The burning of rice residues in the northwest region of India caused an estimated loss of 9.2 million tons of C per year (Sing 2018). Proper management and incorporation of residues could result in the accumulation of SOC and lower the emission of CO_2 to the atmosphere. Farming systems with the capability of producing high biomass could return the biomass to the soil to enhance SOC. These residues will over time be converted into humus compounds, which have more residence time as compared to the original residue contents. Humic substances are also helpful in increasing soil aggregation and the retention of moisture contents in the soil (Šimanský 2016).

23.9.7 Fertilizer Application

Agricultural production systems depend mainly upon nitrogenous fertilizers. The application of nitrogenous fertilizers results in the production of higher biomass and increases SOC contents. One kg N fertilizer can cause 1 kg C emission from the soil; therefore, any C balance in soil depends upon N fertilizer application. However, in general, C sequestration increases with the application of N fertilizers, especially when the 4Rs of nutrient stewardship are properly implemented. The application of fertilizers increases SOC significantly in soils with initially low SOC levels as compared to soils initially enriched with C. Qin et al. (2014) found a significant positive correlation between nitrogen and SOC.

23.9.8 Crop Rotation

Crop rotation has an important role in the increase of SOC. Monoculture systems are not much effective in increasing C contents of soil as compared to diversified cropping systems. The addition of legume crops in the rotation was found to help increase C contents in the soil as compared to where no legume crop was included (Becker et al. 2017). Bolinder et al. (2012) evaluated different crop rotations for 6 years. Rotation A (under sown barely and forage) showed higher soil C [31.18 g kg^{-1}], as compared to rotation B (under sown barely, forage, green fodder, and fodder rape) [27.0 g kg^{-1}], rotation C (under sown Barely, forage, winter rye, green fodder, and root crop) [23.8 g kg^{-1}], and rotation D (under sown barely, green manure, winter rye, peas, and root crop [21.9 g kg^{-1}].

23.10 Conclusion

The benefits of SOC sequestrations are countless to mitigate climate change effects, but there are some hurdles to cross to achieving the full potential of C sequestrations. There is a need to standardize best agronomic management practices according to on-site soil type and climatic conditions. It is also important to understand the additional requirements of essential nutrients and water for SOC sequestration in soils. Further, stable C sources such as biochar are helpful to increase the soil C pool, but the energy input in the preparation, as well as application costs, may undermine its wide adoption. There is also a need to standardize soil C change measurement to better understand the effectiveness of different management practices. Recommended management practices for C sequestration must be effective by being tailored to unique ecosystems and it must not be cost-prohibitive.

References

Agevi H, Onwonga R, Kuyah S, Tsingalia M (2017) Carbon stocks and stock changes in agroforestry practices: A review. Trop Subtrop Agroecosystems 20:101–109.

Akram R, Turan V, Hammad HM, et al (2018) Fate of Organic and Inorganic Pollutants in Paddy Soils. In: Hashmi MZ, Varma A (eds) Environmental Pollution of Paddy Soils, Soil Biology, vol 53. Springer International Publishing AG, Basel, Switzerland, pp 197–214.

Amin A, Nasim W, Fahad S, Ali S, Ahmad S, Rasool A, Saleem N, Hammad HM, Sultana SR, Mubeen M, Bakhat HF, Ahmad N, Shah GM, Adnan M, Noor M, Basir A, Saud S, Habib ur Rahman M, Paz JO (2018) Evaluation and analysis of temperature for historical (1996–2015) and projected (2030–2060) climates in Pakistan using SimCLIM climate model: Ensemble application. Atmos Res 213:422–436. https://doi.org/10.1016/j.atmosres.2018.06.021

Awale R, Emeson MA, Machado S (2017) Soil organic carbon pools as early indicators for soil organic matter stock changes under different tillage practices in Inland Pacific Northwest. Front Ecol Evol 5:https://doi.org/10.3389/fevo.2017.00096.

Baldocchi D, Chu H, Reichstein M (2018) Inter-annual variability of net and gross ecosystem carbon fluxes: A review. Agric For Meteorol 249:520–533. https://doi.org/10.1016/j.agrformet.2017.05.015

Becker JN, Gütlein A, Sierra Cornejo N, Kiese R, Hertel D, Kuzyakov Y (2017) Legume and non-legume trees increase soil carbon sequestration in Savanna. Ecosystems 20:989–999. https://doi.org/10.1007/s10021-016-0087-7

Blanco-Canqui H, Stone LR, Schlegel AJ, et al (2009) No-till Induced Increase in Organic Carbon Reduces Maximum Bulk Density of Soils. Soil Sci Soc Am J 73:1871–1879. https://doi.org/10.2136/sssaj2008.0353

Blanco-Moure N, Gracia R, Bielsa AC, López MV (2016) Soil organic matter fractions as affected by tillage and soil texture under semiarid Mediterranean conditions. Soil Tillage Res 155:381–389. https://doi.org/10.1016/j.still.2015.08.011

Blanco-Canqui H, Ruis SJ (2018) No-tillage and soil physical environment. Geoderma 326:164–200. https://doi.org/10.1016/j.geoderma.2018.03.011

Bolinder MA, Kätterer T, Andrén O, Parent LE (2012) Estimating carbon inputs to soil in forage-based crop rotations and modeling the effects on soil carbon dynamics in a Swedish long-term field experiment. Can J Soil Sci 92:821–833. https://doi.org/10.4141/CJSS2012-036

Busari MA, Kukal SS, Kaur A, Bhatt R, Dulazi AA (2015) Conservation tillage impacts on soil, crop and the environment. Int Soil Water Conserv Res 3:119–129. https://doi.org/10.1016/j.iswcr.2015.05.002

Chan KYA, Van Zwieten, L., Meszaros, I., Downie, A., Joseph, S.D. (2008) Using poultry litter biochars as soil amendments. Aust J of Soil Res 46:437–444

Conant RT, Cerri CEP, Osborne BB, Paustian K (2017) Grassland management impacts on soil carbon stocks: A new synthesis: A. Ecol Appl 27:662–668. https://doi.org/10.1002/eap.1473

De Stefano A, Jacobson MG (2018) Soil carbon sequestration in agroforestry systems: a meta-analysis. Agrofor Syst 92:285–299. https://doi.org/10.1007/s10457-017-0147-9

Demisie W, Liu Z, Zhang M (2014) Effect of biochar on carbon fractions and enzyme activity of red soil. Catena 121:214–221. https://doi.org/10.1016/j.catena.2014.05.020

Drewry JJ, Cameron KC, Buchan GD (2008) Pasture yield and soil physical property responses to soil compaction from treading and grazing – A review. Aust J Soil Res 46:237–256. https://doi.org/10.1071/SR07125

Duval ME, Galantini JA, Capurro JE, Martinez JM (2016) Winter cover crops in soybean monoculture: Effects on soil organic carbon and its fractions. Soil Tillage Res 161:95–105. https://doi.org/10.1016/j.still.2016.04.006

Filser J, Faber JH, Tiunov A V., Brussaard L, Frouz J, De Deyn G, Uvarov A V., Berg MP, Lavelle P, Loreau M, Wall DH, Querner P, Eijsackers H, Jiménez JJ (2016) Soil fauna: Key to new carbon models. Soil 2:565–582. https://doi.org/10.5194/soil-2-565-2016

Hammad R, Elbagory M (2019) Using Plant Growth-promoting Fungi (PGPF), as a Biofertilizer and Biocontrol Agents against Tetranychus cucurbitacearum on Nubian Watermelon (Citrullus lanatus L.). J Adv Microbiol 16:1–15. https://doi.org/10.9734/jamb/2019/v16i230119

Hammad HM, Khaliq A, Abbas F, et al (2020) Comparative Effects of Organic and Inorganic Fertilizers on Soil Organic Carbon and Wheat Productivity under Arid Region. Commun Soil Sci Plant Anal 51:1406–1422. https://doi.org/10.1080/00103624.2020.1763385

Hammond J, Shackley S, Sohi S, Brownsort P (2011) Prospective life cycle carbon abatement for pyrolysis biochar systems in the UK. Energy Policy 39:2646–2655. https://doi.org/10.1016/j.enpol.2011.02.033

Hashimoto K (2019) Global Temperature and Atmospheric Carbon Dioxide Concentration. In: Global Carbon Dioxide Recycling. SpringerBriefs in Energy. Springer, Singapore, pp 5–17.

Higashi T, Yunghui M, Komatsuzaki M, Miura S, Hirata T, Araki H, Kaneko N, Ohta H (2014) Tillage and cover crop species affect soil organic carbon in Andosol, Kanto, Japan. Soil Tillage Res 138:64–72. https://doi.org/10.1016/j.still.2013.12.010

Hussain S, Mubeen M, Akram W, et al (2020) Study of land cover/land use changes using RS and GIS: a case study of Multan district, Pakistan. Environ Monit Assess 192:1–15. https://doi.org/10.1007/s10661-019-7959-1

Kibet LC, Blanco-Canqui H, Jasa P (2016) Long-term tillage impacts on soil organic matter components and related properties on a Typic Argiudoll. Soil Tillage Res 155:78–84. https://doi.org/10.1016/j.still.2015.05.006

Kim HS, Kim KR, Yang JE, Ok YS, Owens G, Nehls T, Wessolek G, Kim KH (2016) Effect of biochar on reclaimed tidal land soil properties and maize (Zea mays L.) response. Chemosphere 142:153–159. https://doi.org/10.1016/j.chemosphere.2015.06.041

Kuo S, Sainju UM, Jellum EJ (1997) Winter Cover Crop Effects on Soil Organic Carbon and Carbohydrate in Soil. Soil Sci Soc Am J 61:145–152. https://doi.org/10.2136/sssaj1997.03615995006100010022x

Lal R, Wilson GF, Okigbo BN (1978) No-till farming after various grasses and leguminous cover crops in tropical alfisol. I. Crop performance. F Crop Res 1:71–84. https://doi.org/10.1016/0378-4290(78)90008-4

Lal R (2011) Sequestering carbon in soils of agro-ecosystems. Food Policy 36:S33–S39. https://doi.org/10.1016/j.foodpol.2010.12.001

Lal R (2015) Restoring soil quality to mitigate soil degradation. Sustain 7:5875–5895. https://doi.org/10.3390/su7055875

Lal R (2016) Soil health and carbon management. Food Energy Secur 5:212–222. https://doi.org/10.1002/fes3.96

Matthews R, Malcolm, H., Buys G, Henshall, P., Moxley J., Morris, A., Mackie E. (2014) Changes to the representation of forest land and associated land-use changes in the 1990--2012 UK Greenhouse Gas Inventory. For Res Cent Ecol Hydrol (DECC Contract GA0510, CEH Contract no NEC0376

Minasny B, McBratney AB (2016) Digital soil mapping: A brief history and some lessons. Geoderma 264:301–311. https://doi.org/10.1016/j.geoderma.2015.07.017

Naeth MA, Chanasyk DS, Rothwell RL, Bailey AW (1991) Grazing impacts on soil water in mixed prairie and fescue grassland ecosystems of Alberta. J Range Manag 44:313–325. https://doi.org/10.4141/cjss91-031

Nayak AK, Rahman MM, Naidu R, Dhal B, Swain CK, Nayak AD, Tripathi R, Shahid M, Islam MR, Pathak H (2019) Current and emerging methodologies for estimating carbon sequestration in agricultural soils: A review. Sci Total Environ 665:890–912. https://doi.org/10.1016/j.scitotenv.2019.02.125

Newey A (2005) Decomposition of Plant Litter and Carbon Turnover as a Function of Soil Depth. Ph.D. Thesis, Australian National University, Canberra, Australia.

Palosuo T, Heikkinen J, Regina K (2016) Method for estimating soil carbon stock changes in Finnish mineral cropland and grassland soils. Carbon Manag 6:207–220. https://doi.org/10.1080/17583004.2015.1131383

Paustian K, Collier S, Baldock J, Burgess R, Creque J, DeLonge M, Dungait J, Ellert B, Frank S, Goddard T, Govaerts B, Grundy M, Henning M, Izaurralde RC, Madaras M, McConkey B, Porzig E, Rice C, Searle R, Seavy N, Skalsky R, Mulhern W, Jahn M (2019) Quantifying carbon for agricultural soil management: from the current status toward a global soil information system. Carbon Manag 10:567–587. https://doi.org/10.1080/17583004.2019.1633231

Poeplau C, Don A (2015) Carbon sequestration in agricultural soils via cultivation of cover crops: A meta-analysis. Agric Ecosyst Environ 200:33–41. https://doi.org/10.1016/j.agee.2014.10.024

Pries CEH., Sulman BN, West C, O'Neill C, Poppleton E, Porras RC, Castanha C, Zhu B, Wiedemeier DB, Torn MS (2018) Root litter decomposition slows with soil depth. Soil Biol Biochem 125:103–114. https://doi.org/10.1016/j.soilbio.2018.07.002

Qi R, Li J, Lin Z, Li Z, Li Y, Yang X, Zhang J, Zhao B (2016) Temperature effects on soil organic carbon, soil labile organic carbon fractions, and soil enzyme activities under long-term fertilization regimes. Appl Soil Ecol 102:36–45. https://doi.org/10.1016/j.apsoil.2016.02.004

Qin Y, Xin Z, Yu X, Xiao Y (2014) Influence of vegetation restoration on topsoil organic carbon in a small catchment of the loess hilly region, China. PLoS One 9. https://doi.org/10.1371/journal.pone.0094489

Sala OE, Lauenroth WK, Parton WJ (1992) Long Term Soil Water Dynamics. Ecology 73:1175–1181.

Sanjari G, Ghadiri H, Ciesiolka CAA, Yu B (2008) Comparing the effects of continuous and time-controlled grazing systems on soil characteristics in Southeast Queensland. Aust J Soil Res 46:348–358. https://doi.org/10.1071/SR07220

Schipper LA, Parfitt RL, Fraser S, Littler RA, Baisden WT, Ross C (2014) Soil order and grazing management effects on changes in soil C and N in New Zealand pastures. Agric Ecosyst Environ 184:67–75. https://doi.org/10.1016/j.agee.2013.11.012

Schmidt MWI, Torn MS, Abiven S, Dittmar T, Guggenberger G, Janssens IA, Kleber M, Gel-Knabner IK, Lehmann J, Manning DAC, Nannipieri P, Rasse DP, Weiner S, Trumbore SE (2015) Persistence of soil organic matter as an ecosystem property. Nature 528:49–56 . https://doi.org/10.1038/nature15744

Sintim HY, Bary AI, Hayes DG, et al (2020) In situ degradation of biodegradable plastic mulch films in compost and agricultural soils. Sci Total Environ 727:138668. https://doi.org/10.1016/j.scitotenv.2020.138668

Shahzad K, Abid M, Sintim HY, Hussain S, Nasim W (2019) Tillage and biochar effects on wheat productivity under arid conditions. Crop Sci 59:1191–1199. https://doi.org/10.2135/cropsci2018.08.0485

Shenbagavalli S, Mahimairaja S (2012) Characterization and effect of biochar on nitrogen and carbon dynamics in soil. Int J Adv Biol Res 2:249–255.

Šimanský V (2016) Changes in soil organic matter parameters during the period of 18 years under different soil management practices. Agriculture 62:149–154. https://doi.org/10.1515/agri-2016-0015

Sing JS (2018) Crop residues management in agro-environmental sustainability. Clim Chang 4:653–660.

Six J, Feller C, Denef K, et al (2002) Soil organic matter, biota and aggregation in temperate and tropical soils – Effects of no-tillage. Agronomie 22:755–775. https://doi.org/10.1051/agro

Stockmann U, Adams MA, Crawford JW, Field DJ, Henakaarchchi N, Jenkins M, Minasny B, McBratney AB, de Courcelles VR, Singh K, Wheeler I, Abbott L, Angers DA, Baldock J, Bird M, Brookes PC, Chenu C, Jastrow JD, Lal R, Lehmann J, O'Donnell AG, Parton WJ, Whitehead D, Zimmermann M (2013) The knowns, known unknowns and unknowns of sequestration of soil organic carbon. Agric Ecosyst Environ 164:80–99. https://doi.org/10.1016/j.agee.2012.10.001

Thomas GW, Haszler GR, Blevins RL (1995) The Effects of Organic Matter and Tillage on Maximum Compactibility. In: 1995 Southern Conservation Tillage Conference for Sustainable Agriculture. Jackson, Mississippi, pp 34–36.

Young A (1976) Tropical Soils and Soil Survey. Cambridge University Press, Cambridge, United Kingdom.

Wang J, Xiong Z, Kuzyakov Y (2016) Biochar stability in soil: Meta-analysis of decomposition and priming effects. GCB Bioenergy 8:512–523. https://doi.org/10.1111/gcbb.12266

Willaarts BA, Oyonarte C, Muñoz-Rojas M, Ibáñez JJ, Aguilera PA (2016) Environmental factors controlling soil Organic carbon stocks in two ccontrasting Mediterranean Climatic Areas of Southern Spain. L Degrad Dev 27:603–611 . https://doi.org/10.1002/ldr.2417

Zhang F, Li M (2019) Impacts of ocean warming, sea level rise, and coastline management on storm surge in a semienclosed bay. J Geophys Res Ocean 124:6498–6514. https://doi.org/10.1029/2019jc015445

Zhao P, Palviainen M, Köster K, Berninger F, Bruckman VJ, Pumpanen J (2019) Effects of biochar on fluxes and turnover of carbon in boreal forest soils. Soil Sci Soc Am J 83:126–136. https://doi.org/10.2136/sssaj2018.04.0149

Zhou ZY, Li FR, Chen SK, Zhang HR, Li G (2010) Dynamics of vegetation and soil carbon and nitrogen accumulation over 26 years under controlled grazing in a desert shrubland. Plant Soil 341:257–268. https://doi.org/10.1007/s11104-010-0641-6

Index

W. N. Jatoi et al. (eds.), *Building Climate Resilience in Agriculture*,
https://doi.org/10.1007/978-3-030-79408-8

Z